Methods for Euclidean Geometry

Methods for Euclidean Geometry

Owen Byer
Eastern Mennonite University

Felix Lazebnik
University of Delaware

Deirdre L. Smeltzer
Eastern Mennonite University

Published and Distributed by
The Mathematical Association of America

CLASSROOM RESOURCE MATERIALS

Classroom Resource Materials is intended to provide supplementary classroom material for students—laboratory exercises, projects, historical information, textbooks with unusual approaches for presenting mathematical ideas, career information, etc.

101 Careers in Mathematics, 2nd edition edited by Andrew Sterrett

Archimedes: What Did He Do Besides Cry Eureka?, Sherman Stein

Calculus Mysteries and Thrillers, R. Grant Woods

Conjecture and Proof, Miklós Laczkovich

Creative Mathematics, H. S. Wall

Environmental Mathematics in the Classroom, edited by B. A. Fusaro and P. C. Kenschaft

Excursions in Classical Analysis: Pathways to Advanced Problem Solving and Undergraduate Research, by Hongwei Chen

Exploratory Examples for Real Analysis, Joanne E. Snow and Kirk E. Weller

Geometry From Africa: Mathematical and Educational Explorations, Paulus Gerdes

Historical Modules for the Teaching and Learning of Mathematics (CD), edited by Victor Katz and Karen Dee Michalowicz

Identification Numbers and Check Digit Schemes, Joseph Kirtland

Interdisciplinary Lively Application Projects, edited by Chris Arney

Inverse Problems: Activities for Undergraduates, Charles W. Groetsch

Laboratory Experiences in Group Theory, Ellen Maycock Parker

Learn from the Masters, Frank Swetz, John Fauvel, Otto Bekken, Bengt Johansson, and Victor Katz

Math Made Visual: Creating Images for Understanding Mathematics, Claudi Alsina and Roger B. Nelsen

Methods for Euclidean Geometry, Owen Byer, Felix Lazebnik, Deirdre L. Smeltzer

Ordinary Differential Equations: A Brief Eclectic Tour, David A. Sánchez

Oval Track and Other Permutation Puzzles, John O. Kiltinen

A Primer of Abstract Mathematics, Robert B. Ash

Proofs Without Words, Roger B. Nelsen

Proofs Without Words II, Roger B. Nelsen

She Does Math!, edited by Marla Parker

Solve This: Math Activities for Students and Clubs, James S. Tanton

Student Manual for Mathematics for Business Decisions Part 1: Probability and Simulation, David Williamson, Marilou Mendel, Julie Tarr, and Deborah Yoklic

Student Manual for Mathematics for Business Decisions Part 2: Calculus and Optimization, David Williamson, Marilou Mendel, Julie Tarr, and Deborah Yoklic

Teaching Statistics Using Baseball, Jim Albert

Visual Group Theory, Nathan C. Carter

Writing Projects for Mathematics Courses: Crushed Clowns, Cars, and Coffee to Go, Annalisa Crannell, Gavin LaRose, Thomas Ratliff, Elyn Rykken

MAA Service Center
P.O. Box 91112
Washington, DC 20090-1112
1-800-331-1MAA   FAX: 1-301-206-9789

To the memory of

Isaak Moiseevich Yaglom (1921–1988),
whose work inspired so many others.

Preface

Approach

Geometry is a rich and beautiful subject with a long history, offering an ideal platform for studying mathematics. It is a foundational source of a very large part of modern mathematics; as recently as the nineteenth century, mathematicians were referred to as "geometers." The problems that arise in Euclidean geometry lend themselves to solutions embedded in a wide variety of mathematical areas. We wish to review and deepen readers' understanding of Euclidean geometry, and emphasize some techniques developed after Euclid, many of which are not well-known or widely used by non-professionals. Our desire is to rekindle the relationship between geometry and broader mathematics.

Several excellent college-level geometry texts have appeared during the last forty years. However, whereas geometry was once perceived as a system of properties of geometric objects (like polygons, circles, polyhedra, spheres), many recent treatments have turned the subject into an arena for introducing formal logic and developing proof-writing skills.

The main goals of this book are different. Instead of carefully building geometries from axiom sets, we will focus on using a variety of mathematical methods to solve problems in Euclidean geometry. Many of these methods arose when people tried to solve problems in old ways and failed. In several cases, the new ideas used in solving specific problems later were developed into areas of mathematics in their own right. Our book is first and foremost one for learning mathematics, with geometry providing the content, and our hope is that studying geometry along the lines we suggest will develop students' appreciation of the subject and of mathematics as a whole.

One important aspect of this book is its thorough treatment of the coordinate method and its applications. Despite the fact that it has been in mathematics for four centuries, it is usually not a tool a student considers using when faced with a geometry problem. One reason for this may be the fear of complicated algebraic manipulations, but we try to illustrate that when the method is used cleverly, the involved algebra is relatively easy. We provide a supplement to the book in which we explain how to write simple procedures in Maple to help with difficult algebraic manipulations (see comments on Chapter 14 below). We believe that this presentation does not detract from mastering geometry, but rather provides a tool similar to that of the calculator for performing mundane numerical computations.

The book contains many problems of varying difficulty; nearly half its pages are devoted to problem statements, hints, and complete solutions. Much can be learned by comparing various solutions. Supplemental problems may be found at the end of each of Chapters 3–13, for which we do not provide solutions in the text itself. Some problems appear repeatedly throughout the book, and we suggest trying to solve them using different methods and comparing the merits of each approach. We place paramount importance on the problems and their solutions, as many of them continue the discussion from the chapter. We try to avoid "baroque" problems (such as the nine-point circle), choosing instead to raise more natural questions, which are often just as challenging. Each set of problems is divided by horizontal lines into three subsets of (generally) increasing difficulty.

We attempt to bring to the book the spirit exercised by professional mathematicians. For example, when we pose the question of describing the set of all points in the plane with a given sum of distances to two fixed lines, we follow up by asking similar questions about the difference, product, and ratio; or, we may press the reader to consider whether a fact remains valid in more general cases.

We try to incorporate some surprising or not well-known facts that we have encountered while teaching the subject and writing the book.

- For example, we highlight that all parabolas are similar (but not all ellipses or all hyperbolas). Students have been told about "narrower" and "wider" parabolas, but while their graphs suggest such differences, it simply isn't the case. This fact surprised Kepler, who wrote about it with excitement in 1609.

- Another little-known fact is that the trajectory of a projectile is not a parabola, but an ellipse, and that Newton pointed this out in his *Principia*.

- Is there a nonregular convex pentagon in which every diagonal is parallel to a side? (Yes.)

- For any $n \geq 6$, is it possible to have n points such that any five of them lie on an ellipse, but no six do? (Yes.) What about on a hyperbola? (Yes.) What about on a parabola? (No!)

In addition to being a text for a college geometry course, the book can be utilized in other ways. Most of the chapters in the second half of the book could be used independently as resources for other high school or college courses, for enrichment materials, or for an independent study. The book could serve as the basis for a capstone course in mathematics. The problems and solutions would be an excellent resource for a problem-solving group.

When choosing the content necessary to achieve our goals, we encountered the painful decision as to which topics or methods to exclude. Among those are

- physics-based concepts, such as center of mass;
- the structure of the groups of isometries (congruences), similarities, and affine transformations;
- non-Euclidean geometries, other than a brief treatment of finite geometries in Chapter 2;
- elements of arithmetic algebraic geometry;
- topics from combinatorial geometry;
- projective transformations and polarities with respect to a conic; and
- the use of eigenvalues of matrices in studying conics.

One could argue that many of the above topics do not belong to the category of plane geometry proper, but the boundaries are constantly changing.

The omission of straight edge and compass constructions bothers us less. Despite their aesthetic value and role in the development of mathematics, we did not see the need for them in our book, no matter how close they are to our hearts.

Solid (3-dimensional) geometry is another matter. Many topics in this book are indispensable for solving problems of solid geometry, and the fact that the Euclidean plane can be embedded in Euclidean space can be extremely helpful for solving problems in the plane. Perhaps another book will be forthcoming to treat this relationship.

When productions of *Hamlet* are discussed, the details matter most. We believe that there are as many Pythagorean theorems as there are *Hamlets*: although thousands of actors each year repeat the same words, the viewers are impressed by only very few productions. We hope the audience will remember ours.

Content

Because of our approach, we will make several assumptions that would not be made in a geometry course that develops axiomatically. We will assume that students have had high school courses in geometry and algebra and that they have been exposed to proofs and the rules of logic. We will use the School Mathematics Study Group (SMSG) axioms. This axiom set has the advantage of permitting us to begin proving meaningful theorems right away about topics with which students are quite familiar.

Since the geometry courses students have in high school differ a great deal, we try to present with proof most basic facts, but not all. Without doubt, some statements of geometry are more exciting than others. For example, the fact that opposite sides of a parallelogram are congruent is not particularly exciting; most people would "see" this fact without studying any geometry. On the other hand, the Pythagorean theorem, though very familiar, is not obvious at all. The reader will find that we concentrate most on some of these interesting, non-obvious facts. They are the pillars of our presentation.

The first two chapters of the text give a short introduction to the development of the subject. Chapter 1 provides a brief history of early geometry, highlighting the accomplishments of the Egyptians, Babylonians, and Greeks. Chapter 2 follows up with a discussion of the axiomatization of geometry that began with Euclid and continued into modern mathematics.

Chapters 3 through 6 introduce the Euclidean geometry of the plane. While some of this content may be familiar to students, our treatment goes beyond that of a typical high school course and will provide a rich source of problems for all. Chapters 3 and 4 cover lines, polygons, and circles. Chapter 5 provides a treatment of length and area, including the Wallace-Bolyai-Gerwien Theorem. Chapter 6 begins with a general discussion of loci of curve properties and concludes with a brief introduction to various geometric properties of the conic sections, without going into their algebraic descriptions; readers of all levels will likely encounter something new in this chapter. Virtually all material presented through Chapter 6 will utilize methods accessible to Euclid's contemporaries and their predecessors.

Chapters 7 through 13 bring out the distinctive flavor of our approach. Each of the chapters discusses a different modern mathematical technique for solving problems in Euclidean geometry (modern in the sense that they were not known to Euclid). These topics include trigonometry, the coordinate method, complex numbers, vectors, affine transformations, and inversions.

Chapter 14 contains a collection of problems for which algebraic software becomes useful to assist with the coordinate method. There is no doubt that the actual software chosen will change as time passes, but we used the symbolic algebra package Maple. As a supplement to the text, we provide a CD with a Maple worksheet showing how to use Maple effectively with the coordinate method, but we also go much further. We use the procedural capabilities of Maple to streamline many of the repetitive constructions that arise while solving the problems we present. Many of the examples and problems on the worksheet will have been solved with traditional methods earlier in the text. However, we believe that learning to use software like Maple to write procedures that solve problems is an essential part of maturing as a mathematician and that revisiting these problems will be beneficial.

Acknowledgments

While writing this book we utilized the many excellent sources listed in the bibliography. We tried to give credit as appropriate, but it is often difficult to trace certain facts or problems to their origins. Certainly, some books affected our views of the subject and our writing more than others. They include the great texts by I. M. Yaglom ([**69**], [**71**], [**72**]), H. Eves ([**19**]), and V. V. Prasolov ([**47**]). An excellent recent book, which complements and extends our text, is the one by Brannan, Esplen, and Gray ([**7**]), but its goals are different and it assumes greater mathematical maturity. Alexander Bogomolny's excellent web site `http://www.cut-the-knot.org/` has been an enduring source of inspiration.

We give special thanks to the following individuals for their feedback on various portions of the book: Patricia Ahlborn, Stuart Anderson, Willard Baxter, David Bellamy, Alla Bogomolnaya, Josh Bowman, Gary Ebert, Briana Gascho, Branko Grünbaum, Greg Inman, Kim Jongerius, Sarah Kalichman, Valery Kanevsky, Elena Koublanova, Marianne Lios, Donna Mark, Alex McAllister, Vladimir Naroditsky, Igor Pak, Irina Pankratova, Mark Roche, Paula Shorter, Jeff Suzuki, Michelle Swartley, Elena Tartakovskaya, Rachel Todd, Eve Torrence, Glen Van Brummelen, Ray Viglione, Kenneth Wantz, and many students from our geometry courses. We are grateful to Daniel Piker for making available to us in high resolution the electronic image "The Astronomer" in the Inversion chapter. We thank Doug Meade for suggesting MAA as a publisher, and Don Albers, Jerry Bryce, Elaine Pedreira, and Beverly Ruedi at MAA for overseeing the entire production process. The greatest measure of appreciation goes to our families, for their enduring support and patience throughout this project—in particular, our spouses, Barbara, Luba, and Sherwyn.

Contents

1

Early History

Certain things just have their epoque, when they are found at different places, just as in spring when violets come into the light everywhere.

— F. W. Bolyai (1775–1856)

In the first two chapters we provide a brief history of geometry in ancient civilizations and discuss how axiomatic assumptions have changed as we have moved into the modern era. Of course, many books have been written on the history of mathematics, and much of ancient history involves the beautiful subject of geometry. We hope this introduction will help develop your appreciation for the accomplishments of the early geometers and whet your appetite to explore the subject in more detail as we proceed through the remainder of the book.

The origins of geometry are debatable; scholars since the ancient Greeks have offered various opinions on the time, location, and purpose of the first use of geometry. Geometry as we know it today certainly results from the combined influence of artifacts, oral tradition, stone and clay tablets, papyri, great schools and libraries, and the unquenchable human thirst for knowledge. This rich subject has emerged as the collective effort of several distinct people groups, each of which had its own motivations for studying the subject.

Part of the difficulty in pinpointing the origin of geometry lies in the definition of the term. The word *geometry* comes from the Greek word *geometrein* (*geo*: earth, and *metrein*: to measure). The Greek historians Herodotus (5th century B.C.) and Proclus (5th century A.D.) no doubt had this definition in mind when they each claimed that geometry originated with the ancient Egyptians, who needed to resurvey their land each year following the annual flooding of the Nile. Aristotle (4th century B.C.) partly agreed, but he had a different explanation for the Egyptian pursuit of mathematics, as evidenced in this quote from his *Metaphysics*: "The mathematical sciences originated in the neighborhood of Egypt, because there the priestly class was allowed leisure."

Indeed, if one takes a broader view of geometry than that of measurement (which Aristotle evidently did), other scenarios for its origins become plausible. Neolithic peoples showed an awareness of similarity, congruence, and spatial relationships with their designs on pottery and woven baskets. In ancient India, workers used simple geometric relationships known as the *Sulvasutras*, or "rules of the cord," to construct altars and temples.

Some have advanced the thesis that these early uses of geometry were rooted in primitive religious rituals. Unfortunately, because the early Indians and Chinese used perishable items like bamboo and bark for their writings, we have very little physical record of what they knew, when they knew it, and why they studied it. Therefore, we will focus our discussion on the accomplishments of the Egyptians and subsequent civilizations for which there exist ample written records.

1.1   The Egyptians

Egypt was united by Menes around 3000 B.C., and it flourished under a relatively peaceful existence for about three millennia until it entered a gradual decline, ultimately being conquered by the Romans in 31 B.C. Egyptian mathematics served a maintenance role, as opposed to a system of thought that helped advance the society technologically or academically. In this regard, Egyptian mathematicians were best known for their accuracy in measurements, both in the surveying of property and in the construction of their pyramids. Their surveyors were later called "rope-fasteners" by the Greeks because their primary measuring device was a rope with carefully placed knots at equal intervals. The Great Pyramid at Gizeh, built by Khufu around 2600 B.C., stands as a lasting tribute to the preciseness and determination of a masterfully organized group of people. This pyramid stood 480 feet tall, with a square (the side lengths differed by less than $4\frac{1}{2}$ inches) base area of 13 acres. The four sides of the pyramid were intentionally aligned with the directions north, south, east, and west, within a fraction of a degree!

Most of our information about Egyptian mathematics comes to us from the Rhind Papyrus (1650 B.C.) and the Moscow Papyrus (1850 B.C.). These papyri contain many specific calculations of areas and volumes of standard geometric objects, but the Egyptians clearly were much more interested in empirical results than they were in finding accurate general formulas. For instance, examples abound that they used the formula $A = \frac{1}{4}(a + c)(b + d)$ to calculate the area of a quadrilateral with consecutive side lengths of a, b, c, and d. Although this formula always overestimates the actual area,[1] it provides a reasonable approximation if the quadrilateral is somewhat rectangular. (See Problem 1.3.)

In contrast to this errant formula, Problem 14 of the Moscow Papyrus demonstrates (again, using a specific problem) that the Egyptians knew how to calculate the volume of a truncated square pyramid with altitude h and base and top side lengths of a and b, respectively:

$$V = \frac{h}{3}(a^2 + ab + b^2).$$

This formula is not an easy one to guess, but neither is it clear how they found it; some consider the result to be the crowning achievement of the Egyptians, surpassing even their construction of the pyramids.

Several of the geometric problems in the Rhind Papyrus involved finding the amounts of grain that could fit in a cylindrical bin. The Egyptians therefore recognized the need for finding the area of a circle; in fact, Problem 50 reads as follows:

> *Example of a round field of a diameter 9 khet. What is its area? Take away $\frac{1}{9}$ of the diameter, namely 1; the remainder is 8. Multiply 8 times 8; it makes 64. Therefore it contains 64 setat of land.*

---

[1] Burton noted that this overestimation was to the advantage of tax collectors, who collected taxes based on the amount of land determined by surveyors. ([**10**], page 62.)

This method translates into the area formula $A = \frac{256}{81}r^2$ for a circle of radius r, which gives the reasonable approximation of $\pi \approx 3.1605$. While the above area formula isn't stated explicitly, a diagram accompanying Problem 48 of the papyrus suggests that it might have come from approximating a circle as an octagon inscribed in a square.

We summarize Egyptian geometry by concluding that it essentially consisted of applied arithmetic used for specific problems. The calculations were invariably practical and no attempt was made to write general rules that followed from deductive logic. We can safely say that their geometry provided nothing of theoretical value that became part of modern day mathematics. To find contemporaries of the Egyptians who practiced more theoretical (and thus more enduring) mathematics, we must turn to the Babylonians.

1.2   The Babylonians

The "Babylonians" of the Mesopotamian valley actually consisted of many different people groups, including the Sumerians, Chaldeans, Akkadians, and the Persians. In contrast to the isolation of Egypt, the Fertile Crescent was a desirable land containing many caravan routes, which made it vulnerable to repeated invasion. Invariably, the invaders would absorb and become part of the culture, learning and contributing to Babylonian mathematics in the process. Most of what we know about this mathematics comes from relatively recent interpretations of cuneiform texts (clay tablets) from two distinct periods of Babylonian history: 2100 B.C. to 1600 B.C. and 600 B.C. to 300 A.D.

The Babylonians are well known for calculations in their sexagesimal (base 60) positional number system. The Babylonians also originated the division of a circle into 360 equivalent parts. One credible explanation for this division is that the Babylonian "time-mile," used in early Sumeria as a measurement of the amount of time it took to walk a "Babylonian mile,"[2] led the Babylonians to divide the circle into 360 equivalent parts.([18], Chapter 2) It is plausible that the selection of the number 360 was tied "to their astronomy, with its 360-day year."([22])

The Babylonians' consistent explanations of how to carry out certain calculations is indicative of their attempt to categorize theoretical problems, and in this sense they moved beyond the Egyptian focus on surveying and bookkeeping. While the Cairo Mathematical Papyrus (interpreted in 1962) finally gave evidence that the Egyptians were actually aware of several Pythagorean Triples,[3] a group of tablets discovered in Susa in 1936 had already established that the Babylonians used the Pythagorean theorem in general settings. The Babylonians equalled or surpassed the Egyptians in other areas of geometry as well: their approximations of π included $3\frac{1}{8} = 3.125$ (see Problem 1.5); they knew general formulas for areas of triangles and parallelograms and volumes of certain prisms and parallelepipeds[4]; they recognized the similarity of noncongruent figures; and they knew that an angle inscribed in a semicircle was a right angle. Much of Babylonian geometry was algebraic in flavor, as many of their problems reduced to solving a quadratic equation or a system of equations in several unknowns.

Although the Babylonians utilized general arithmetic and geometric rules, their language was still quite concrete. Moreover, no evidence was given that they employed methods of deductive logic or that they were even concerned with proving their results. They did not differentiate clearly

---

[2] A "Babylonian mile" was a distance equal to approximately seven of our miles.

[3] Three positive integers form a Pythagorean Triple if they represent lengths of sides of a right triangle.

[4] Interestingly, the Babylonians used two different formulas for the volume of the frustrum of a square pyramid, but only one of them was correct—and thus equivalent to the one obtained by the Egyptians (See [6], page 39, and Problem 1.6).

between calculations that were exact and those that were simply approximations. All in all, while they did pose many problems that were recreational in nature, their mathematics, like that of the Egyptians, was mainly utilitarian. The first civilization to become interested in abstract mathematics and to recognize the importance of a proof grounded in deductive logic was the Greeks.

1.3 The Greeks

As the second millennium B.C. came to a close, the Egyptian and Babylonian civilizations began to lose world prominence. A new civilization of independent city states emerged in trading towns along the coast of Asia Minor, southern Italy, and mainland Greece. In the 4th century B.C., Alexander the Great unified Greece and helped spread Greek culture throughout Europe and western Asia. The next few hundred years after Alexander the Great's death brought to an end a brilliant historical period of intellectual, scientific, and cultural progress. It was during this time that scholars began to ask "why" rather than "how," and mathematics as we know it today was born.

Although mathematical documents from this time period have not survived, oral tradition has generally given credit to Thales of Miletus (640 B.C. to 546 B.C.) for initiating the formal study of demonstrative mathematics. Legend has it that Thales dabbled in commercial ventures, and evidently his travels to Egypt introduced him to some of the geometric calculations used by the early Egyptians. He subsequently discovered many other geometric propositions himself, but in contrast to the intuition and experimental methods utilized by the Egyptians, Thales used deductive reasoning to validate his discoveries. The following theorems have been attributed to him by Herodotus and Proclus.

(1) An angle inscribed in a semicircle is a right angle. (This theorem was recognized by the Babylonians 1400 years earlier.)
(2) A circle is bisected by its diameter.
(3) The base angles of an isosceles triangle are equal.
(4) The vertical angles formed by two intersecting angles are equal.
(5) Two triangles are equal if they have one side and two adjacent angles, respectively, equal. (Some have conjectured that Thales used this proposition to determine the distance from a ship to shore.)

The next significant figure in the string of Greek mathematicians is Pythagoras (572–480 B.C.), who very well could have been a student of the much older Thales. We know little about the life of Pythagoras, because many of the stories that have been passed down are shrouded in mysticism as fables about him and his followers; however, it is understood that around the age of 50 he founded a distinctive school with political, philosophical, and religious aims. The Pythagoreans believed that all that was meaningful (i.e., religion, matter, music, mankind) could be described via whole numbers, and as a result, the "brotherhood" of students focused their study on number theory, music, geometry, and astronomy. These four subjects were collectively coined the "quadrivium" during the Middle Ages. Very little mathematics can be attributed specifically to Pythagoras himself, in no small part because his followers customarily gave him credit for *all* of their discoveries. Indeed, at least two authors suggest that he could not even prove his namesake proposition, the Pythagorean theorem.

Although the instruction in the school was predominantly oral in nature and the Pythagoreans supposedly swore to keep their findings secret, scholars generally believe that the Pythagoreans extended Thales' formalization of geometry and set the direction of future Greek mathematics. The following quote gives an apt summary of what we can conclude about Pythagoras:

> *It is indeed difficult to separate history and legend concerning the man, for he means so many things to the populace—the philosopher, the astronomer, the mathematician, the abhorrer of beans, the saint, the prophet, the performer of miracles, the magician, the charlatan. That he was one of the most influential figures in history is difficult to deny, for his followers, whether deluded or inspired, spread their beliefs over most of the Greek world.* ([6], page 49)

While the Pythagoreans did not keep an accurate written record of their work, the geometry teacher Hippocrates (460–380 B.C.), who was undoubtedly influenced by them, did. Although copies of his text *Elements of Geometry* have not survived, references to it from authors after him suggest that he was the first person to lay out propositions in a logical sequence and to use letters of the alphabet to represent points and lines in geometric figures. In these ways, his work foreshadowed Euclid's *Elements* by nearly a century. Hippocrates is best known for his work on the first two of the three construction problems from antiquity: use a straight edge and compass to square the circle, double the cube, and trisect the angle.

The philosopher Plato (427–348 B.C.) also contributed greatly to the development and spread of mathematical knowledge of the Hellenic Age. Around 388 B.C., he established the Academy in Athens, which is alleged to have had the front gate inscription "Let no man ignorant of Geometry enter here." Almost all of the mathematical advances of the 4th century B.C. were achieved by friends or students of Plato, so his Academy served as an important link between the early Pythagoreans and the later scholars at the Museum in Alexandria (founded by Ptolemy I around 300 B.C.). Plato is not known for doing much original mathematics; rather, his major influence on the subject stemmed from his enthusiastic conviction that the study of mathematics was invaluable for training of the mind. At Plato's Academy, mathematics became a pure science, as the mathematicians were not at all concerned with practical applications of their work. In fact, the philosophical Greeks of this time were pursuing intellectual activity in many academic areas for its intrinsic value.

Eudoxus (400–347 B.C.) of Cnidos was a student at Plato's Academy, and he is considered by some to be the most brilliant mathematician to precede Archimedes. He formalized the axiomatic method initiated by Plato's most celebrated student, Aristotle, but Eudoxus is best known for his rigorous definition of two magnitudes being *in proportion*. Although his definition applied to both commensurable and incommensurable quantities (what we would now call irrational ratios), which was somewhat disconcerting to the Pythagoreans, they did not subsequently acknowledge that he had discovered the irrational numbers.[5]

Another major contribution of Eudoxus resulted from his formulation of the *continuity lemma*, which states that if one magnitude can be multiplied by an integer to exceed another magnitude, the two magnitudes have a non-zero ratio. He used this lemma to validate his *Method of Exhaustion*, a calculus-type procedure later refined by Archimedes and used to find areas and volumes.

From the founding of Ptolemy's Museum in about 300 B.C. until its destruction by the Arabs in 641 A.D., the Museum and the adjacent Alexandrian library served as a central location for a

---

[5] Because of our familiarity with the real number line, in modern mathematics we freely associate, say, the diagonal of a unit square with a number representing its length. However, though the Pythagoreans did consider ratios of whole numbers, they made no such connection between "geometric magnitudes" and numbers, as their number set consisted of whole numbers only.

long line of scholars and their manuscripts. Of these scholars, the most important was undoubtedly Euclid (330–270 B.C.). His *Elements of Geometry* was a 13-book treatise containing a list of 467 propositions of plane and solid geometry that encompassed all of the known geometry at the time. Euclid drew upon much of the work of his predecessors: books I and II seem to have come from the Pythagoreans, books III and IV from Hippocrates, books V, VI, and XII from Eudoxus, and books X and XIII from Theaetetus (c. 417–369 B.C.).

Euclid's careful arrangement of axioms, definitions, and theorems formed an axiomatic system that became the standard curriculum for geometry students for over 2000 years. However, Euclid's work wasn't perfect and during the modern era mathematicians have clarified and modified the assumptions he used in the *Elements*; in the process, they discovered new types of geometries that share logical structures with the Euclidean (plane) geometry, but that result in theorems that do not hold in it. We further explore Euclid's assumptions and the subsequent modifications in the next chapter.

We must mention at least two other great Greek mathematicians who lived after Euclid. Archimedes (287–212 B.C.) of Syracuse, Sicily, when it was a Greek city-state, is often considered to be the greatest of the Greek mathematicians. In his mathematics, Archimedes developed methods very similar to the coordinate systems of analytic geometry and the limiting process of integral calculus. Apollonius of Perga (c. 262–190 B.C.) was a Greek geometer and astronomer noted for his research and writings on conic sections. His innovative methodology and terminology in the field of conics influenced many scholars for thousands of years to come.

What happened to the whole Greek culture, and to its mathematics in particular, in the last several centuries B.C. is still hard to comprehend. We wish to close this section with a quotation from Paul Valéry's *Crisis of the Mind (1919)*, [First Letter]:

> *Greece founded geometry. It was a mad undertaking: we are still arguing about the pos-sibility of such a folly. What did it take to bring about that fantastic creation? Consider that neither the Egyptians nor the Chinese nor the Chaldeans nor the Hindus managed it. Consider what a fascinating adventure it was, a conquest a thousand times richer and actually far more poetic than that of the Golden Fleece. No sheepskin is worth the golden thigh of Pythagoras. This was an enterprize requiring gifts that, when found together, are usually the most incompatible. It required argonauts of the mind, tough pilots who refused to be either lost in their thoughts or distracted by their impressions. Neither the frailty of the premises that supported them, nor the infinite number and subtlety of the inferences they explored could dismay them.*

1.4 Problems

Below we provide a few problems that we hope will help you reflect on the chapter and appreciate some of the accomplishments of our mathematical ancestors. Feel free to use any mathematics you have learned previously to help you with them.

1.1 Name seven great mathematicians born before the year 200 B.C. Where was each one from? Describe a main achievement of each one of them. Feel free to use a library or the internet.

1.2 In your view, what deep mathematical results were known to the ancient Egyptians? to the Babylonians? to the Greeks? Feel free to use a library or the internet.

1.3 Show that the Egyptian formula used for calculating the area of a quadrilateral, $A = \frac{1}{4}(a + c)(b + d)$ where a, b, c, and d are consecutive side lengths of the quadrilateral, is correct if used for finding the area of a rectangle. Will the approximation overestimate or underestimate area for a non-rectangular quadrilateral?

1.4 An isosceles triangle with side lengths 50, 50, and 60 is inscribed in a circle. Determine the radius of the circle. (The solution to this problem appears on a tablet discovered by the French at Susa in 1936, and gives evidence that the Babylonians knew the Pythagorean theorem.)

1.5 An Old Babylonian tablet gives the ratio of the perimeter of a regular hexagon to the circumference of the circumscribed circle. When converted from the base-60 Babylonian numeration system, the ratio given on the tablet is $24/25$. Show that this leads to $3\frac{1}{8}$ as an approximation of π.

1.6 As we have noted, the Babylonians used two different formulas to calculate the volume of the frustrum of a square pyramid. The two formulas are given below. Determine which is correct (and, thus, equivalent to the formula used by the Egyptians).

(a) $V = h(\frac{a+b}{2})^2$

(b) $V = h((\frac{a+b}{2})^2 + \frac{1}{3}(\frac{a-b}{2})^2)$

1.7 Can you find a way to prove the correct result given in Problem 1.6?

2

Axioms: From Euclid to Today

*Never undertake to prove things that are so evident in
themselves that one has nothing clearer by which to
prove them.*

— Blaise Pascal (1623–1662)

Euclid's *Elements of Geometry* is undoubtedly one of the most significant writings ever produced,
often cited as being second only to the Bible in total number of copies printed. Written around
300 B.C., this voluminous work had a tremendous impact on the historical development of mathematics. With his explicit and thorough use of the axiomatic method, Euclid brought to prominence
an approach that has come to dominate all of what is now known as "pure mathematics." His work
refrains from reliance on physical data or empirical evidence and contains no reference to geometric
applications. As a follower of the Platonic tradition, in which the physical is perceived as merely
the shadow of a conceptual ideal, Euclid was drawn to what he viewed as the more perfect world of
thought.

2.1   Axiomatic Systems

The significance of Euclid's work lies both in his recognition of the need for reducing mathematics
to a set of basic assumptions, building more complex results in logical fashion from that basis,
and in the systematic way in which he carried out this process. In geometry, as in the rest of
mathematics, conjectures arise most often as a result of experimentation, trial and error, or intuition.
The validity of a conjecture, though, is not conclusively established until a proof exists. While
intuition and experimentation may be convincing, they do not serve as proof in mathematics. Euclid
recognized the insufficiency of experimentation, approximation, and intuition for establishing truth,
and attempted instead to lay a thorough axiomatic foundation upon which proofs were built.

Greek mathematicians thought of geometry as a representation of the physical world, with terms
such as *point*, *line*, and *plane* corresponding to observable realities and interacting with each other
in ways that seemed obvious. In writing his *Elements*, Euclid aimed to provide definitions of basic
geometric objects, to clearly state (as axioms) the apparent truths about these objects, and then
to apply the rules of logic to derive other statements of geometric truth. ([**64**], p. 32) Within a
mathematical system, the collection of foundational truths forms an axiom set from which all other

truths within the system will be derived. For this reason, then, it is important that the axioms cover all of the essential concepts of the intuition being represented. The Euclidean axioms were seen as "self-evident truths" insofar as they were simply abstractions of what was "obvious" in the physical world.

2.1.1 Components of an axiomatic system

Any axiomatic system must contain the following elements: defined terms, axioms, rules of inference, and undefined terms. Let us examine each of these components more closely, in the context of Euclidean geometry. Certainly, the objects we wish to discuss, as well as the features of and relationships between these objects, must be defined as clearly as possible; this provides us with the requisite shared vocabulary for our discussion. Nevertheless, we will find that some terms must remain undefined, a reality which apparently eluded Euclid and his contemporaries, as his *Elements of Geometry* attempts to define even such basic terms as *point* and *line*. In Euclid's *Elements*, a *point* is defined as "that which has no part"—a definition that is not sufficient unless *part* is also defined. In the same vein, a *line* is defined as "breadthless length" and *straight line* as "that which lies evenly with the points on itself."

Just as there must be objects or relationships that are left undefined within a mathematical system, there must be statements that are accepted but left unproven. These statements, the axioms within an axiomatic system, should specify the essential assumptions upon which the system will be built. An axiom set should supply a sufficient number of clear statements so as to remove the ambiguities that often accompany the intuitive processes, but not so many as to become redundant or inefficient. Within a specified set, an individual axiom is **independent** of the other axioms if it cannot be proven from the others. If each individual axiom in a set is independent of the other axioms, then the entire set is said to be independent.

Theorems (or propositions) form the largest part of any interesting axiomatic system. Theorems are statements that are established through proof, making use of the defined and undefined terms, axioms, and previously proven theorems. Theorems are used to give general results rather than specific ones (e.g., the Pythagorean theorem establishes that $a^2 + b^2 = c^2$ where a, b, and c are the side lengths of <u>any</u> right triangle), and the validity of a theorem is determined using reasoning alone, rather than empirical data. The mechanism for constructing a proof is a set of rules of inference, or logic. That is, the proof of a theorem will consist of a sequence of statements, each of which follows (according to the laws of inference) from the ones preceding it. The proof must begin with an axiom, a definition, or the statement of a theorem that has already been proven, and it must end with the statement being proved. In this text, we will assume that the reader has knowledge of the standard rules of logic.

Obviously, an axiomatic system is worthless if it is possible to prove, using valid rules of logic and the axioms provided, a pair of contradictory statements. We call an axiomatic system **consistent** if deducing a pair of contradictory statements from the axiom set is impossible. Because consistency is considered to be such an important property for axiomatic systems, mathematicians require any worthwhile axiomatic system to be consistent. Note that, in contrast, a set of axioms need not be independent.

2.1.2 Models

One way of assessing both consistency and independence of an axiomatic system is through the use of *models*. In order to create a model for an axiomatic system, each undefined term must be associated with an object in such a way that the axioms hold. If the objects used to interpret the undefined terms are taken from the real world, then the model is called **concrete**; on the other hand,

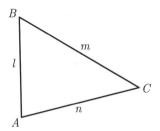

FIGURE 2.1. The three-point geometry.

if the interpretation of the undefined terms comes from another axiomatic system, then the model is called **abstract** (or **ideal**).

Example 1. This is an example of a **finite geometry**, in which there are finitely many points and lines. This particular finite geometry is often referred to as the three-point geometry. Consider the following set of axioms, in which *point*, *line*, and *on* are taken as undefined terms.

(1) There exist exactly three distinct points.

(2) Each two distinct points are on exactly one line.

(3) Not all the points are on the same line.

(4) Each two distinct lines are on at least one point.

   Model 1. Let $\{a, b, c\}$ be the set of *points*, and $\{\{a, b\}, \{a, c\}, \{b, c\}\}$ be the set of *lines*. We say that a point is *on* the line, or a line is *on* the point, if the point is an element of the line. This gives us an abstract model for the three-point geometry within set theory. The reader should verify that all axioms hold for this model.

   Model 2. Although *point* and *line* are undefined terms, the familiar interpretation of these terms leads us to the model given in Figure 2.1 for the three-point geometry within the Euclidean plane (an abstract model). The reader should verify that all axioms hold for this model.

   Model 3. Suppose each *point* is interpreted as a flavor of ice cream in your freezer (chocolate, vanilla, and pistachio) and the lines are ice cream cones. We say that a point is *on* the line, or a line is *on* the point, if one scoop of the given flavor is in the cone. Consider three cones, each having two scoops of ice cream:

   The first cone has one scoop of chocolate and one of vanilla;
   The second has one scoop of chocolate and one of pistachio;
   The third has one scoop of vanilla and one of pistachio.

With this interpretation, we can easily verify that a concrete model for the three-point geometry has been created.

A concrete model for an axiomatic system confirms the **absolute consistency** of the system. If contradictory statements could be proven from a set of axioms, then any concrete model for the system would exhibit corresponding "real life" contradictions, which we assume to be impossible. An abstract model verifies that the system has **relative consistency**. In this case, the consistency of one system is linked to the consistency of the other.

Finally, we call an axiom set **complete** if it is not possible to introduce an additional axiom to the set without also adding new undefined terms such that the new set of axioms is both consistent and independent. It could be said that maintaining independence of axioms places limits on the number of axioms included in a set, while completeness assures that the set of axioms is large enough. The questions of whether or not a set of axioms is complete, or independent, or consistent, can be very difficult to answer, and we will not examine them in depth here. The discussion of these fundamental questions can be found in books on mathematical logic. An excellent presentation of the notions discussed in this section appears in [**23**].

2.2 Euclidean and Non-Euclidean Geometries

> *His definitions do not always define, his axioms are not always indemonstrable, his demonstrations require many axioms of which he is quite unconscious.*
> — Bertrand Russell (1872–1970)

> *Almost from the time of its writing and lasting almost to the present, the Elements has exerted a continuous and major influence on human affairs. It was the primary source of geometric reasoning, theorems, and methods at least until the advent of non-Euclidean geometry in the 19th century.*
> — B. L. van der Waerden (1903–1996)

2.2.1 Euclid's axiomatic system

Euclid's *Elements of Geometry* begins with a list of 23 definitions, followed by Euclid's axiom set. As we have noted, *Elements of Geometry* mistakenly attempts to define <u>all</u> of the terms that are used. We will discover similar flaws in the axiom set. Nevertheless, Euclid's efforts at systematically recording all basic assumptions and communicating with clarity are laudable.

Euclid divided his axioms into two categories: the "Common Notions" listed basic assumptions that applied to all of mathematics, while the "Postulates" provided statements that dealt specifically with geometry. Both are listed below, with parenthetical commentary given for some of the postulates.

The Common Notions

(1) Things that are equal to the same things are equal to one another.

(2) If equals be added to equals, the wholes are equal.

(3) If equals be subtracted from equals, the remainders are equal.

(4) Things that coincide with one another are equal to one another.

(5) The whole is greater than the part.

The Postulates

(1) To draw a straight line from any point to any point. (*That is, two points determine a (unique) line.*[1])

(2) To produce a finite straight line continuously in a straight line. (*That is, any line segment may be extended to a line segment that contains it.*)

---

[1] Although not stated, Euclid apparently assumes that the straight line between two points is unique. Likewise, the circle specified in Postulate (3) is assumed to be unique. Also, it is generally believed that in Euclid's postulates, lines are not infinite.

(3) To describe a circle with any center and radius. (*Given a specified point that will serve as center, and a specified distance to serve as radius, a (unique) circle can be created.*)

(4) That all right angles are equal to one another. (*This statement of congruence of all right angles establishes a useful and natural uniformity of angle measurement.*)

(5) That, if a straight line falling on two straight lines makes the interior angles on the same side less than two right angles, the straight lines, if produced indefinitely, meet on that side on which are the angles less than the two right angles.

It should be immediately apparent that the flavor of the fifth postulate (now known as the Euclidean parallel postulate) differs from that of the first four, as it is not as intuitively "obvious" – in addition to being much longer! Euclid himself seemed uncomfortable with the fifth postulate, as evidenced by the fact that he delayed explicit use of this statement in a proof until the 29th proposition in his *Elements* ([23], p. 148), even though avoiding Postulate (5) sometimes required writing a much longer proof than would otherwise be necessary.

The more familiar form of this fifth postulate is an equivalent statement known as Playfair's axiom.[2] Playfair's axiom requires a definition of parallel lines. Euclid wrote: "parallel straight lines are straight lines that, being in the same plane and being produced indefinitely in both directions, do not meet one another in either direction." That is, two distinct co-planar lines are defined as **parallel** if they do not intersect. Using this definition, consider Playfair's Axiom:

For every line l, and for every point P that does not lie on l, there exists a unique line m through p that is parallel to l.

2.2.2 Non-Euclidean geometries

For nearly 2000 years, beginning with the Greeks of Euclid's time, many mathematicians suspected that the fifth postulate was not a postulate at all, but a theorem. Hundreds of mathematicians attempted to find a proof that would settle the question, and many an errant proof was accepted as valid for years before a flaw was discovered. Most commonly, the flaw involved some form of circular reasoning, in which an assumed statement turned out to be equivalent to the Euclidean parallel postulate.

Why was this strange activity taking place? One explanation is that people have always had a strong instinct *to know* whether they can answer certain questions and were always confident that the answers existed and could be found. In many cases, a practical outcome to the quest mattered little or not at all.

The turning point in the pursuit of a proof for Euclid's fifth postulate came with the work of Italian Jesuit mathematician Girolomo Saccheri (1667–1733). Saccheri's plan was to consider an axiom set consisting of the *negation* of the fifth postulate, the other four postulates, and the common notions, and show that this axiom set was inconsistent. Doing so would establish the fifth postulate as a theorem.

Taking the Euclidean parallel postulate as stated by Playfair, we see that the negation is: For some line *l*, there is a point *P* not on *l* such that either (a) no line through *P* is parallel to *l*, or (b) more than one line through *P* is parallel to *l*. Saccheri considered each of these options, hoping to find a contradiction to one of the first four postulates or to one of the first twenty-eight

---

[2] This form of the Euclidean parallel postulate is the namesake of Scottish mathematician John Playfair (1748–1819). As with many mathematical namesakes, however, Playfair was not the originator of this statement. In fact, it was given as early as the fifth century by the Greek mathematician Proclus (411–485). See [18], page 496.

propositions given by Euclid (all of which were purported to hold without reference to the parallel postulate). Alternative (a) led to a contradiction of Euclid's Proposition 27, allowing Saccheri to turn his attention to the second option. Saccheri invested considerable time and energy examining the modified axiom set with a postulate of multiple parallels. In this investigation, he discovered many theorems whose conclusions were vastly different from those of Euclidean geometry. In Saccheri's words, the statements he deduced were "repugnant to the nature of a straight line." ([**64**], p. 66; [**23**], p. 155) That is, Saccheri was able to prove results that seemed quite wrong to a mathematician whose experience was rooted in Euclidean geometry. These results seemed intuitively incorrect to Saccheri.

Saccheri published his work in a book whose complete title is *"Euclides ab omni naevo vindicatus: sive conatus geometricus quo stabiliuntur prima ipsa universae Geometriae principia." (Euclid Freed of Every Flaw: A geometrical work in which are established the fundamental principles of a universal geometry.)* ([**23**], p. 154) Because the results that Saccheri discovered were contradictory to what he believed to be the only valid geometry, he concluded that he had, indeed, discovered an inconsistency and "freed Euclid of his flaws." In fact, Saccheri never succeeded in finding the inconsistency he hoped for, although he failed to recognize his own flaw! While Saccheri believed that he had established Euclid's fifth postulate as a theorem, he had instead stumbled upon the first statements of a non-Euclidean geometry, now known as hyperbolic geometry.

By the early nineteenth century, other mathematicians were beginning to acknowledge the possibility of alternative, non-Euclidean geometries, specifically that of the "multiple parallels" geometry unwittingly introduced by Saccheri. Almost concurrently, in 1829 and 1831, formal presentations of a new geometry were produced by Nicolai Lobachevsky (1792–1856) of Russia and János Bolyai (1802–1860) of Hungary. It is also true that the German mathematician Karl Friedrich Gauss (1777–1855) discovered many results of hyperbolic geometry years earlier, but he did not publish them. Not being able to find a contradiction – i.e., two contradictory statements in their axiomatic system— made these mathematicians believe that no such contradiction existed. But how could one be sure of this? Maybe a contradiction would be found with just one more day, or month, or year of searching.

In 1868, the Italian mathematician Eugenio Beltrami (1835–1900) proved that the Lobachevsky and Bolyai geometry is consistent as long as the Euclidean geometry is, by suggesting its model within the Euclidean geometry. Hence, any contradiction in hyperbolic geometry implies a contradiction in Euclidean geometry. Do we know that Euclidean geometry is consistent? What we know at this time is that it is as consistent as the arithmetic of integers . . .[3]

Now the door was opened for other geometries to be considered and by 1870 many different geometries were constructed. The ultimate recognition of alternative geometries—perhaps less intuitive but nonetheless consistent—caused a shift in the perception of the role of axioms. While axioms had previously been viewed as written expressions of an obvious reality, the modern emphasis is primarily on consistency. The mathematical perception of axioms became more abstract with the discovery of non-Euclidean geometries, which led to profound changes in people's understanding of geometry and of mathematics as a whole.

2.2.3   More on Euclid's "Elements"

Amidst the flurry of activity around non-Euclidean geometries, the work of Euclid himself was re-examined with a more critical eye. With a more rigorous understanding of axiomatic systems and logic, what was previously accepted as adequate was now seen as lacking. We have already observed Euclid's failure to include undefined terms. In addition, note that nowhere in Euclid's list

---

[3] To be more explicit, the coordinate method demonstrates that Euclidean geometry is as consistent as the arithmetic of the real numbers, which can be shown to be as consistent as the arithmetic of the rational numbers, which, in turn, is as consistent as the arithmetic of the integers.

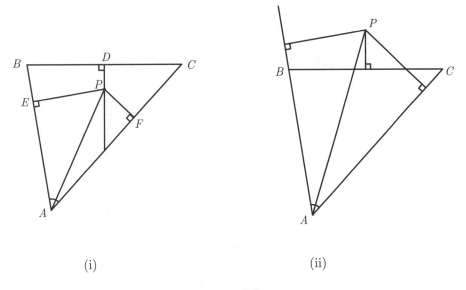

(i)   (ii)

FIGURE 2.2.

of axioms does he provide a statement that implies the existence of <u>any</u> points or lines! Similarly, there is no axiom specifying that at least two distinct points lie on each line, nor any that precludes all points from being collinear. Because Euclid's axioms rely heavily on his own understanding of plane geometry, he failed to explicitly state some of his most basic assumptions, which resulted in his axiom set being incomplete.

A further critique of Euclid's methods focuses on his dependence upon diagrams in proofs. Because Euclid's diagrams generally adhered to the implicit assumptions of Euclidean geometry (including those that Euclid failed to state as axioms), the majority of Euclid's theorems are valid. Nevertheless, not all of his "proofs" are logically sound. The following well-known "proof" (<u>not</u> attributed to Euclid) that all triangles are isosceles demonstrates the kind of erroneous reasoning that can result from over-reliance on diagrams. The example and the paragraph following include terms that have not yet been introduced in this text, although they are ones that many readers will recognize from a high school geometry class.

Example 2. "*Prove*": Any triangle is isosceles (i.e., has two sides of the same length).[4]

"*Proof*": Given a triangle ABC, construct the angle bisector of the interior angle at A and the perpendicular bisector of segment $\overline{BC}$ at point D. If the angle bisector at A and the perpendicular bisector of $\overline{BC}$ are parallel (in fact, coincident), we can immediately conclude that $\triangle ABC$ is isosceles. On the other hand, if the angle bisector at A and the perpendicular bisector of $\overline{BC}$ are not parallel, then they intersect at a point which we label P. Let E and F be the feet of the perpendiculars from P to $\overline{AB}$ and $\overline{AC}$, respectively, as shown in Figure 2.2(i). Now $\triangle AEP \cong \triangle AFP$ (AAS) and $\triangle BDP \cong \triangle CDP$ (SAS), from which it follows that $PE = PF$ and $PB = PC$. Since $\triangle BEP$ and $\triangle CFP$ are right triangles, the third sides must also be congruent, so that $\triangle BEP \cong \triangle CFP$ (SSS). Thus, $AB = AE + EB = AF + FC = AC$, and $\triangle ABC$ is isosceles. □

The logical fallacy in this argument is cultivated by an inaccurate diagram. In fact, when the angle bisector at A and the perpendicular bisector of $\overline{BC}$ intersect at a point, P, that point must lie outside

---

[4] From [**64**], p. 37.

of $\triangle ABC$ (in fact, on its circumcircle). (See Figure 2.2(ii).) Precisely one of E or F, however, must lie inside $\triangle ABC$ while the other lies in the triangle's exterior, resulting in a figure like the one below. Hence, the fallacy comes in our assumption that $AB = AE + EB$ and $AC = AF + FC$, highlighting the importance of a clear understanding of the concept of "betweenness."

2.3 Alternate Axiom Sets

2.3.1 Hilbert

As noted in the previous section, the acknowledgement that more than one consistent geometry exists, and the realization that Euclid's axiom system had flaws, took root by the late 1800's. ([23], p. 185) As a response to the latter realization, a string of mathematicians undertook the quest for a more rigorous and complete foundation for Euclidean geometry. One of the earliest and most notable attempts came from the German mathematician David Hilbert (1862–1943). Considered one of the world's leading mathematicians in the early twentieth century, Hilbert's mathematical interests were tremendously diverse. His interest in geometry was one facet of a larger goal to establish consistent, independent, and complete axiomatic systems for various branches of mathematics. Although it was eventually shown (by Kurt Gödel in 1931) that such a goal is unattainable, Hilbert's work in geometry represents an important turning point in the formalization of mathematics.

David Hilbert's systematization of Euclidean geometry built on the work of Moritz Pasch, who published the first rigorous treatise on geometry in 1882. ([56], p. 8) Hilbert's goals were to expand Euclid's axiom set to one that was complete and to introduce a level of rigor and formal logic beyond that used by Euclid. ([64], pp. 45–46) Hilbert took as undefined terms *point*, *line*, *plane*, *on*, *between*, and *congruent*. His axiom set includes 16 statements, each placed in one of five sets. The heading for each group of axioms gives some indication of the Euclidean concept being addressed: Axioms of connection (or incidence), Axioms of order, Axioms of congruence, Axiom of parallels (stated in the form of Playfair's axiom), and Axioms of continuity. Particularly notable is that the Axioms of continuity identify a clear connection between Euclidean geometry and formal systems of arithmetic.

While Hilbert's presentation of Euclidean geometry satisfactorily addresses most of the problems found in *Elements of Geometry*, the axioms are stated in a way that resembles Euclid's axiom set, making them quite intuitive for the mathematical community of Hilbert's time. In addition, Hilbert maintained a level of rigor that became the standard for subsequent work in mathematics. The complete set of Hilbert's axioms can be viewed at the end of the chapter.

2.3.2 Birkhoff

Hilbert's treatment of Euclidean geometry appeared in *Foundations of Geometry*, published in 1899. Several other versions of Euclidean axiom sets appeared soon after, including those of Oswald Veblen in 1904, Edward (E.V.) Huntington in 1913, and Henry Forder in 1927. ([56], pp. 8–9) In 1932, American mathematician George D. Birkhoff (1884–1944) published a Euclidean axiom set with a notably different flavor than those of Hilbert and Euclid. Birkhoff's remarkably concise axiom set contains only four statements and features measurement as the unifying concept. The powerful first postulate puts the points of any line in Euclidean geometry in one-to-one correspondence with the real numbers; in this way, all the properties of the real numbers are associated with the points of any line. Conspicuously absent from Birkhoff's axiom set is a statement of parallelism. In fact, Birkhoff's axioms are designed so that the Euclidean parallel postulate can be proven as a theorem.

The undefined terms in Birkhoff's system are *point*, *line*, *distance*, and *angle*, with the concept of *between* being defined in terms of distance. Birkhoff's postulate set is also given at the end of the chapter.

2.4   The SMSG Postulates

In the late 1950s, a growing concern that the United States was lagging behind the Soviet Union in mathematics and science education and research spurred the creation of the National Science Foundation. The early initiatives of the NSF included the establishment of the School Mathematics Study Group (SMSG), directed by Edward G. Begle and charged with a mandate to revise the U.S. secondary mathematics curriculum. Operating under the scope of this mandate, the SMSG produced a new axiom set for Euclidean geometry. Although many of the materials put forth by the SMSG were criticized and ultimately discarded, the SMSG axiom set was largely well-received and has continued to be widely used in high school and college geometry courses. ([64], p. 58)

The SMSG hoped to produce a Euclidean axiom set that is understandable and usable by high school students with little prior exposure to geometry and little experience with constructing proofs. In their quest for accessibility, the SMSG chose to relinquish independence of the axioms, recognizing that independence often necessitates an initial investment of time in proving "obvious" theorems. ([64], p. 59) As a result, the SMSG axiom set contains some redundancy but is highly readable. Because these are the axioms that we intend to reference throughout the text, we devote the remainder of this section to a more thorough discussion of the SMSG postulate set, patterning our treatment after the one found in [64].

The undefined terms identified in the SMSG postulate set are *point*, *line*, and *plane*. Interestingly, the word *contain* (along with several others) is not given as an undefined term, although it presumably plays the same role as the undefined term *on* used by Hilbert. Elements of the axiom sets of both Hilbert and Birkhoff are retained in the SMSG postulates. The twenty-two axioms in the SMSG set can be sub-divided into the following categories: Axiom of incidence, axioms about linear measure, axioms about space relationships, axioms of separation, axioms of angular measure, axiom of congruence, axiom of parallelism, and axioms about area and volume measure. We now state the postulates and give a brief accompanying discussion.

The SMSG Postulates for Euclidean Geometry

> **Postulate 2.1.**   *Given any two distinct points, there is exactly one line that contains them.*

The first axiom is equivalent to both Hilbert's Postulate I(1) and Birkhoff's Postulate (2) and is the single axiom addressing incidence between points and lines. With the next three postulates, the concept of distance is established, in the manner of Birkhoff's Postulate (1).

> **Postulate 2.2.**   *(The Distance Postulate.) To every pair of distinct points there corresponds a unique positive number. This number is called the distance between the two points.*

> **Postulate 2.3.** *(The Ruler Postulate.) The points of a line can be placed in a correspondence with the real numbers such that*
>
> (1) *To every point of the line there corresponds exactly one real number,*
> (2) *To every real number there corresponds exactly one point of the line, and,*
> (3) *The distance between two distinct points is the absolute value of the difference of the corresponding real numbers.*

> **Postulate 2.4.** *(The Ruler Placement Postulate.) Given two points P and Q of a line, the coordinate system can be chosen in such a way that the coordinate of P is zero and the coordinate of Q is positive.*

Postulate 2.3 allows any line to be viewed as a "number line," or ruler. The establishment of a one-to-one correspondence between the points of any line and the set of real numbers permits a natural understanding of the concept of *distance*, as indicated in Postulate 2.2. Furthermore, with Postulate 2.4, the freedom exists to coordinatize any line in such a way that both the point corresponding to 0 and the direction designated as positive can be chosen arbitrarily. ([**64**], p. 61)

The next four postulates (2.5–2.8) create a foundation for the familiar understanding of three-dimensional geometry. These four postulates, along with Postulate 2.1, formalize the notion of *incidence*: any two distinct points determine a unique line, a line and the points on that line are incident with the same plane(s), and any three non-collinear points determine a unique plane. The two subsequent postulates (2.9 and 2.10) specify a partitioning of a plane by any line in that plane, and of space by any plane. All of these axioms are in the style of Hilbert.

> **Postulate 2.5.** *(a) Every plane contains at least three non-collinear points. (b) Space contains at least four non-coplanar points.*

> **Postulate 2.6.** *If two points lie in a plane, then the line containing these points lies in the same plane.*

> **Postulate 2.7.** *Any three points lie in at least one plane, and any three non-collinear points lie in exactly one plane.*

> **Postulate 2.8.** *If two distinct planes intersect, then that intersection is a line.*

Postulate 2.9. *(The Plane Separation Postulate.) Given any line and a plane containing it, the points of the plane that do not lie in the line form two sets such that*

(1) *each of the sets is convex, and*

(2) *if P is in one set and Q is in the other, then segment $\overline{PQ}$ intersects the line.*

Postulate 2.10. *(The Space Separation Postulate.) The points of space that do not lie in a given plane form two sets such that*

(1) *each of the sets is convex, and*

(2) *if P is in one set and Q is in the other, then segment $\overline{PQ}$ intersects the plane.*

The next four postulates (2.11–2.14) are categorized as axioms of angle measure.

Postulate 2.11. *(The Angle Measurement Postulate.) To every angle there corresponds a real number between 0 and 180.*

Postulate 2.12. *(The Angle Construction Postulate.)[5] Let $\overrightarrow{AB}$ be a ray on the edge of the half-plane H. For every r between 0 and 180 there is exactly one ray $\overrightarrow{AP}$ with P in H such that $m\angle PAB = r°$.*

Postulate 2.13. *(The Angle Addition Postulate.) If D is a point in the interior of $\angle BAC$, then $m\angle BAC = m\angle BAD + m\angle DAC$.*

Postulate 2.14. *(The Supplement Postulate.) If two angles form a linear pair, then they are supplementary.*

Postulates 2.13 and 2.14 are examples of the redundancy found in the SMSG postulates; each of these could be proven using the preceding axioms (along with a definition of *supplementary angles* as being angles whose measure sums to 180°).

---

[5] Postulates 2.11 and 2.12 are comparable to Birkhoff's Postulate (3), establishing a one-to-one correspondence between angle measures and the real numbers between 0 and 180. Note that while Birkhoff allows angle measures between 0° and 360°, the SMSG axiom set restricts the range of angle measures to 0° to 180°. ([**64**], p. 62)

Postulate 2.15 gives the familiar SAS congruence condition for triangles. A more thorough discussion of the concept of congruence is found in Chapter 3.

Postulate 2.15. *(The SAS Postulate.) Given a one-to-one correspondence between two triangles (or between a triangle and itself), if two sides and the included angle of the first triangle are congruent to the corresponding parts of the second triangle, then the correspondence is a congruence.*

In his *Elements of Geometry*, Euclid gives the SAS Postulate as a theorem. In order to prove this theorem, Euclid makes the assumption that figures in the plane can be moved from one position to another. Although such a transformation is indeed possible, as we will see in Chapter 12, Euclid never formally establishes its validity. Consequently, within the confines of his manuscript, making use of such a technique is not legitimate.[6]

Postulate 2.16 states the Euclidean parallel postulate, given in the familiar form of Playfair's Postulate.

Postulate 2.16. *(The Parallel Postulate.) Through a given external point there is at most one line parallel to a given line.*

The next four postulates (2.17–2.20) address areas of plane regions.

Postulate 2.17. *To every polygonal region there corresponds a unique positive real number called its area.*

Postulate 2.18. *If two triangles are congruent, then the triangular regions have the same area.*

Postulate 2.19. *Suppose that the region R is the union of two regions R_1 and R_2. If R_1 and R_2 intersect at most in a finite number of segments and points, then the area of R is the sum of the areas of R_1 and R_2.*

---

[6] Hilbert apparently recognized the problem with Euclid's proof of the SAS congruence condition, but he chose not to incorporate the use of transformations into his presentation of Euclidean geometry. Instead, Hilbert states a slightly weaker condition in Postulate III(5):

If for two triangles, $\triangle ABC$ and $\triangle A'B'C'$, the congruences $\overline{AB} \equiv \overline{A'B'}$, $\overline{AC} \equiv \overline{A'C'}$, and $\angle BAC \equiv \angle B'A'C'$ are valid, then the congruence $\angle ABC \equiv \angle A'B'C'$ is also satisfied.

Using this weaker statement, Hilbert proved the stronger SAS congruence condition. Again, the decision of the SMSG was that there was little instructional value to be gained by giving the weaker condition and requiring a proof of the full SAS. (In fact, it is interesting to note that some high school geometry books include the ASA and SSS congruence conditions for triangles as postulates also.)

> **Postulate 2.20.** *The area of a rectangle is the product of the length of its base and the length of its altitude.*

These postulates formally assert the Euclidean properties associated with *area*, some of which Euclid himself assumed but did not state. Specifically, each plane region has an associated unique positive real number, called its "area"; congruent triangles—and, consequently, other congruent polygons—encompass equal areas; and areas within a partitioned region can be treated additively. Postulate 2.20 gives a formula for determining the area of a rectangle. This area formula, in combination with Postulate 2.19, allows formulas for other planar figures to be determined.

The concluding SMSG postulates (2.21 and 2.22) turn again to three-dimensional geometry. While we do not treat solid geometry in this book, we include these postulates for completeness. Postulate 2.21 extends the formula for the area of a rectangle given in Postulate 2.20 to a formula for the volume of a rectangular parallelepiped.

> **Postulate 2.21.** *The volume of a rectangular parallelepiped is equal to the product of the length of its altitude and the area of its base.*

> **Postulate 2.22.** *(Cavalieri's Principle.) Given two solids and a plane, if for every plane that intersects the solids and is parallel to the given plane the two intersections determine regions that have the same area, then the two solids have the same volume.*

These SMSG axioms will be referred to in upcoming chapters. Although this initial discussion has been cursory, a deeper understanding of these foundational concepts will be developed as we proceed.

We hope that our readers know that mathematics is more than logic, and that very few mathematicians spend much of their time thinking about axiomatic systems. We agree with Gian-Carlo Rota (1932–1999), who wrote:

> *The facts of mathematics are verified and presented by the axiomatic method. One must guard, however, against confusing the presentation of mathematics with the content of mathematics. An axiomatic presentation of a mathematical fact differs from the fact that is being presented as medicine differs from food. It is true that this particular medicine is necessary to keep mathematicians from self-delusions of the mind. Nonetheless, understanding mathematics means being able to forget the medicine and enjoy the food.*[7]

And this is what we intend to do in the rest of the book.

---

[7] From *Indiscrete Thoughts*, Gian-Carlo Rota, Birkhauser, 1997, p. 96.

2.5   Problems

> *Mathematicians are like lovers ... Grant a mathematician the least principle, and he will draw from it a consequence which you must also grant him, and from this consequence another.*
>
> — B. Fontenelle[8] (1657–1757)

2.1 (a) Give an example of an inconsistent axiomatic system.

    (b) Give an example of a dependent axiomatic system.

        A geometry in which there are only a finite number of points is called a **finite geometry**.

2.2 Consider the following axiom set for a finite geometry. The undefined terms are *point*, *line*, and *on*.

    Axiom A: Every two distinct points are on a line.

    Axiom B: Every two distinct lines are on a point.

    Axiom F: There exists exactly 10 points.

    Is the set of axioms consisting of Axioms A, B, and F consistent? Is it independent?

2.3 Consider the following axiom set for a finite geometry known as the Fano plane, in honor of Italian mathematician Gino Fano (1871–1952). The undefined terms are *point*, *line*, and *on*.

    Axiom 1: There exists at least one line.

    Axiom 2: There are exactly three points on every line.

    Axiom 3: Not all points are on the same line.

    Axiom 4: There is exactly one line on any two distinct points.

    Axiom 5: There is at least one (common) point on any two distinct lines.

    (a) Construct a model for the Fano plane.

    (b) Is this set of axioms consistent?

    (c) Is this set of axioms independent?

    (d) Does the parallel postulate hold in this axiomatic system?

2.4 Using the axioms of the Fano plane (see Problem 2.3), provide a proof of each of the following statements.

    (a) In the Fano plane, any two distinct lines have exactly one point in common.

    (b) The Fano plane contains exactly seven points and seven lines.

2.5 Consider the following axiom set. The undefined terms are *point*, *line*, and *on*.

    Axiom A: Every two distinct points are on a line.

    Axiom B: Every two distinct lines are on a point.

    Axiom C: There exist four points such that no three of them are together on a line.

    Axiom D: There is a line with exactly three points on it.

    (a) Prove that the set of five axioms from Problem 2.3 is equivalent to the set of four axioms presented in this problem in the following sense: each set of axioms implies every axiom from the other set.

    (b) Suppose *Axiom D* is replaced by the following.

        Axiom E: There is a line with exactly 4 points on it.

        Is the set of axioms consisting of Axioms A, B, C, and E consistent? Is it independent?

---

[8] From *The World of Mathematics*, Volume 3, James R. Newman, Simon and Shuster, 1956.

2.6   Chapter 2 Appendix

Hilbert's Axioms (as stated in [64])

Group I: Axioms of Incidence (Connection)

(1) For every two points A and B, there exists a line that contains each of the points A and B.

(2) For every two points A and B, there is no more than one line that contains each of the points A and B.

(3) There exist at least two points on a line. There exist at least three points that do not lie on a line.

(4) For any three points A, B, and C that do not lie on the same line, there exists a plane, α, that contains each of the points A, B, and C. For every plane, there exists a point that it contains.

Group II: Axioms of Order

(1) If point B is between points A and C, then A, B, and C are distinct points on the same line and B is between C and A.

(2) For any two distinct points A and C, there is at least one point B on the line $\overleftrightarrow{AC}$ such that C is between A and B.

(3) If A, B, and C are three points on the same line, then no more than one is between the other two.

(4) Let A, B, and C be three points that are not on the same line, and let l be a line in the plane containing A, B and C that does not meet any of the points A, B or C. Then, if l passes through a point of the segment $\overline{AB}$, it will also pass through a point of segment $\overline{AC}$ or a point of segment $\overline{BC}$.

Group III: Axioms of Congruence

(1) If A and B are two points on a line a, and if A' is a point on the same or on another line a', then it is always possible to find a point B' on a given side of the line a' such that $\overline{AB}$ and $\overline{A'B'}$ are congruent.

(2) If a segment $\overline{A'B'}$ and a segment $\overline{A''B''}$ are congruent to the same segment $\overline{AB}$, then segments $\overline{A'B'}$ and $\overline{A''B''}$ are congruent to each other.

(3) On a line a, let $\overline{AB}$ and $\overline{BC}$ be two segments that, except for B, have no point in common. Furthermore, on the same or another line a', let $\overline{A'B'}$ and $\overline{B'C'}$ be two segments that, except for B', have no points in common. In that case, if $\overline{AB}$ is congruent to $\overline{A'B'}$ and $\overline{BC}$ is congruent to $\overline{B'C'}$, then $\overline{AC}$ is congruent to $\overline{A'C'}$.

(4) If $\angle ABC$ is an angle and if $\overrightarrow{B'C'}$ is a ray, then there is exactly one ray $\overrightarrow{B'A'}$ on each "side" of $\overleftrightarrow{B'C'}$ such that $\angle A'B'C'$ is congruent to $\angle ABC$. Furthermore, every angle is congruent to itself.

(5) If, for two triangles $\triangle ABC$ and $\triangle A'B'C'$, the congruences $\overline{AB} \cong \overline{A'B'}$, $\overline{AC} \cong \overline{A'C'}$, and $\angle BAC \cong \angle B'A'C'$ are valid, then the congruence $\angle ABC \cong \angle A'B'C'$ is also satisfied.

Group IV: Axiom of Parallels

(1) Let a be any line and A a point not on it. Then there is at most one line in the plane that contains a and A that passes through A and does not intersect a.

Axioms of Continuity

(1) *Archimedes' Axiom.* If $\overline{AB}$ and $\overline{CD}$ are any segments, then there exists a number n such that n copies of $\overline{CD}$ constructed contiguously from A along the ray $\overrightarrow{AB}$ will pass beyond the point B.

(2) *Axiom of Line Completeness.* An extension of a set of points on a line with its order and congruence relations that would preserve the relations existing among the original elements, as well as the fundamental properties of line order and congruence that follow from Axioms I-3 and V-1, is impossible.

Birkhoff's Postulates (as stated in [64])

Definitions

(1) A point B is *between* A and C ($A \neq C$) if $d(A, B) + d(B, C) = d(A, C)$.

(2) The points A and C together with all points B between A and C form *line segment* $\overline{AC}$.

(3) The *half-line m'* with *endpoint O* is defined by two points, O and A in line m ($A \neq O$) as the set of all points A' of m such that O is not between A and A'.

(4) If two distinct lines have no points in common, they are *parallel*. A line is always regarded as parallel to itself.

(5) Two half-lines m and n through O are said to form a *straight angle* if $m\angle mOn = (\pm)\pi$. Two half-lines m and n through O are said to form a *right angle* if $m\angle mOn = (\pm)\pi/2$, in which case we also say that m is *perpendicular* to n.

(6) If A, B and C are three distinct points, the three segments $\overline{AB}$, $\overline{BC}$, and $\overline{CA}$ are said to form a *triangle*, $\triangle ABC$, with sides $\overline{AB}$, $\overline{BC}$, and $\overline{CA}$ and vertices A, B, and C. If A, B and C are collinear, then $\triangle ABC$ is said to be degenerate.

(7) Any two geometric figures are *similar* if there exists a one-to-one correspondence between the points of the two figures such that all corresponding distances are in proportion and corresponding angles have equal measures (except, perhaps, for their sign). Any two geometric figures are *congruent* if they are similar with a constant of proportionality $k = 1$.

Postulates

(1) *Postulate of Line Measure.* The points A, B, $\ldots$, of any line m can be placed into a one-to-one correspondence with the real numbers r so that $|r_B - r_A| = d(A, B)$ for all points A and B.

(2) *Point-line Postulate.* One and only one line m contains two given points P and Q ($P \neq Q$).

(3) *Postulate of Angle Measure.* The half-lines $m, n, \ldots$, through any point O can be placed into a one-to-one correspondence with the real numbers a (mod 2π) so that if $A \neq O$ and $B \neq O$ are points of m and n, respectively, the difference $(a_n - a_m)$ (mod 2π) is $m\angle AOB$.

(4) *Postulate of Similarity.* If $\triangle ABC$ and $\triangle A'B'C'$ and, for some positive constant k, $d(A', B') = kd(A, B)$, $d(A', C') = kd(A, C)$, and also $m\angle BAC = (\pm)m\angle B'A'C'$, then $d(B', C') = kd(B, C)$ and $m\angle A'B'C' = (\pm)m\angle ABC$ and $m\angle A'C'B' = (\pm)m\angle ACB$.

3

Lines and Polygons

3.1   Introduction

In the beginning everything is self-evident, and it is hard to see whether one self-evident proposition follows from another or not. Obviousness is always the enemy of correctness.

— Bertrand Russell (1872–1970)[1]

In this brief section we present some definitions and facts concerning lines, segments, angles, and polygons that will be utilized in subsequent sections. Though we will be employing the SMSG Postulates in this text, the reader is welcome to use any definition or axiom from any other axiomatic system—even Euclid's definition of a (straight) line as "that which lies evenly with the points on itself," if that helps! Our goal is to review basic facts and quickly pass to more interesting (or less "obvious") theorems and problems, rather than to provide careful proofs of these facts from the axioms. Many of the concepts discussed in this introductory section are likely to be familiar to the reader, in which case this section can serve primarily as a reference to be consulted as needed.

3.1.1   Segments, angles, and polygons

Let's consider Postulates 2.1–2.4. Suppose A and B are two distinct points in the geometry specified by the postulates. By Postulate 2.1, there is exactly one line, which we denote $\overleftrightarrow{AB}$, containing both A and B. By Postulate 2.4, we can identify point A with the number 0 and point B with any positive number.[2] Postulate 2.3 asserts that the remaining points of $\overleftrightarrow{AB}$ can be placed in one-to-one correspondence with the real numbers. From this, we conclude that there are infinitely many points on $\overleftrightarrow{AB}$, one for each real number.

Furthermore, Postulate 2.3 defines the **distance between two points** to be the absolute value of the difference between the numbers corresponding to those points, and by Postulate 2.2, this is the <u>unique</u> distance between two points. We will denote the distance between points P and Q by PQ or

---

[1] From *Beyond Reductionism*, as quoted in *Contemporary Abstract Algebra* by Joseph A. Gallian

[2] The ambiguity of the assignment of a value to B should not be troubling; the difference between assigning B a value of 1 and assigning B a value of 12, for example, could be thought of as equivalent to the difference between measuring a fixed length in feet or in inches. We just choose the scale.

FIGURE 3.1. The segment $\overline{AB}$, with $A - C - B$.

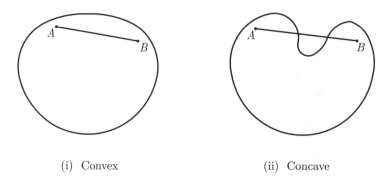

(i) Convex   (ii) Concave

FIGURE 3.2.

by $d(P, Q)$. So, if point P corresponds to the real number p and point Q corresponds to the number q, then $PQ = d(P, Q) = |p - q| = |q - p|$.

We say that a point C is **between** points A and B if A, B, and C are distinct points on the same line and $AC + CB = AB$; for brevity, we will denote this relationship as $A - C - B$. Given two distinct points A and B, we define the **line segment** $\overline{AB}$ to be the set of points containing A, B, and all points between A and B. Points A and B are called the **endpoints** of $\overline{AB}$, and any other point of the segment is called an **interior** point of $\overline{AB}$. The distance AB is also called the **length** of $\overline{AB}$.

A **figure** in the plane is just a subset of the points in the plane. A figure is called **convex** if it contains all interior points of the segment joining any two of its points. Otherwise, a figure is called **non-convex**, or **concave**. (See Figure 3.2 for examples of convex and concave figures.)

For distinct points A and B, the **ray** $\overrightarrow{AB}$ is the line segment $\overline{AB}$ together with all points D such that $A - B - D$. Rays BA and BC are called **opposite** if $A - B - C$. Two rays originating from the same point form two **angles**. The common endpoint of the rays is called the **vertex** of each angle, and the rays are called the **sides** of each angle. If the rays coincide, we say that we have one angle of measure $0°$ (**zero** angle), and another of measure $360°$. If the union of the rays is a line, we say that each angle has measure $180°$, and we call each of the two angles **straight**. (See Figure 3.3 (iv).)

In all other cases, by Postulate 2.12, one angle has a unique measure (in degrees) between 0 and 180. If $\overrightarrow{QP}$ and $\overrightarrow{QR}$ are the rays, we denote this angle by $\angle PQR$, and its measure by $m\angle PQR$; the measure of the second angle is defined as $360 - m\angle PQR$.[3] We take for granted that in this case the sides of angle PQR partition the plane into two subsets, one convex and the other concave, neither of which contain the points of the rays. The convex region is called the **interior** of the angle; the concave region is called its **exterior**. An angle with degree measure in the interval $(0°, 90°)$ is called **acute**; an angle with measure equal to $90°$ is called **right**; an angle with degree measure in the interval $(90°, 180°)$ is called **obtuse**. (See Figure 3.3.)

When the sum of the measures of two angles is $90°$, they are called **complementary** angles, and when the sum is $180°$, they are called **supplementary** angles. Two angles PQR and RQT form a

---

[3] We do not introduce a special notation for this angle, and when there is no danger of ambiguity, we refer to it as to the first one by writing $\angle PQR$.

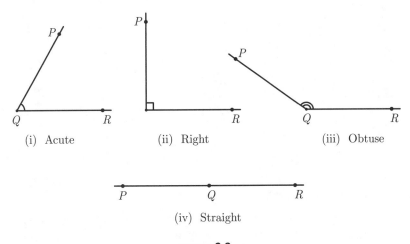

(i) Acute   (ii) Right   (iii) Obtuse

(iv) Straight

FIGURE **3.3.**

linear pair if P, Q, and T are collinear with $P - Q - T$. By Postulate 2.14, two angles that form a linear pair are supplementary angles.

Points, lines, and segments are used to define other geometric objects. One such object is the **polygon**, whose name is derived from the Greek words "poly" (many) and "gonia" (angle). For brevity, we will often simply call an n-sided polygon an n-gon. To define the term, let $A_1, A_2, \ldots, A_n$ be distinct points in a plane such that no three consecutive points are collinear, and suppose that no two of segments $\overline{A_1 A_2}$, $\overline{A_2 A_3}, \ldots, \overline{A_{n-1} A_n}$, and $\overline{A_n A_1}$ share an interior point. The n-**gon** $A_1 A_2 \ldots A_n$ is defined as the union of these n segments. Figure 3.4 shows three examples of 5-gons. The points $A_1, A_2, \ldots, A_n$ are the **vertices** of the polygon and the segments are the **sides** of the polygon.

Observe that a polygon partitions the plane into two sets – the **interior** of the polygon and the **exterior** of the polygon – with no point of the polygon in either of them. We say that a **polygon is convex** if its interior is a convex figure. Otherwise, we call it **non-convex** or **concave**. A polygon is called **regular** if all of its sides are congruent and all of its angles are congruent. (Again, see Figure 3.4.) A 3-gon is also called a **triangle**. All interior angles of a triangle have measure in $(0°, 180°)$. A 4-gon is called a **quadrilateral**, a 5-gon is a **pentagon**, and a 6-gon is a **hexagon**. For $n \geq 4$, an interior angle of an n-gon may have measure in $(180°, 360°)$.

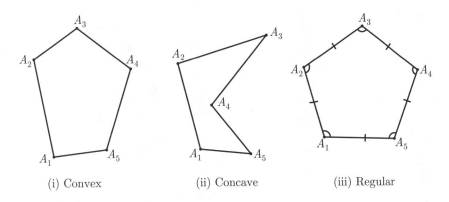

(i) Convex   (ii) Concave   (iii) Regular

FIGURE **3.4.** Polygons.

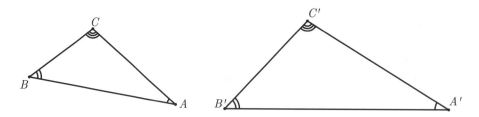

FIGURE 3.5. Similar triangles.

3.1.2 Similarity and congruence

Loosely speaking, two geometric figures are similar if they have the same shape, and they are congruent if they are similar and have the same size. More precisely, let Φ and Φ' be two figures in the plane. We say that Φ is **similar** to Φ', denoted here as $\Phi \sim \Phi'$, if there exists a positive real number k and an onto function $f : \Phi \to \Phi'$ mapping every point $A \in \Phi$ to a point $f(A) = A' \in \Phi'$ such that for every two points $A, B \in \Phi$, $A'B' = k\,AB$. The number k is called the **coefficient of similarity**. It can be shown that two polygons are similar if and only if there is a one-to-one correspondence between their vertices such that each pair of corresponding sides is in the same proportion and each pair of corresponding angles is congruent. Figure 3.5 illustrates similar triangles ABC and $A'B'C'$.

Clearly, every figure is similar to itself: take $k = 1$ and the identity function $f(x) = x$. If $\Phi \sim \Phi'$ with similarity coefficient $k = 1$, we say that Φ is **congruent** to Φ'. Intuitively, congruent figures are "exactly the same" even though they may be different sets of points. (See Figure 3.6.) Euclidean geometry is concerned only with those properties of figures that hold in all congruent figures. We write $\Phi \cong \Phi'$ to denote that Φ is congruent to Φ'.

Establishing similarity or congruence of figures by using the definition may not be a simple matter, as, strictly speaking, we have to determine a number k and a function f that satisfy the required properties. For similarity and congruence of two triangles, see, for example, Postulate 2.15, as well as several other statements from the following section on triangles. Some of the following facts are easy to prove, and others are taken as axioms:

- Any two infinite lines are congruent, as well as any two rays.

- Any two segments are similar, and they are congruent if and only if they are of equal length.

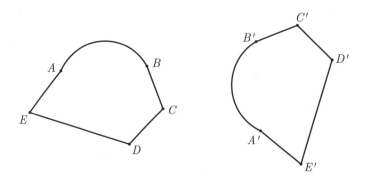

FIGURE 3.6. $ABCDE \cong A'B'C'D'E'$.

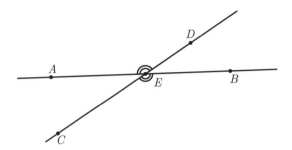

FIGURE 3.7. Vertical Angles are Congruent.

- Any two circles are similar, and they are congruent if and only if their radii are of equal length.
- Two angles are congruent if and only if they have equal measures.

Two congruent segments or two congruent angles are often depicted on diagrams by means of identical marks: a slash or equal number of slashes for segments, and an arc or equal number of arcs for angles. For example, this can be seen in Figure 3.4(iii), where the five congruent sides are each marked with a slash, and in Figure 3.5, where pairs of angles of the same measure have the same number of arcs.

3.1.3 Special angles and parallel lines

When two distinct lines intersect they form four angles having a common vertex. (See Figure 3.7.) The angles in each pair $\{\angle AEC, \angle BED\}$ and $\{\angle CEB, \angle DEA\}$ are called **vertical** angles.

Theorem 3.1. (Vertical angles are congruent.) *Suppose point E lies on the distinct segments $\overline{AB}$ and $\overline{CD}$. Then the vertical angles DEA and CEB are congruent, as are the angles AEC and BED. (See Figure 3.7.)*

Proof. Since $\angle AEC$ and $\angle CEB$ form a linear pair, they are supplementary. Likewise, $m\angle DEA + m\angle AEC = 180°$. Substitution gives $m\angle DEA = m\angle CEB$, as desired. The proof that the vertical angles AEC and BED are congruent is identical. $\square$

We define two lines to be **parallel** if they coincide or do not intersect. To indicate that lines AB and PQ are parallel, we write $\overleftrightarrow{AB} \parallel \overleftrightarrow{PQ}$. If lines AB and PQ are not parallel, they intersect at a unique point, say C. In this case we will write $\overleftrightarrow{AB} \cap \overleftrightarrow{PQ} = C$, rather than the technically correct but cumbersome $\overleftrightarrow{AB} \cap \overleftrightarrow{PQ} = \{C\}$.

When two distinct lines, l and m, not necessarily parallel, are intersected by a third line t, called a **transversal** (see Figure 3.8), then the following terminology is used to refer to the pairs of angles formed by them.

The angles in each pair $\{1, 5\}$, $\{2, 6\}$, $\{3, 7\}$, and $\{4, 8\}$ are called **corresponding** angles;
The angles in each pair $\{3, 6\}$ and $\{4, 5\}$ are called **alternate interior** angles;
The angles in each pair $\{1, 8\}$ and $\{2, 7\}$ are called **alternate exterior** angles;
The angles in each pair $\{3, 5\}$ and $\{4, 6\}$ are called **same side interior** angles.

The following theorem relates the congruence of some of these angles with parallelism of lines l and m. It provides the main tools for proof of two lines being parallel.

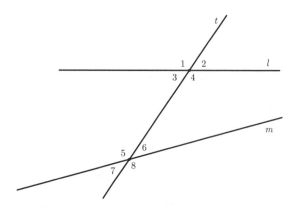

FIGURE 3.8.

Theorem 3.2. *Let l and m be distinct lines, and let t be a transversal of l and m. Then the following statements are equivalent.*

(1) *Lines l and m are parallel.*

(2) *Any two corresponding angles are congruent.*

(3) *Any two alternate interior angles are congruent.*

(4) *Any two alternate exterior angles are congruent.*

(5) *Any two same side interior angles are supplementary.*

Proof. We prove the equivalence of the statements (1) and (3) only. Equivalences between (3), (2), (4), and (5) are easier to demonstrate, and we leave those to the reader. Let t intersect l and m at points A and B, respectively, to form $\angle 1$, $\angle 2$, $\angle 3$, and $\angle 4$, as shown in Figure 3.9(i). We want to show that l and m are parallel if and only if $m\angle 1 = m\angle 4$ (and $m\angle 2 = m\angle 3$).

($\Rightarrow$) Assume that l is parallel to m. By Postulate 2.14, $m\angle 1 + m\angle 2 = 180°$ and $m\angle 3 + m\angle 4 = 180°$. Assume, for sake of argument, that $m\angle 1 > m\angle 4$. Then there exists a ray, $\overrightarrow{AC}$, on the opposite side of t from $\angle 4$, such that $m\angle CAB = m\angle 4$. (See Figure 3.9(ii).)

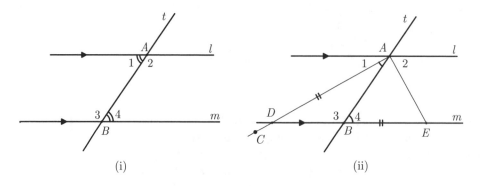

FIGURE 3.9. Congruent Alternate Interior Angles.

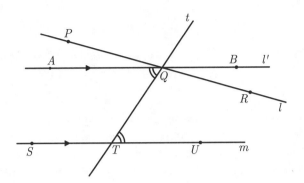

FIGURE 3.10.

Since C does not lie on l, by the Parallel postulate, line AC must intersect m at some point, D. Construct point E on line m such that $D - B - E$ and $BE = AD$. By the SAS postulate, $\triangle DAB$ is congruent to $\triangle EBA$; thus, $\angle 3 \cong \angle BAE$. Substituting congruent angles, $m\angle DAB + m\angle BAE = 180°$, making these a linear pair; therefore, D, A, and E are collinear, which implies that A lies on m, a contradiction.

($\Leftarrow$) Assume that l is not parallel to m. Let Q and T be the points of intersection of l and m, respectively, with t. Let P and R be points on l such that $P - Q - R$; let S and U be points on m such that $S - T - U$. (See Figure 3.10.)

By the Parallel postulate, there is a line l' through Q that is parallel to m. Let A and B be points on l' such that $A - Q - B$. Since l' is not coincident with l, we can assume without loss of generality that A is interior to $\angle PQT$. Hence, $m\angle PQA > 0$. By the first part of this proof, $\angle AQT \cong \angle QTU$. By the Angle Addition postulate (2.13), $m\angle PQT = m\angle PQA + m\angle AQT > m\angle AQT = m\angle UTQ$. Thus, $\angle PQT$ is not congruent to $\angle QTU$. $\qquad\square$

Two intersecting lines, l_1 and l_2, are said to be **perpendicular**, written $l_1 \perp l_2$, if the measure of each angle formed by the lines is $90°$. It follows from the Parallel postulate that given a line l and a point P not on l, there is a unique line m that contains P and is perpendicular to l. Let Q be the intersection of m and l. It can be shown that PQ is smaller than PR for any point R on l distinct from Q. (See Figure 3.11.) The length of $\overline{PQ}$ is called the **distance** from P to l, and it is denoted by $d(P, l)$.

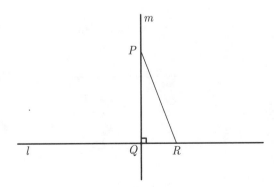

FIGURE 3.11.

3.2 Triangles

You, who wish to study great and wondrous things,...
must read these theorems about triangles...
— Regiomontanus[4] (1436–1476)

Triangles are the simplest and the most thoroughly studied polygons. Through the centuries people have discovered numerous structural properties about the triangle. Here we review only a few of the better known properties; additional facts can be found in the problems. Proofs in this section will use only "traditional" methods of Euclidean geometry. In subsequent chapters we will present alternative proofs of many statements from this section as illustrations of other mathematical ideas and techniques.

If A, B, and C are the vertices of a triangle, we denote the triangle by $\triangle ABC$. We say that sides $\overline{BC}$, $\overline{AC}$, and $\overline{AB}$ are **opposite** the angles A, B, and C, respectively, and vice versa. When convenient, the lengths of the sides of a triangle opposite angles A, B, and C will be denoted by a, b, and c, respectively. That is, $BC = a$, $AB = c$, and $AC = b$.

A triangle is called **acute** if each of its angles is acute (that is, having measure less than $90°$), **right** if one of its angles is right (having measure equal to $90°$), and **obtuse** if one of its angles is obtuse (having measure greater than $90°$).

3.2.1 Angle relations in polygons

We begin with the following theorem, which is certainly one of the simplest and most well-known facts about triangles, yet is not obvious at all!

Theorem 3.3. *The sum of the measures of the interior angles of a triangle is* $180°$.

Proof. Consider a line l through B parallel to $\overline{AC}$. (See Figure 3.12.) Let D and F be two points on l such that $D - B - F$. Then $m\angle A = m\angle DBA$ and $m\angle C = m\angle FBC$ as measures of alternate interior angles. Therefore,

$$m\angle A + m\angle B + m\angle C = m\angle DBA + m\angle ABC + m\angle FBC = 180°. \qquad \square$$

This theorem allows, almost immediately, generalizations to convex polygons with more than three sides.

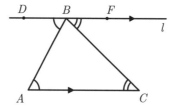

FIGURE 3.12.

---

[4] From *Regiomontanus on Triangles*, translated by Barnabas Hughes, Madison, University of Wisconsin Press, 1967.

Theorem 3.4. *The sum of the measures of the interior angles of a convex n-gon is $180(n-2)°$.*

Proof 1. Let $A_1 A_2 \ldots A_n$ be a convex polygon. Segments $A_1 A_3$, $A_1 A_4, \ldots, A_1 A_{n-1}$ lie inside of the polygon due to convexity, and therefore partition the polygon into $n-2$ triangles. (See Figure 3.13(i).) This partitions $\angle A_2 A_1 A_n$ of the polygon into $n-2$ angles, one from each triangle. For $3 \leq i \leq n-1$, the interior angle A_i of the polygon is partitioned into two angles by the triangles that share side $\overline{A_1 A_i}$. Therefore, the measure sum of all interior angles of the polygon is equal to the measure sum of the interior angles of all $n-2$ triangles, which is $180(n-2)°$. □

Proof 2. Let $A_1 A_2 \ldots A_n$ be a convex polygon and O be an interior point. (See Figure 3.13(ii).) Then all segments $A_i O$, $1 \leq i \leq n$, lie inside of the polygon and partition it into n triangles. For each i, the interior angle A_i of the polygon is partitioned into two angles by the pair of triangles that share side $\overline{O A_i}$. Therefore, the sum of the measures of these angles is equal to the sum of the measures of the interior angles of all these triangles, which is $180n°$, decreased by the sum of the measures of the angles of these triangles having O as a vertex, which is $360°$. Therefore, we have

$$m\angle A_1 + m\angle A_2 + \cdots + m\angle A_n = 180°n - (m\angle A_1 O A_2 + \cdots + m\angle A_n O A_1)$$
$$= (180n - 360)° = 180(n-2)°,$$

the desired result. □

While the interior angles of a polygon are practically self-defined, care must be taken when describing the exterior angles, as two (congruent ones) are present at each vertex. In our discussion and figures, when we refer to *the exterior angle* of a polygon, we will mean $\angle DBA$, where D is a point on the ray $\overrightarrow{CB}$ and points A, B, and C occur in clockwise order along the polygon. (See Figure 3.14.)

Corollary 3.5. (The Exterior Angle Theorem) *The measure of an exterior angle of a triangle is equal to the sum of the measures of its two remote interior angles.*

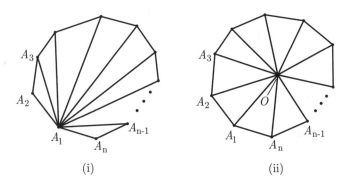

(i) (ii)

FIGURE 3.13. Convex Polygons.

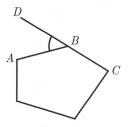

FIGURE 3.14. The exterior angle $\angle DBA$ of a polygon.

Proof 1. Let $\angle DCB$ be an exterior angle of $\triangle ABC$. (See Figure 3.15(i).) Since $m\angle A + m\angle B + m\angle C = 180°$ and $m\angle C + m\angle DCB = 180°$, we find that $m\angle A + m\angle B = m\angle DCB$. $\qquad\square$

Proof 2. Let $\angle DCB$ be an exterior angle of $\triangle ABC$. Let $\overrightarrow{CF}$ be a ray parallel to $\overline{AB}$. (See Figure 3.15(ii).) We accept without proof that $\overrightarrow{CF}$ lies in the interior of $\angle BCD$. Then $m\angle A = m\angle FCD$ (corresponding angles), and $m\angle FCB = m\angle B$ (alternate interior angles). Therefore,

$$m\angle A + m\angle B = m\angle FCD + m\angle FCB = m\angle BCD.$$

$\qquad\square$

The claim of the next corollary may be surprising: though the sum of measures of all interior angles in a convex n-gon $(= 180(n-2)°)$ depends on n, the sum of measures of the exterior angles (one at each vertex) does not.

> **Corollary 3.6.** *The sum of the measures of the exterior angles of a convex n-gon, one at each vertex, is* $360°$.

Proof. The sum of the measures of an interior angle and the exterior angle at a vertex is $180°$, since the angles form a linear pair. Adding the sums for all n pairs (one for each vertex) gives $180n°$. But the measure sum of all interior angles from these pairs gives $180(n-2)°$ by Theorem 3.4. Therefore, the measure sum, in degrees, of the exterior angles from these pairs is

$$180n - 180(n-2) = 180n - (180n - 360) = 360.$$

$\qquad\square$

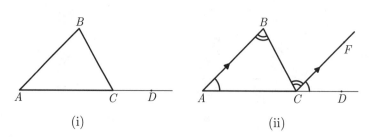

FIGURE 3.15.

3.2.2 Congruence of triangles

Suppose we have two triangles and we wish to prove that they are congruent. How can we do it? As we noted earlier, the question is not easy if we wish to use the definition of congruence: we have to define a distance-preserving bijection (one-to-one, onto function) between their points. In fact, seldom does anyone do it this way, though one such example may be seen in the footnote to our proof of Lemma 5.10. Rather, the following theorem is used, which we will accept without proof.[5] Essentially, when we wish to prove that two triangles are congruent, we will simply prove one of the (equivalent) statements listed in Theorem 3.7. (Note that SMSG Postulate 2.15 states that $\triangle ABC$ and $\triangle A'B'C'$ are congruent if and only if statement (3) holds.) Clearly, our intuition and experience with "moving" figures in a plane or in space are in agreement with this theorem.

> **Theorem 3.7.** (Conditions for Congruency of Triangles) *The following statements are equivalent.*
>
> (1) $\triangle ABC \cong \triangle A'B'C'$.
> (2) Side-Side-Side (SSS): *Three sides in $\triangle ABC$ are congruent to the corresponding three sides in $\triangle A'B'C'$.*
> (3) Side-Angle-Side (SAS): *Two sides and the angle formed by them in $\triangle ABC$ are congruent to the corresponding two sides and the angle formed by them in $\triangle A'B'C'$.*
> (4) Angle-Side-Angle (ASA): *Two angles and their common side in $\triangle ABC$ are congruent to the corresponding two angles and a common side in $\triangle A'B'C'$.*
> (5) Angle-Angle-Side (AAS): *Two angles and the side opposite to one of them in $\triangle ABC$ are congruent to the corresponding two angles and the side in $\triangle A'B'C'$.*

Theorem 3.8 provides one more statement which is useful for establishing congruence of right triangles. Its proof follows immediately from the Pythagorean theorem, which we will prove later in this section.

> **Theorem 3.8.** (HL Theorem) *Two right triangles are congruent if and only if the hypotenuse and a leg of one triangle are congruent to the hypotenuse and a leg of the other triangle.*

These theorems provide the standard techniques in Euclidean geometry for proving congruency, not just of two triangles, but also of two segments or two angles. Usually such proofs proceed as follows:

- We identify the pairs of corresponding segments (angles) of two triangles that we suspect are congruent.
- The congruency of the triangles is established by one of the tests: SSS, SAS, ASA, AAS, or HL.
- The segments (angles) are claimed to be congruent as the corresponding parts of congruent triangles.

In the next section we will employ the above approach in numerous examples. The claim cited in the last step is important enough for us to highlight:

---

[5] A proof of the statements in Theorem 3.7 can be found in many geometry texts; see [**64**], for example.

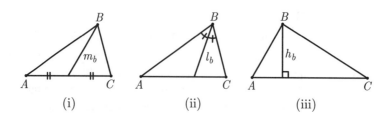

FIGURE 3.16. (i) Median, (ii) Angle Bisector, and (iii) Altitude.

CPCTC Principle: Corresponding Parts of Congruent Triangles are Congruent.

3.2.3 Median, bisector, and altitude

A **median** of a triangle is a segment connecting a vertex with the midpoint of the opposite side. The lengths of medians containing vertices A, B, and C are denoted by m_a, m_b, and m_c, respectively. The median from B in $\triangle ABC$ is shown in Figure 3.16(i).

A **bisector** of a triangle is a segment connecting a vertex with the point of intersection of the bisector of that angle with the opposite side. The lengths of bisectors containing vertices A, B, and C are denoted by l_a, l_b, and l_c, respectively. The bisector of $\angle B$ in $\triangle ABC$ is shown in Figure 3.16(ii).

An **altitude** of a triangle is a segment connecting a vertex with the base of the perpendicular dropped from that vertex onto a line containing the opposite side of the triangle. This base point can be inside the triangle, coincide with an endpoint of the side, or be outside the triangle. The lengths of altitudes containing vertices A, B, and C are denoted by h_a, h_b, and h_c, respectively. Figure 3.16(iii) shows the altitude from B in $\triangle ABC$.

3.2.4 Isosceles triangles

A triangle is called **isosceles** if two of its sides are congruent. The common vertex of the congruent sides is called the **vertex** of the isosceles triangle, and the side opposite of the vertex is called the **base**. In the next theorem we collect several characteristic properties of isosceles triangles. The list can be easily extended, but that is not our goal here.

Theorem 3.9. *For every triangle ABC, the following statements are equivalent.*

(1) $a = c$. *(The triangle is isosceles.)*

(2) $\angle A \cong \angle C$.

(3) $m_a = m_c$.

(4) $l_a = l_c$.

(5) $h_a = h_c$.

(6) *The bisector, median, and altitude at B coincide.*

Proof. To prove the equivalence of the six statements, we must establish that each statement implies each of the other statements. Thus, our proof consists of a series of proofs of implications; in this case, statement (1) is shown to be equivalent to each of the other statements.

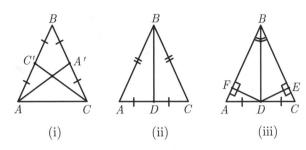

FIGURE 3.17.

$(1) \Rightarrow (2)$: $\triangle ABC \cong \triangle CBA$ by SAS $(\overline{AB} \cong \overline{BC}, \overline{CB} \cong \overline{AB}, \angle ABC \cong \angle CBA)$. Hence, $\angle A \cong \angle C$ by CPCTC.

$(2) \Rightarrow (1)$: $\triangle ABC \cong \triangle CBA$ by AAS $(\angle ABC \cong \angle CBA, \angle BAC \cong \angle BCA, \overline{AC} \cong \overline{CA})$. Hence, $AB = CB$ ($c = a$) by CPCTC.

$(1) \Rightarrow (3)$: Let A' and C' be the midpoints of $\overline{BC}$ and $\overline{BA}$, respectively. (See Figure 3.17(i).) Then $AC' = CA'(= a/2 = c/2)$, $AC = CA$, and $\angle C'AC \cong \angle A'CA$ (because $(1) \Rightarrow (2)$). Hence, $\triangle AA'C \cong \triangle CC'A$ by SAS, and $m_a = AA' = CC' = m_c$ by CPCTC.

$(3) \Rightarrow (1)$: We postpone the proof to Problem S3.2.7.

$(1) \Rightarrow (4)$: Since $(1) \Rightarrow (2)$, it is sufficient to show that $(2) \Rightarrow (4)$. This is immediate by applying ASA to two triangles defined by the base $\overline{AC}$ and the bisectors.

$(4) \Rightarrow (1)$: We postpone the proof to Problem 3.2.24.

$(1) \Rightarrow (5)$: Since $(1) \Rightarrow (2)$, it is sufficient to show that $(2) \Rightarrow (5)$. This is immediate by applying AAS to two right triangles defined by the base $\overline{AC}$ and the altitudes. Note that the argument does not depend on the measure of $\angle B$.

$(5) \Rightarrow (1)$: Since $(2) \Rightarrow (1)$, it is sufficient to show that $(5) \Rightarrow (2)$. The two right triangles containing the base $\overline{AC}$ and one of the altitudes are congruent by HL. Thus, by CPCTC, $\angle A \cong \angle C$.

$(1) \Rightarrow (6)$: Let D be the midpoint of $\overline{AC}$. (See Figure 3.17(ii).) Then $\overline{BD}$ is a median, and $\triangle ABD \cong \triangle CBD$ by SSS. This implies that $\angle ABD \cong \angle CBD$ by CPCTC; hence $\overline{BD}$ is the bisector of $\angle B$. Also, the congruency of the triangles implies that $\angle BDA \cong \angle BDC$. Since these angles form a linear pair, they are both right angles; hence $\overline{BD}$ is the altitude from B.

$(6) \Rightarrow (1)$: We are going to prove a stronger statement, namely that the coincidence of any two of these three segments implies that the triangle is isosceles.

If the bisector and the altitude at B coincide, then $\triangle ABD \cong \triangle CBD$ by ASA, and therefore $a = c$.

If the median and the altitude at B coincide, then $\triangle ABD \cong \triangle CBD$ by SAS, and therefore $a = c$.

Let $\overline{BD}$ be the median and the bisector at B. (See Figure 3.17(iii).) Let F and E be the bases of the perpendiculars dropped from D to the sides AB and CB, respectively: $\overline{DF} \perp \overline{AB}$ and $\overline{DE} \perp \overline{CB}$. Then $DF = DE$, since D is on the bisector of $\angle B$. This implies that $\triangle DFA \cong \triangle DEC$ by HL. Hence, $\angle DAF \cong \angle DCE$ by CPCTC. This implies (2), which gives us (1). □

Usually it is clear from the flow of a proof when a conclusion is reached based on the CPCTC Principle. Therefore, we will now stop mentioning it.

From now on, all references to the CPCTC Principle will be omitted!

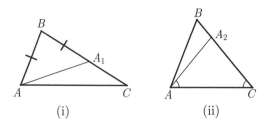

FIGURE 3.18.

3.2.5 Basic triangle inequalities

The ubiquitous "triangle inequality," which appears in various forms throughout mathematics, has a natural role to play in geometry. For real numbers a and b, the triangle inequality says

$$|a + b| \le |a| + |b|.$$

When applied to a triangle ABC in the plane, the triangle inequality gives

$$a + b > c, \quad b + c > a, \quad c + a > b,$$

where $a = BC$, $b = AC$, and $c = AB$, and we accept it without proof. Also this particular version of the inequality explains the use of "triangle" in the term. This implies $-c < a - b < c$, so $|a - b| < c$. Hence we observe that the absolute value of the difference of lengths of two sides of a triangle is smaller than the length of the third side:

$$|a - b| < c, \quad |b - c| < a, \quad |c - a| < b.$$

While the triangle inequality addresses relationships between side lengths in a triangle, the next theorem establishes an important relationship between the angle measures and side lengths in a triangle.

Theorem 3.10. *For every triangle ABC, $a > c$ if and only if $m\angle A > m\angle C$.*

Proof. ($\Rightarrow$) Let $a > c$ and let A_1 be the point on $\overline{BC}$ such that $BA_1 = AB = c$. (See Figure 3.18(i).) Then $m\angle BAA_1 = m\angle BA_1A$ as the measures of angles at the base of isosceles triangle ABA_1. At the same time, $m\angle BA_1A = m\angle A_1AC + m\angle C$ as the measure of the exterior angle in a triangle. Hence,

$$m\angle A > m\angle BAA_1 = m\angle BA_1A = m\angle A_1AC + m\angle C > m\angle C.$$

Thus, $m\angle A > m\angle C$.

($\Leftarrow$) Let $m\angle A > m\angle C$ and let A_2 be a point on $\overline{BC}$ such that $m\angle A_2AC = m\angle C$. (See Figure 3.18(ii).) Then $\triangle AA_2C$ is isosceles, so $AA_2 = A_2C$. Thus, $c = AB < BA_2 + A_2A$ by the triangle inequality. Hence

$$c < BA_2 + A_2A = BA_2 + A_2C = BC = a.$$

This proves $a > c$, as desired. $\square$

A polygon is called **equilateral** if all of its sides are congruent. A polygon is called **equiangular** if all of its interior angles are congruent. For a triangle these two notions are equivalent, as stated in the following immediate corollary of Theorem 3.10.

Corollary 3.11. *A triangle is equilateral if and only if it is equiangular.*

3.2.6 Similarity of triangles

We treat similarity of triangles in the same manner as congruence of triangles. Suppose we wish to prove that two triangles are similar. As before, using the definition of similarity requires defining a bijection between their points which changes all distances in the same ratio. In reality, such a high-level approach is almost never used. Rather, we utilize the following analog of Theorem 3.7. Again, we accept this theorem without proof.[6]

Theorem 3.12. (Conditions for Similarity of Triangles) *The following statements are equivalent.*

(1) $\triangle ABC \sim \triangle A'B'C'$.

(2) Angle-Angle (AA): *Two angles of one triangle are congruent to the corresponding two angles of the other triangle.*

(3) Side-Angle-Side (SAS): *An angle of one triangle is congruent to the angle of the other, and the pairs of sides that form these angles are in proportion.*

(4) Side-Side-Side (SSS): *Three sides of one triangle are in proportion with the corresponding three sides of the other triangle.*

Theorem 3.12 provides the basis for proving similarity of triangles in Euclidean geometry. It is often used to show that two pairs of segments are in proportion or that two angles are congruent. Usually such proofs proceed as follows. Note that this method is analogous to the one used for proving the congruence of triangles.

- We identify the pairs of corresponding segments (angles) of two triangles which we suspect are similar.
- The similarity of the triangles is established by one of the tests: AA, SAS, or SSS.
- The segments (angles) are claimed to be in proportion (congruent) as the corresponding parts of similar triangles.

The following theorem has many applications.

Theorem 3.13. *Let m, n, and p be parallel lines and l_1 and l_2 be transversals of these lines. Let $M_i = m \cap l_i$, $N_i = n \cap l_i$, and $P_i = p \cap l_i$, $i = 1, 2$. (See Figure 3.19.) Then,*

$$M_1 N_1 / N_1 P_1 = M_2 N_2 / N_2 P_2.$$

Proof. Let points N_3 on n and P_3 on p be such that $\overleftrightarrow{M_2 N_3} \parallel l_1 \parallel \overleftrightarrow{N_2 P_3}$. Let angles $1, 2, \ldots, 8$ be as shown in Figure 3.19. Then $\angle 1 \cong \angle 2$ and $\angle 3 \cong \angle 4$ as alternate interior angles. Therefore,

---

[6] The interested reader is directed to [64] for further discussion of similarity of triangles.

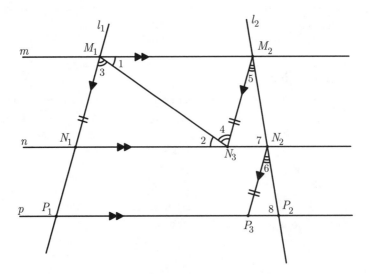

FIGURE 3.19.

$\triangle N_1 M_1 N_3 \cong \triangle M_2 N_3 M_1$ by ASA; hence $M_1 N_1 = N_3 M_2$. Similarly we can show that $N_1 P_1 = N_2 P_3$. This implies

$$\frac{M_1 N_1}{N_1 P_1} = \frac{M_2 N_3}{N_2 P_3}. \tag{3.1}$$

We also have $\angle 5 \cong \angle 6$ and $\angle 7 \cong \angle 8$ as corresponding angles. Therefore, $\triangle M_2 N_3 N_2 \sim \triangle N_2 P_3 P_2$ by AA. Hence their corresponding sides are in proportion:

$$\frac{M_2 N_2}{N_2 P_2} = \frac{M_2 N_3}{N_2 P_3}. \tag{3.2}$$

Equalities (3.1) and (3.2) give the desired result. $\square$

One important specific case of Theorem 3.13 occurs when line m passes through the (possible) point of intersection of l_1 and l_2, as shown in Figure 3.20(i). Corollary 3.14 states this fact in a slightly stronger form.

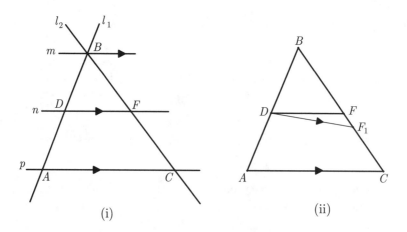

FIGURE 3.20.

> **Corollary 3.14.** *Consider* $\triangle ABC$ *and points D and F on $\overline{AB}$ and $\overline{BC}$, respectively. Then $\overline{DF} \parallel \overline{AC}$ if and only if points D and F divide $\overline{AB}$ and $\overline{BC}$ in the same proportion:*
>
> $$BD/DA = BF/FC.$$

Proof. ($\Rightarrow$) This part is a direct corollary of Theorem 3.13. ($\Leftarrow$) Suppose points D and F are on $\overline{AB}$ and $\overline{BC}$, respectively, and $BD/DA = BF/FC$. Consider a line through D parallel to line AC, and let F_1 be the intersection of this line with line BC. (See Figure 3.20(ii). We omit the proof that F_1 lies on $\overline{BC}$.) By the previous part of the statement, $BD/DA = BF_1/F_1C$. For the sake of contradiction, suppose $F_1 \neq F$. It is clear that if $B - F - F_1$, then $BD/DA = BF_1/F_1C > BF/FC$, as $BF_1 > BF > 0$ and $0 < F_1C < FC$. Similarly, if $B - F_1 - F$, then $BD/DA = BF_1/F_1A < BF/FC$. In each case we obtain a contradiction. This proves that $F_1 = F$, so $\overline{DF} \parallel \overline{AC}$. $\square$

A **midline** of a triangle is a segment joining the midpoints of two of its sides. Immediate applications of Corollary 3.14 are found in the proofs of Theorems 3.15 and 3.16.

> **Theorem 3.15.** (Triangle Midline Theorem) *Each midline of a triangle is parallel to a side of the triangle and has length half the length of that side.*

Proof. Let $\overline{DF}$ be a midline in $\triangle ABC$, as illustrated in Figure 3.21. Then $BD/AD = BF/CF = 1$, and $\overline{DF} \parallel \overline{AC}$ by Corollary 3.14. Triangles DBF and ABC share $\angle B$, while $BD/BA = BF/BC = 1/2$. Hence $\triangle DBF \sim \triangle ABC$ by SAS, so $DF/AC = 1/2$. $\square$

> **Theorem 3.16.** (Triangle Bisector Property) *Given $\triangle ABC$, let B_1 be an interior point of $\overline{AC}$. Then $\overline{BB_1}$ is a bisector of $\angle B$ if and only if $AB_1/B_1C = AB/BC = c/a$.*

Proof. ($\Rightarrow$) Suppose $\overline{BB_1}$ is the bisector of $\angle B$. Consider the line passing through C and parallel to $\overline{BB_1}$. Let C_1 be the intersection of this line with line AB. (See Figure 3.22(i).)

Since $\overline{BB_1} \parallel \overline{CC_1}$, we have that $\angle 1 \cong \angle 4$ as corresponding angles and $\angle 2 \cong \angle 3$ as alternate interior angles. But $\angle 1 \cong \angle 2$, since $\overline{BB_1}$ is the bisector of $\angle B$. Therefore, $\angle 3 \cong \angle 4$ and $\triangle CBC_1$ is isosceles. Hence $a = BC = BC_1$. Since $\overline{BB_1} \parallel \overline{CC_1}$, $AB_1/B_1C = AB/BC_1 = c/a$ by Corollary 3.14. Thus, we have proved that the bisector of $\angle B$ divides the opposite side in ratio c/a.

($\Leftarrow$) Let B_1 be a point on $\overline{AC}$ such that $AB_1/B_1C = c/a$. We wish to show that $\overline{BB_1}$ is the bisector of $\angle B$. Let B_2 be a point on $\overline{AC}$ such that $\overline{BB_2}$ is the bisector of $\angle B$. (See Figure 3.22(ii).)

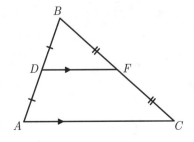

FIGURE 3.21.

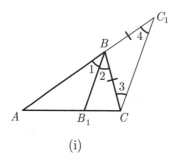 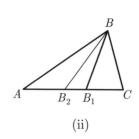

(i)   (ii)

FIGURE 3.22.

Then, as was just established, $AB_2/B_2C = c/a$. Hence we have

$$AB_1/B_1C = AB_2/B_2C.$$

Adding 1 to both sides gives the following equivalent statements:

$$AB_1/B_1C + 1 = AB_2/B_2C + 1,$$
$$(AB_1 + B_1C)/B_1C = (AB_2 + B_2C)/B_2C,$$
$$AC/B_1C = AC/B_2C,$$
$$B_1C = B_2C.$$

Since B_1 and B_2 are on the same side of C on $\overrightarrow{AC}$, we conclude that $B_1 = B_2$. Thus, $\overline{BB_1}$ is the bisector of $\angle B$. □

If the addition of 1 to both sides of the equality near the end of this proof seems too mysterious (see a generalization in Problem 3.2.4), we note that the conclusion $B_1 = B_2$ could also be easily obtained by an argument similar to one used in the proof of Corollary 3.14. If $B_1 \neq B_2$, then B_2 is either on $\overrightarrow{AB_1}$ or on $\overrightarrow{B_1C}$. In the first case, $AB_1/B_1C > AB_2/B_2C$, since the numerator of the first fraction is greater than that of the second, while the denominator of the first fraction is smaller than the one of the second. In the second case, $AB_1/B_1C < AB_2/B_2C$ for similar reasons. In either case we contradict the fact $AB_1/B_1C = AB_2/B_2C$, and we conclude that $B_1 = B_2$.

Three or more lines or segments are said to be **concurrent** if they all intersect in a common point. A **cevian** of a triangle is a segment from a vertex of the triangle to the opposite side. The following beautiful theorem was first published by Italian mathematician Giovanni Ceva (1647–1734) in 1678. It provides necessary and sufficient conditions for cevians of a triangle to be concurrent.

> **Theorem 3.17.** (Ceva's Theorem). Let $\overline{AD}$, $\overline{BE}$, and $\overline{CF}$ be cevians of a triangle ABC. (See Figure 3.23) Then $\overline{AD}$, $\overline{BE}$, and $\overline{CF}$ are concurrent if and only if
> $$\frac{AE}{EC} \cdot \frac{CD}{DB} \cdot \frac{BF}{FA} = 1.$$

Proof. Suppose that $\overline{AD}$, $\overline{BE}$, and $\overline{CF}$ are concurrent at a point P. Draw line l through B parallel to $\overline{AC}$. Let $A' = \overrightarrow{AD} \cap l$, and let $C' = \overrightarrow{CF} \cap l$.

Triangles ACD and $A'BD$ are similar by AA, as are triangles ACF and $BC'F$. This gives

$$\frac{CD}{DB} = \frac{AC}{BA'} \quad \text{and} \quad \frac{BF}{FA} = \frac{BC'}{CA}.$$

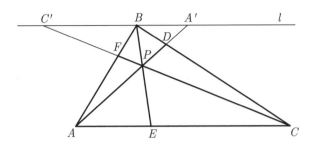

FIGURE 3.23.

Thus,

$$\frac{CD}{DB} \cdot \frac{BF}{FA} = \frac{AC}{BA'} \cdot \frac{BC'}{CA} = \frac{BC'}{A'B}.$$

Furthermore, $\triangle AEP \sim \triangle A'BP$ and $\triangle CEP \sim \triangle C'BP$, yielding

$$\frac{A'B}{AE} = \frac{BP}{EP} = \frac{BC'}{EC}, \text{ which implies } \frac{AE}{EC} = \frac{A'B}{BC'}.$$

By multiplying equal expressions we see that

$$\frac{AE}{EC} \cdot \frac{CD}{DB} \cdot \frac{BF}{FA} = \frac{A'B}{BC'} \cdot \frac{BC'}{A'B} = 1.$$

Conversely, suppose that D, E, and F are points on $\triangle ABC$ with $A - F - B$, $B - D - C$, and $C - E - A$, such that $\frac{AE}{EC} \cdot \frac{CD}{DB} \cdot \frac{BF}{FA} = 1$. Let X be the point of intersection of $\overline{AD}$ and $\overline{BE}$. Let F' be the intersection of $\overrightarrow{CX}$ with $\overline{AB}$. Then, by the first part of the proof,

$$\frac{AE}{EC} \cdot \frac{CD}{DB} \cdot \frac{BF'}{F'A} = 1.$$

This implies that $BF'/F'A = BF/FA$, from which it follows that $F' = F$. We conclude that $\overline{AD}$, $\overline{BE}$, and $\overline{CF}$ are concurrent. $\qquad\square$

There are a multitude of methods for proving this remarkable theorem. See our solutions to Problems 5.17, 7.23, 8.16, and 12.11.

Ceva's theorem can be used to prove concurrency of all altitudes, or all medians, or all bisectors of a triangle.

Corollary 3.18. *In a triangle ABC, the three medians are concurrent, the three bisectors are concurrent, and the three lines containing its altitudes are concurrent.*

Proof. Here we prove only that the altitudes are concurrent. Proofs of the other two statements are left to Problems 3.2.11 and 3.2.15.

The statement is trivial for any right triangle, where the altitudes meet at the vertex of the right angle. Suppose $\triangle ABC$ is acute. Then each altitude lies inside the triangle, as in Figure 3.24(i). Let $\overline{AD}$, $\overline{BE}$, and $\overline{CF}$ be the altitudes.

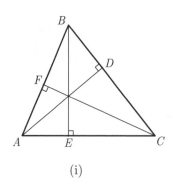

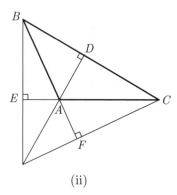

(i)   (ii)

FIGURE 3.24.

The right triangles ACD and BCE are similar by AA. Consequently,

$$\frac{CD}{EC} = \frac{DA}{EB}.$$

Likewise, $\triangle CAF \sim \triangle BAE$ and $\triangle ABD \sim \triangle CBF$, so

$$\frac{AE}{FA} = \frac{EB}{FC} \quad \text{and} \quad \frac{BF}{DB} = \frac{FC}{DA}.$$

Thus,

$$\frac{AE}{EC} \cdot \frac{CD}{DB} \cdot \frac{BF}{FA} = \frac{AE}{FA} \cdot \frac{CD}{EC} \cdot \frac{BF}{DB} = \frac{EB}{FC} \cdot \frac{DA}{EB} \cdot \frac{FC}{DA} = 1.$$

By Ceva's theorem, the altitudes are concurrent.

A similar argument proves the statement when $\triangle ABC$ is obtuse; see Figure 3.24(ii). We leave the details of this case to the reader.   □

The points of concurrency of the altitudes, medians, and angle bisectors are called the **orthocenter**, the **centroid**, and the **incenter** of the triangle, respectively.

3.2.7 Homotheties

A very useful type of similarity is the homothety, also known as a dilation or central similarity. Given a point O and a number $k \neq 0$, we consider the function H, from the plane to itself, mapping every point X to a point $X' = H(X)$ such that

(a) $OX' = |k| \cdot OX$ and

(b) rays OX and OX' are equal for $k > 0$, and opposite for $k < 0$.

Such a map is called a **homothety**, and we denote it by $H(O, k)$. Point O is called the **center of homothety** and k is called the **coefficient of homothety**. Note that the center of homothety is the only point of the plane fixed by it: $H(O) = O$. For the special case where $k = -1$, observe that we have symmetry with respect to O; i.e., O will be the midpoint of $\overline{XX'}$.

The SAS similarity axiom implies that a homothety $H = H(O, k)$ is a similarity of the plane with coefficient $|k|$. It is easy to check that H maps lines to lines, parallel lines to parallel lines, segments to segments, and n-gons to n-gons. In particular, in Figure 3.21, $\triangle ABC = H(B, 2)(\triangle DBF)$, and $\triangle DBF = H(B, \frac{1}{2})(\triangle ABC)$. Another example is provided in Figure 3.25, where the image of triangle T is shown for three different homotheties.

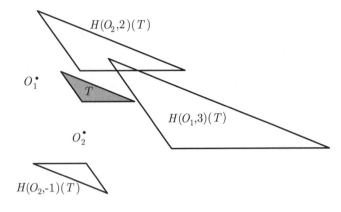

FIGURE 3.25.

3.2.8 Right triangles

A study of right triangles is an essential part of any discussion of triangles, and the following theorem is fundamental to this study.

Theorem 3.19. *Given a triangle ABC with $m\angle C = 90°$ and altitude $\overline{CD}$ (as pictured in Figure 3.26),*

(1) $\triangle ACD \sim \triangle CBD \sim \triangle ABC$, and

(2) $CD = h_c = (AD \cdot BD)^{1/2}$, $DA = b^2/c$, and $BD = a^2/c$.

Proof. Since $\triangle ABC$, $\triangle ACD$, and $\triangle CBD$ are right triangles, $\angle A$ and $\angle B$ are complementary, $\angle A$ and $\angle ACD$ are complementary, and $\angle BCD$ and $\angle B$ are complementary. Hence $\angle A \cong \angle BCD$ and $\angle B \cong \angle ACD$, while $\triangle ABC \sim \triangle ACD \sim \triangle CBD$ by AA. This proves (a).

Since corresponding sides in similar triangles are in proportion, we have:

$$\triangle ACD \sim \triangle CBD \ \Rightarrow\ \frac{AD}{h_c} = \frac{h_c}{BD} \ \Leftrightarrow\ CD = h_c = (AD \cdot BD)^{1/2};$$

$$\triangle ACD \sim \triangle ABC \ \Rightarrow\ \frac{AD}{b} = \frac{b}{c} \ \Leftrightarrow\ AD = \frac{b^2}{c};$$

$$\triangle CBD \sim \triangle ABC \ \Rightarrow\ \frac{BD}{a} = \frac{a}{c} \ \Leftrightarrow\ BD = \frac{a^2}{c}.$$

$\square$

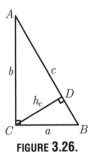

FIGURE 3.26.

And now we come to the most celebrated theorem of Euclidean geometry, which dates around 500 B.C. This theorem generalized empirical data collected centuries before by the Babylonians and Egyptians. The theorem, though familiar, is strikingly nonobvious. Over the years, a multitude of proofs of the Pythagorean theorem have been written; at this point, we present just one short proof. Note that the Pythagorean theorem is often stated only as an implication: "If $\triangle ABC$ is a right triangle, then $a^2 + b^2 = c^2$." We prove both this implication and its converse.

Theorem 3.20. (Pythagorean Theorem) *In a triangle ABC, $m\angle C = 90°$ if and only if $a^2 + b^2 = c^2$.*

Proof. ($\Rightarrow$) Let $m\angle C = 90°$, and let $\overline{CD}$ be the altitude at C. Point D lies on $\overline{CD}$. Hence $BD + DA = AB$. Using Theorem 3.19(2), we obtain

$$BD + DA = AB \iff$$
$$\frac{a^2}{c} + \frac{b^2}{c} = c \iff$$
$$a^2 + b^2 = c^2.$$

($\Leftarrow$) Let $\triangle ABC$ be a triangle with side lengths a, b, and c, such that $a^2 + b^2 = c^2$. Let $\triangle DEF$ be a right triangle with legs having lengths a and b. By the first part of this proof, the hypotenuse of $\triangle DEF$ has length $\sqrt{a^2 + b^2}$; that is, the hypotenuse of $\triangle DEF$ has length c. By SSS, $\triangle DEF \cong \triangle ABC$, so $\triangle ABC$ is a right triangle with $m\angle C = 90°$. $\square$

One can view numbers a, b, and c as lengths of three corresponding elements (each a hypotenuse) in similar triangles CBD, ACD, and ABC. Note that the coefficients of similarity of triangles CBD and ACD with triangle ABC are given by the ratios $k_a = a/c$ and $k_b = b/c$, respectively. Another way to state the Pythagorean theorem is that the sum of the squares of these coefficients is 1.

$$\boxed{k_a^2 + k_b^2 = 1}$$

Indeed, $k_a^2 + k_b^2 = (a/c)^2 + (b/c)^2 = (a^2 + b^2)/c^2 = c^2/c^2 = 1$.

This leads to the following extension of one direction of the Pythagorean theorem.

Theorem 3.21. (Extended Pythagorean Theorem) *Consider a triangle ABC with $m\angle C = 90°$. Let $\overline{CD}$ be the altitude of the triangle and r, s, and t be the lengths of three corresponding segments in similar triangles CBD, ACD, and ABC, respectively. Then $r^2 + s^2 = t^2$.*

Proof. Since $\triangle CBD \sim \triangle ABC$ with coefficient $k_a = a/c$, $r = k_a t$. Similarly, $\triangle ACD \sim \triangle ABC$ with coefficient $k_b = b/c$, so $s = k_b t$. Hence,

$$r^2 + s^2 = k_a^2 t^2 + k_b^2 t^2 = (k_a^2 + k_b^2)t^2 = (1)t^2 = t^2.$$

$\square$

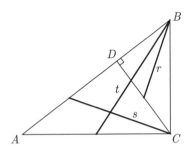

FIGURE 3.27.

The phrase "three corresponding segments" in the statement of Theorem 3.21 may refer to angle bisectors, altitudes, medians, or any other interesting segment of a right triangle. The following example illustrates use of the theorem for one such set of corresponding segments.

Example 3. Consider $\triangle ABC$ with $m\angle C = 90°$ and altitude $\overline{CD}$. Suppose the length of the median of $\triangle BCD$ at B is $r = 11$ and the length of the median of $\triangle ABC$ at B is $t = 18$. (See Figure 3.27.) What is the length s of the median of $\triangle ACD$ at C?

Solution: From the Extended Pythagorean theorem, $s = \sqrt{t^2 - r^2} = \sqrt{203}$. □

3.2.9 Problems

A great discovery solves a great problem, but there is a grain of discovery in the solution of any problem.
— George Polya (1887–1985)

In all problems below, the letters a, b, and c denote the lengths of sides of a triangle ABC opposite to the angles $\angle A$, $\angle B$, and $\angle C$, respectively. Let h_a, l_a, and m_a denote the lengths of the altitude, bisector, and median at A, respectively, (and likewise for B and C). Let $p = (a + b + c)/2$ be the semiperimeter of the triangle.

3.2.1 Consider $\triangle ABC$ with $m\angle C = 90°$; let $\overline{CD}$ be its altitude at C. Suppose the length of the angle bisector of $\triangle ACD$ at C is 5 and the length of the angle bisector of $\triangle CBD$ at B is 8. What is the length of the angle bisector of $\triangle ABC$ at B?

3.2.2 Consider $\triangle ABC$ with $a = 8, b = 3, c = 6$. Let $\overline{AD}$ be the altitude at A, $\overline{AF}$ be the bisector at A, and $\overline{AM}$ be the median at A.
(a) Find CD, CF, and CM.
(b) Find h_a, l_a, and m_a.

3.2.3 Suppose $a \leq b \leq c$ in $\triangle ABC$. Prove that if $a^2 + b^2 > c^2$, then the triangle is acute, and if $a^2 + b^2 < c^2$, then it is obtuse.

3.2.4 (Derived proportions) Prove that if $a_1/a_2 = b_1/b_2$, then

$$\frac{a_1 + ca_2}{a_1 + da_2} = \frac{b_1 + cb_2}{b_1 + db_2}$$

for any numbers c and d, assuming that all denominators are non-zero.

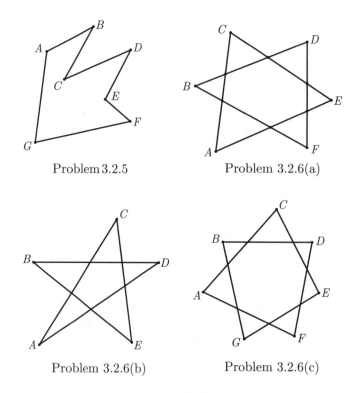

Problem 3.2.5

Problem 3.2.6(a)

Problem 3.2.6(b)

Problem 3.2.6(c)

FIGURE 3.28.

3.2.5 Find the sum of measures of all interior angles of the polygon $ABCDEFG$ shown in Figure 3.28.

3.2.6 (a) What is the sum $m\angle A + m\angle B + m\angle C + m\angle D + m\angle E + m\angle F$ in the "6-star" in Figure 3.28?

    (b) What is the sum $m\angle A + m\angle B + m\angle C + m\angle D + m\angle E$ in the "5-star" in Figure 3.28?

    (c) What is the sum $m\angle A + m\angle B + m\angle C + m\angle D + m\angle E + m\angle F + m\angle G$ in the "7-star" in Figure 3.28? Can you generalize?

3.2.7 Prove that a convex n-gon cannot have more than three acute interior angles.

3.2.8 Is there a triangle with all altitudes shorter than 1 inch, but with each side longer than 1 mile?

3.2.9 Consider a line l and two points A and B on the same side of l.

    (a) Prove that there exists a point C on l such that the acute angles formed by $\overline{AC}$ with l and $\overline{BC}$ with l are congruent.

    (b) Prove that point C described in (a) has the following remarkable property: for every other point X on l,

$$AC + BC < AX + BX.$$

In other words, show that point C minimizes the sum of the distances from a point of l to A and B.

    (c) Think about a possible application of (b).

    (d) If A and B are on opposite sides of l, where is the point C on l that minimizes $AC + CB$?

3.2.10 Prove that any two medians of a triangle intersect at a point which divides their length in ratio 2:1 (starting from the vertex).

3.2.11 Prove that the three medians in any triangle are concurrent.

3.2.12 Given two distinct points in the plane, A and B, prove that a point P is on the perpendicular bisector of $\overline{AB}$ if and only if P is equidistant from A and B (i.e., $AP = PB$).

3.2.13 Given an angle ABC in the plane, such that $0° < m\angle ABC < 180°$, prove that a point P is on the bisector of $\angle ABC$ if and only if P is equidistant from the rays $\overrightarrow{BA}$ and $\overrightarrow{BC}$.

3.2.14 Prove that the perpendicular bisectors to three sides of a triangle are concurrent.

3.2.15 Prove that the three angle bisectors in any triangle are concurrent.

3.2.16 Consider a triangle ABC with $m\angle C = 90°$. Let $\overline{CD}$ be an altitude. Suppose the length x of the median of $\triangle CBD$ at B is 4, and the length z of the median of $\triangle ABC$ at B is 7. What is the length y of the median of $\triangle ACD$ at C?

3.2.17 Prove that in a right triangle with an angle $30°$, the length of the side opposite this angle is equal to half of the length of the hypotenuse.

3.2.18 Given $\triangle ABC$, show that h_a, l_a, and m_a can be expressed by the following formulae:

$$h_a = \frac{2}{a}\sqrt{p(p-a)(p-b)(p-c)},$$
$$l_a = \frac{2}{b+c}\sqrt{bcp(p-a)}, \text{ and}$$
$$m_a = \frac{1}{2}\sqrt{2(b^2+c^2)-a^2}.$$

3.2.19 Prove that in any triangle ABC,
   (a) $m_a < \dfrac{b+c}{2}$,
   (b) $p < m_a + m_b + m_c < 2p$.

3.2.20 Consider a line l and two points A and B. Describe all points C on l such that $|AC - BC|$ is maximal in each of the following cases.
   (a) A and B are on the same side of l and $\overline{AB}$ is not parallel to l.
   (b) $\overline{AB}$ is parallel to l.
   (c) $\overline{AB}$ intersects l in one point.

3.2.21 Let O be an interior point of a triangle ABC. Prove that $p < OA + OB + OC < 2p$.

3.2.22 Consider a triangle ABC with $a^3 + b^3 = c^3$. Can angle C be acute? right? obtuse?

3.2.23 (Triangle Exterior Angle Bisector Property) Given $\triangle ABC$, let $\overline{BB_1}$ be a segment joining B with a point B_1 on the ray $\overrightarrow{AC}$ such that $A - C - B_1$. Prove that $\overline{BB_1}$ is a bisector of the exterior angle at B if and only if $AB_1/B_1C = AB/BC = c/a$.

3.2.24 Prove that if two angle bisectors of a triangle have the same length, then the triangle is isosceles.

3.2.25 Prove that in any non-isosceles triangle, the bisector of an angle lies between the altitude and the median at the same vertex.

3.2.26 Let $\overline{AA_1}$, $\overline{BB_1}$, and $\overline{CC_1}$ be the altitudes of an acute triangle ABC, where A_1, B_1, and C_1 are on $\overline{BC}$, $\overline{AC}$, and $\overline{AB}$, respectively. Prove that $\overline{B_1B}$ bisects $\angle C_1B_1A_1$.

3.2.27 Modify the previous problem for an obtuse triangle. Is a similar statement correct in this case?

3.2.28 Given an isosceles triangle ABC with $a = c$, prove that the sum of distances from every point of $\overline{AC}$ to $\overline{AB}$ and $\overline{CB}$ is constant.

3.2.29 Given an equilateral triangle ABC, prove that the sum of the three distances from every interior point or boundary point to its sides is the same.

3.2.30 Prove that the sum of measures of the interior angles of any polygon (convex or not) is $180(n - 2)°$.

3.2.10 Supplemental Problems

*S*3.2.1 Is every equilateral 4-gon equiangular? Is every equiangular 4-gon equilateral? Answer similar questions for n-gons with $n > 4$.

*S*3.2.2 A carpenter has a board that is $9\frac{5}{16}$ inches wide. He needs to cut it into three strips of equal width. He has a 12-inch ruler that unfortunately only has markings at each inch. Describe a way in which he can create the strips he needs. How would he adapt the method if he wished to create five strips? Assume that the saw blade has no width.

*S*3.2.3 Consider a right triangle ABC, $m\angle C = 90°$. Let $\overline{CD} \perp \overline{AB}$, $D \in \overline{AB}$. If the perimeters of $\triangle ABC$ and $\triangle BCD$ are 8 and 5, respectively, what is the perimeter of $\triangle ACD$?

*S*3.2.4 Consider a triangle ABC with right angle at C, with $AC = 3$ and $BC = 4$. What are the side lengths of a square $MNPQ$ with P and Q on $\overline{AB}$, M on $\overline{AC}$, and N on $\overline{BC}$?

*S*3.2.5 For a triangle ABC, prove that $\angle C$ is a right angle if and only if $m_c = c/2$.

---

*S*3.2.6 Let $H = H(O, k)$ be a homothety and l be a line. Prove that $l' = H(l)$ is a line parallel to l.

*S*3.2.7 Prove that if two medians of a triangle have the same length, then the triangle is isosceles.

*S*3.2.8 Consider a right triangle ABC, with $m\angle C = 90°$. Let $\overline{CD} \perp \overline{AB}$, $D \in \overline{AB}$. Let p and q be the distances from D to the sides AC and BC, respectively. Find AB.

*S*3.2.9 Prove that for any triangle ABC, another triangle can be built whose sides are congruent to the medians of $\triangle ABC$. Does the statement hold if the medians are replaced by the altitudes? By the bisectors?

*S*3.2.10 Consider a right triangle ABC with the right angle at C. Prove that the bisector at C also bisects the angle between the median and the altitude at C.

*S*3.2.11 Let m, n, and p be the distances from a point in the interior of an equilateral triangle to its sides. Find the length of the side of the triangle.

*S*3.2.12 Given n red and n blue points in a plane, $n \geq 1$, prove that there are n segments such that each segment joins one red and one blue point and no two segments intersect.

*S*3.2.13 Given $\triangle ABC$, let $A' \in \overline{BC}$ and $B' \in \overline{AC}$, and let N be the point of intersection of $\overline{AA'}$ and $\overline{BB'}$. Find $AB'/B'C$, if $AN = 4NA'$ and $BN = 3NB'$.

---

*S*3.2.14 Four points in a plane determine six distances. Prove that the ratio of the longest of them to the shortest is at least $\sqrt{2}$.

*S*3.2.15 Let $\overline{AD}$ and $\overline{CE}$ be the medians in $\triangle ABC$, and let $m\angle BAD = m\angle BCE = 30°$. Prove that $\triangle ABC$ is equilateral.

*S*3.2.16 Given a finite set of points in a plane, not all collinear, show that there exists a line passing through only two of these points.

3.3 Parallelograms and Trapezoids

3.3.1 Parallelograms

A **parallelogram** is a quadrilateral with two pairs of parallel sides. Thus, $ABCD$ is a parallelogram if $\overline{AB} \parallel \overline{CD}$ and $\overline{BC} \parallel \overline{AD}$. Parallelograms are quite exceptional quadrilaterals and, undoubtedly, they have received more attention from geometers than any other polygon besides triangles. The following theorem summarizes properties of parallelograms. Each of properties (2)–(5) can be used to prove that a quadrilateral is a parallelogram.

Theorem 3.22. *For every convex quadrilateral $ABCD$, the following statements are equivalent.*

(1) *$ABCD$ is a parallelogram.*

(2) *$\triangle ABC \cong \triangle CDA$ and $\triangle BAD \cong \triangle DCB$; i.e., each diagonal divides $ABCD$ into two congruent triangles.*

(3) *Pairs of opposite sides of $ABCD$ are congruent; i.e., $\overline{AB} \cong \overline{CD}$ and $\overline{BC} \cong \overline{AD}$.*

(4) *Diagonals AC and BD bisect each other.*

(5) *Pairs of opposite angles of $ABCD$ are congruent; i.e., $\angle A \cong \angle C$ and $\angle B \cong \angle D$.*

Proof. We will show that $(1) \Rightarrow (2) \Rightarrow (3) \Rightarrow (4) \Rightarrow (5) \Rightarrow (1)$. The convexity requirement is used to claim that diagonals lie in the interior of $ABCD$ and that the sum of the measures of the interior angles is $360°$ (though the latter is true for all non-intersecting quadrilaterals).

$(1) \Rightarrow (2)$: Let $ABCD$ be a parallelogram and consider $\triangle ABC$ and $\triangle CDA$. Since lines AB and CD are parallel, $\angle CAB \cong \angle DCA$ as alternate interior angles; likewise, since lines BC and AD are parallel, $\angle BCA \cong \angle DAC$. Hence, $\triangle ABC \cong \triangle CDA$ by ASA. The proof that $\triangle BAD \cong \triangle DCB$ is analogous.

$(2) \Rightarrow (3)$: The congruency of triangles implies the congruency of the corresponding sides.

$(3) \Rightarrow (4)$: If pairs of opposite sides of $ABCD$ are congruent, then each diagonal divides $ABCD$ into two triangles which are congruent by SSS. Therefore, the corresponding angles CAB and ACD are congruent, and the corresponding angles ABD and CDB are congruent. Let O be the intersection of the diagonals. Since sides AB and CD are congruent, triangles AOB and COD are congruent by ASA. This implies that $AO = OC$ and $BO = OD$; i.e., the diagonals bisect each other.

$(4) \Rightarrow (5)$: Let O be the intersection of the diagonals of $ABCD$. Then $\angle AOB \cong \angle COD$ as vertical angles. Hence, $\triangle AOB \cong \triangle COD$ by SAS. Similarly $\triangle BOC \cong \triangle DOA$. Then

$$m\angle A = m\angle BAO + m\angle DAO = m\angle DCO + m\angle BCO = m\angle C.$$

Therefore angles A and C are congruent. Similarly, $\angle B \cong \angle D$.

$(5) \Rightarrow (1)$: Since $m\angle A + m\angle B + m\angle C + m\angle D = 360°$, $m\angle A = m\angle C$, and $m\angle B = m\angle D$, we see that $m\angle A + m\angle B = 180°$. Since $\angle A$ and $\angle B$ are supplementary, Theorem 3.2 implies that $\overline{AD} \parallel \overline{BC}$. Likewise, $\overline{AB} \parallel \overline{CD}$. $\qquad \square$

The point O of intersection of the diagonals of a parallelogram is called its **center**. It is easy to see (Problem S3.3.3) that this is the center of symmetry of the parallelogram: every line passing through O cuts two sides of the parallelogram at two points such that O is the midpoint of the segment connecting them.

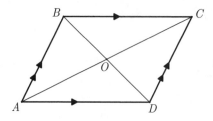

FIGURE 3.29.

The following theorem states an important relation between the lengths of sides and diagonals in a parallelogram. It is easy to prove that the given equation holds in a parallelogram. The converse statement allows us to prove that a convex quadrilateral is a parallelogram by just doing a simple computation with six segment lengths.

> **Theorem 3.23.** *ABCD is a parallelogram if and only if* $AC^2 + BD^2 = AB^2 + BC^2 + CD^2 + DA^2$.

Proof. ($\Rightarrow$) One can easily justify (Problem 3.3.5) the existence of a vertex with the property that the base of a perpendicular dropped from it to a line defined by another side is in the interior of that side. This justifies Figure 3.30, where $E \in \overline{AD}$ and $\overline{BE} \perp \overline{AD}$.

Then the base of a perpendicular CF dropped from C on line AD lies in the exterior of $\overline{AD}$. Since $ABCD$ is a parallelogram, $\angle BAD \cong \angle CDF$ as corresponding angles, and $\overline{AB} \cong \overline{CD}$ as opposite sides. Hence, right triangles AEB and DFC are congruent by AAS. Let $AB = CD = m$, $BC = AD = n$, $AE = DF = k$, and $BE = CF = h$. Then by the Pythagorean theorem, $h^2 = m^2 - k^2$, $AC^2 = (n+k)^2 + h^2$ and $BD^2 = (n-k)^2 + h^2$. Therefore,

$$AC^2 + BD^2 = ((n+k)^2 + h^2) + ((n-k)^2 + h^2)$$
$$= 2(n^2 + k^2 + h^2) = 2(n^2 + k^2 + (m^2 - k^2))$$
$$= 2(m^2 + n^2) = AB^2 + BC^2 + CD^2 + DA^2.$$

($\Leftarrow$) We postpone proving the converse until we develop superior techniques for doing so. See Problem 11.8. $\square$

A parallelogram with all interior angles congruent is called a **rectangle**.
A parallelogram with all sides congruent is called a **rhombus**.
A **square** is a parallelogram that is both a rectangle and a rhombus.

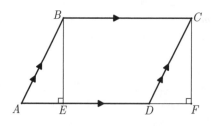

FIGURE 3.30.

Note that if the word "parallelogram" is changed to "quadrilateral" in the definition of rectangle, rhombus, and square, one gets an equivalent definition due to Theorem 3.22.

The following theorem summarizes properties of rectangles. Each of properties (2)–(4) can be used to prove that a quadrilateral is a rectangle.

Theorem 3.24. *For every parallelogram $ABCD$, the following statements are equivalent.*

(1) *$ABCD$ is a rectangle.*

(2) *$ABCD$ has a right angle.*

(3) *$\triangle ABD \cong \triangle DCA$.*

(4) *Diagonals AC and BD are congruent.*

The following theorem will summarize properties of rhombi. Each of properties (2)–(4) can be used to prove that a quadrilateral is a rhombus.

Theorem 3.25. *For every parallelogram $ABCD$, the following statements are equivalent.*

(1) *$ABCD$ is a rhombus.*

(2) *A diagonal bisects an interior angle.*

(3) *Each diagonal bisects two interior angles.*

(4) *The diagonals are perpendicular.*

Proofs of Theorems 3.24 and 3.25 are left to the reader.

3.3.2 Trapezoids

A **trapezoid** is a quadrilateral with exactly one pair of parallel sides. The parallel sides are called the **bases** of the trapezoid, and the nonparallel sides are called the **lateral** sides of the trapezoid. According to this definition, a trapezoid is never a parallelogram.[8] A segment joining the midpoints of the lateral sides of a trapezoid is called its **midline**.

Theorem 3.26. *Given a trapezoid $ABCD$ with bases $\overline{AD} \parallel \overline{BC}$, let E and F be the midpoints of the lateral sides AB and CD, respectively. Then*

(1) *the midline $\overline{EF}$ is parallel to the bases, and*

(2) *$EF = \frac{1}{2}(AD + BC)$.*

Proof. (1) Consider a line l through E parallel to the bases of the trapezoid. Let F' be the point of intersection of l with line CD. (See Figure 3.31.) Theorem 3.13 implies that $AE/EB = DF'/F'C$. Hence, F' is the midpoint of $\overline{CD}$; i.e., $F' = F$.

---

[8] In some texts a trapezoid is defined as a quadrilateral with at least one pair of parallel sides. Such a definition implies that every parallelogram is a trapezoid.

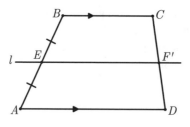

FIGURE **3.31.**

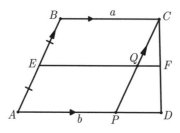

FIGURE **3.32.**

(2) Assume that $a = BC < b = AD$, for otherwise the proof is similar. Let $P \in \overline{AD}$ such that lines AB and CP are parallel. Let $Q = \overline{CP} \cap \overline{EF}$. (See Figure 3.32.) Then $EBCQ$ and $ABCP$ are parallelograms, so $BC = EQ = AP$. Hence, Q is the midpoint of $\overline{CP}$ (by Theorem 3.13 again), so $\overline{QF}$ is a midline of $\triangle PCD$ with $PD = b - a$. By Theorem 3.15, $QF = (b - a)/2$. Hence,

$$EF = EQ + QF = a + \frac{b - a}{2} = \frac{a + b}{2}.$$

□

A trapezoid is called **isosceles** if its lateral sides are congruent.

Theorem 3.27. *Given a trapezoid $ABCD$ with bases $\overline{AD} \parallel \overline{BC}$, the following statements are equivalent.*

(1) *$ABCD$ is isosceles.*

(2) *$\angle A \cong \angle D$ and $\angle B \cong \angle C$; i.e., angles at each base are congruent.*

(3) *$\overline{AC} \cong \overline{BD}$; i.e., the diagonals are congruent.*

The proof of Theorem 3.27 is left to the reader.

3.3.3 Problems

> *If you can't solve a problem, then there is an easier problem you can solve: find it . . .*
> — George Polya (1887–1985)

3.3.1 Prove Theorem 3.24.

3.3.2 Prove Theorem 3.27.

3.3.3 Given four points A, B, C, and D in a plane, no three on a line, prove that the midpoints of segments AB, BC, CD, and DA are vertices of a parallelogram.

3.3.4 Let $\overline{BO}$ be a median of $\triangle ABC$, with $BO = m_b$. Let D be a point on ray BO such that O is a midpoint of $\overline{BD}$. Prove that $ABCD$ is a parallelogram. Use this fact to show that

$$m_b = \frac{1}{2}\sqrt{2a^2 + 2c^2 - b^2}.$$

3.3.5 Prove that every parallelogram has a vertex such that the base of the perpendicular dropped from that vertex to the opposite side lies on that side.

3.3.6 Prove that the four rays that bisect interior angles of a parallelogram form a rectangle.

3.3.7 Let $ABCD$ be a trapezoid with $\overline{BC} \parallel \overline{AD}$, $AD = a > BC = b$. Let M and N be the midpoints of the respective diagonals AC and BD. Prove that $MN = (a - b)/2$.

3.3.8 Prove that the lines defined by the three altitudes of a triangle are concurrent.

3.3.9 Let $ABCD$ be a parallelogram, and let $F \in \overline{AD}$ be such that $AF = AD/5$. Let E be the intersection of segments BF and AC. Prove that $AE = AC/6$. Generalize the problem.

3.3.10 Let $ABCD$ and $A_1B_1C_1D_1$ be parallelograms with $A_1B_1C_1D_1$ inscribed in $ABCD$, i.e., vertices A_1, B_1, C_1, and D_1 lie on $\overline{AB}$, $\overline{BC}$, $\overline{CD}$, and $\overline{DA}$, respectively. Prove that the centers of these parallelograms coincide.

3.3.11 Let $ABCD$ be a trapezoid with $\overline{BC} \parallel \overline{AD}$, $AD = a$, $BC = b$, and let O be the intersection of the diagonals AC and BD. A line through O that is parallel to the bases intersects lateral sides AB and CD at points E and F, respectively.

(a) Show that $EO = OF$.

(b) Express EF in terms of a and b.

3.3.12 Let l be a line passing through the vertex M of parallelogram $MNPQ$ and suppose l intersects lines NP, PQ, and NQ in points R, S, and T, respectively. Prove that

$$1/MR + 1/MS = 1/MT.$$

3.3.13 Let $ABCD$ be a trapezoid. Let M be the point of intersection of the bisectors of angles A and B, and let N be the point of intersection of the bisectors of angles C and D. Prove that

$$MN = \frac{1}{2}(AD + BC - AB - CD).$$

3.3.14 Prove that the line joining the point of intersection of the extensions of the nonparallel sides of a trapezoid to the point of intersection of its diagonals bisects each base of the trapezoid.

3.3.15 The midpoints of the sides AB and CD, and of sides BC and ED of a convex pentagon (5-gon) $ABCDE$ are joined by two segments. The midpoints H and K of these segments are joined. Prove that $\overline{HK} \parallel \overline{AE}$ and $HK = \frac{1}{4}AE$.

3.3.16 Two pirates decide to hide a stolen treasure on a desert island, which has a well (W), a birch tree (B), and a pine tree (P). To bury the treasure, one of the pirates starts at W and walks to B and then turns right $90°$ and walks the same distance as WB, reaching the point Q. The second pirate starts at W and walks to P, after which he turns left $90°$ degrees and walks the same distance as WP, reaching the point R. Then they bury the treasure halfway between Q and R. Some months later the two pirates return to dig up the treasure only to discover that the well was gone. Can they find the treasure?

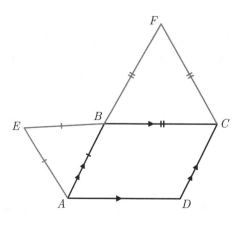

FIGURE 3.33.

3.3.4 Supplemental Problems

*S*3.3.1 Prove Theorem 3.25.

*S*3.3.2 Show that in a triangle ABC, angle A is acute, right, or obtuse depending on whether a is, respectively, less than, equal to, or greater than $2m_a$, where m_a is the length of the median of the triangle at A.

*S*3.3.3 Let O be the center of a parallelogram $ABCD$. Let a line passing through O intersect sides BC and DA at points E and F, respectively. Prove that O is the midpoint of $\overline{EF}$, that $BE = DF$, and that $EC = FA$.

*S*3.3.4 Point P lies in the interior of a rectangle $ABCD$. Prove that $PA^2 + PC^2 = PB^2 + PD^2$.

*S*3.3.5 Consider the four bisectors of the angles formed by the diagonals of a parallelogram. Prove that the points of intersection of the bisectors with the sides of the parallelogram are vertices of a rhombus.

---

*S*3.3.6 Let $ABCD$ be a parallelogram. Two equilateral triangles, ABE and BCF, are constructed in such a way that E is on the opposite side of line AB from C and D, and F is on the opposite side of line BC from A and D. Prove that $\triangle DEF$ is equilateral. (See Figure 3.33.)

*S*3.3.7 Given the lengths of all sides of a trapezoid, determine the length of its diagonals.

*S*3.3.8 A rectangle is inscribed in a triangle with side lengths 13, 14, and 15, in such a way that two of its vertices are on the longest side of the triangle, and two other vertices are on other sides of the triangle. If one of the diagonals of the rectangle is parallel to the shortest side, find the lengths of the sides of the rectangle.

*S*3.3.9 Given that all angles of a convex hexagon $ABCDEF$ are congruent, prove that

$$|BC - EF| = |DE - AB| = |AF - CD|.$$

*S*3.3.10 Let $ABCD$ be a trapezoid, and let P and Q be the midpoints of the bases BC and AD, respectively. Take a point M on ray AC such that M is outside of the trapezoid. Let lines MP and MQ intersect lateral sides AB and CD at points H and K, respectively. Prove that $\overline{HK}$ is parallel to the bases.

*S*3.3.11 Consider a parallelogram, $ABCD$. Let E and F be the midpoints of sides AB and CD, respectively. Let S, R, Q, and T be the respective points of intersection of $\overline{AC}$ and $\overline{DE}$, $\overline{DB}$ and $\overline{AF}$, $\overline{AC}$ and $\overline{BF}$, and $\overline{BD}$ and $\overline{CE}$. Prove that $SRQT$ is a parallelogram.

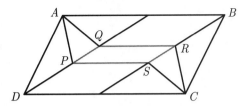

FIGURE 3.34.

*S*3.3.12 A perpendicular segment is dropped from every vertex of a convex quadrilateral to its diagonal. Prove that the bases of these perpendiculars are vertices of a quadrilateral similar to the original (in the sense that their corresponding angles are congruent and the ratios of lengths of the corresponding sides are equal).

*S*3.3.13 If the angles A and C of parallelogram $ABCD$ are trisected and the angles B and D are bisected, and if the bisector of D meets the trisectors of A at P and Q, and if the bisector of B meets the trisectors of C at R and S, prove that $PQRS$ is a parallelogram in which two sides are parallel to two sides of $ABCD$. (See Figure 3.34.)

---

*S*3.3.14 Let $ABCD$ be a convex quadrilateral, and let M and N be the midpoints of the diagonals AC and BD. Assume that $M \neq N$ and let M_1 and N_1 be the points of intersection of line MN with the (possibly extended) sides AB and CD. Prove that if $M_1M = NN_1$, then $\overline{BC} \parallel \overline{AD}$.

*S*3.3.15 Each vertex of a quadrilateral is joined to the centroid of the triangle formed by the other three vertices. Prove that the four segments are concurrent. This problem also appears as Problem S8.2.

*S*3.3.16 A convex polygon is inside a square with unit sides. Prove that the sum of squares of lengths of its sides is at most 4.

4

Circles

The eye is the first circle; the horizon which it forms is the second; and throughout nature this primary figure is repeated without end. It is the highest emblem in the cipher of the world.
— Ralph Waldo Emerson[1] (1803–1882)

The circle, perhaps the simplest of all geometric shapes, is the focus of the third book of Euclid's *Elements*. Frequently used as a symbol of infinity and wholeness, the circle possesses a simplicity and symmetry that is inherently aesthetically pleasing. Despite this simplicity, however, study of the circle has given rise to many profound theorems and mathematical insights.

4.1 Definitions and Properties

The definition of a circle becomes obvious when using a compass to draw one on paper or a chalkboard. Given a point O in the plane and a positive real number r, the **circle** $\mathcal{C}(O, r)$ is the set of all points P in the plane such that the distance from P to O is r. Point O is called the **center** of the circle and r is the **radius**. We say that all points of the circle are **equidistant** from O. It is sometimes helpful to think of a point as a circle of radius zero.

The following theorem will be useful in proving important properties of circles, as well as in future chapters. Part (1) appeared as Problem 3.2.12 and part (2) appeared as Problem 3.2.13.

Theorem 4.1.

(1) *Given two distinct points in the plane, A and B, a point P is on the perpendicular bisector of $\overline{AB}$ if and only if P is equidistant from A and B (i.e., $AP = PB$).*

(2) *Given an angle ABC in the plane, $0° < m\angle ABC < 180°$, a point P is on the bisector of $\angle ABC$ if and only if P is equidistant from the rays $\overrightarrow{BA}$ and $\overrightarrow{BC}$.*

---

[1] From *Essays and English Traits: Circles*, 1841.

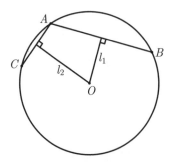

FIGURE 4.1.

Given a single point or a pair of points, infinitely many circles contain the point or the pair. Is the same true for a set of three points? If there are not infinitely many circles passing through three points, how many are there? Must there always be at least one?

> **Theorem 4.2.** *Given three non-collinear points, there exists a unique circle passing through them. No three distinct collinear points lie on a circle.*

Proof. Let A, B, and C be non-collinear points. Construct the perpendicular bisectors, l_1 and l_2, to $\overline{AB}$ and $\overline{AC}$. (See Figure 4.1.) Because A, B, and C are non-collinear, the lines $\overleftrightarrow{AB}$ and $\overleftrightarrow{AC}$ cannot be parallel; consequently, the lines l_1 and l_2 are not parallel either and must intersect at a point O. By Theorem 4.1 (1), $AO = BO$ and $AO = CO$. Thus, the circle $\mathcal{C}(O, AO)$ passes through all three points, A, B, and C.

For uniqueness of the circle, let O_1 be the center of a circle through A, B, and C. Then O_1 is equidistant from A and C, and from A and B. By Theorem 4.1, O_1 belongs to both l_1 and l_2; hence, $O_1 = O$, and the circle with center at O_1 is the same as the circle $\mathcal{C}(O, OA)$. Note that this proof also establishes that there is no circle passing through three distinct collinear points.   □

In the preceding proof, since O is equidistant from B and C, the perpendicular bisector to $\overline{BC}$ passes through O. Thus, we see that the three perpendicular bisectors to the sides of $\triangle ABC$ are concurrent at O, and O can be determined by finding the point of intersection of any two of these perpendicular bisectors. One obvious consequence of Theorem 4.2 is that any triangle can be *circumscribed* by a circle.

The following corollary specifies the possible intersections between a circle and a line. Note that the first statement of the theorem is a corollary to Theorem 4.2. We leave the proof of the second statement to the reader.

> **Corollary 4.3.** *A line and a circle cannot intersect at more than two points. Moreover, a line l intersects a circle $\mathcal{C}(O, r)$ in 0, 1, or 2 points if and only if $d(O, l) > r$, $d(O, l) = r$, or $d(O, l) < r$, respectively.*

Further definitions are required in order to proceed with our study of circles. A **chord** of a circle is any line segment joining two points of the circle. A chord that passes through the center of the circle is a **diameter** of the circle. If the vertex of an angle falls at the center of a circle, then the angle is called a **central angle** of the circle. (For example, note the central angle, $\angle AOB$, in Figure 4.2.)

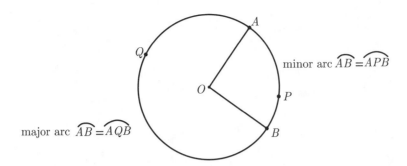

FIGURE 4.2. Arcs of a Circle.

Let A and B be two points on $\mathcal{C} = \mathcal{C}(O, r)$. The points divide $\mathcal{C}$ into two **arcs**. If $\overline{AB}$ is a diameter of $\mathcal{C}$, then the arcs are congruent and we call them **semicircles**. Otherwise, A and B are the endpoints of a **minor arc** (which is in the interior of $\angle AOB$) and a **major arc**. (See Figure 4.2.) We say that the chord AB **subtends** each of the arcs.

We will use the notation $\overset{\frown}{AB}$ to denote an arc with endpoints at A and B. Generally, $\overset{\frown}{AB}$ will refer to the minor arc; when ambiguity may be an issue, we will use the notation $\overset{\frown}{APB}$, where P is a point of the intended arc, as in Figure 4.2. The **sector** AOB is the region bounded by central angle AOB and $\overset{\frown}{AB}$.

The relationship between central angles and arcs suggests a natural way to impose *measurement* on arcs. Recall that according to our postulates, any angle is assigned a value (in degrees) between 0 and 180. Therefore, the measure of any central angle corresponds to a value between 0 and 180; however, in order to accommodate the measure of both major and minor arcs, we need to allow a greater range of values. We will define the **angular measure** (or just **measure**) of an arc, $\overset{\frown}{AB}$, of a circle $\mathcal{C}(O, r)$ to be $m\angle AOB$ if $\overset{\frown}{AB}$ is a minor arc; $180°$ if $\overline{AB}$ is a diameter; or $360° - m\angle AOB$ if $\overset{\frown}{AB}$ is a major arc. Observe that the measure of an arc does not depend on the radius of the circle. We assume without proof that two arcs in a given circle (or in congruent circles) are congruent if their angular measures are equal.

Theorem 4.4. *Let segments AB and CD be two chords of a circle $\mathcal{C} = \mathcal{C}(O, r)$. (See Figure 4.3.) The following statements are equivalent.*

(1) $\overline{AB} \cong \overline{CD}$

(2) $m\angle AOB = m\angle COD$

(3) $\overset{\frown}{AB} \cong \overset{\frown}{CD}$

(4) $d(O, \overline{AB}) = d(O, \overline{CD})$.

Proof. If one of the chords is a diameter of $\mathcal{C}$, the claim is obvious, so we assume that neither chord is a diameter.

(1) $\Rightarrow$ (2): Assume that $AB = CD$, as shown in Figure 4.3(i). Since A, B, C, and D are all points on the circle, $OA = OB = OC = OD$. By SSS, $\triangle AOB \cong \triangle COD$. Therefore, $\angle AOB \cong \angle COD$.

(2) $\Rightarrow$ (1): Assume that $\angle AOB \cong \angle COD$, as shown in Figure 4.3(ii). As before, $OA = OB = OC = OD$. Thus, by SAS, $\triangle AOB \cong \triangle COD$, so $\overline{AB} \cong \overline{CD}$.

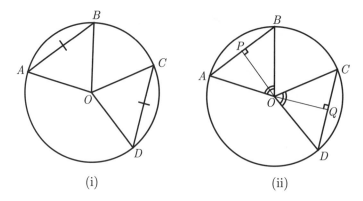

FIGURE 4.3.

(1) ⇒ (4): As shown above, under the assumption that $AB = CD$, we conclude that $\triangle AOB \cong \triangle COD$, with both being isosceles triangles. Construct the perpendicular from O to $\overline{AB}$; label the point of intersection P. Similarly, construct the perpendicular from O to $\overline{CD}$ with foot at Q. (See Figure 4.3(ii).) By Theorem 3.9, $AP = BP$ and $CQ = DQ$; thus, by the HL theorem, $\triangle APO$, $\triangle BPO$, $\triangle CQO$, and $\triangle DQO$ are all congruent. Hence, $OP = OQ$, and (4) is proven.

(4) ⇒ (1): The distance $d(O, \overline{AB})$ is the length of a perpendicular line segment from O to $\overline{AB}$. Let P be the foot of this perpendicular line segment, so that $d(O, \overline{AB}) = OP$. Similarly, let $d(O, \overline{CD}) = OQ$. Suppose that $d(O, \overline{AB}) = d(O, \overline{CD})$ (or, equivalently, $OP = OQ$). Since A, B, C, and D are all points on the circle, $OA = OB = OC = OD$. Thus, by the HL theorem, $\triangle AOP \cong \triangle BOP \cong \triangle COQ \cong COQ$, so $AB = CD$.

(2) ⇔ (3): By the definition of arc measure, $\angle AOB \cong \angle COD$ if and only if $m(\widehat{AB}) = m(\widehat{CD})$.   □

Theorem 4.5. *Let segments AB and CD be two chords of a circle, $\mathcal{C}(O, r)$. (See Figure 4.4.) The following statements are equivalent.*

(1) $AB > CD$

(2) $m\angle AOB > m\angle COD$

(3) $m(\widehat{AB}) > m(\widehat{CD})$

(4) *the distance from O to $\overline{AB}$ is smaller than the distance from O to $\overline{CD}$.*

We leave the proof of the previous theorem to the reader. The next theorem gives conditions under which a diameter of a circle is perpendicular to a chord.

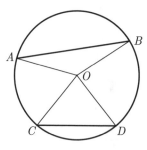

FIGURE 4.4.

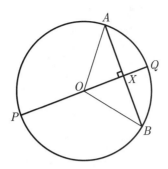

FIGURE 4.5.

Theorem 4.6. *Let C be a circle with center O; let $\overline{AB}$ be a chord of C that is not a diameter; let $\overline{PQ}$ be a diameter of C.*

(1) *$\overline{PQ}$ is perpendicular to $\overline{AB}$ if and only if $\overline{PQ}$ passes through the midpoint of $\overline{AB}$.*

(2) *$\overline{PQ}$ is perpendicular to $\overline{AB}$ if and only if $\overline{PQ}$ divides an arc subtended by $\overline{AB}$ into two congruent arcs. (The point that divides an arc of a circle into two congruent arcs is called the **midpoint** of the arc.)*

Proof. We refer to Figure 4.5 throughout the proof.

(1) ($\Rightarrow$) Assume $\overline{PQ} \perp \overline{AB}$ and X is their intersection. Since $\overline{OA}$ and $\overline{OB}$ are radii of the circle, $OA = OB$. Then, by the HL theorem, $\triangle OXA \cong \triangle OXB$; thus, $AX = BX$, and X is the midpoint of $\overline{AB}$.

($\Leftarrow$) Assume that $\overline{PQ}$ passes through the midpoint, X, of a chord of the circle, $\overline{AB}$; then $AX = BX$. As before, $OA = OB$, since both are radii. Hence, by SSS congruence, $\triangle OXA \cong \triangle OXB$, so $\angle OXA \cong \angle OXB$. Since these angles are supplementary, they must be right. We conclude that $\overline{PQ} \perp \overline{AB}$.

(2) The proof is left to the reader. $\qquad\qquad\square$

We say that points A and B are **symmetric** with respect to a line l if $A = B \in l$ or if l is the perpendicular bisector of $\overline{AB}$. The one-to-one correspondence of the plane that maps each point to a symmetric point with respect to l is called the **symmetry** with respect to l, or **reflection** with respect to l. A figure is called **symmetric** with respect to a line if the figure is congruent to its image under the reflection with respect to the line. Observe that in Figure 4.5, point A is mapped to point B under reflection about $\overleftrightarrow{PQ}$. Since A and B are simply the endpoints of an arbitrary chord that is perpendicular to $\overline{PQ}$, we conclude that $\overset{\frown}{PAQ}$ is mapped to $\overset{\frown}{PBQ}$ under reflection about $\overleftrightarrow{PQ}$. This observation is stated formally but without proof in Corollary 4.7.

Corollary 4.7. *A line passing through the center of a circle is an axis of symmetry for the circle.*

The orthogonality relation between diameters and chords described in Theorem 4.6 holds for chords of arbitrarily small length, as well as for lines containing these chords. Intuitively, this suggests that a line intersecting a circle at one point should be perpendicular to the diameter containing the point. This is indeed the case, as specified by Corollary 4.8.

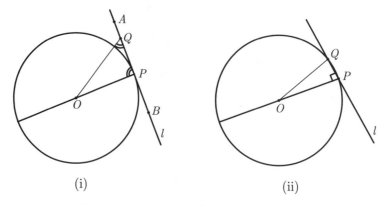

(i)   (ii)

FIGURE 4.6.

Corollary 4.8.

(1) *A line intersects a circle at exactly one point if and only if it is perpendicular to the diameter of the circle containing that point.*

(2) *For any point P on a circle, there exists a unique line passing through P and intersecting the circle only at this point.*

Proof.

(1) Consider a circle $\mathcal{C} = \mathcal{C}(O, r)$ with point P on the circle.

($\Rightarrow$) Let l be a line intersecting $\mathcal{C}$ only at P. Suppose that l is not perpendicular to $\overline{OP}$. Choose points A and B on l such that $A - P - B$. Exactly one of $\angle OPA$ or $\angle OPB$ must be an acute angle; assume $m\angle OPA < 90°$. Find point $Q\ (\neq P)$ on $\overrightarrow{PA}$ such that $m\angle OPA = m\angle OQP$, as shown in Figure 4.6(i). (Why must such a point exist?) Then $\triangle OPQ$ is an isosceles triangle; hence $OP = OQ$, so Q is a point on both l and $\mathcal{C}$, contradicting our assumption. We conclude that $m\angle OPA = 90°$.

($\Leftarrow$) Let l be a line passing through point P such that $l \perp \overline{OP}$. Suppose that l intersects $\mathcal{C}$ at a second point, Q. Now $\overline{QP}$ is a chord of $\mathcal{C}$. (See Figure 4.6(ii).) By Theorem 4.6, since $\overline{OP}$ is perpendicular to $\overline{QP}$ at P, P must be the midpoint of $\overline{QP}$ – clearly a contradiction if $Q \neq P$.

(2) There is a unique line, l, through P that is perpendicular to $\overline{OP}$. Using part (1), l intersects $\mathcal{C}$ at one point, so l is the desired line. $\square$

The unique line passing through point P on a circle $\mathcal{C}$ and intersecting the circle only at P (whose existence has been established by the preceding theorem) is called the **tangent** to the circle at P. A line intersecting a circle at *two* points (i.e., containing a chord of the circle) is called a **secant** to the circle. The tangent to a circle at point P can be viewed as a limiting position of a sequence of secants $\overleftrightarrow{PB}$, created by fixing P and letting B move along the circle, getting closer and closer to P. For example, in Figure 4.7, the tangent line, l, to the circle at P is the limiting position of the sequence of secant lines $\overleftrightarrow{PB}, \overleftrightarrow{PB_1}, \overleftrightarrow{PB_2}, \dots$.

The idea of a tangent to a circle has been around for a long time; Euclid included this concept in his work, as did other Greek mathematicians. However, making a useful definition of a tangent to other curves took many centuries. In fact, clarifying this notion was essential for the rigorous development of calculus. In calculus, the tangent line is viewed as the limiting position of secant

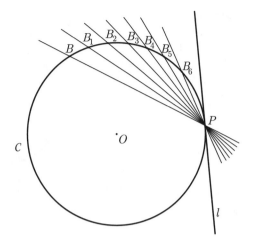

FIGURE **4.7.**

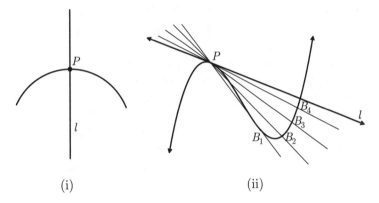

(i)   (ii)

FIGURE **4.8.**

lines, not as a line sharing exactly one point with the curve. In general, requiring a tangent line to share exactly one point with the curve is neither sufficient (as shown in Figure 4.8(i)) nor necessary (as shown in Figure 4.8(ii), where the tangent l at P intersects the curve twice).

Theorem 4.9.   *Two tangent segments to a circle that intersect at a point external to the circle, as shown in Figure 4.9, are congruent.*

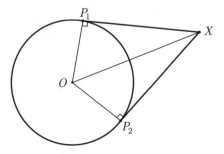

FIGURE **4.9.**

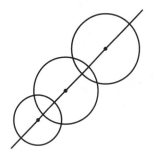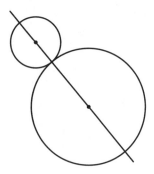

FIGURE 4.10.

Proof. Consider a circle, $C = C(O, r)$, and a point X that is exterior to C; let $\overline{XP_1}$ and $\overline{XP_2}$ be distinct tangent segments to C.

By Corollary 4.8, $\angle OP_1X$ and $\angle OP_2X$ are both right angles. In addition, as radii of $C(O, r)$, $\overline{OP_1} \cong \overline{OP_2}$. Hence, by the HL theorem, $\triangle OP_1X \cong \triangle OP_2X$, so $XP_1 = XP_2$. $\square$

Corollary 4.7 implies that the line passing through the centers of a set of circles is the axis of symmetry of their union; that is, such a line is an axis of symmetry for each of the circles in the set. (See Figure 4.10 for examples.) If there are more than two circles in the set, there is no guarantee that such a line exists, of course.

The axis of symmetry of the union of circles will help us explore the possibilities for the intersection of two circles, which we do next. Two circles are called **tangent** if they are simultaneously tangent to a line at the same point. The circles can be externally tangent or internally tangent, as illustrated in Figure 4.11.

Theorem 4.10.   *Two distinct circles intersect at exactly one point if and only if they are tangent.*

Proof. Consider two distinct circles, $C_1 = C(O_1, r_1)$ and $C_2 = C(O_2, r_2)$. (See Figure 4.11.)

($\Rightarrow$) Suppose that C_1 and C_2 intersect at exactly one point, P. Line O_1O_2 is an axis of symmetry for both circles, so the reflection of P about $\overline{O_1O_2}$ will also be a point P' that lies on both circles; hence $P = P'$, and P lies on $\overline{O_1O_2}$.

By Corollary 4.8, there is a unique line, l, through P that is tangent to C_1; likewise, there is a unique line, m, through P that is tangent to C_2. We must show that $l = m$. By definition of tangency, $l \perp \overline{O_1P}$ and $m \perp \overline{O_2P}$; since P lies on $\overline{O_1O_2}$, l must equal m, as there is a unique perpendicular line to $\overline{O_1O_2}$ through P.

($\Leftarrow$) Suppose that C_1 and C_2 are tangent at P. Then they are tangent to some line l at P. By Corollary 4.8, $\overline{O_1P} \perp l$ at P and $\overline{O_2P} \perp l$ at P. Hence, O_1, P, and O_2 are all collinear, and the line, m, containing these three points is an axis of symmetry for the union of C_1 and C_2. If the circles share another point, $X \neq P$, then they would necessarily share a point X' that is symmetric to X with respect to $\overline{O_1O_2}$. In this case, the three points X, P, and X' would all lie on each of the circles. By Theorem 4.2, this is an impossibility since C_1 and C_2 are distinct circles. $\square$

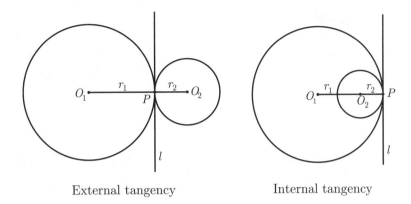

External tangency   Internal tangency

FIGURE 4.11.

Two distinct circles can intersect at 0, 1, or 2 points, as established in the following theorem.

> **Theorem 4.11.** *Let $\mathcal{C}(O_1, r_1)$ and $\mathcal{C}(O_2, r_2)$ be two distinct circles in the plane. Denote the distance $O_1 O_2$ between the centers as d. Then*
>
> (1) *the circles do not intersect if and only if the maximum of $\{r_1, r_2, d\}$ is greater than the sum of the other two values;*
> (2) *the circles intersect at one point if and only if the maximum of $\{r_1, r_2, d\}$ is equal to the sum of the other two values;*
> (3) *the circles intersect at two points if and only if the maximum of $\{r_1, r_2, d\}$ is less than the sum of the other two.*

Proof. Without loss of generality, assume that $r_1 \geq r_2$.

(1) ($\Rightarrow$) Assume that the circles do not intersect. This could occur in either of two ways: $\mathcal{C}(O_2, r_2)$ lies within $\mathcal{C}(O_1, r_1)$, or $\mathcal{C}(O_2, r_2)$ is entirely outside of $\mathcal{C}(O_1, r_1)$.

Suppose that $\mathcal{C}(O_2, r_2)$ lies within $\mathcal{C}(O_1, r_1)$. (One version of this is illustrated in Figure 4.12(i).) If $d = 0$, then certainly $r_1 > r_2 = r_2 + d$, and we are done. If $d \neq 0$, then construct $\overrightarrow{O_1 O_2}$ and label points of intersection of $\overrightarrow{O_1 O_2}$ with $\mathcal{C}(O_1, r_1)$ and $\mathcal{C}(O_2, r_2)$ as P_1 and P_2, respectively, so that $O_1 - O_2 - P_2 - P_1$. Then $O_1 O_2 = d$, $O_2 P_2 = r_2$, and $O_1 P_1 = r_1$. Since $r_1 = d + r_2 + P_2 P_1$ and $P_2 P_1 > 0$, $r_1 > d + r_2$. The case when $\mathcal{C}(O_2, r_2)$ is entirely outside of $\mathcal{C}(O_1, r_1)$ (as shown in Figure 4.12(ii)) is similarly proven.

($\Leftarrow$) Suppose that the circles intersect at some point P. If O_1, O_2, and P are collinear, then $max\{r_1, r_2, d\}$ is equal to the sum of the other two values. If O_1, O_2, and P are not collinear, then these three points are the vertices of a triangle. In this case, the maximum of the set $\{r_1, r_2, d\}$ is less than the sum of the other two by the triangle inequality. Thus, by the contrapositive, if the maximum of the set $\{r_1, r_2, d\}$ is greater than the sum of the other two values, there can be no point of intersection for the circles.

We leave the proofs of (2) and (3) to Problem 4.5. $\qquad\square$

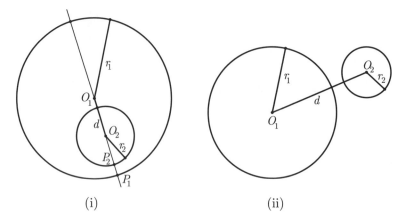

FIGURE 4.12.

The following example finds a distance related to the tangents to two circles. An extension of this example may be found in Problem 4.15.

Example 4. *Consider two externally tangent circles with respective centers O_1 and O_2 and radii $r_1 < r_2$. Find the distance from the intersection of their common exterior tangent lines to O_1.*

Solution: Certainly the intersection point, M, lies on $\overrightarrow{O_2 O_1}$. We need to find $a = MO_1$. Suppose one of the common tangent lines intersects the circles at points P_1 and P_2, respectively, as shown in Figure 4.13. Since line $P_1 P_2$ is perpendicular to both $\overline{O_1 P_1}$ and $\overline{O_2 P_2}$, $\triangle MP_1O_1 \sim \triangle MP_2O_2$. Therefore,

$$\frac{a}{r_1} = \frac{a + r_1 + r_2}{r_2},$$

from which it follows that the desired distance is $a = \frac{r_1(r_2+r_1)}{r_2-r_1}$. $\square$

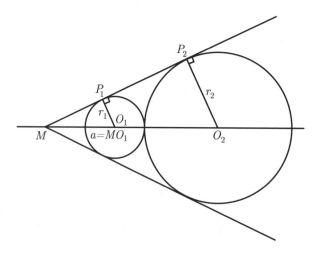

FIGURE 4.13.

4.2 Inscribed Angles

> *He [Walt Disney] made the maximum use of circles be-*
> *cause they were the easiest to draw fast and he insisted*
> *that his helpers do the same.*
>
> — Paul Johnson[2]

An understanding of central angles and their associated arcs has provided us with one means of exploring properties of circles. We now introduce another class of angles that will prove useful. Let P, A, and B be three points on a circle C. We say that $\angle APB$ is **inscribed** in C. (See Figure 4.14.) In this case, we say that the arc $\overset{\frown}{ARB}$ is **intercepted by** $\angle APB$.

The main fact about inscribed angles is given in the following theorem.

Theorem 4.12. *The measure of an inscribed angle is half the measure of the arc it intercepts (or, equivalently, half the measure of the central angle corresponding to the intercepted arc).*

Proof. Consider first an inscribed angle, $\angle APB$, such that $\overline{PB}$ is a diameter of the circle $C(O, r)$. As shown in Figure 4.15, $\overset{\frown}{AQB}$ is the intercepted arc for this inscribed angle, and $\angle AOB$ is the corresponding central angle.

Let $m\angle AOB = x$. Since $\overline{OP}$ and $\overline{OA}$ are both radii, $\triangle OPA$ is isosceles. Let $m\angle OPA = m\angle OAP = y$. By the Exterior Angle theorem (Corollary 3.5), $x = 2y$.

There are two other cases to consider: the case when the center, O, lies in the interior of $\angle APB$ and the case when O lies in the exterior of $\angle APB$. Suppose that O is in the interior of $\angle APB$,

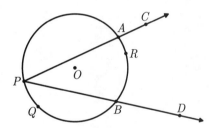

FIGURE 4.14.

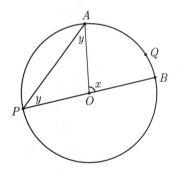

FIGURE 4.15. Inscribed angle with diameter PB.

---

[2] From *Inventors*, p. 268–269.

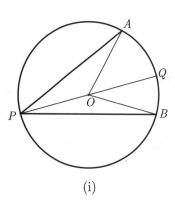
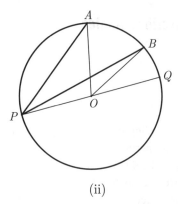

(i)   (ii)

FIGURE 4.16.

as in Figure 4.16(i). Let $\overline{PQ}$ be the diameter of $\mathcal{C}(O, r)$ having P as an endpoint. As just shown, $m\angle APQ = \frac{1}{2}m\angle AOQ$ and $m\angle QPB = \frac{1}{2}m\angle QOB$. By angle addition,

$$m\angle APB = m\angle APQ + m\angle QPB = \frac{1}{2}(m\angle AOQ + m\angle QOB) = \frac{1}{2}m\angle AOB.$$

The remaining case is similarly proven and the details are left to the reader. □

The proof of Theorem 4.12 utilizes one of the favorite tools of mathematicians: reduction to a previous case. Observing that the argument can be split into two cases involving the diameter as one ray of the inscribed angle greatly reduces the work required in proving the new case. Watch for opportunities to use this technique in your own proofs.

Corollary 4.13. *Consider a circle, $\mathcal{C} = \mathcal{C}(O, r)$.*

(1) *Two non-intersecting chords of $\mathcal{C}$, $\overline{AB}$ and $\overline{CD}$, are parallel if and only if the arcs $\overset{\frown}{AC}$ and $\overset{\frown}{BD}$ are congruent. (See Figure 4.17(i).)*

(2) *If $\overline{AB}$ is a chord of $\mathcal{C}(O, r)$, C is a point on the circle, and l is a line tangent to $\mathcal{C}$ at C, then l is parallel to $\overline{AB}$ if and only if the arcs $\overset{\frown}{AC}$ and $\overset{\frown}{BC}$ are congruent. (See Figure 4.17(ii).)*

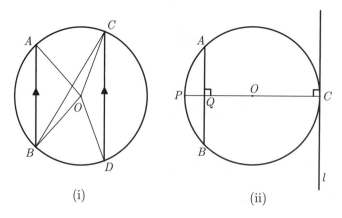

(i)   (ii)

FIGURE 4.17.

Proof.

(1) Assume that $\overline{AB}$ and $\overline{CD}$ are non-intersecting chords of C. Then $\overset{\frown}{AC} \cong \overset{\frown}{BD}$ if and only if $m\angle AOC = m\angle BOD$, which, by Theorem 4.12, occurs if and only if $m\angle ABC = m\angle BCD$. However, as they are alternate interior angles, $\angle ABC \cong \angle BCD$ if and only if $\overline{AB}$ and $\overline{CD}$ are parallel.

(2) Assume that $\overline{AB}$ is a chord of C, the point C is on the circle, and l is a line tangent to C through C. Let $\overline{PC}$ be the diameter of the circle having one endpoint at C. By Corollary 4.7, $\overline{PC}$ is an axis of symmetry for C.

($\Rightarrow$) Suppose that l is parallel to $\overline{AB}$. By Corollary 4.8, l is perpendicular to the diameter through C, $\overline{PC}$; hence $\overline{AB} \perp \overline{PC}$ also. Consequently, A is mapped to B under reflection about $\overline{PC}$, and C, of course, is fixed. Thus, $\overset{\frown}{AC}$ is mapped to $\overset{\frown}{BC}$ under reflection about $\overline{PC}$ and the two arcs are congruent.

($\Leftarrow$) Suppose now that $\overset{\frown}{AC}$ is congruent to $\overset{\frown}{BC}$. Let Q be the point of intersection of $\overline{AB}$ and $\overline{PC}$. Because the arcs are congruent and a diameter of the circle passes through C, B must map to A under reflection about $\overline{PC}$. Thus, $\overline{AQ}$ maps to $\overline{BQ}$ under reflection about $\overline{PC}$, so $\overline{AB} \perp \overline{PC}$. As before, l is perpendicular to $\overline{PC}$ by Corollary 4.8, and we conclude that l is parallel to $\overline{AB}$. $\qquad\square$

Note that the second part of Corollary 4.13 can be viewed as a "limiting case" of part (1) as CD becomes arbitrarily small. Likewise, the next corollary can be viewed as a "limiting case" of Theorem 4.12, where the angle between a chord and a tangent at the endpoint of the chord is a "limiting position" of an inscribed angle.

Corollary 4.14. *The measure of an angle formed by an intersecting chord and tangent line is half the measure of the arc subtended by the chord (or, equivalently, half the measure of the corresponding central angle).*

Proof. Let $\overline{AB}$ be a chord of a circle C and l be the tangent line at B. Construct a line m through A that is parallel to l; let D be the second point of intersection of m with C, as shown in Figure 4.18. (Such a point will exist unless $\overline{AB}$ is a diameter of the circle, in which case the statement is clearly true.)

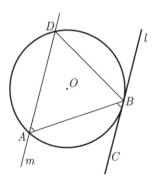

FIGURE 4.18.

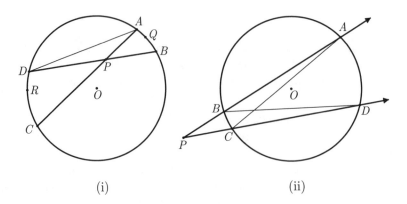

FIGURE 4.19.

We wish to determine the measure of $\angle ABC$. By Theorem 4.12, $m\angle DAB = \frac{1}{2}m(\widehat{DB})$. By Corollary 4.13(2), $\widehat{AB} \cong \widehat{DB}$ since l is parallel to m. Using alternate interior angles, $m\angle ABC = m\angle DAB = \frac{1}{2}m(\widehat{DB}) = \frac{1}{2}m(\widehat{AB})$. □

Theorem 4.15. *Consider a circle, $\mathcal{C} = \mathcal{C}(O, r)$.*

(1) *The measure of the vertical angles formed by two intersecting chords of $\mathcal{C}$ is half of the sum of the measures of the two intercepted arcs.*

(2) *The measure of the angle formed by two externally intersecting secant lines to $\mathcal{C}$ is half of the positive difference of the measures of the two intercepted arcs.*

Proof.

(1) Consider circle $\mathcal{C}(O, r)$ with chords $\overline{AC}$ and $\overline{BD}$ that intersect at point P, as shown in Figure 4.19(i). We wish to determine the measure of the vertical angles $\angle APB$ and $\angle DPC$. Note that by Theorem 4.12, $m\angle ADP = \frac{1}{2}m(\widehat{AQB})$. Similarly, $m\angle DAP = \frac{1}{2}m(\widehat{DRC})$. However, $\angle APB$ is an exterior angle to $\triangle APD$. Therefore,

$$m\angle APB = m\angle ADP + m\angle DAP$$
$$= \frac{1}{2}m(\widehat{AQB}) + \frac{1}{2}m(\widehat{DRC}) = \frac{1}{2}\left(m(\widehat{AQB}) + m(\widehat{DRC})\right).$$

(2) Consider circle $\mathcal{C}(O, r)$ with secants $\overleftrightarrow{AB}$ and $\overleftrightarrow{CD}$ that intersect at point P in the exterior of the circle. Assume $m(\widehat{AD}) > m(\widehat{BC})$. By the Exterior angle theorem for $\triangle PBD$,

$$m\angle BPD = m\angle ABD - m\angle BDP = \frac{1}{2}m(\widehat{AD}) - \frac{1}{2}m(\widehat{BC})$$
$$= \frac{1}{2}(m(\widehat{AD}) - m(\widehat{BC})).$$

□

Theorem 4.12 can be thought of as a "limiting case" of Theorem 4.15: when P approaches the circle, one of the arcs collapses to a point. Nevertheless, we used Theorem 4.12 heavily to prove Theorem 4.15.

The following corollary will prove useful in future chapters of this text, particularly Chapter 6.

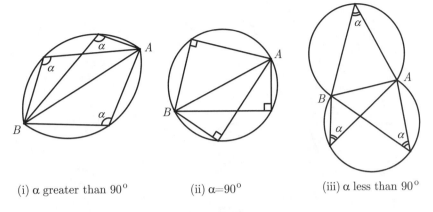

(i) α greater than 90°   (ii) α=90°   (iii) α less than 90°

FIGURE **4.20.**

Corollary 4.16. *Let a segment $\overline{AB}$ and a constant $\alpha \in (0°, 180°)$ be given. The set of all points X in the plane such that $m\angle AXB = \alpha$ is the union of two congruent circular arcs with the points A and B deleted. In particular, if $\alpha = 90°$, the set is a circle with $\overline{AB}$ as a diameter and with points A and B deleted. (See Figure 4.20.)*

Proof. Let X be a point such that $m\angle AXB = \alpha$. The points A, X, and B determine a circle, $\mathcal{C} = \mathcal{C}(O, r)$. By Theorem 4.12, $m\angle AXB = \frac{1}{2}m(\widehat{ADB})$, where $\widehat{ADB}$ is the arc of $\mathcal{C}$ with A and B as endpoints not containing X. (See Figure 4.21(i).) Let Y be any other point of $\widehat{AXB}$. Then $m\angle AYB = m\angle AXB$, since they intercept the same arc.

Reflect $\widehat{AXB}$ about $\overline{AB}$. The resulting arc ($\widehat{AX'B}$ in Figure 4.21) is congruent to $\widehat{AXB}$ and any angle inscribed in this arc must be congruent to an angle inscribed in $\widehat{AXB}$. Thus, any point of either arc serves as the vertex of an angle with rays passing through A and B and having angle measure α.

It remains to show that this is the complete set of points having the desired property. Let P be a point in the exterior of the region bounded by $\widehat{AXB}$ and $\widehat{AX'B}$. (See Figure 4.21(ii).) Let S and

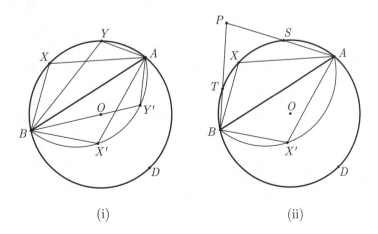

(i)   (ii)

FIGURE **4.21.**

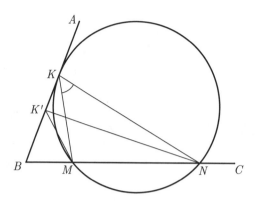

T be the points of intersection of $\overline{AP}$ and $\overline{BP}$ with $\mathcal{C}$ respectively.[3] Then by Theorem 4.15(2), the measure of $\angle APB$ is one-half the positive difference of the measures of two intercepted arcs, one of which has measure 2α. (In Figure 4.21(ii), $m\angle APB = \frac{1}{2}(m(\widehat{ADB}) - m(\widehat{ST})) = \alpha - \frac{1}{2}m(\widehat{ST}) < \alpha$.) Thus, it is not possible that the measure of $\angle APB$ is α, so P is not in our set.

The case when P is in the interior of the region bounded by $\widehat{AXB}$ and $\widehat{AX'B}$ can be handled similarly. For all such points, $m\angle APB > \alpha$. $\qquad\square$

Example 5. Given an angle ABC and two points M and N on $\overrightarrow{BC}$, let K be a point of $\overrightarrow{BA}$ such that $m\angle MKN$ is the greatest. Describe the position of K in geometrical terms.

Solution: Consider a circle through M and N that is tangent to $\overrightarrow{BA}$. (See Figure 4.22.) Such a circle clearly exists and is unique: its center is the intersection of the bisector of the angle with the perpendicular bisector of $\overline{MN}$.

The point of tangency of the circle with $\overrightarrow{BA}$ is the point K such that $m\angle MKN$ is maximized. By Theorem 4.12, $m\angle MKN = 1/2m(\widehat{MN})$. Every other point K' on $\overrightarrow{BA}$ is outside of the circle. Hence, by Theorem 4.15(2), $m\angle MK'N$ is half of the positive difference of the measures of the arcs of the circle cut by its sides. Since the larger of the two arcs determined by any such angle $MK'N$ is $\widehat{MN}$, $m\angle MK'N < m\angle MKN$. $\qquad\square$

This example has an interesting "application." Suppose that a group of people is sitting on a "hill," watching a movie on a large vertical screen $\overline{MN}$. (See Figure 4.23.) From what point K on the hill will the view be the best; i.e., where will $m\angle MKN$ be the greatest? Based on the previous discussion, consider all circles through M and N that touch the hill. Among these circles choose the one of the smallest radius and let K be a point of tangency of the circle to the hill. (There might be more than one.) The view from any such point K will be the best, in the sense of providing the maximum viewing angle.

4.3   Lengths of Tangents and Chords

In this short section we explore some relationships between the lengths of chords and tangent segments to a circle.

---

[3] If P is in the interior of $\mathcal{C}$, consider instead the points of intersection of $\overline{AP}$ and $\overline{BP}$ with $\mathcal{C}'$, where $\mathcal{C}'$ is the circle determined by A, X', and B.

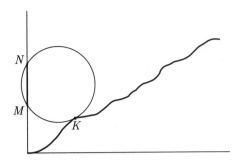

FIGURE 4.23.

Theorem 4.17. *Let $C = C(O, r)$ be a circle and let A be a point not on C. Consider two secant lines, l and l', to C through A; let the points of intersection of these lines with the circle be B and C, and B' and C', respectively. (See Figure 4.24.) Then*

$$AB \cdot AC = AB' \cdot AC'.$$

Proof. First, let A be in the exterior of the circle, as shown in Figure 4.24(i). Consider triangles ACB' and $AC'B$. They share a common angle at A. Also, $\angle BCB' \cong \angle BC'B'$, as they are both inscribed in arc BB'.

Consequently, $\triangle ACB' \sim \triangle AC'B$ by AA. Hence,

$$\frac{AB}{AB'} = \frac{AC'}{AC}$$

and the result follows.

Now, suppose A lies inside the circle, as shown in Figure 4.24(ii). Consider $\overline{BB'}$ and $\overline{CC'}$. Again using congruence of inscribed angles, $\triangle ABB' \sim \triangle ACC'$. The proportion

$$\frac{AB}{AC'} = \frac{AB'}{AC}$$

again yields the desired result. □

Clearly the following corollary is a "limiting case" of Theorem 4.17.

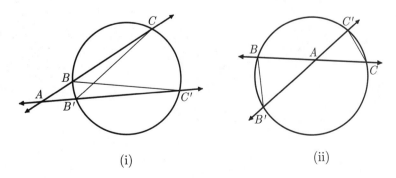

(i)   (ii)

FIGURE 4.24.

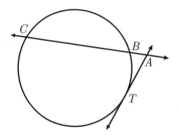

FIGURE 4.25.

Corollary 4.18. *Let A be a point in the exterior of a circle, let $\overline{AT}$ be a tangent to the circle, and let a secant through A intersect the circle at points B and C (so that $A - B - C$, as shown in Figure 4.25). Then*

$$AT^2 = AB \cdot AC.$$

Proof. By Corollary 4.14, $\angle ACT \cong \angle ATB$. Therefore, $\triangle ATB \sim \triangle ACT$ by AA. Therefore,

$$\frac{AT}{AB} = \frac{AC}{AT},$$

and the result follows.   □

The following example provides a nice application of Corollary 4.18.

Example 6.   Let circles C_1 and C_2 intersect at points A and B. Prove that the lengths of tangent segments to these circles from any point of line AB, but not of segment AB, are equal.

Solution: Let P lie on line AB, not on segment AB, and let D_1 and D_2 be points on C_1 and C_2, respectively, such that $\overline{PD_1}$ and $\overline{PD_2}$ are tangent to the two circles. (See Figure 4.26.) Then from Corollary 4.18, $(D_1P)^2 = PA \cdot PB = (D_2P)^2$, so $D_1P = D_2P$.   □

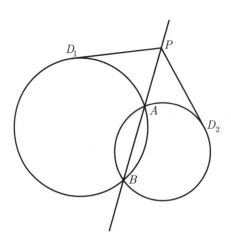

FIGURE 4.26.

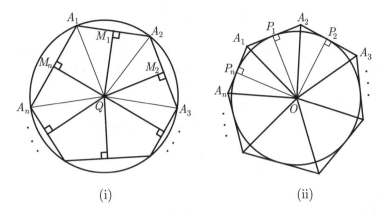

FIGURE 4.27.

4.4 Incircle and Circumcircle

As previously noted, Theorem 4.2 demonstrates that any triangle can be circumscribed by a circle. Is the same true for other polygons? Certainly not. Since three points uniquely determine a circle, the additional vertices of an n-sided polygon ($n > 3$) will need to be constrained in order for the polygon to be circumscribable.

Likewise, we might ask when a circle can be *inscribed* in a triangle. How about in other polygons? These are the kinds of questions we wish to address in this section.

If there exists a circle C that is tangent to all sides of the polygon $\mathcal{P}$, then C is **inscribed** in $\mathcal{P}$ and $\mathcal{P}$ is **circumscribed** around C; alternatively, we say that C is the **incircle** of $\mathcal{P}$, its center is the **incenter** of $\mathcal{P}$, and the length of its radius is the **inradius** of $\mathcal{P}$. Similarly, if all of the vertices of $\mathcal{P}$ lie on a circle C, then $\mathcal{P}$ is **inscribed** in C and C is **circumscribed** around $\mathcal{P}$; in this case, C is the **circumcircle** of $\mathcal{P}$, its center is the **circumcenter** of $\mathcal{P}$, the length of its radius is the **circumradius** of $\mathcal{P}$, and we say that $\mathcal{P}$ is a **cyclic polygon**. These notions are illustrated in Figure 4.27. Theorem 4.19 provides necessary and sufficient conditions for the existence of the incircle and circumcircle of a polygon and assures the uniqueness of these circles if they exist.

Theorem 4.19. *Let $\mathcal{P}$ be a polygon.*

(1) *There exists a point Q that is equidistant from all vertices of $\mathcal{P}$ if and only if the perpendicular bisectors of all sides of $\mathcal{P}$ are concurrent. If such a point exists, it is unique, and it is the circumcenter of $\mathcal{P}$.*

(2) *There exists a point O that is equidistant from all sides of $\mathcal{P}$ if and only if the bisectors of all interior angles of $\mathcal{P}$ are concurrent. If such a point exists, it is unique, and it is the incenter of $\mathcal{P}$.*

Proof. Let $\mathcal{P}$ be a polygon with vertices $A_1, A_2, \ldots, A_n, A_{n+1} = A_1$.

(1) Label the midpoint of side $\overline{A_i A_{i+1}}$ as M_i for $i = 1, 2, \ldots, n$.

($\Rightarrow$) Suppose that Q is a point that is equidistant from all vertices of $\mathcal{P}$; that is, $QA_1 = QA_2 = \cdots = QA_n$. By SSS, $\triangle A_i M_i Q \cong \triangle A_{i+1} M_i Q$. Thus, $\angle A_i M_i Q$ and $\angle A_{i+1} M_i Q$ are congruent angles that form a linear pair, and we conclude that $\overline{M_i Q}$ is the perpendicular bisector of $\overline{A_i A_{i+1}}$. Hence, the perpendicular bisectors of the sides of $\mathcal{P}$ are all concurrent.

($\Leftarrow$) Suppose all the perpendicular bisectors of the sides of $\mathcal{P}$ are concurrent at a point Q. By the HL theorem, $\triangle A_i M_i Q \cong \triangle A_{i+1} M_i Q$; consequently, $\overline{QA_i} \cong \overline{QA_{i+1}}$. By transitivity of the congruence relation, all vertices are equidistant from Q.

(2) ($\Rightarrow$) Suppose that O is a point that is equidistant from all sides of $\mathcal{P}$. The distance from O to the side $\overline{A_i A_{i+1}}$ will be the length of a line segment perpendicular to $\overline{A_i A_{i+1}}$ from O. Let P_i denote the point of intersection of the line through O that is perpendicular to $\overline{A_i A_{i+1}}$; then $OP_1 = OP_2 = \cdots = OP_n$. By the HL theorem, $\triangle P_n A_1 O \cong \triangle P_1 A_1 O$, making $\angle P_n A_1 O$ and $\angle P_1 A_1 O$ congruent. Likewise, $\triangle P_i A_{i+1} O \cong \triangle P_{i+1} A_{i+1} O$ for $1 \leq i \leq n$, so $\angle P_i A_{i+1} O \cong \angle P_{i+1} A_{i+1} O$ for $1 \leq i \leq n$. Thus, all angle bisectors of $\mathcal{P}$ meet at O.

($\Leftarrow$) Suppose the bisectors of all interior angles of $\mathcal{P}$ are concurrent, meeting at a point O. By AAS, $\triangle P_n A_1 O \cong \triangle P_1 A_1 O$, which implies that $OP_n = OP_1$. Likewise, $\triangle P_i A_{i+1} O \cong \triangle P_{i+1} A_{i+1} O$ for $1 \leq i \leq n$, so $OP_i = OP_{i+1}$ for $1 \leq i \leq n$. By transitivity, $OP_1 = OP_2 = \cdots = OP_n$. $\square$

Corollary 4.20. *Every triangle has both an incircle and circumcircle.*

Proof. The existence of the circumcircle of a triangle was established in Theorem 4.2. The existence of the incircle follows from the fact that the interior angle bisectors of a triangle are concurrent; see Problem 3.2.15. An alternative proof of this fact follows from Theorem 4.1(2). $\square$

Quadrilaterals with circumcircles or incircles have interesting properties, as we see in the following theorems.

Theorem 4.21. *A quadrilateral $ABCD$ is cyclic (has a circumcircle) if and only if*

$$m\angle A + m\angle C = m\angle B + m\angle D.$$

Proof. ($\Rightarrow$) Suppose $ABCD$ is inscribed in a circle. Then the chord AC simultaneously subtends $\overset{\frown}{ABC}$ and $\overset{\frown}{ADC}$. (See Figure 4.28(i).) Since the measure of an inscribed angle is half the measure of the intercepted arc (by Theorem 4.12), the sum of the measures of the angles at B and D must be $180°$.

($\Leftarrow$) Assume that $m\angle A + m\angle C = m\angle B + m\angle D = 180°$. Consider just the points A, B, and C; these three points determine a circle, $\mathcal{C}$. If D is on $\mathcal{C}$, we are done. Suppose that D is inside $\mathcal{C}$. Join D to C and A and extend $\overline{CD}$ until it crosses $\mathcal{C}$; call the point of intersection D_1. (See Figure 4.28(ii).) Join D_1 to A. By assumption, $m\angle B + m\angle D = 180°$; furthermore, $m\angle B + m\angle D_1 = 180°$ since $ABCD_1$ is inscribed in $\mathcal{C}$. Thus, $m\angle D = m\angle D_1$. However, $\angle ADC$ is exterior to $\triangle D_1 AD$, which implies that $m\angle ADC > m\angle D_1$. We have reached a contradiction, and we conclude that it is impossible for D to be inside $\mathcal{C}$.

Assuming that D is outside of $\mathcal{C}$ leads to a contradiction in a similar way. Thus, we conclude that $ABCD$ is circumscribed by $\mathcal{C}$. $\square$

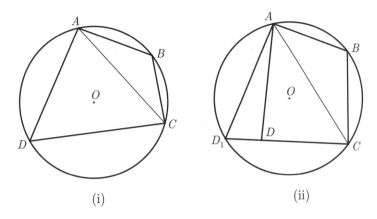

FIGURE 4.28.

We leave the proof of the following corollary to Problem 4.12.

Corollary 4.22. *A quadrilateral $ABCD$ with a pair of parallel sides is cyclic if and only if $ABCD$ is an isosceles trapezoid or a rectangle.*

The next theorem is attributed to the Greek mathematician Claudis Ptolemaeus, commonly known as Ptolemy, whose work dates from the second century A.D. Ptolemy is best known as the author of an influential thirteen-volume series on trigonometry and astronomy. This compilation is entitled *Mathematical Syntaxis*, although it came to be known as *Almagest* (from Arabic for "the greatest") in order to distinguish it from another collection of books which was produced at about the same time. Ptolemy's theorem implies the Pythagorean theorem (see Problem S4.4) and familiar formulas for the sine and cosine of the sum or difference of two angles (see Chapter 7).

Theorem 4.23. *(Ptolemy's theorem) Let $ABCD$ be a cyclic quadrilateral. Then*
$$AB \cdot CD + AD \cdot BC = AC \cdot BD.$$

Proof. Assume that $ABCD$ is a quadrilateral for which a circumcircle exists. Locate point E on $\overline{BD}$ such that $m\angle BCA = m\angle ECD$ (see Figure 4.29); such a point certainly exists since $m\angle BCD > m\angle BCA$. Since $\angle BAC$ and $\angle BDC$ are both inscribed angles on $\overset{\frown}{BC}$, they are congruent. Thus, $\triangle ABC \sim \triangle DEC$, so $AB/DE = AC/CD$. Again using inscribed angles, $\angle DAC \cong \angle DBC$. Noting that additionally $\angle DCA \cong \angle ECB$, we see that $\triangle ADC \sim \triangle BEC$, so $AC/BC = AD/BE$. From the equalities above we obtain

$$AB \cdot CD + BC \cdot AD = AC \cdot DE + AC \cdot BE = AC(BE + DE) = AC \cdot BD.$$

$\square$

Theorem 4.24. *Let $ABCD$ be a convex quadrilateral. There exists an incircle of $ABCD$ if and only if $AB + CD = BC + AD$.*

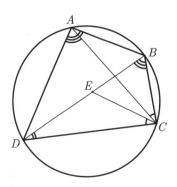

FIGURE 4.29.

Proof. Consider quadrilateral $ABCD$.

($\Rightarrow$) Let $\mathcal{C} = \mathcal{C}(O, r)$ be the incircle of $ABCD$. Let P, Q, R, and S be the points of tangency of $\mathcal{C}$ with the sides $\overline{AB}$, $\overline{BC}$, $\overline{CD}$, and $\overline{AD}$, respectively, as shown in Figure 4.30(i).

By Theorem 4.9, $AP = AS$. Similarly, $BP = BQ$, $CR = CQ$, and $DR = DS$. The desired sum, $AB + CD = BC + AD$, follows from the addition of these four equalities.

($\Leftarrow$) Assume that $AB + CD = BC + AD$. Suppose first that $ABCD$ is a parallelogram. Then $AB = CD$ and $BC = AD$. Since $AB + CD = BC + AD$, we conclude that $ABCD$ is a rhombus. Because the diagonals of a rhombus are bisectors of its interior angles (Theorem 3.25), $ABCD$ has an incircle by Theorem 4.19(2).

Now, consider the case when $ABCD$ is not a parallelogram. In particular, assume that $\overrightarrow{AB}$ and $\overrightarrow{CD}$ intersect at a point X, as shown in Figure 4.30(ii). Let $\mathcal{C}$ be the incircle of $\triangle AXD$. Assume that $\mathcal{C}$ is not an incircle of $ABCD$; first, suppose that $\overline{BC}$ intersects $\mathcal{C}$ but is not tangent to $\mathcal{C}$. Let l be a line tangent to $\mathcal{C}$ that is parallel to $\overline{BC}$; let B' and C' be the points of intersection of l with $\overrightarrow{AB}$ and $\overrightarrow{DC}$, respectively. Then $\mathcal{C}$ is an incircle for $AB'C'D$, so $AB' + C'D = B'C' + AD$. However, $B'C' < BC$ (by similar triangles), $AB' > AB$, and $C'D > CD$. Therefore,

$$AB + CD < AB' + C'D = B'C' + AD < BC + AD,$$

contradicting our assumption.

The case when $\mathcal{C}$ is not an incircle of $ABCD$ but $\overline{BC}$ does not intersect $\mathcal{C}$ is handled similarly. $\square$

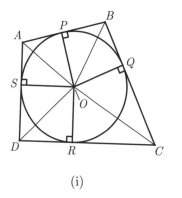

(i)

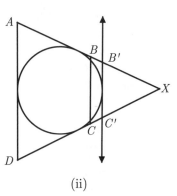

(ii)

FIGURE 4.30.

4.5 Problems

Intuition–trial–error–speculation–conjecture–proof is a sequence for understanding of mathematics.
— Saunders Mac Lane (1909–2005)[4]

4.1 Consider a circle with points A, B, C, and D lying on the circle, and point P lying off the circle.

    (a) Assume $P - A - B$ and $P - C - D$, where $PA = 4$, $PB = 12$, and $PC = CD$. Find PD.

    (b) Assume $A - P - B$ and $C - P - D$, where $PA = 4$, $PB = 12$, and $PC = CD$. Find PD.

4.2 Prove that for any regular n-gon, there exists a unique incircle, a unique circumcircle, and that their centers coincide. (The common center is called the **center** of the regular n-gon.)

4.3 Given an angle ABC and a circle that is tangent to its sides, let D be a point on a smaller arc defined by the points of tangency. Let a line tangent to the circle at D intersect the sides of the angle at points E and F. Prove that the perimeter of $\triangle EBF$ does not depend on the position of point D.

4.4 Suppose two circles with centers O_1 and O_2 are externally tangent at a point A. Consider a line passing through A and intersecting the circles at points B_1 and B_2, respectively. Prove that lines O_1B_1 and O_2B_2 are parallel.

4.5 Prove part (2) or (3) of Theorem 4.11, your choice.

4.6 Let acute triangle ABC with orthocenter H be inscribed in circle $\mathcal{C}$. Prove that the reflection of H across any side of $\triangle ABC$ lies on $\mathcal{C}$.

4.7 Suppose two circles, $\mathcal{C}_1$ and $\mathcal{C}_2$, intersect at points A and B. Consider a line passing through A, distinct from line AB, and intersecting the circles at points D_1 and D_2, respectively, such that $D_1 - A - D_2$.

    (a) Prove that $m\angle D_1BD_2$ does not depend on the line through A.

    (b) Where is D_1D_2 the greatest? the smallest?

4.8 Given two circles, $\mathcal{C}(O_1, r_1)$ and $\mathcal{C}(O_2, r_2)$, that are externally tangent at point A,

    (a) find the length of their common external tangent segment.

    (b) find the radius of a circle that is tangent to both circles and to their common external tangent.

4.9 Let two circles be externally tangent at a point A. Consider their common external tangent line $\overleftrightarrow{BC}$, where B and C are the points of tangency. Prove that $m\angle BAC = 90°$.

4.10 Consider three mutually externally tangent circles. Prove that the lines passing through the point of tangency of two circles and the center of the third are concurrent.

4.11 Given 10^6 points in a plane, prove that for any positive integer n, there exists a circle in the plane that contains exactly n, $1 \leq n \leq 10^6$, of the given points in its interior.

---

[4] From *A Mathematical Autobiography*

4.12 Prove Corollary 4.22.

4.13 Let A, B, C, and D be points (in that order) on a circle, and let A_1, B_1, C_1, and D_1 be the midpoints of arcs AB, BC, CD, and DA, respectively. Prove that $\overline{A_1C_1} \perp \overline{B_1D_1}$.

4.14 Suppose two lines meet at a point A and that n circles $\mathcal{C}_1, \mathcal{C}_2, ..., \mathcal{C}_n$ are placed consecutively between the two lines in such a way that each circle is tangent to each line, and circle $\mathcal{C}_i$ is tangent to circle $\mathcal{C}_{i+1}$. Assuming $\mathcal{C}_1$ has radius 1 and $\mathcal{C}_2$ has radius r, find the radius of $\mathcal{C}_n$.

4.15 Consider two nonoverlapping circles, $\mathcal{C}_1 = \mathcal{C}(O_1, R_1)$ and $\mathcal{C}_2 = \mathcal{C}(O_2, R_2)$, $R_1 < R_2$. Find the distance from the intersection of their common exterior tangent lines to O_1.

4.16 Given two nonintersecting circles, each in the exterior of the other, consider a segment with one endpoint on each circle. Where is the length of the segment the smallest? (In other words, using which points on the circle would we measure the *distance* between the two circles?) Where is the length of the segment the greatest?

4.17 Prove that in any triangle the angle bisector at any vertex lies between the altitude and the median at the same vertex.

4.18 Let $A - B - C - D - E - F$ be points on a circle such that $\overline{AB} \parallel \overline{DE}$ and $\overline{BC} \parallel \overline{EF}$. Prove that $\overline{AF} \parallel \overline{CD}$.

4.19 Let $ABCDEF$ be a hexagon circumscribed around a circle with center at point O. Suppose diagonals $\overline{AD}$ and $\overline{BE}$ intersect at O. Prove that $\overline{CF}$ also passes through O.

4.20 Prove that in any triangle, a line passing through the feet of two of the altitudes is perpendicular to the line passing through the third vertex and the circumcenter of the triangle.

4.21 Let point M be a point on the circumcircle of an equilateral triangle ABC. Prove that one of the lengths MA, MB, or MC is the sum of two others.

4.22 Four intersecting lines form four triangles as shown in Figure 4.31. Prove that the circumcircles (two of them are shown in the figure) around these triangles share a common point.

4.23 Let two circles be internally tangent at a point M. Let $\overline{AB}$ be a chord of the larger circle which is tangent to the smaller circle at a point T. Prove that $\overrightarrow{MT}$ is a bisector of $\angle AMB$.

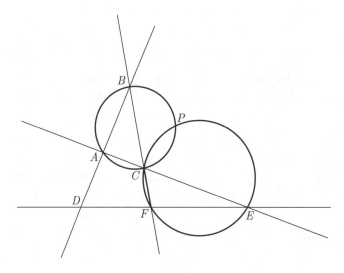

FIGURE 4.31.

4.24 Consider three circles with equal radii that intersect at a point. For each pair of these circles, consider the other point of their intersection. Prove that these three points lie on a circle of the same radius.

4.25 Consider a triangle ABC with sides of lengths a, b, and c. Let $p = (a + b + c)/2$ be the semiperimeter of the triangle. Find the length

(a) r of the radius of its incircle. Show that the answer can be given in the form
$$r = \frac{\sqrt{p(p-a)(p-b)(p-c)}}{p}.$$

(b) R of the radius of its circumcircle. Show that the answer can be given in the form
$$R = \frac{abc}{4\sqrt{p(p-a)(p-b)(p-c)}}.$$

4.26 Given three circles of radii r_1, r_2, and r_3, such that each is externally tangent to the other two, find the radius of a circle that passes through the points of tangency of these three circles.

4.27 Let G be the centroid (point of intersection of the medians) of a $\triangle ABC$, H be the orthocenter (the point of intersection of the altitudes of the triangle), and O be the center of the circumcircle of the triangle. Prove that G is on $\overline{HO}$ and $HG : GO = 2$.[5]

4.28 Let A, B, C, and D be four consecutive vertices of a regular 7-gon. Prove that
$$\frac{1}{AB} = \frac{1}{AC} + \frac{1}{AD}.$$

4.6 Supplemental Problems

S4.1 Consider all chords of a circle that pass through a fixed point A inside the circle. Prove that the set of midpoints of the chords form a circle, and describe the circle.

S4.2 Suppose $\overline{MN}$ and $\overline{MP}$ are tangent to a circle with center O, and $m\angle NMP = \alpha$. Let $\overline{BC}$ be another tangent to the circle, with $B \in \overline{MN}$ and $C \in \overline{MP}$. Find $m\angle BOC$.

S4.3 Consider a circular billiard table with center O. A point (ball) located at the circumference is hit in an arbitrary direction and moves along an (infinite) trajectory reflecting from the circular edge. Suppose that the reflection law is such that for every three consecutive reflection points A, B, and C on the trajectory, $\angle ABO \cong \angle CBO$. Prove that there is a circle inside the billiard such that the point's trajectory never crosses it.

S4.4 Use Ptolemy's theorem to prove the Pythagorean theorem.

S4.5 Prove that if two diagonals of a regular pentagon intersect each other, the longer segments of each diagonal will be congruent to the sides of the pentagon.

S4.6 Let P be the center of the square constructed on the hypotenuse $\overline{AC}$ of a right triangle ABC. Prove that $\overline{BP}$ bisects $\angle ABC$.

S4.7 (a) Prove that the image of a circle under a homothety is a circle.

(b) Consider two circles $\mathcal{C}_1$ and $\mathcal{C}_2$ of distinct radii, each in the exterior of the other. Prove that there are two homotheties mapping $\mathcal{C}_1$ to $\mathcal{C}_2$.

(c) Prove that for any two tangent circles there exists a homothety mapping one to the other.

(d) Given two circles, one in the interior of the other, is there always a homothety mapping one circle to the other?

---

[5] This theorem is often attributed to Euler, and the line containing points O, G, and H is referred to as the *Euler line*.

*S*4.8 Let $\overline{PA}$ and $\overline{PB}$ be two tangent segments from a point P to a circle $\mathcal{C} = \mathcal{C}(O, r)$, with A and B being the points of tangency. If $OP = p$, find the radius of the circle tangent to the segments and the minor arc AB.

*S*4.9 Let $ABCD$ be an isosceles trapezoid with $\overline{BC} \| \overline{AD}$, $AB = CD = 5$, $BC = 1$, and $AD = 7$. Find the radius of its circumcircle.

*S*4.10 Join each vertex of a triangle to the point where the incircle of the triangle touches the side opposite of the vertex. Prove that the three segments are concurrent.

*S*4.11 Fix a diameter of a circle, and let K be a point on the diameter. Let $\overline{AB}$ be a chord of the circle which passes through K and forms a $45°$ angle with the diameter. Prove that the sum $AK^2 + KB^2$ does not depend on the position of K on the diameter.

*S*4.12 Let $ABCD$ be a convex quadrilateral. Suppose that incircles of $\triangle ABC$ and $\triangle ADC$ are tangent to each other. Prove that the incircles of $\triangle ABD$ and $\triangle CBD$ are also tangent to each other.

*S*4.13 Four circles are built on the sides of a convex quadrilateral as on diameters. Prove that they completely cover the quadrilateral.

*S*4.14 Let $\overline{AF}$ be the bisector of $\angle A$ in $\triangle ABC$, and let it intersect the incircle of the triangle at points D and E such that $A - D - E - F$. Prove that $AD > EF$.

---

*S*4.15 Consider two circles in the plane such that each circle lies in the exterior of the other. Let A_1 and B_1 be the points of intersection of the first circle with two tangents from its center to the second circle, and A_2 and B_2 be the points of intersection of the second circle with two tangents from its center to the first circle. Prove that $A_1 B_1 = A_2 B_2$. (This problem will also appear as Problem S8.8.)

*S*4.16 For a fixed point in the interior of a circle, how should one draw two perpendicular chords through the point in order to maximize the sum of the chords' lengths?

*S*4.17 A triangle is inscribed in a circle. From any point on the circumference, perpendiculars are drawn to the lines containing the sides of the triangle. Prove that the feet of these perpendiculars are collinear.

5

Length and Area

We are not very pleased when we are forced to accept a mathematical truth by virtue of a complicated chain of formal conclusions and computations, which we traverse blindly, link by link, feeling our way by touch. We want first an overview of the aim and of the road; we want to understand the <u>idea</u> of the proof, the deeper context.

— Hermann Weyl[1] (1885–1955)

The notions of length, area, and volume are so familiar that we generally use and discuss them without concern for being misunderstood. We do not hesitate to measure the length of an oval skating ring, the area of a continent, or the volume of a strangely shaped container. The result of any such measurement is a non-negative real number; our main (unspoken) requirement is that if a figure is cut into pieces, the measure of the whole is the sum of the measures of the parts.

On the other hand, it is not easy to define these notions rigorously, nor to describe the classes of figures which have length, or area, or volume. After all, a figure in geometry is just an arbitrary set of points. Does a point have length, and if it does, what is it? Everyone will agree that if a point has length, it must be zero, no matter which units we use. Now, every segment is a union of points, and it is reasonable to think that its length should be the sum of lengths of all its points. This means the length of a segment is an infinite sum. But if we are adding zeros, even infinitely many, surely the result must be zero as well. Hence the length of a segment is zero. Something is wrong ...

Actually, many things are wrong with the above argument. The only reason we include it here is to alert the reader that one has to be careful even when speaking about a concept as "simple" as the length of a segment.

The truth is that it is impossible to discuss rigorously the notions of length, area, and volume for most figures within the context of elementary Euclidean geometry. The branch of mathematics in which this is done is called Measure Theory, which is a part of a larger area called Mathematical Analysis. One comes close to a serious discussion of these notions in calculus courses, where powerful methods for computing lengths, areas, and volumes are developed, but those courses

---

[1] From *Unterrichtsblätter für Mathematik und Naturwissenschaften*, **38**, 177–188 (1932). Translation by Abe Shenitzer appeared in *The American Mathematical Monthly*, v. 102, no. 7, p. 646.

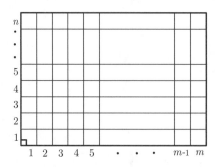

FIGURE 5.1.

usually lack general definitions and rigor. In this section our goals are modest: to discuss areas of polygonal regions and the area and circumference of a circle.

5.1 Area of a Polygonal Region

> *Err*
> *and err*
> *and err again*
> *but less*
> *and less*
> *and less.*
>
> — Piet Hein (1905–1996)

To simplify the language, we will refer to the area of a polygonal region as just the **area of a polygon**. The main axioms for the area of polygons are Postulates 2.17–2.20. We begin with a few comments about Postulate 2.20, which states that the area of a rectangle is equal to ah, where a is the length of the base, and h is the length of the altitude. We wish to motivate its inclusion in our postulate system.

In many other axiomatic systems this postulate is replaced by a statement that the area of a unit square is 1. Then the formula for the area is derived along the following lines. Using induction, Postulate 2.19 can easily be generalized to any finite number of polygonal regions. Then, if the lengths of the base and the altitude of a given rectangle are positive integers, say m and n, dividing these sides into m and n congruent segments, respectively, and drawing corresponding vertical and horizontal lines, we divide the rectangle into $m \times n$ congruent unit squares. (See Figure 5.1.) This *proves* that its area is $m \times n$.

If the lengths of the base and the altitude are positive rational numbers, say p/q and r/t, we can again *prove* that its area is $p/q \times r/t$. In order to see this, we first divide the side of length p/q into p congruent segments, then divide each of the p segments into t congruent segments; similarly, divide the side of length r/t into r congruent segments, then divide each of the r segments into q congruent segments. For each of these divisions, draw the corresponding vertical and horizontal lines. As $p/q = pt/qt$ and $r/t = qr/qt$, this construction divides the rectangle into $pt \times qr$ congruent squares, with each side length equal $1/qt$. Dividing each side of the unit square into qt congruent segments, and drawing corresponding vertical and horizontal lines, we partition the unit square into $(qt)^2$ smaller congruent squares, each having sides of length $1/qt$. Clearly, the area of each small square is $1/(qt)^2$. Therefore, as we claimed, the area of our rectangle is

$$(pt \times qr) \times \frac{1}{(qt)^2} = \frac{p}{q} \times \frac{r}{q}.$$

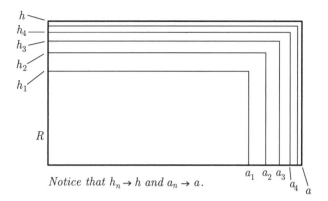

Notice that $h_n \to h$ and $a_n \to a$.

FIGURE **5.2.**

In order to prove the result for all rectangles, we must consider the case when a or b (or both) are positive irrational numbers. In order to do this, we employ the notion of a limit of a sequence.[2] It is known that every real number is a limit of a sequence of rational numbers, and the latter can be chosen in infinitely many ways. Consider two sequences of rational numbers a_n and h_n, such that $a_n \to a$ and $h_n \to h$ when $n \to \infty$. For every n, the area of a rectangle R_n with sides a_n and h_n is $a_n h_n$, since we already established the result. It is intuitively clear that when n grows, the rectangles R_n become "closer and closer" to our rectangle R. (See Figure 5.2.)

Since $a_n h_n \to ah$ when $n \to \infty$, it is reasonable to *define* the area of R as ah. What is left to prove (but will not be proven here) is that this definition is independent of the choice of rational sequences a_n and h_n, and that the area defined in this way satisfies the other postulates related to area.

The preceding arguments provide the motivation for Postulate 2.20 and demonstrate how many difficulties can be avoided if we just accept it. Once we know how to compute the area of a rectangle, we can easily derive the formulae for the areas of other polygons.

Theorem 5.1.

(1) *The area of a parallelogram $ABCD$ is equal to the product of the lengths of a side and corresponding altitude:*

$$Area(ABCD) = ah_a. \ \text{(See Figure 5.3.)}$$

(2) *The area of a triangle ABC is equal to one half of the product of the length of any side and corresponding altitude:*

$$Area(\triangle ABC) = \frac{1}{2}ah_a = \frac{1}{2}bh_b = \frac{1}{2}ch_c. \ \text{(See Figure 5.4(i).)}$$

(3) *The area of a trapezoid $ABCD$ is equal to the product of the lengths of its midline and altitude:*

$$Area(ABCD) = \frac{1}{2}(a+b)h. \ \text{(See Figure 5.4(ii).)}$$

---

[2] The reader not familiar with the notion of the limit can skip the rest of this paragraph and accept the formula as fact when a and b are irrational.

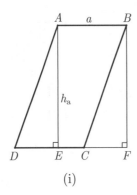
(i)

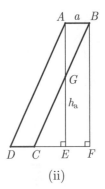
(ii)

FIGURE **5.3.**

Proof.

(1) Let $ABCD$ be a parallelogram with $\overline{AB} \parallel \overline{CD}$ and $\overline{AD} \parallel \overline{BC}$.

Consider first the case in which the altitude from A intersects $\overline{CD}$ at E, as shown in Figure 5.3(i). Let $\overline{BF}$ be perpendicular to $\overrightarrow{DC}$ at F. Then, $\triangle AED \cong \triangle BFC$, so the area of $ABCD$ must be the same as the area of the rectangle $ABFE$.

Next, suppose the altitude from A intersects $\overrightarrow{DC}$ at E so that $D - C - E$, as shown in Figure 5.3(ii). Let $\overline{BF}$ be perpendicular to $\overrightarrow{DC}$ at F, and let G be the intersection of $\overline{AE}$ and $\overline{BC}$. As before, $\triangle AED \cong \triangle BFC$. Making use of Postulate 2.19, $\text{Area}(ABCD) = \text{Area}(\triangle ADE) + \text{Area}(\triangle BGA) - \text{Area}(\triangle CGE)$. Similarly, $\text{Area}(ABFE) = \text{Area}(\triangle BCF) + \text{Area}(\triangle ABG) - \text{Area}(\triangle CGE)$. It follows that the areas of $ABCD$ and $ABFE$ are the same.

(2) The result follows quickly by viewing $\triangle ABC$ as one half of a parallelogram having the same base $\overline{CB}$ and the same altitude at A.

(3) Let $ABCD$ be a trapezoid with parallel sides $\overline{AB}$ and $\overline{CD}$, as shown in Figure 5.4(ii). Assume that $ABCD$ is not a parallelogram. Draw a line through B parallel to $\overline{AD}$; let E be the point of intersection of this line with $\overleftrightarrow{CD}$. (This intersection point may or may not actually lie on $\overline{CD}$.) Construct the altitude from B, $\overline{BF}$. Now $ABED$ and BCE are a parallelogram and a triangle,

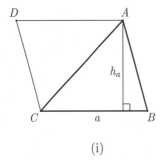
(i)

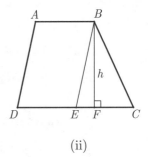
(ii)

FIGURE **5.4.**

respectively, each having $\overline{BF}$ as an altitude. Using Postulate 2.19, the area of $ABCD$ is the sum of the areas of $ABED$ and BCE. Hence,

$$Area(ABCD) = DE \cdot h + \frac{1}{2}EC \cdot h \qquad (5.1)$$

$$= \frac{1}{2}(AB + DE) \cdot h + \frac{1}{2}EC \cdot h \qquad (5.2)$$

$$= \frac{1}{2}(AB + CD) \cdot h. \qquad (5.3)$$

$\square$

There are many problems not related to area for which one can use an area argument to obtain a beautiful solution. We provide one such example here by giving another proof of the Triangle Bisector Property (Theorem 3.16). Several other examples occur in the problem set for this chapter.

Theorem 5.2. (Triangle Bisector Property) *Given $\triangle ABC$, let B_1 be an interior point of $\overline{AC}$. Then $\overline{BB_1}$ is a bisector of $\angle B$ if and only if $AB_1/B_1C = AB/BC = c/a$.*

Proof. Let h be the length of the altitude at B. Then the area of $\triangle AB_1B$ is $h \cdot AB_1/2$ and the area of $\triangle CB_1B$ is $h \cdot B_1C/2$. On the other hand, by using $\overleftrightarrow{AB}$ and $\overleftrightarrow{BC}$ as bases, we see that the areas of these triangles can also be written as $d(B_1, \overleftrightarrow{AB}) \cdot AB/2$ and $d(B_1, \overleftrightarrow{BC}) \cdot BC/2$, respectively. Therefore,

$$\frac{AB_1}{B_1C} = \frac{d(B_1, \overleftrightarrow{AB}) \cdot AB}{d(B_1, \overleftrightarrow{BC}) \cdot BC}.$$

The result now follows, because $d(B_1, \overleftrightarrow{AB}) = d(B_1, \overleftrightarrow{BC})$ if and only if $\overleftrightarrow{BB_1}$ bisects angle B, by Problem 3.2.13. $\square$

The following theorem provides alternate ways for computing areas for triangles.

Theorem 5.3. *For any triangle ABC with semiperimeter $p = (a + b + c)/2$,*

(1) *$Area(\triangle ABC) = \sqrt{p(p-a)(p-b)(p-c)}$. (Heron's Formula)*
(2) *$Area(\triangle ABC) = pr$, where r is the inradius of $\triangle ABC$.*
(3) *$Area(\triangle ABC) = abc/(4R)$ where R is the circumradius of $\triangle ABC$.*

Proof.

(1) Given $\triangle ABC$, let h_a be the length of the altitude $\overline{AD}$, with $C - D - B$ (at least one of the altitudes of a triangle is in its interior). Let $CD = x$. (See Figure 5.5.) By the Pythagorean theorem, $b^2 - x^2 = h_a^2 = c^2 - (a - x)^2$. Thus, $b^2 - c^2 + a^2 = 2ax$, so $x = (b^2 - c^2 + a^2)/(2a)$.

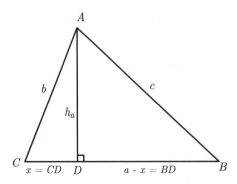

FIGURE 5.5.

We now show that p, $p - a$, $p - b$, and $p - c$ are factors of $h_a{}^2$.

$$h_a{}^2 = b^2 - \left(\frac{b^2 - c^2 + a^2}{2a}\right)^2 = \frac{4a^2b^2 - (b^2 - c^2 + a^2)^2}{4a^2}$$

$$= \frac{\left(2ab - b^2 + c^2 - a^2\right)\left(2ab + b^2 - c^2 + a^2\right)}{4a^2}$$

$$= \frac{\left(c^2 - (a - b)^2\right)\left((a + b)^2 - c^2\right)}{4a^2}$$

$$= \frac{(c - a + b)(c + a - b)(a + b - c)(a + b + c)}{4a^2}$$

$$= \frac{4}{a^2}\left(\frac{a + b + c}{2} - a\right)\left(\frac{a + b + c}{2} - b\right)\left(\frac{a + b + c}{2} - c\right)\left(\frac{a + b + c}{2}\right)$$

$$= \frac{4p(p - a)(p - b)(p - c)}{a^2}.$$

So, $h_a = \dfrac{2}{a}\sqrt{p(p - a)(p - b)(p - c)}$. The desired result follows from Theorem 5.1(2).

(2) Let O be the incenter of $\triangle ABC$. Let D, E, and F be the points of tangency of the incircle with $\overline{BC}$, $\overline{AC}$, and $\overline{AB}$, respectively. Join O with the vertices, A, B, and C, and with the points of tangency, D, E and F, as shown in Figure 5.6. Since at each point of tangency, the radius is perpendicular to the tangent, $\overline{OD}$, $\overline{OE}$, and $\overline{OF}$ are altitudes of $\triangle COB$, $\triangle AOC$, and $\triangle BOA$, respectively.

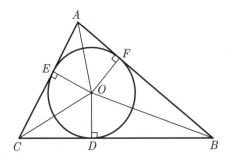

FIGURE 5.6.

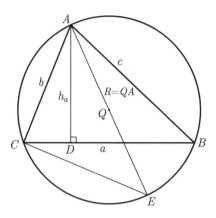

FIGURE 5.7.

Let $r = OD$. Then Area$(\triangle ABC) = \dfrac{1}{2}rBC + \dfrac{1}{2}rAC + \dfrac{1}{2}rAB = pr$.

(3)   In any triangle, one altitude must lie inside the triangle. Assume that for $\triangle ABC$, the altitude $\overline{AD}$ is inside the triangle. Let Q be the center of the circumcircle of $\triangle ABC$. Let $\overline{AE}$ be a diameter of the circumcircle of $\triangle ABC$, and join E with C, as shown in Figure 5.7. Then $m\angle ACE = 90°$.

Now, $\triangle ADB \sim \triangle ACE$ by the AA theorem. Thus, $h_a/c = b/2R$, which implies the statement. $\qquad\square$

The following theorem provides a useful method for finding the area of a quadrilateral with perpendicular diagonals.

Theorem 5.4.   *Let $ABCD$ be a convex quadrilateral with perpendicular diagonals AC and BD. Then*

$$Area(ABCD) = \frac{1}{2}AC \cdot BD.$$

In particular, the area of a rhombus is equal to one half of the product of the lengths of its diagonals.

Proof.   Consider convex quadrilateral $ABCD$ with perpendicular diagonals $\overline{AC}$ and $\overline{BD}$. (See Figure 5.8.) Let E be the point of intersection of $\overline{AC}$ and $\overline{BD}$. The area of $ABCD$ is the sum of the areas of $\triangle ADC$ and $\triangle ABC$. Thus,

$$\text{Area}(ABCD) = \frac{1}{2}AC \cdot DE + \frac{1}{2}AC \cdot EB = \frac{1}{2}AC \cdot BD. \qquad\square$$

In order to find the area of an arbitary polygon, we can simulate the method already used in this section, i.e., we divide it into triangles and sum their areas. This division may be achieved by using some of the diagonals of the polygon, and the sides of each triangle will be either sides of the polygon or one of its diagonals. The existence of such a *triangulation* may appear obvious, even for nonconvex polygons like the one shown in Figure 5.9, but a proof is not trivial. One way to prove

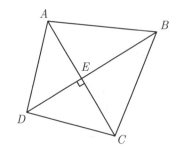

FIGURE 5.8.

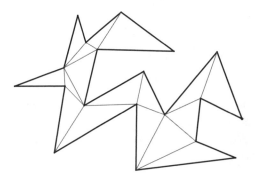

FIGURE 5.9.

this result is to use the fact that any n-gon for $n \geq 4$ must have a diagonal that lies completely in the interior of the n-gon, and then use induction. (We proved this fact in our solution of Problem 3.2.30.) We leave the details to the reader.

5.2   Area and Circumference of a Circle

Let us examine two common approaches to determining the area of some bounded figures. To simplify the language, we will refer to the area of a polygonal region as just the area of a polygon. We say that a figure in the plane is **bounded** if it is contained in the interior of a convex polygon. Let Φ be a bounded convex figure with a smooth boundary.[3]

5.2.1   A first approach to area

Consider a sequence A_n of convex polygons inscribed in Φ, $n \geq 2$, such that the set of vertices of A_n is a subset of the set of vertices of A_{n+1} for $n \geq 3$. It can be shown that the sequence can be chosen in such a way that the length of a longest side of A_n approaches 0 when $n \to \infty$. Let $a_n = \text{Area}(A_n)$. Then the sequence (a_n) is increasing and bounded from above by the area of any polygon which contains Φ in its interior. Therefore (a_n) converges.[4] It also can be shown that the limit will be the same for any other sequence of convex polygons inscribed in Φ and having the same properties as A_n.

---

[3] Other approaches can be used to define the areas of non-convex figures with more general boundaries.

[4] The fact that a monotone and bounded sequence converges is discussed in most calculus texts.

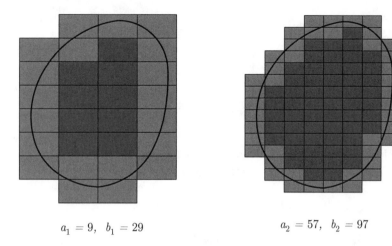

$$a_1 = 9, \quad b_1 = 29 \qquad\qquad a_2 = 57, \quad b_2 = 97$$

FIGURE **5.10.**

This limit is called the **area of** Φ. The computation of area by taking the limit of (a_n) as we did here is sometimes called "the method of exhaustion."

5.2.2 A second approach to area

View Φ as a figure in the plane, and superimpose on the plane a grid of congruent rectangles, each with an area of one unit. Count the number of rectangles that are inside Φ and call this number a_1. Now count the number of rectangles that intersect Φ, and add this value to a_1; call the sum b_1. Refine the grid by dividing the sides of each rectangle into two congruent segments and drawing corresponding vertical and horizontal lines. In this way, each of the original rectangles is divided into four smaller congruent rectangles, with each of the smaller rectangles having area $1/4$. Again counting the number of rectangles inside Φ and the number of rectangles intersecting Φ, we get two numbers a_2 and b_2. The areas of the corresponding figures are $a_2 \cdot 2^{-2}$ and $b_2 \cdot 2^{-2}$. In Figure 5.10, note that the four areas are $a_1 = 9$, $a_2/4 = 57/4 = 14.25$, $b_2/4 = 97/4 = 24.25$, and $b_1 = 29$. Observe that $a_1 < a_2/4 < b_2/4 < b_1$.

Continuing in this manner, we obtain two sequences for the areas of the corresponding figures, $(a_n \cdot 2^{-2(n-1)})$ and $(b_n \cdot 2^{-2(n-1)})$, such that for every $n \geq 1$,

$$a_1 < a_2 \cdot 2^{-2} < \cdots < a_n \cdot 2^{-2(n-1)} < b_n 2^{-2(n-1)} < b_{n-1} 2^{-2(n-2)} < \cdots < b_1.$$

As in the first approach, each sequence is monotone and bounded, and thus convergent. Again, it can be shown that their limits are equal and that their common value does not depend on the choice of the original grid. The **area of** Φ is defined as this limit.

When Φ is a circle of radius r, both of these approaches lead to the value πr^2, where $\pi = 3.141592\ldots$, a non-repeating decimal. When the first approach is used, one can take A_n as the sequence of inscribed regular 2^n-gons, where $n \geq 2$.

5.2.3 Circumference of a circle

What is the distance around (perimeter of) a circle of radius r? Though at first glance the question seems to be a perfectly legitimate one for us to ask, we do not yet have tools for answering it. So far we know nothing about lengths of any curves other than segments and their finite unions. Nevertheless the question has a clear intuitive meaning: if a piece of rope is bent, in order to find

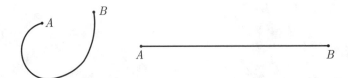

FIGURE 5.11.

its length we only need to straighten it out and measure the length of the obtained "segment." (See Figure 5.11.) It seems that "cutting" the circle and straightening it out should give us the value of the length of the circle.

This intuitive idea cannot easily be converted into a rigorous argument, though. First of all, a curve is not a physical object, such as a rope; secondly, there is ambiguity as to what "straighten" means. A solution to this difficulty is again obtained by defining the length of a curve in a way that both allows us to compute it and which agrees with our intuition. Here is how we proceed with doing so for a circle.

Let A_n be a sequence of regular 2^n-gons inscribed in the circle, $n \geq 2$, and let p_n denote the perimeter of A_n. It is easy to argue that the sequence (p_n) is increasing and bounded from above by the perimeter of any convex polygon containing the circle in its interior. Hence, (p_n) converges. The value of its limit, C, is called the **circumference** of the circle. The details of the computation can be found in many calculus texts. It turns out that $C = 2\pi r$.[5]

If one accepts that $C = 2\pi r$, then the area of the interior of a circle (which we call the **area of the circle**) may be found via the following non-rigorous argument. Join the center O of the circle with vertices of A_n. We obtain n congruent isosceles triangles, each with altitude h_n. The sum of the lengths of the bases c_n of these triangles gives the perimeter of A_n for all n. (See Figure 5.12.)

When n becomes large, the length of the altitude h_n approaches r, and the sum of the areas of all triangles, i.e., the area of A_n, approaches the area of the circle. Hence

$$\text{Area}(A_n) = n\left(\frac{1}{2}c_n h_n\right) = \frac{1}{2}\left(nc_n\right)h_n.$$

As $n \to \infty$, this last expression converges to

$$\frac{1}{2}C \cdot r = \frac{1}{2}\left(2\pi r\right)r = \pi r^2.$$

We summarize the preceding results in the following theorem.

Theorem 5.5. *The circumference of a circle of radius r is $2\pi r$, and its area is πr^2.*

It is common for the number π to be *defined* as the ratio of the circumference of a circle to its diameter:

$$\pi = \frac{C}{d} = \frac{C}{2r}.$$

---

[5] Such an approach requires that π be defined in advance and independently of the notion of circumference. Another relation explained in calculus is that the derivative of the area function is circumference, i.e., $\frac{d}{dr}(\pi r^2) = 2\pi r$.

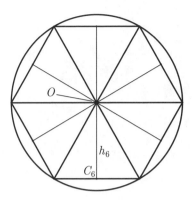

FIGURE 5.12. A_6 inscribed in $\mathcal{C}(O, r)$.

If this approach is taken, then the fact that $C = 2\pi r$ comes as a definition. However, in order for this definition to make sense, one must have (i) the circumference already defined, and (ii) the assurance that this ratio is independent of the circle. The second statement is often derived from the fact that all circles are similar, and hence the lengths of the corresponding linear elements are in proportion. Clearly both this approach and the alternative approach we presented have their own merits and drawbacks.

The following corollary is obvious and we omit the proof, simply reminding the reader that both radians and degrees are valid units for measuring angles. By definition

$$1 \text{ radian} = \frac{180°}{\pi} \approx 57.2957 \text{ degrees.}$$

Corollary 5.6. *Let α represent the measure in degrees of a central angle AOB in a circle with center O and radius r. (See Figure 5.13.)*
Then

$$\text{length of arc } AB = \frac{\alpha}{180}\pi r, \quad \text{and} \quad \text{area of sector } AOB = \frac{\alpha}{360}\pi r^2.$$

If β is the measure of $\angle AOB$ in radians,

$$\text{length of arc } AB = \beta r, \quad \text{and} \quad \text{area of sector } AOB = \frac{\beta}{2}r^2.$$

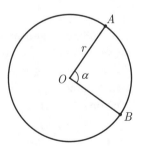

FIGURE 5.13.

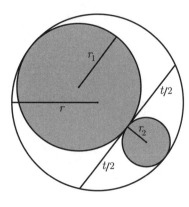

FIGURE 5.14.

As we see, one of the primary benefits of using radians to measure the central angle is that one can then simplify the formulae for the length of the corresponding arc and the area of the corresponding sector.[6]

Example 7. (Arbelos - The Shoemaker's Knife) Consider two circles tangent to each other, inscribed in a third circle, and notice the nonshaded region outside of the two smaller circles and inside the larger circle. (See Figure 5.14.) Archimedes called each half of this shape an *arbelos*, which literally means "a shoemaker's knife." In Proposition 4 of his Book of Lemmas, Archimedes found the area of each piece as a function of t, the length of the chord of the large circle which is tangent to two smaller circles at their point of tangency.

Archimedes' Solution: Let r_1 and r_2 denote the radii of the two small circles. The radius of the larger circle will then be $r = r_1 + r_2$. The tangent chord is perpendicular to the diameter of the larger circle passing through the centers of the smaller circles, and, hence, is bisected by it. By Theorem 4.17, $2r_1 \cdot 2r_2 = (t/2)^2$. Using the area formula for circles, we have that the area of each arbelos is

$$\frac{1}{2}\left(\pi r^2 - \pi r_1^2 - \pi r_2^2\right) = \frac{\pi}{2}\left((r_1 + r_2)^2 - r_1^2 - r_2^2\right) = \pi(r_1 r_2) = \pi(t^2/16).$$

$\square$

5.3 Wallace-Bolyai-Gerwien Theorem

Recall the importance of Postulate 2.19 in allowing us to find formulae for the areas of parallelogram, triangle, and trapezoid in Theorem 5.1 by sliding congruent regions in the plane. The Wallace-Bolyai-Gerwien theorem generalizes this approach, demonstrating that a polygon in the plane can be "transformed" into any other polygon in the plane having the same area. Before stating the theorem more formally, we introduce some terminology.

Technically a polygon consists of the union of segments, but to simplify the presentation in this section, we will also refer to polygonal regions as polygons. Let $\mathcal{P}$ and $\mathcal{P}'$ be two polygons in a

---

[6] Another reason comes from calculus, where measuring angles in radians simplifies the formulae for the derivatives of trigonometric functions.

plane. Suppose that $\mathcal{P}$ can be cut by straight lines into finitely many pieces, and the pieces can then be reassembled in such a way that $\mathcal{P}'$ is created; conceptually, one can imagine $\mathcal{P}$ being cut into polygonal puzzle pieces and the pieces then put together, maybe differently, so that $\mathcal{P}'$ is constructed. When this is possible, we write $\mathcal{P} \equiv \mathcal{P}'$, and we say that $\mathcal{P}$ is **equidecomposable** to or **congruent by addition** with $\mathcal{P}'$.

Lemma 5.7. *The relation $\equiv$ described above is an equivalence relation.*[7]

Proof.

(1) Clearly, $\mathcal{P} \equiv \mathcal{P}$ for any polygon $\mathcal{P}$, so the relation is reflexive.

(2) Likewise, it is obvious that if $\mathcal{P} \equiv \mathcal{Q}$, then $\mathcal{Q} \equiv \mathcal{P}$. Thus, $\equiv$ is symmetric.

(3) For transitivity, suppose that $\mathcal{P} \equiv \mathcal{Q}$ and $\mathcal{Q} \equiv \mathcal{R}$. Then $\mathcal{Q} \equiv \mathcal{P}$ also. On $\mathcal{Q}$, draw all of the lines that cut the polygon $\mathcal{Q}$ into pieces from which $\mathcal{P}$ can be assembled. In addition, draw all of the lines that cut $\mathcal{Q}$ into polygons from which $\mathcal{R}$ can be assembled. From this refined partition of $\mathcal{Q}$, both $\mathcal{P}$ and $\mathcal{R}$ can be assembled. Thus, $\mathcal{P} \equiv \mathcal{R}$.

$\square$

The following theorem is quite remarkable: it states that every two plane polygons with equal areas are equidecomposable. The theorem is attributed variously to the mathematicians W. Wallace, F. Bolyai, P. Gerwien, and a Mr. Lowry. Lowry purportedly provided a simple proof of the theorem in 1814 in response to a problem posed by Wallace. Wallace himself, it is believed, had a solution to the problem at the time he posed it; he presented a full proof of the theorem in 1831. Independently, Bolyai and Gerwien provided proofs in 1833 and 1835, respectively.

Theorem 5.8. *(Wallace-Bolyai-Gerwien Theorem) Any two polygons in a plane having equal area are equidecomposable.*

The proof of Theorem 5.8 is preceded by several lemmas.

Lemma 5.9. *Every triangle is equidecomposable with some rectangle.*

Proof. Consider $\triangle ABC$; assume that $\overline{AB}$ is the longest side in $\triangle ABC$. Then the altitude at C lies in the interior of the triangle. Let M and N be the midpoints of sides $\overline{AC}$ and $\overline{BC}$, respectively. Let D be the foot of the altitude from C. Let $\overleftrightarrow{MN}$ and $\overleftrightarrow{CD}$ be lines dividing $\triangle ABC$ into polygonal regions, as shown in Figure 5.15. By Corollary 3.14, $\overleftrightarrow{MN} \parallel \overleftrightarrow{AB}$.

By rotating triangle 1 about the point M and triangle 2 about the point N, $\triangle ABC$ can be transformed into the rectangle $AEFB$.

$\square$

---

[7] A binary relation $\equiv$ on a set A is said to be an equivalence relation if (a) $x \equiv x$ for all $x \in A$, (b) $x \equiv y$ implies $y \equiv x$ for all $x, y \in A$, and (c) $x \equiv y$ and $y \equiv z$ together imply $x \equiv z$ for all $x, y, z \in A$. For example, the similarity relation on the set of all triangles is an equivalence relation.

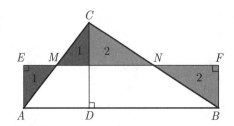

FIGURE 5.15.

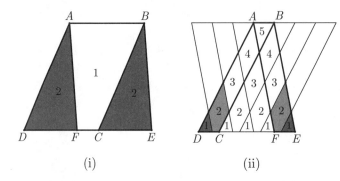

(i)   (ii)

FIGURE 5.16.

Lemma 5.10. *Two parallelograms of equal area and sharing a common side are equidecomposable.*

Proof. Consider parallelograms $ABCD$ and $ABEF$. Suppose first that $\overline{CD}$ and $\overline{EF}$ intersect, as shown in Figure 5.16(i). Since $\overline{AD} \parallel \overline{BC}$ and $\overline{AF} \parallel \overline{BE}$, $\triangle ADF \cong \triangle BCE$. Thus, $ABCD \equiv ABEF$.

Now, suppose that $\overline{CD}$ and $\overline{EF}$ do not intersect, as illustrated in Figure 5.16(ii). Note that since $ABCD$ and $ABEF$ are parallelograms, $AB = CD = EF$. Construct a sequence of parallelograms that are congruent to $ABCD$, with the first one sharing side $\overline{BC}$ with $ABCD$, and each subsequent parallelogram in the sequence similarly sharing a side with the previous one. The sequence should be continued until parallelogram $ABEF$ is covered. (This will be possible, because $\overline{CD}$ and $\overline{EF}$ must lie on the same line.) The intersection of these parallelograms with $ABCD$ results in a polygonal decomposition of $ABCD$, labeled as regions 1 to 5 in the figure. Similarly, construct a sequence of parallelograms that are congruent to $ABEF$, with the first one sharing side $\overline{AF}$ with $ABEF$, continuing until parallelogram $ABCD$ is covered. The intersection of these parallelograms with $ABEF$ results in a polygonal decomposition of $ABEF$. Moreover, the corresponding polygons of these decompositions clearly must be congruent,[8] so $ABCD \equiv ABEF$. $\qquad \square$

---

[8] We claim, for example, that the pentagons enclosing the regions labeled 2 are "clearly congruent," as each of them is mapped to another by a translation in the direction of the line AB, by a distance AB.

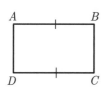
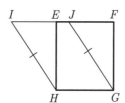

FIGURE **5.17.**

Lemma 5.11. *Two rectangles of equal area are equidecomposable.*

Proof. Consider rectangles $ABCD$ and $EFGH$ having equal area. Construct parallelogram $GHIJ$ so that $\overline{IJ}$ is on $\overleftrightarrow{EF}$ and $IH = AB$. (See Figure 5.17.)

Since $\triangle IEH \cong \triangle JFG$, rectangles $EFGH$ and $GHIJ$ have the same area. By Lemma 5.10, $EFGH \equiv GHIJ$. However, $GHIJ$ could in the same way have been constructed to share a side with $ABCD$; so, again by Lemma 5.10, $ABCD \equiv GHIJ$. Thus, by Lemma 5.7, $ABCD \equiv EFGH$. □

Lemma 5.12. *Any polygon in the plane is equidecomposable with a rectangle.*

Proof. Let polygon $\mathcal{P}$ be given. As we noted at the end of Section 5.1, any polygon may be triangulated, so $\mathcal{P}$ can be divided into a finite number of triangles, $\triangle_1, \triangle_2, \ldots, \triangle_n$, as demonstrated in Figure 5.18.

To decompose $\mathcal{P}$ into a rectangle, start with an arbitrary line segment, $\overline{AB}$. Construct lines $\overleftrightarrow{CA}$ and $\overleftrightarrow{DB}$ with $\overleftrightarrow{CA} \perp \overline{AB}$ and $\overleftrightarrow{DB} \perp \overline{AB}$. (See Figure 5.19.) Choose point A_1 on $\overrightarrow{AC}$ and corresponding point B_1 on $\overrightarrow{BD}$ so that Area(AA_1B_1B) = Area($\triangle_1$). Choose point A_2 on $\overrightarrow{A_1C}$ and corresponding point B_2 on $\overrightarrow{B_1D}$ so that Area($A_1A_2B_2B_1$) = Area($\triangle_2$). Continue this process for all n triangles in the triangulation of $\mathcal{P}$.

By Lemma 5.9, each of $\triangle_1, \triangle_2 \ldots, \triangle_n$ is equidecomposable with some rectangle, say $R_1, R_2, \ldots, R_n$, respectively. By Lemma 5.11, $AA_1B_1B \equiv R_1$, $A_1A_2B_2B_1 \equiv R_2$, $\ldots$, $A_{n-1}A_nB_nB_{n-1} \equiv R_n$. Furthermore, by Lemma 5.7, $\triangle_1 \equiv AA_1B_1B$, $\triangle_2 \equiv A_1A_2B_2B_1$, $\ldots$, and $\triangle_n \equiv A_{n-1}A_nB_nB_{n-1}$. Thus, $\mathcal{P} \equiv AA_nB_nB$. □

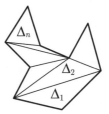

FIGURE **5.18.**

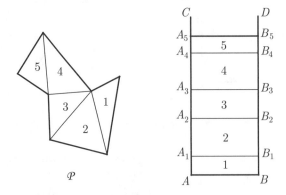

FIGURE 5.19.

We now have the necessary tools for proving the Wallace-Bolyai-Gerwien theorem.

Proof. (Wallace-Bolyai-Gerwien Theorem)

Consider polygons $\mathcal{P}$ and $\mathcal{P}'$ such that $\mathrm{Area}(\mathcal{P}) = \mathrm{Area}(\mathcal{P}')$. By Lemma 5.12, $\mathcal{P} \equiv ABCD$ and $\mathcal{P}' \equiv A'B'C'D'$, where $ABCD$ and $A'B'C'D'$ are rectangles. Since $\mathrm{Area}(\mathcal{P}) = \mathrm{Area}(\mathcal{P}')$, $\mathrm{Area}(ABCD) = \mathrm{Area}(A'B'C'D')$. By Lemma 5.11, $ABCD \equiv A'B'C'D'$. By Lemma 5.7, $\mathcal{P} \equiv \mathcal{P}'$. $\qquad\square$

We have seen in this chapter that determining areas of polygons in the plane can be reduced to determining areas of triangles, and the latter can be reduced to determining areas of rectangles. Areas of rectangles can be found by a simple formula, but a continuity argument was needed to establish it in complete generality (for rectangles with both rational and irrational side lengths). Therefore one may say that our theory of areas of polygons requires an application of continuity argument only once, and the rest can be accomplished with dissecting and reassembling tricks.

At the same time, when attempting to generalize this approach and develop the theory of volumes of polyhedra in 3-space, one inevitably uses a continuity argument at least twice: first, for the formula for the volume of a rectangular parallelepiped with irrational side lengths, and again for the formula for the volume of a tetrahedron, $V = \frac{1}{3}$(Area of the base)(height), for which the calculus technique of integration is usually used. Note that the "analogous" question of determining the area of a triangle in the plane can be done by simple dissection and reassembling, as we did in the proof of Lemma 5.9.

This raises the question: is our inability to dissect and reassemble an arbitrary tetrahedron into, say, a cube due to our lack of imagination, or is the difficulty inherent? David Hilbert suspected it was the latter, and he included a question on this topic among his now famous "Hilbert's problems," a list of twenty-three questions posed by Hilbert in 1900 as challenges to the mathematical community as they entered a new century. He suggested that it should be possible to find two tetrahedra of equal volume that are not equidecomposable (cannot be decomposed into finitely many congruent parts). This was the first of the twenty-three problems to be solved, when in the same year Max Dehn proved that a regular tetrahedron cannot be cut into finitely many polyhedral pieces which can be reassembled to yield a cube.

We finish by mentioning two other famous results concerning areas, dissections and rearrangements. In what we discussed above, all "dissections" could be done by lines (in a plane) or planes (in 3-space). Sometimes, instead of "dissection," the word "cutting" is used. Every piece obtained after such a dissection has an area or a volume. In the results we are going to mention, the dissection means much less: just a partitioning of a set into (pairwise disjoint) subsets. The meaning of reassembling is the same, and it corresponds to motions of these subsets in a plane or in space.

The **Banach-Tarski paradox** is one of the most famous of these results. It states that a solid ball can be dissected into five pieces and reassembled into *two* solid balls, each congruent to the original. When first stated in 1924, the Banach-Tarski paradox claimed such a result with a dissection of a solid sphere into six pieces; in 1947, Robinson showed that only five pieces are required. It should be noted that the pieces involved in this dissection defy the accepted notion of volume.

We conclude with another, more recent, result. In 1925, Tarski asked whether a circular disc could be cut into pieces which could then be rearranged to form a square of the same area. In 1989, Miklòs Laczkovich provided a proof that a circular disk could be partitioned into about 10^{50} subsets that could be reassembled (by rotations or translations only!) to form a square with no gaps or overlapping pieces. We note again that the pieces involved in this dissection defy the accepted notion of area.

5.4 Problems

> *There is a crack in everything. That's how the light gets in.*
>
> — Leonard Cohen (1934–)

5.1 Given $\triangle ABC$ with $a = 8, b = 3, c = 6$, find

    (a) Area($\triangle ABC$).

    (b) h_a.

    (c) r (inradius) and R (circumradius).

5.2 (a) Given an isosceles triangle ABC with $AB = BC$, prove that the sum of the two distances from any point of $\overline{AC}$ to the lateral sides AB and BC is the same.

    (b) Given an equilateral triangle ABC, prove that the sum of the three distances from any interior or boundary point to the sides is the same. Generalize the statement to any regular n-gon.

5.3 (a) Prove that the three medians of a triangle divide it into six triangles of equal area.

    (b) Let X be a point inside a triangle ABC such that

$$\text{Area}(\triangle AXB) = \text{Area}(\triangle BXC) = \text{Area}(\triangle CXA).$$

Prove that X is the centroid (intersection of medians) of the triangle.

5.4 Given a convex quadrilateral $ABCD$, consider another quadrilateral whose vertices are the midpoints of the sides of $ABCD$. Prove that its area is half of the area of $ABCD$.

5.5 Consider three semicircles built on the sides of a right triangle, where each diameter coincides with a side of the triangle. Prove that the area of the semicircle on the hypotenuse is equal to the sum of the areas of two other semicircles. Can you generalize this problem?

5.6 For a given triangle ABC, let A_1 and C_1 be points on $\overline{BC}$ and $\overline{AB}$, respectively. Let $BC_1/BA = \lambda$ and $BA_1/BC = \mu$. Prove that

$$\frac{\text{Area}(\triangle C_1 B A_1)}{\text{Area}(\triangle ABC)} = \lambda\mu.$$

5.7 Suppose a rope is tied along the Earth's equator forming a ring. Extend this rope just by 3 feet and lift it uniformly to the same height in the air. Can a cat crawl under it?

5.8 Prove that in any triangle ABC,

$$\frac{1}{h_a} + \frac{1}{h_b} + \frac{1}{h_c} = \frac{1}{r},$$

where h_a, h_b, and h_c are altitude lengths, and r is the radius of the incircle.

5.9 Prove that a diagonal of a convex quadrilateral divides its area in half if and only if it bisects the other diagonal.

5.10 (a) Out of all rectangles with a given perimeter, find the one with greatest area. What about one with smallest area?

   (b) Out of all rectangles with a given area, find the one with smallest perimeter. What about one with greatest perimeter?

5.11 Three circles, each of radius r are externally tangent to each other. Find the area of a figure bounded by them (the region between the circles, bounded by three arcs).

5.12 Let $ABCD$ be a trapezoid with bases $\overline{AD}$ and $\overline{BC}$, $AD = a$ and $BC = b$. Let $E \in \overline{AB}$, $F \in \overline{CD}$ and $\overline{EF} \parallel \overline{BC} \parallel \overline{AD}$. Find EF if the areas of the two trapezoids into which $\overline{EF}$ divides $ABCD$ are equal.

5.13 Consider three mutually externally tangent circles, centered at A, B, and C, with respective radii a, b, and c. Find the area of triangle ABC. What if two of the circles are externally tangent to each other, with both internally tangent to the third?

5.14 Let A_1, B_1, C_1, and D_1 be the midpoints of sides $\overline{CD}$, $\overline{DA}$, $\overline{AB}$, and $\overline{BC}$, respectively, of a square $ABCD$ of area 1. Find the area of the quadrilateral bounded by segments AA_1, BB_1, CC_1, and DD_1. Can you generalize the problem to the case where a point other than the midpoint is chosen on each side? Suppose that $\overline{DA_1}/DC = k$, and similarly for the other three sides.

5.15 Consider three semicircles built on the three sides of a right triangle ABC such that the semicircles on the legs are built in the exterior of the triangle, and the one on the hypotenuse is built on the same side as the triangle. (See Figure 5.20.) Prove that the sum of the areas bounded by these semicircles (that is, the two nonshaded crescent regions in Figure 5.20) is equal to the area of the triangle.

5.16 Let $ABCD$ be a trapezoid with bases $\overline{AD}$ and $\overline{BC}$, and let O be the intersection of its diagonals. Let Area($\triangle AOD$) $= S_1$ and Area($\triangle BOC$) $= S_2$. Find Area($ABCD$).

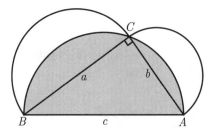

FIGURE 5.20.

5.17 Let A_1, B_1, and C_1 be points chosen on the sides $\overline{BC}$, $\overline{CA}$, and $\overline{AB}$ of $\triangle ABC$, respectively. Suppose that the segments $\overline{AA_1}$, $\overline{BB_1}$ and $\overline{CC_1}$ are concurrent at point O.

(a) Prove that

$$\frac{\text{Area}(\triangle AOB)}{\text{Area}(\triangle BOC)} = \frac{AB_1}{B_1C}.$$

(b) Prove that

$$\frac{AC_1}{C_1B} \cdot \frac{BA_1}{A_1C} \cdot \frac{CB_1}{B_1A} = 1 \qquad \text{(Ceva's theorem)}.$$

5.18 Four circles of radius 1 are centered at the vertices of a square with side length 1. Find the area of the figure formed by the intersection of these circles.

5.19 Let M be a point in the interior of an angle with vertex B, and let A and C be the points of intersection of a line through M with the sides of the angle. Describe the line for which $\text{Area}(\triangle ABC)$ is the smallest (the greatest).

5.20 Two triangles, each of area greater than 1, are inside a circle of radius 1. Prove that the triangles must share a common point.

5.21 (a) Out of all isosceles triangles with a given perimeter, find the one with greatest area. What about one with smallest area?

(b) Out of all isosceles triangles of a given area, find the one with smallest perimeter. What about one with largest perimeter?

5.22 Consider an equilateral triangle ABC. Let A_1, B_1, and C_1 be points on $\overline{BC}$, $\overline{CA}$, and $\overline{AB}$, respectively, such that $BA_1/A_1C = CB_1/B_1A = AC_1/C_1B = 1/2$. Let $A_2 = \overline{BB_1} \cap \overline{CC_1}$, $B_2 = \overline{CC_1} \cap \overline{AA_1}$, and $C_2 = \overline{BB_1} \cap \overline{AA_1}$. Find

$$\frac{\text{Area}(\triangle A_2B_2C_2)}{\text{Area}(\triangle ABC)}.$$

5.23 Show how to dissect two given squares into several parts and assemble them into another square.

5.24 Consider a convex n-gon $A_1 \ldots A_n$. Let $A_1' \ldots A_n'$ be another n-gon obtained by drawing n lines parallel to the sides of $A_1 \ldots A_n$, at distance 1 from them and in the exterior of $A_1 \ldots A_n$. (See Figure 5.21.)

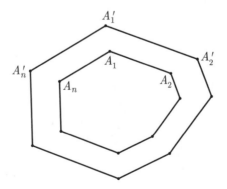

FIGURE 5.21.

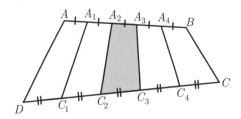

FIGURE 5.22.

Prove that

$$\text{Area}(A'_1 \ldots A'_n) > \text{Perimeter}(A_1 \ldots A_n) + \pi.$$

(This is a numerical inequality. We assume, of course, that numerical values of the area and the perimeter come from corresponding units.)

5.25   (a) Out of all triangles with a given perimeter, find the one with greatest area.

(b) Out of all triangles with a given area, find the one with smallest perimeter.

5.26   Each of two opposite sides of a convex quadrilateral is divided into five congruent segments and the corresponding points on opposite sides are joined. (See Figure 5.22.) Prove that the "middle" quadrilateral (shaded) has area equal to one fifth the area of the original one.

5.27   Prove that in any triangle of area A and inradius r, $A \geq 3\sqrt{3}r^2$.

5.28   Suppose n coins, each of radius r are placed on a round table of radius R such that no two coins overlap and no other coin of the same radius r can be placed on the table without coins overlapping. Prove that

$$\frac{1}{2}\left(\frac{R}{r} - 1\right) < \sqrt{n} < \frac{R}{r}.$$

5.29   Among all triangles inscribed in a circle, prove that the equilateral triangle has the greatest area.

5.30   *Carnot's Theorem.*   The following theorem is attributed to Lazare Nicolas Margurite Carnot (1753–1823).[9]

In any triangle ABC, the (algebraic) sum of the distances (suitably signed, as described in the next paragraph) from the circumcenter O to the sides, is $R + r$, the sum of the circumradius and the inradius. In symbols,

$$OM_A + OM_B + OM_C = R + r.$$

In acute triangles, the circumcenter O is always located inside the triangle. In this case, all three segments OM_A, OM_B, and OM_C lie entirely inside the triangle, as in Figure 5.23(i). If one of the angles is obtuse, the circumcenter falls outside the triangle, as in Figure 5.23(ii). One of the segments (the one that corresponds to the side opposite the obtuse angle) lies entirely outside the triangle and its length will be given a negative sign in the sum. The other two segments lie only partially outside the triangle, and their lengths will be given a positive sign in the sum. Segments OM_A, OM_B, and OM_C serve as altitudes of triangles OBC, OAC, and OAB with the bases on the sides a, b, and c. The sign convention guarantees that the

---

[9] The statement, proof, and discussion of Carnot's theorem came from the webpage `http://www.cut-the-knot` `.org/proofs/carnot.shtml`, which cites [**30**].

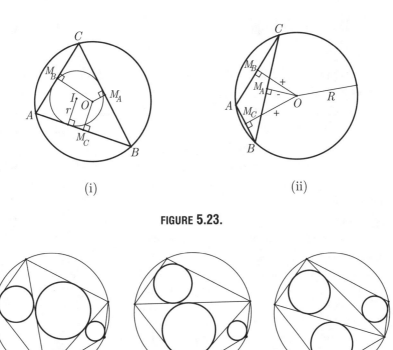

FIGURE 5.23.

(i) (ii)

FIGURE 5.24.

areas of these triangles (taken with the proper signs) add up to the area of $\triangle ABC$, which is the important step in proving Carnot's theorem.

Let a convex polygon, inscribed in a circle, be triangulated by drawing various noninter- secting diagonals between the vertices, and let the inscribed circle be drawn in each of the triangles. Use Carnot's theorem to prove that the sum of the radii of all these circles is constant (independent of the triangulation used).[10] An illustration of the theorem is shown in Figure 5.24 where congruent pentagons are inscribed in three congruent circles but triangulated in three different ways. The sum of the radii of the inscribed circles will be the same for all three triangulations.

5.5 Supplemental Problems

$S5.1$ Each of three circles in the plane touches two other externally. If their radii lengths are a, b, and c, find the area of the triangle with vertices at the centers of the circles.

$S5.2$ Centers of each of two circles of equal radii lengths r lie on the other circle. Find the radius of a circle that is internally tangent to each of the circles and to the line passing through the centers of the circles.

$S5.3$ The midpoints of the sides of one convex quadrilateral coincide with the midpoints of sides of another convex quadrilateral. Prove that the areas of the quadrilaterals are equal.

---

[10] This problem was taken from the website *http://www.cut-the-knot.org/proofs/jap.shtml*. Again, a reference to Hons- berger's book [**30**] is given.

*S*5.4 Is there a triangle with two altitudes longer than 10 inches and with area smaller than 40 square inches?

---

*S*5.5 Three vertices of a triangle lie on the sides of a parallelogram. Prove that the area of the triangle is at most half the area of the parallelogram.

*S*5.6 Let M and N be the midpoints of sides $\overline{BC}$ and $\overline{AD}$ of a convex quadrilateral $ABCD$, respectively. Let $P = \overline{AM} \cap \overline{BN}$, and let $Q = \overline{MD} \cap \overline{CN}$. Prove that the area of the quadrilateral $MQPN$ is the sum of the areas of $\triangle ABP$ and $\triangle CDQ$.

*S*5.7 Consider a circle C circumscribed around a square with a unit side length. Consider another circle C_1 built on one side of a square as on the diameter. Find the area of the part of C_1 that is in the exterior of C.

*S*5.8 A hexagon $ABCDEF$ is inscribed into a circle such that its diagonals $\overline{AD}$, $\overline{BE}$ and $\overline{CF}$ are diameters. Prove that the area of the hexagon is twice the area of $\triangle ACE$.

*S*5.9 Consider a point P in the interior of a triangle ABC. Draw three lines passing through the point and parallel to the sides of the triangle. They divide the triangle into three parallelograms and three triangles. Let S_1, S_2 and S_3 be the areas of these smaller triangles. Find the area of $\triangle ABC$.

*S*5.10 Let $\overline{AB}$ be the hypothenuse of a right triangle, and let K be the point of tangency of $\overline{AB}$ with the incircle of the triangle. Prove that the area of the triangle is equal to $AK \cdot KB$.

*S*5.11 Consider two non-overlapping square regions in a plane. Show that one of them can be cut into four pieces such that they, together with the second square, can be assembled into one square.

*S*5.12 (a) Suppose the medians of a triangle have lengths 4, 5, and 6. What is the area of the triangle?
      (b) Suppose the altitudes of a triangle have lengths 4, 5, and 6. What is the area of the triangle?

*S*5.13 Consider a parallelogram, $ABCD$. Let E and F be the midpoints of sides AB and CD, respectively. Let $S, R, Q,$ and T be the respective points of intersection of $\overline{AC}$ and $\overline{DE}$, $\overline{DB}$ and $\overline{AF}$, $\overline{AC}$ and $\overline{BF}$, and $\overline{BD}$ and $\overline{CE}$. Find the ratio of the areas of $SRQT$ and $ABCD$.

---

*S*5.14 Prove that for every convex polygon there exists two perpendicular lines that cut it into four parts with equal areas.

*S*5.15 In a trapezoid with perpendicular diagonals, the length of a diagonal is 5, and the length of the altitude of the trapezoid is 4. Find the area of the trapezoid.

*S*5.16 Prove that if a line divides both the perimeter and the area of a triangle in half, then it passes through the center of the incircle of the triangle.

6

Loci

The diversity of the phenomena of nature is so great, and the treasures hidden in the heavens so rich, precisely in order that the human mind shall never be lacking in fresh nourishment.

— Johannes Kepler[1] (1571–1630)

Having completed a survey of lines, polygons, circles, and angles, we come to another collection of well-known figures in the plane: ellipses, parabolas, and hyperbolas. In what situations do these figures appear? What is our motivation for studying them?

One way in which these figures arise quite naturally is when we try to find answers to questions of the type, "What is the set of all points (loci) of a plane that satisfy a given property?" Another is when we wish to understand the trajectory of a moving point. Yet a third situation occurs when we seek to describe the intersection of two surfaces in space. We begin this section with a general discussion of loci, which will lead us to the curves mentioned above.

6.1 Locus of a Property

The set of all points in a plane that satisfy a given property is called the **locus** of that property. That is, a locus corresponding to a specified property is a figure in the plane such that each point of the figure satisfies the property and any point of the plane not in the figure does not satisfy the property.

We have already seen some examples of loci.

- In the previous section, a circle with center O and radius $r > 0$ was defined as the set of all points X in the plane that are distance r from O; in other words, $\mathcal{C}(O, r)$ is the locus of points X satisfying $OX = r$.

- The locus of all points equidistant from two distinct points, A and B, is the perpendicular bisector of $\overline{AB}$.

- The locus of all points equidistant from the sides of a given angle is the bisector of the angle.

---

[1] *from Mysterium Cosmographicum*, 1596.

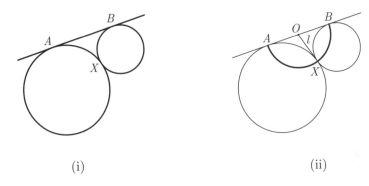

(i)   (ii)

FIGURE 6.1.

- For a pair of distinct points, A and B, and a given real number α, $0 < \alpha < \pi$, the locus of all points C such that $m\angle ACB = \alpha$ is the union of two circular arcs. (See Corollary 4.16.)

Let us see a few more examples. Note that we will often use the phrase **locus of** X to refer to the locus of points satisfying the defining attributes of X.

Example 8. Given distinct points A and B, consider a pair of circles on one side of $\overleftrightarrow{AB}$, tangent to the line at points A and B, and tangent externally to each other at point X. (See Figure 6.1(i).) Find the locus of X.

Solution: By Corollary 4.8, there is a unique line, l, through X that is perpendicular to the radii of both circles at X. Let O be the point of intersection of l and $\overline{AB}$. (See Figure 6.1(ii).) By Theorem 4.9, $OA = OX = OB$. Thus, the locus of points X with the desired property lies on a semicircle of $\mathcal{C} = \mathcal{C}(O, AB/2)$, where O is the midpoint of $\overline{AB}$. By varying the size of the two tangent circles in a continuous fashion, it is easy to see that all points of the semicircle except A and B will be obtained. The complete locus is shown in bold in Figure 6.1(ii).   □

Example 9. Let A and B be two fixed points on a circle. For an arbitrary point C on the open[2] major arc AB, let X be a point on the ray $\overrightarrow{AC}$ such that $A - C - X$ and $CX = BC$. (See Figure 6.2(i).) Find the locus of the points X satisfying this property.

Solution: For a point $C \neq B$ on the circle, the points X satisfying $CX = BC$ form a circle with center at C and radius BC, but only one of these points will lie on $\overrightarrow{AC}$ with $A - C - X$, as illustrated in Figure 6.2(ii).

Now, $\angle ACB$ is an exterior angle to isosceles triangle BCX, which contains congruent base angles at B and X. Therefore, $m\angle AXB = (m\angle ACB)/2$. Furthermore, by Theorem 4.12, the measure of angle ACB is half of the measure of the minor arc AB, from which it follows that $m\angle AXB = (m\widehat{AB})/4$.

Therefore, by Corollary 4.16, since $m\angle AXB$ takes on a single value, X must lie on an arc of a circle containing A and B, with points A and B deleted. The resulting locus is depicted as the open arc DB in Figure 6.2(iii), where D lies on the line tangent to the original circle at A.   □

---

[2] By an **open arc**, we mean the set of all points on the arc, excluding the endpoints.

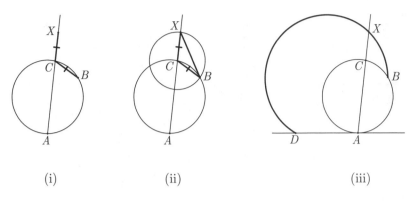

(i)   (ii)   (iii)

FIGURE 6.2.

In Problem 6.12, the reader may explore what happens in the case where fewer restrictions are placed on C and X.

Example 10. Let A and C be acute angles in triangle ABC. Consider all rectangles $DEFG$ inscribed in the triangle in such a way that D is on $\overline{AB}$, E is on $\overline{BC}$, and F and G are on $\overline{AC}$. Find the locus of all centers of these rectangles.

Solution: Experiments suggest that the locus is the interior of the segment joining the midpoint M of $\overline{AC}$ with the midpoint N of the altitude BH from B. (Two 'degenerate cases', when the rectangle becomes a segment, give these points.) Let K be the midpoint of diagonal $\overline{EG}$. Our plan will be to show that K lies on $\overline{MN}$. The reader is invited to try it for herself before proceeding with the solution if she wishes.

Note that the center of $DEFG$ is the midpoint K of its diagonal $\overline{EG}$, and let L be the base of the perpendicular from K to $\overline{AC}$. (See Figure 6.3.) To show that $K \in \overline{MN}$, we will show that the ratio ML/LK ($= MH/HN$) is independent of K.

For simplicity, label $BH = h$, $AH = b_1$, $AC = b$, and $DG = d$. We assume that $b_1 \geq b/2$. From similar right triangles AGD and AHB we obtain $AG/AH = d/h$, so $AG = b_1 d/h$. Also, from similar triangles DBE and ABC, $DE/AC = (h-d)/h$. Hence $FG = DE = (h-d)b/h$. Therefore,

$$ML = AL - AM = (AG + FG/2) - b/2$$

$$= \left(\frac{db_1}{h} + \frac{(h-d)b}{2h}\right) - \frac{b}{2} = \frac{d(2b_1 - b)}{2h}.$$

Since $KL = d/2$, we have shown $ML/KL = (2b_1 - b)/h$, which is independent of d, and thus independent of the distance from K to $\overline{AC}$. This proves that $K \in \overline{MN}$. Now, in order for all possible rectangles to be obtained, D will move from A to B, which will cause d to vary on $(0, h)$. Thus, K covers the interior of $\overline{MN}$. □

The preceding examples all demonstrate loci that are comprised of lines, circles, or parts of lines or circles. However, many simple properties lead to other figures, and in the following section we will consider several such loci, each of which is likely to be familiar to the reader.

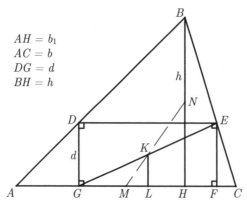

$AH = b_1$
$AC = b$
$DG = d$
$BH = h$

FIGURE 6.3.

6.2   Three Notable Loci

> *Geometry existed before the creation.*
> — Johannes Kepler (1571–1630)

The curves we explore in this section are usually referred to as conic sections because they can be constructed by intersecting a cone with a plane. (See Figure 6.18 in Section 6.2.4.) Often students encounter the conic sections for the first time in an algebraic setting. Historically, however, these curves were known for 2000 years before the Cartesian coordinate system finally provided the proper setting for algebraic description.

It turns out that the same curves can be defined as certain loci.[3] In this section, we present these definitions and study some properties of the curves by using methods of plane geometry that we have developed so far. In Chapter 8 we will show how one can use the coordinate method to study these curves by algebraic means.

6.2.1   Ellipse

Given two points, F_1 and F_2, and a positive number s, suppose we wish to find the locus $\mathcal{E}$ of all points C such that $CF_1 + CF_2 = s$. In other words, we wish to find all points in the plane the sum of whose distances to the points F_1 and F_2 is a fixed value, s. Such a figure $\mathcal{E}$ is called an **ellipse** and will be denoted as $\mathcal{E}(F_1, F_2; s)$. Each point F_1 and F_2 is called a **focus** of $\mathcal{E}$; together, they are the **foci** of $\mathcal{E}$. For any point C on $\mathcal{E}$, the segments CF_1 and CF_2 are called **focal radii** of $\mathcal{E}$.

It can happen that $\mathcal{E}$ will be empty. By the triangle inequality,

$$s = CF_1 + CF_2 = F_1C + CF_2 \geq F_1F_2.$$

Hence, $\mathcal{E}$ is empty if $s < F_1F_2$, and $\mathcal{E}$ is the segment F_1F_2 if $s = F_1F_2$.

The interesting case occurs when $s > F_1F_2$. Some points of $\mathcal{E}$ are easy to find in this situation. For example, let C be the point on $\overrightarrow{F_1F_2}$ such that $CF_2 = (s - F_1F_2)/2$. (In Figure 6.4(i), this point is designated as C_1.) Then

$$CF_1 + CF_2 = (CF_2 + F_1F_2) + CF_2 = 2CF_2 + F_1F_2 = s,$$

so C is on $\mathcal{E}$. It is easy to see that C has a counterpart, C' on $\overrightarrow{F_2F_1}$ with $C'F_1 = (s - F_1F_2)/2$.

---

[3] Often one uses arguments from spatial geometry to demonstrate the equivalence, either so-called Dandelin spheres or the coordinate method in R^3.

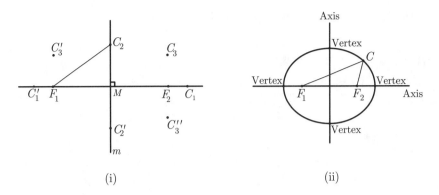

FIGURE **6.4.**

Next, consider the midpoint, M, of $\overline{F_1 F_2}$. Let the line m be the perpendicular bisector of $\overline{F_1 F_2}$ at M. Since $s/2 > F_1 F_2/2$, there must be a point C on m such that $CF_1 = s/2$; in particular, C is one of two points on m such that $CM = \sqrt{(s/2)^2 - (F_1 F_2/2)^2}$. (This is C_2 in Figure 6.4(i).) Then, since $CF_1 = CF_2$ (by the Pythagorean theorem), $CF_1 + CF_2 = s/2 + s/2 = s$.

It is clear that for any point C on $\mathcal{E}$, the points C' and C'', symmetric to C with respect to m and $\overleftrightarrow{F_1 F_2}$, respectively, are also on $\mathcal{E}$. (This is illustrated with the points C_3, C_3' and C_3'' in Figure 6.4(i).) Hence, $\mathcal{E}$ will certainly possess two lines of symmetry, m and $\overleftrightarrow{F_1 F_2}$. Therefore, in order to fully describe $\mathcal{E}$ it suffices to determine the portion of $\mathcal{E}$ lying in the interior of one of the four right angles defined by m and $\overleftrightarrow{F_1 F_2}$.

Letting $s = s_1 + s_2$, with $0 < s_2 < s_1$ and $s_1 - s_2 < F_1 F_2$, we can find C such that $CF_1 = s_1$ and $CF_2 = s_2$. Doing this in different ways (i.e., with different choices of s_1 and s_2) will give us different points of one fourth of $\mathcal{E}$ (located in the first quadrant if the x- and y-axes are the axes of symmetry for $\mathcal{E}$). Due to the symmetric nature of the ellipse, this one-fourth of the ellipse can be reflected about axes of symmetry to obtain the remainder of $\mathcal{E}$. The complete ellipse is shown in Figure 6.4(ii).

It is helpful to visualize the ellipse as being created by tacking the ends of a piece of string of length s to the two foci and then allowing a pencil pulled taut against the string to trace out a path as it moves around. (See Figure 6.5.) Obviously this "string property" will yield points on the curve because the sum of the distances from any such point to the two foci will be constant.

We define the **interior** of an ellipse $\mathcal{E}(F_1, F_2; s)$ as the loci of all points C such that

$$CF_1 + CF_2 < s,$$

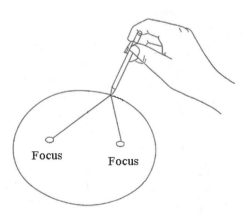

FIGURE **6.5.**

and the **exterior** of $\mathcal{E}$ as the loci of all points C such that

$$CF_1 + CF_2 > s.$$

For now we accept without proof that

i. an ellipse divides the plane into two parts, its **interior** and its **exterior**, with foci in the interior,

ii. the union of an ellipse and its interior is a convex figure, i.e., given two points A and B from the union, segment AB will lie in the union, and

iii. every line through an interior point intersects the ellipse in two points.

The lines of symmetry are called the **axes** of the ellipse and the intersection of these axes with the ellipse yields its four **vertices**.

Example 11. Consider four linked "rods", $\overline{AB}$, $\overline{BC}$, $\overline{CD}$, and $\overline{DA}$, such that $AB = CD$ and $BC = DA$. By a "rod", we mean a line segment of fixed length. (The four points A, B, C, and D can be pictured as vertices of the quadrilateral $ACDB$, where $\overline{AD}$ and $\overline{BC}$ are diagonals of the quadrilateral.) Let F be the intersection of $\overline{AD}$ and $\overline{BC}$. Find the locus of the points F if the placement of $\overline{AB}$ is fixed while the placements of the other rods can vary.

Two of the infinitely many possible arrangements are shown in Figure 6.6. Notice that in order for $\overline{AD}$ and $\overline{BC}$ to intersect, it must be that $AD > AB$. (An opportunity to explore a similar question for the case when $AD < BC$ is given in Problem 6.3.)

Solution: Let $\mathcal{L}$ be the desired locus. Notice that if A, B, C, and D are collinear (which happens when $A - B - D - C$ or $D - C - A - B$), then $\overline{AD}$ and $\overline{BC}$ lie on the same line ($\overleftrightarrow{AB}$) and their intersection is a segment of length $AD - AB$.

Assume that the points are not all collinear. By SSS, $\triangle ADB \cong \triangle CBD$ and $\triangle BCA \cong \triangle DAC$. Therefore, $\angle FDB \cong \angle FBD$, so $\triangle BFD$ is isosceles; similarly, $\triangle CFA$ is isosceles.[4] Thus, $FA = FC$ and $FD = FB$. Therefore,

$$FA + FB = AF + FD = AD,$$

where $AD = s$ is a fixed value. Therefore, F is on the ellipse $\mathcal{E}(A, B; s)$.

The preceding paragraph demonstrates that every point in $\mathcal{L}$ is a point on $\mathcal{E}$. To complete the proof, we must show that any point of $\mathcal{E}$ is also a point in the locus. To show this, suppose that F' is any point on $\mathcal{E}(A, B; s)$. By definition of an ellipse, if F' is on $\mathcal{E}$, then $F'A + F'B = AD = BC$. Position the rod $\overline{AD}$ so that $A - F' - D$. Then $AF' + F'B = AF' + F'D$, so $F'B = F'D$.

For this positioning of $\overline{AD}$, there is a unique positioning of C so that $AB = CD$ and $BC = DA$. Let $E = \overline{AD} \cap \overline{BC}$. Assume that $E \neq F'$ and $A - F' - E - D$, as shown in Figure 6.7. (A similar argument will work if $A - E - F' - D$.) Then,

$$F'B < F'E + EB \implies$$
$$F'B < F'E + ED \implies$$
$$F'B < F'D,$$

a contradiction. Thus $F' = E$, and we conclude that F' is on $\mathcal{L}$. $\square$

Let us continue the discussion of the shape of a general ellipse, $\mathcal{E}(F_1, F_2; s)$. Observe that since $|CF_1 - CF_2| \leq F_1 F_2$ for any point C on the ellipse, if $F_1 F_2$ is very small compared to s, then CF_1 is very close to CF_2. Since their sum is s, both CF_1 and CF_2 are close to $s/2$, and $\mathcal{E}$ is "close" to a circle

---

[4] In fact, since $\angle AFC$ and $\angle DFB$ are vertical angles, all four of the angles FDB, FBD, FAC, and FCA are congruent.

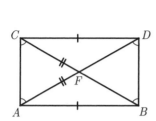
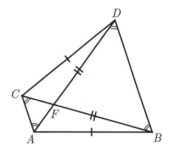

FIGURE 6.6.

of radius $s/2$ with center at the midpoint of $\overline{F_1 F_2}$. Thus, we see what happens in the extreme case where $F_1 F_2 = 0$ (i.e., $F_1 = F_2$): $\mathcal{E}$ is the circle of radius $s/2$ centered at F_1 with $CF_1 = CF_2 = s/2$. As noted before, the other extreme case occurs when $s = F_1 F_2$, yielding the degenerate "ellipse" consisting of the line segment $F_1 F_2$.

6.2.2 Hyperbola

In the preceding section, we fixed two points, F_1 and F_2, and a positive number s, and then we found the locus of all points C such that $CF_1 + CF_2 = s$. Consider now the locus of points in the plane such that the *difference* of the distances from F_1 and F_2 is a specified value, d. Let distinct points F_1 and F_2 be given, and assume $0 < d < F_1 F_2$. We wish to find the locus $\mathcal{H}$ of all points C such that $|CF_1 - CF_2| = d$. $\mathcal{H}$ is called a **hyperbola** and each of the points F_1 and F_2 is called a **focus** of $\mathcal{H}$. We will denote the hyperbola with foci F_1 and F_2 and difference d as $\mathcal{H}(F_1, F_2; d)$.

Whereas the ellipse, $\mathcal{E}$, contains four easily identifiable points (its vertices), we only can find two such points on $\mathcal{H}$. Let V be the point on $\overline{F_1 F_2}$ such that $VF_2 = (F_1 F_2 - d)/2$. Then

$$|VF_1 - VF_2| = |(F_1 F_2 - VF_2) - VF_2| = |F_1 F_2 - 2VF_2| = d.$$

Similarly, the point V' on $\overline{F_1 F_2}$ such that $V'F_1 = (F_1 F_2 - d)/2$ will also be on $\mathcal{H}$. These two points V and V' are called the **vertices** of the hyperbola $\mathcal{H}$.

The symmetry observed in $\mathcal{E}$ is also present in $\mathcal{H}$: for any point C on $\mathcal{H}$, the points C' and C'' obtained by reflection of C about the perpendicular bisector of $\overline{F_1 F_2}$ and about $\overleftrightarrow{F_1 F_2}$, respectively, are also on $\mathcal{H}$. As with the ellipse, these lines of symmetry are called the **axes** for the hyperbola, and the intersection of the line $F_1 F_2$ with the curve yields the vertices. An example of a hyperbola is shown in Figure 6.8. While there are not obvious "interior" and "exterior" regions for a hyperbola in the way there are for an ellipse, we define them similarly as the locus of all points C such that

$$|CF_1 - CF_2| > d$$

for the **interior** of a hyperbola $\mathcal{H}(F_1, F_2; d)$ and the locus of all points C such that

$$|CF_1 - CF_2| < d$$

for its **exterior**. The interior of a hyperbola contains its foci (see Problem 6.9), and consists of two regions. The remaining region forms the exterior.

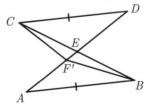

FIGURE 6.7.

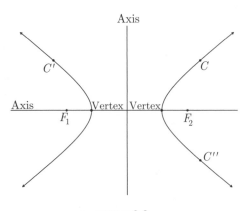

FIGURE 6.8.

Example 12. Given two non-intersecting circles, find the locus of the centers of all circles that are tangent to both of them.

Solution: Let $C_1 = C(O_1, r_1)$ and $C_2 = C(O_2, r_2)$ be given circles and let $C = C(X, r)$ be tangent to both of them. There are various possibilities for the relative position of the circles with respect to each other; we consider two of them in detail and leave the others to the reader. (See Problem S6.3.)

If C_1 is in the exterior of C_2, and C is tangent to both of them externally, then $XO_i = r + r_i$, as seen in Figure 6.9(i). Then $|XO_1 - XO_2| = |r_1 - r_2|$. If $r_1 \neq r_2$, the locus is one branch of the hyperbola with $d = |r_1 - r_2|$ and foci at O_1 and O_2; if $r_1 = r_2$, then the locus is the perpendicular bisector of $\overline{O_1 O_2}$.

If C_1 is in the interior of C_2, and C is tangent to C_1 externally and to C_2 internally, then $XO_1 = r + r_1$ and $XO_2 = r_2 - r$, as shown in Figure 6.9(ii). Then $XO_1 + XO_2 = r_1 + r_2$, and the locus is the ellipse with $s = r_1 + r_2$ and foci at O_1 and O_2.

The loci for the two cases considered in Figure 6.9(i) and (ii) are shown in Figure 6.10.

Other cases to consider are: when C_1 is in the exterior of C_2 and both lie internally tangent to C (locus is the other branch of the hyperbola we saw in the first case considered); when C_1 is in the exterior of C_2 and C is tangent to one of them internally and one of them externally (locus is a hyperbola); and when C_1 lies in the interior of C_2, with C_1 lying internally tangent to C while C lies internally tangent to C_2 (locus is an ellipse). $\quad\square$

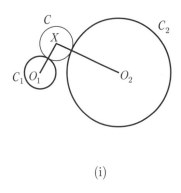

(i)

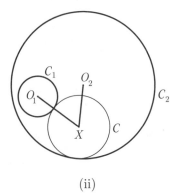

(ii)

FIGURE 6.9.

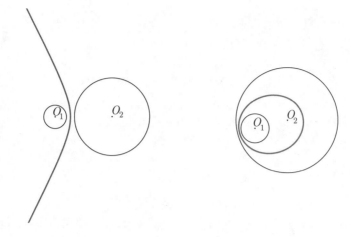

FIGURE **6.10.**

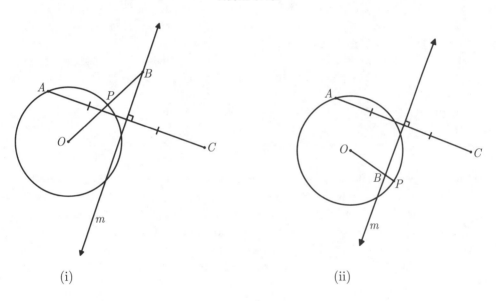

(i)   (ii)

FIGURE **6.11.**

The following lengthy example is optional. It may be skipped without causing future difficulty.

Example 13. Let a circle and a point C outside of the circle be fixed. For an arbitrary point A on the circle, construct the perpendicular bisector m of $\overline{AC}$. Prove that the union of these bisectors is a hyperbola with its exterior.

Solution: We claim that the desired locus is the hyperbola $\mathcal{H}(C, O; r)$ and its exterior, where O and r are the center and radius of the given circle, respectively. First we show that any point B on m lies on $\mathcal{H}$ or its exterior. To do this, we must show that $|BO - BC| \le r$. Since m is a perpendicular bisector of $\overline{AC}$, $BA = BC$. Let P be the intersection of $\overleftrightarrow{BO}$ and the circle. If B lies in the exterior of the circle (as in Figure 6.11(i)), then $BO = BP + PO = BP + r$. In this case, $BO - BC = BP + r - BA \le r$. The last inequality follows from the fact that P is the point on the circle closest to B, giving $BP \le BA$.

If B lies in the interior of the circle (as in Figure 6.11(ii)), then $BO = r - BP$. In this case,

$$BC - BO = BA + BP - r$$
$$\leq AO + BO + BP - r$$
$$= AO + OP - r = 2r - r = r.$$

Next we show that any point B on $\mathcal{H}$ or in the exterior of $\mathcal{H}$ must lie on the perpendicular bisector of segment CA for some point A on the circle. Such a point A is completely characterized by the condition $BA = BC$. We consider two cases, depending on the location of B. In each case, we use the fact that since B lies on or in the exterior of $\mathcal{H}$, $|BO - BC| \leq r$. Since r, BO, and BC are all positive, $|BO - BC| \leq r$ implies

$$|BO - r| \leq BC \leq BO + r.$$

Suppose that B lies in the exterior of the given circle. Then the minimum and maximum distances from B to the circle are $BX = BO - r = |BO - r|$ and $BY = BO + r$, respectively. (Compare with Problem 4.15.) In fact, due to continuity of the distance function, BA will take on all values in the interval $[BO - r, BO + r]$ as A moves around the circle. In particular, BA will assume the value BC.

On the other hand, if B lies in the interior of the given circle, then BA will take on all values in the interval $(r - BO, r + BO)$ as A moves around the circle. In particular, BA will attain the value BC.

We conclude that $\mathcal{H}(C, O; r)$ and its exterior, shown as the shaded region in Figure 6.12, is the desired locus.[5] $\square$

Problem S6.15 considers the analogous situation when the fixed point C is inside the circle.

We conclude this subsection by briefly considering the degenerate "hyperbolas" that arise if we relax our restrictions on d, F_1, and F_2. First, if $d = F_1 F_2$, $\mathcal{H}$ will be $\overleftrightarrow{F_1 F_2}$, excluding the open segment $F_1 F_2$. Second, if $d = 0$ and $F_1 \neq F_2$, then the locus of points satisfying $|CF_1 - CF_2| = d$ will be the perpendicular bisector of $\overline{F_1 F_2}$. If $F_1 = F_2$ and $d = 0$, then the locus will consist of all points in the plane. Note that if $F_1 = F_2$ and $d > 0$, or, more generally, when $F_1 F_2 < d$, then $\mathcal{H}$ is empty.

6.2.3 Parabola

A third well-known curve in the plane is generated by fixing a point F and a line l not containing F and finding the locus $\mathcal{P}$ of points C such that $CF = d(C, l)$; that is, we find the set of all points in the plane that are equidistant from a specified point and a specified line. Such a figure is called a **parabola** and will be denoted as $\mathcal{P}(F, l)$. Point F and line l are called the **focus** and the **directrix** of the parabola $\mathcal{P}$, respectively.

Assuming F is not on l, the distance from F to l is the length of the perpendicular line segment, $\overline{FX}$, from F to l. The midpoint, V, of this segment is certainly on $\mathcal{P}$ and is called the **vertex** of $\mathcal{P}$. If C is any point distinct from V, then C has a counterpart, C', on $\mathcal{P}$, found by reflecting C over $\overleftrightarrow{FX}$. (See Figure 6.13.)

The line FX is called the **axis** for the parabola, and its intersection with the parabola gives the vertex of the parabola. An example of a parabola with its labeled components is shown in Figure 6.14.

---

[5] The image in Figure 6.12 (and all others in the book as well) was created by using Geometer's Sketchpad®. The perpendicular bisector was traced while the point A on the circle was animated around the circle.

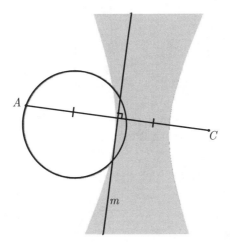

FIGURE **6.12.**

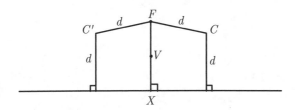

FIGURE **6.13.**

Similarly to the ellipse and hyperbola, we define the **interior** (respectively, **exterior**) of a parabola $\mathcal{P}(F, l)$ as the loci of points C such that $CF < d(C, l)$ (respectively, $CF > d(C, l)$). As before, the focus of a parabola lies in its interior; the directrix lies in its exterior.

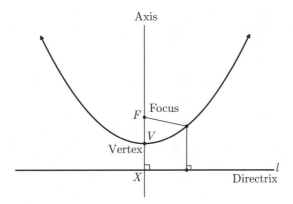

FIGURE **6.14.**

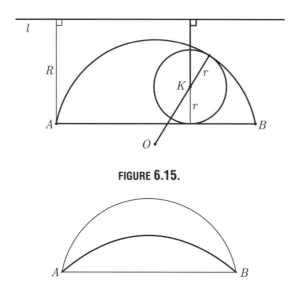

FIGURE **6.15.**

FIGURE **6.16.**

Example 14.   Find the locus of the centers of circles inscribed in a given circular segment (where a "circular segment" consists of a chord of a circle and the minor arc defined by the chord).

Solution: Consider a circular segment defined by a chord $\overline{AB}$ and the minor arc in a circle with center O and radius R. Let K be the center of a circle of radius r inscribed in the circular segment. Then $OK = R - r$. Let l be a line parallel to $\overline{AB}$ at distance R from it. (See Figure 6.15.)

Then $d(K, l) = R - r$, and hence $KO = d(K, l)$. Therefore, K lies on a parabola with focus O and directrix l. The value of r is small precisely when K is close to A or to B. By continuity, the locus is a parabolic arc defined by A and B, with A and B deleted. The complete locus is shown in bold in Figure 6.16. $\qquad\square$

We close this subsection by noting a distinction between the parabola and the two previous curves. The ellipse and hyperbola each had parameters associated with them (F_1 and F_2, along with s for the ellipse and d for the hyperbola). By varying those parameters we could change the shape of the associated curve, indeed almost to degeneracy. However, this does not happen with the parabola, since

all parabolas are similar!

As we know, the parabola is completely defined by its focus and directrix. Given two parabolas $\mathcal{P}_1 = \mathcal{P}(F_1, l_1)$ and $\mathcal{P}_2 = \mathcal{P}(F_2, l_2)$, $\mathcal{P}_1 \sim \mathcal{P}_2$ with coefficient of similarity $d(F_2, l_2)/d(F_1, l_1)$. [6]

In the left side of Figure 6.17 we see a portion of a parabola. By zooming out in both directions by a factor of 2 we obtain the dark portion of the parabola on the right side of Figure 6.17, so certainly

---

[6] Some readers who are familiar with the algebraic representation of parabolas may object here, claiming, for example, that the parabola defined by $y = 2x^2$ is narrower than the parabola defined by $y = x^2$. However, the difference between these two parabolas is in *size* only, as illustrated in the accompanying figures.

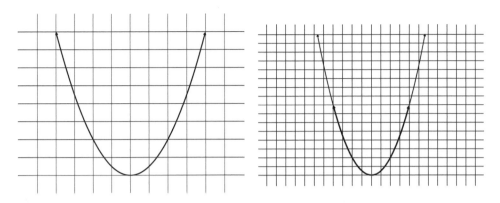

FIGURE 6.17.

the two images are similar (with ratio 2:1). Thus the corresponding parabolas must be similar, even though the second parabola would appear to be narrower than the first if they were superimposed. (Compare the light extension of the parabola on the right with the original parabola in the figure on the left.)

Interestingly, as the focus moves toward the directrix, although the resulting parabolas are indeed similar, the parabolas appear to be narrowing to a ray on the axis, originating at the vertex. The vertex is also approaching the focus. As noted earlier, when the focus and vertex hit the directrix, the locus is the axis.

6.2.4   History and comments

Why, out of infinitely many loci, do we choose to emphasize ellipses, hyperbolas, and parabolas? Maybe it is because of their appearance very early in mathematics as shadows of circles or as conic sections. A conic section is so named because it is the intersection of a double napped right circular cone with a plane, as shown in Figure 6.18.[7] In the special case when the plane contains the vertex of the cone, the intersection is a point, a single line, or a pair of intersecting lines. Suppose then that a plane does not contain the vertex. If the plane is parallel to a line on the surface of the cone, then the intersection will be a parabola. Otherwise, the intersection will be an ellipse or a hyperbola, depending on whether the plane intersects one nap of the cone or both of them.

It is interesting to observe the continuously varying intersections of a plane with the cone as slight adjustments are made to the tilt of the plane. Imagine "hinging" the plane at a point on the cone so that the plane lies parallel to the axis of the cone. Now gradually swivel the plane through an angle of $90°$ until it is perpendicular to the axis of the cone. The resulting intersections will start as hyperbolas and gradually morph into a parabola when the plane is parallel to the edge of the cone. When this angle is passed, the intersections will at first be elongated ellipses that resemble the parabola and will gradually become more circular. Indeed, once the plane is horizontal, the intersection with the cone will be a circle.

This transformation of one curve into another can also be seen directly from the locus definitions of the curves, without appealing to the three-dimensional model. We present this transformation without giving details of a proof. Given an ellipse, fix both a vertex on the major axis and the focus nearest to it, and consider the sequence of ellipses that are formed using these two points while

---

[7] Taken from www.answers.com/topic/conic-section.

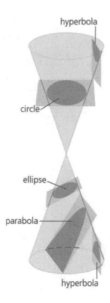

FIGURE 6.18. Intersection of cone and plane.

letting the other focus move farther and farther away. As seen in Figure 6.19, these ellipses approach a parabola sharing the same fixed focus and vertex as the ellipses.

Similarly, a parabola can be viewed as a limiting case of hyperbolas. Fix a vertex and one focus of a hyperbola, and consider the sequence of hyperbolas that are formed using these two points while letting the other focus move farther and farther away. As seen in Figure 6.20, with the right vertex and focus fixed and the left vertex moving further and further to the left, the right branches of the hyperbolas approach a parabola sharing the same fixed focus and vertex with the hyperbolas.

Conic sections were studied extensively by the ancient Greek mathematicians. In his *Commentary*, Proclus states that Menaechmus, a student of Eudoxus, discovered them around 350 B.C. However, it was Apollonius of Perga who gave them their present names and systematized their study in his famous tract, *Conics*. ([**10**], pages 207–208) Centuries later, the great Galileo Galilei (1564–1642) claimed that the trajectory of a rock or a cannon ball would follow a parabolic path in a vacuum. ([**18**], page 321) In 1609, Johannes Kepler (1571–1630) published *Astronomia Nova*, in which he stated the principle that has become known as "Kepler's First Law": the trajectory of each planet is an ellipse with the sun as one focus.

Kepler based his conclusions on rich experimental data obtained largely by Tycho Brahe. However, Isaac Newton (1643–1727) later derived these facts theoretically by assuming the law of gravity and other basic laws of physics. Newton also corrected Galileo's claim about the trajectory

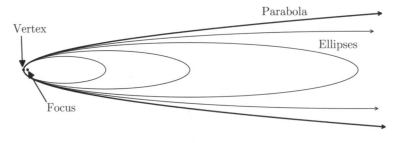

FIGURE 6.19.

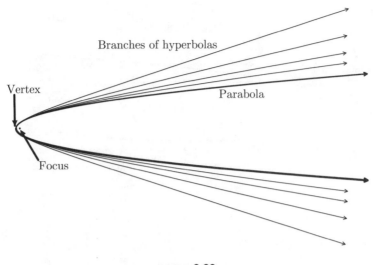

FIGURE 6.20.

of a projectile, realizing that it is elliptical, rather than parabolic. One of the foci of the ellipse is close to the center of the earth and the other is close to the projectile. Of course, near this latter focus, the parabola and ellipse are nearly coincident, as we observed in Figure 6.19. Newton realized this also, as evidenced in this quote from his *Principia*.[8]

> *If the ellipsis, by having its center removed to an infinite distance, degenerates into a parabola, the body will move in this parabola; and the force, now tending to a center infinitely remote, will become equable. Which is Galileo's Theorem.*

Newton's conclusion that a trajectory followed an elliptic arc is based on the assumptions that the earth is a sphere exerting the same gravitational force as a material point (one with the same mass as the earth) at its center and that the direction of the force is radial. The Galileo conclusion can be derived theoretically from the assumption that the earth is flat and the direction of the force of gravity is vertical.

Of course, the calculations used to determine the shape of the earth's orbit relative to the sun apply to other objects in the solar system as well. The total energy E of a comet is the sum of its own kinetic energy and its potential energy (which is negative in a gravity force system) in the gravitational field exerted by the sun. We have

$$E = \frac{mv^2}{2} - \frac{GMm}{r},$$

where v is the velocity of the comet at its closest point to the sun, $G \approx 6.67 \times 10^{-11}$ cubic meter per kilogram-second squared is the universal gravitational constant, and $M = 1.989 \times 10^{30}$ kg is the mass of the sun. The type of orbit of the comet is determined as follows:

(1) When $E < 0$, then $v < \sqrt{2GM/r}$, and the orbit is an ellipse.

(2) When $E = 0$, then $v = \sqrt{2GM/r}$, and the orbit is a parabola.

(3) When $E > 0$, then $v > \sqrt{2GM/r}$, and the orbit is a hyperbola.

---

[8] For some reason this fact, though known long ago, is omitted from modern textbooks. The quote as presented here appears in the article "Ballistic trajectory: Parabola, ellipse, or what?" by L. Burko and R. Price, in *Am. J. Phys* **73** (6) June 2005.

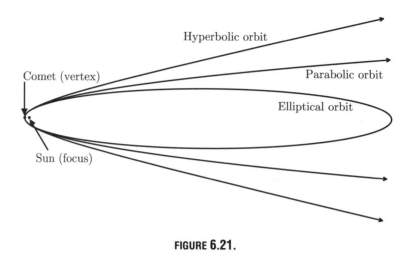

FIGURE 6.21.

In each case above, the center of the sun is a focus of the orbit and the vertex of the orbit occurs where the comet is closest to the sun. The velocities will be approximate and are affected by several factors like air resistance, direction of firing, other objects in space, etc. Nevertheless, there will be an exact velocity, called the 3rd cosmic velocity, which will cause the comet to escape the solar system entirely and continue a parabolic or hyperbolic trajectory in interstellar space (until it is altered by some mass). This will be the situation in the latter two cases above, and occurs when an object is launched from earth with a velocity of at least (approximately) 16.5 km/sec. Interestingly, of the 610 comets identified prior to 1970, 245 have elliptical orbits, 295 have parabolic orbits,[9] and 70 have hyperbolic orbits.[10] We show these three orbits in Figure 6.21.

Now, the same formulas above may be used to classify the orbit of a rocket relative to the Earth (by letting M be the mass of the earth). The speed required to escape directly from the Earth's surface (i.e., to "defy gravity") is referred to as the 1st cosmic velocity, and it is about 7.8 km/sec. At speeds close to this, the rocket becomes a satellite of the Earth and its orbit is close to a circle. When the speed reaches the 2nd cosmic velocity, about 11.2 km/sec, the rocket escapes from the earth and begins an independent elliptical orbit around the sun.

6.3   Reflection Properties and Applications

Various mathematical techniques can be used to study properties of ellipses, hyperbolas, and parabolas, including the coordinate method, transformations, and calculus. Although we will use some of these tools later in the book, recall that here we wish to use only the methods of plane geometry that have previously been introduced. Though the techniques we have at our disposal at this point are inferior to the methods just mentioned, the Greeks were able to work wonders with them.

---

[9] Theoretically, a perfect parabolic orbit is possible, but in practice this is unlikely due to the requirement of an exact velocity. In practice, an orbit is classified as parabolic if there is not conclusive evidence otherwise.

[10] These numbers are from *College Algebra: Concepts and Models*, by Larson, Hostetler, and Hodgkins. Houghton Mifflin (2006).

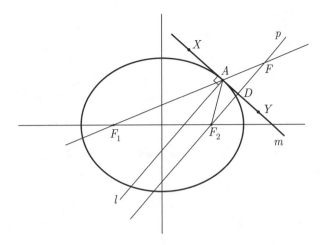

FIGURE 6.22.

We now define the tangent line to an ellipse, $\mathcal{E}$. Line m is **tangent** to $\mathcal{E}$ at a point A if it intersects $\mathcal{E}$ at A and contains no interior points of $\mathcal{E}$.[11] The existence and uniqueness of a tangent line to an ellipse at a specified point follow from the next theorem.

Theorem 6.1. *Let A be a point on $\mathcal{E} = \mathcal{E}(F_1, F_2; s)$, and let line m through A be perpendicular to the bisector of $\angle F_1 A F_2$. Then m is tangent to $\mathcal{E}$ at A.*

Proof. Assume $AF_1 \geq AF_2$, as in Figure 6.22. Choose a point X on m, $X \neq A$. It is sufficient to show that X lies in the exterior of the ellipse. We will do this by showing that $XF_1 + XF_2 > AF_1 + AF_2$; the result then follows immediately from the definition of the exterior of an ellipse, since $AF_1 + AF_2 = s$.

Let p be the line through F_2 that is perpendicular to m, let p and m meet at D, and let p and line $F_1 A$ meet at F. We wish to show that F is the reflection of F_2 across m. Since m and the bisector, l, of $\angle F_2 A F_1$ are perpendicular, $\angle DAF_2 \cong \angle XAF_1 \cong \angle DAF$. Hence, $\triangle FAD \cong \triangle F_2 AD$ by ASA, and it follows that $DF = DF_2$ and $AF = AF_2$. Additionally, $\triangle F_2 DX \cong \triangle FDX$ by SAS, so $XF = XF_2$. Therefore, $F_1 X + XF_2 = F_1 X + XF > F_1 F = F_1 A + AF_2$, as desired.

In order to complete the proof we need to show that m is the unique tangent to $\mathcal{E}$. To this end, suppose that n is a tangent to $\mathcal{E}$ at A. By the definition of tangent to ellipse, every point Q of n distinct from A is in the exterior of $\mathcal{E}$; hence, $QF_1 + QF_2 > AF_1 + AF_2$. Then segments AF_1 and AF_2 form congruent angles with n (see Problem 3.2.9). Therefore n is perpendicular to the bisector of $\angle F_1 A F_2$, and thus, $n = m$. $\qquad\square$

---

[11] We could equivalently define the tangent to $\mathcal{E}$ at A as the line m that intersects $\mathcal{E}$ at A only. However, this definition would need to be modified when we define a tangent line to a parabola. We prefer to use a definition that is as consistent as possible with the definitions of tangent lines to the other conics.

We observed earlier that a tangent line to a circle forms equal angles (each of them 90°) with the radius drawn to the point of tangency; this property characterizes the tangent completely. Theorem 6.1 shows how this characterization generalizes to tangent lines of an ellipse.

Corollary 6.2. *(Reflection Property of an Ellipse) The focal radii of a point on an ellipse form congruent angles with the tangent to the ellipse at this point.*

Corollary 6.2 is often stated as the reflection property of the ellipse: any light ray emanating from one focus of an ellipse will reflect off the ellipse and pass through the second focus. This property has many practical applications. We mention one of them here and leave another for Problem 6.1. In the medical procedure known as lithotripsy, a patient suffering from kidney stones is positioned in a rotational ellipsoid [12] bath in such a way that the affected kidney is at one focus. Shock waves are emitted from the other focus, and thus the stones are broken up and can be passed out of the body without surgery. [8]

Similar reflection properties can be established for the hyperbola and the parabola. In each case these properties are related to tangent lines to the curve, which in turn are defined in terms of interior and exterior points. We will pursue these ideas for the parabola here and leave those for the hyperbola to the reader in Problems 6.9, 6.10, and 6.11.

We define the **tangent line** to a parabola $\mathcal{P}$ at a point A to be the line m intersecting $\mathcal{P}$ at A but containing no points in the interior of $\mathcal{P}$. The existence and uniqueness of the tangent line to $\mathcal{P}$ at A follow from Theorem 6.3 and Problem 3.2.9.

Theorem 6.3. *Let A be a point on a parabola $\mathcal{P}(F, l)$. Let $\overleftrightarrow{AX}$ be a line parallel to the axis of $\mathcal{P}$ such that X is in the interior of $\mathcal{P}$. The line m through A that is perpendicular to the bisector of $\angle FAX$ is tangent to $\mathcal{P}$ at A.*

Proof. Let the angles at A be labeled as in Figure 6.23. Let C be the foot of a perpendicular from A to l. By the definition of a parabola, $AF = AC$. Because m is perpendicular to the angle bisector of $\angle FAX$, $\angle 2 \cong \angle 5$. In addition, $\angle 1$ and $\angle 5$ are congruent vertical angles. Thus, $\angle 2 \cong \angle 1$, so m is the perpendicular bisector of $\overline{FC}$. Then every point on m is equidistant from F and C.

Let $Y \neq A$ be a point on m. Let D be the foot of the perpendicular from Y to l. Since YD is the distance from Y to l, $YD < YC = YF$. By definition, Y is in the exterior of $\mathcal{P}$. Thus, m is the tangent to $\mathcal{P}$ at A. $\square$

Corollary 6.4. *Let A be a point on a parabola $\mathcal{P}(F, l)$, m the tangent to $\mathcal{P}$ at A, and ray AX parallel to the axis of $\mathcal{P}$ in the interior of $\mathcal{P}$. Then $\overrightarrow{AF}$ and $\overrightarrow{AX}$ form congruent angles with m. (In Figure 6.23, this asserts that $\angle 2 \cong \angle 5$.)*

Corollary 6.4 is often stated as the reflection property of a parabola, and it also has many practical applications. If a light source is placed at the focus of a rotational paraboloid,[13] the light will be

---

[12] A **rotational ellipsoid** is defined to be the surface obtained by rotating an ellipse around its major axis.

[13] A **rotational paraboloid** is defined to be the surface obtained by rotating a parabola around its axis.

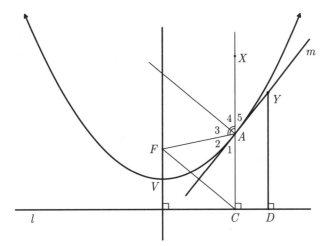

FIGURE 6.23.

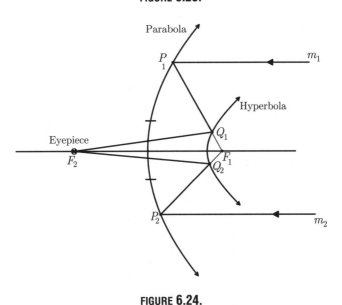

FIGURE 6.24.

reflected off the inner surface as rays that are parallel to the axis. This principle is used in car headlights and flashlights.

The principle can be used in reverse as well: a set of incoming rays all parallel to the line of symmetry will be reflected off the surface of the parabola and pass through the focus. An excellent practical use of this occurs with the Cassegrain[14] telescope. It uses the reflective properties of both the parabola and hyperbola to focus incoming parallel rays to an eyepiece. The cross section is shown in Figure 6.24.

The curve on the right forms the right branch of a hyperbola with foci F_1 and F_2. The curve on the left represents the primary mirror, which is shaped in the form of a parabola with focus at F_1. Incoming light rays m_1 and m_2 are reflected off the mirrored surface of the parabola towards the

---

[14] Laurent Cassegrain (1629–1693). At the time of his death, Cassegrain was working as a teacher giving science classes at the Collège de Chartres, a French lycée (that is, a high school-like institution). See [**68**].

focus F_1. Before they reach F_1, however, they are reflected off of the hyperbolic shaped mirror at the respective points Q_1 and Q_2. We leave it to the reader to prove (see Problems 6.10 and 6.11) that the reflected light rays will now meet to form an image at F_2, where the eyepiece is located. Note that a gap must be placed in the center of the parabolic mirror to allow the light to proceed unimpeded to the eyepiece.

The main advantages of the design of this telescope are the compactness of the tube length, which is due to "folding the optics," and having the focus F_2 removed from the tube. The latter allows easy placement of a camera or other instrument of F_2. Probably the most common type of serious amateur astronomical telescope is a modification of this telescope, called the Schmidt-Cassegrain telescope, which makes use of an additional corrector plate to eliminate chromatic abberation. (See [3] or http://www.astro-tom.com/telescopes/telescopes.htm.)

6.4 Problems

The best way to escape from a problem is to solve it.
— Alan Saporta (1856–1925)

The worst thing you can do to a problem is solve it completely.
— Daniel Kleitman (1934–)

6.1 Science museums often contain an elliptical whispering room ("top half" of a rotational ellipsoid) with two designated locations across the room from each other. One can whisper at one location and the sound will be heard clearly at the other. What two properties of the rotational ellipsoid make this possible?

6.2 Consider a ladder standing vertically against a building. Find the curve traced by the midpoint of the ladder as the base is pulled away from the building.

6.3 Consider four linked "rods," $\overline{AB}$, $\overline{BC}$, $\overline{CD}$, and $\overline{DA}$, such that $\overline{AB}$ intersects $\overline{CD}$ and $AB = CD > BC = AD$. Let F be the intersection of $\overleftrightarrow{AD}$ and $\overleftrightarrow{BC}$. If the placement of $\overline{AB}$ is fixed while the placements of the other rods can vary, show that all such points F lie on a fixed hyperbola or on the axis of symmetry for that hyperbola.

6.4 Fix a segment $\overline{AB}$ and for a given point C, consider the parallelogram $ABCD$. Determine the locus of F as C varies, where F is the base of the perpendicular from B to $\overleftrightarrow{AD}$.

6.5 For the following problems, let l_1 be parallel to l_2 and let $k > 0$ be a real number.

  (a) Find the locus of all points X such that $d(X, l_1) + d(X, l_2) = k$.
  (b) Find the locus of all points X such that $|d(X, l_1) - d(X, l_2)| = k$.
  (c) Find the locus of all points X such that $\dfrac{d(X, l_1)}{d(X, l_2)} = k$.
  (d) Find the locus of all points X such that $d(X, l_1) \cdot d(X, l_2) = k$.

6.6 For the following problems, let l_1 intersect l_2 at O, and let $k > 0$ be a real number.

  (a) Find the locus of all points X such that $d(X, l_1) + d(X, l_2) = k$.
  (b) Find the locus of all points X such that $|d(X, l_1) - d(X, l_2)| = k$.
  (c) Find the locus of all points X such that $\dfrac{d(X, l_1)}{d(X, l_2)} = k$.
  (d) Find the locus of all points X such that $d(X, l_1) \cdot d(X, l_2) = k$.

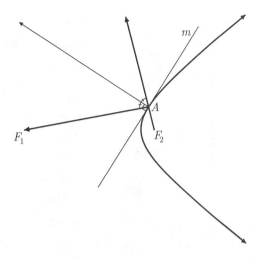

FIGURE **6.25.**

6.7 Let A and C be two fixed points on a circle, $\mathcal{C}$. Let point B move along the circle. Find the locus of all points X where X is

(a) the incenter of $\triangle ABC$,
(b) the circumcenter of $\triangle ABC$,
(c) the centroid of $\triangle ABC$,
(d) the orthocenter of $\triangle ABC$.

6.8 Let two lines, l_1 and l_2, be given. Find the locus of the midpoints of all segments with exactly one endpoint on each line.

6.9 Recall that for an ellipse, we accepted without proof that the interior of the ellipse contained the foci. In this problem we will establish a similar property for the hyperbola. Assume that the hyperbola $\mathcal{H} = \mathcal{H}(F_1, F_2; d)$ divides the plane into three regions and prove that the region of the plane bounded by $\mathcal{H}$ and not containing either focus is the exterior of $\mathcal{H}$ and the remaining two regions of the plane form the interior.

6.10 In this problem we will discuss a reflection property of the hyperbola that is analogous to that of the ellipse. For additional information about the interior and exterior of a hyperbola, see the statement of Problem 6.9. Let $\mathcal{H}(F_1, F_2; d)$ be a hyperbola and let A be a point on $\mathcal{H}$. We define a **tangent** to $\mathcal{H}$ at A to be a line intersecting $\mathcal{H}$ at A but containing no interior points of $\mathcal{H}$. Suppose A is on the branch of $\mathcal{H}$ closest to F_2. Let m be perpendicular to the bisector of the angle made by $\overrightarrow{F_2A}$ and $\overrightarrow{AF_1}$ at A. (See Figure 6.25.) Show that m is the tangent to $\mathcal{H}$ at A.

6.11 Use the result of the previous problem to state a reflective property for the hyperbola.

6.12 Let A and B be two fixed points on a circle. Let C be an arbitrary point on the circle, $C \neq A$, and find the locus of points X on the *line AC* such that $CX = BC$.

6.13 Let points A and B and a positive number $k \neq 1$ be given. Find the locus of all points X such that $XA/XB = k$.

6.14 Let perpendicular rays CA and CB be given. Consider all placements of a rigid right triangle XZY with right angle at Z, such that X is on $\overrightarrow{CA}$, Y is on $\overrightarrow{CB}$, and C and Z are on opposite sides of $\overleftrightarrow{XY}$. (See Figure 6.26.) Find the locus of all such points Z.

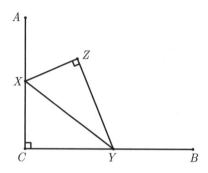

FIGURE 6.26.

6.15 Given a circle, $C = C(O, r)$, and a point $A \neq O$ in the interior of C, consider an arbitrary chord, through A but not through O, and the two tangents to C at the endpoints of the chord. Let X be the intersection of the tangents. Find the locus of all such points X. (One such point X is depicted in Figure 6.27.)

6.16 Let A and B be two points on an ellipse with foci F_1 and F_2. Let the tangents to the ellipse at A and B meet at point S. Prove that $\angle A F_1 S \cong \angle S F_1 B$. In words, the point of intersection of two tangents to an ellipse lies on the bisector of the angle formed by joining a focus to the points of tangency.

6.5   Supplemental Problems

S6.1 Given two points A and B, draw an arbitrary line l through A and construct the line through B perpendicular to l. Let D be the intersection of the two lines. Find the locus of all such points D.

S6.2 Fix a line m and a point K not on m. Let X be the center of a circle through K and tangent to m. Find the locus of all points X.

S6.3 Determine the locus of the centers of the circles that are tangent to each of two given circles for the three cases not considered in Example 12.

S6.4 Let l_1 be parallel to l_2 and let $k > 0$ be a real number. Find the locus of all points X such that $\frac{d(X, l_1)}{d(X, l_2)} = k$.

S6.5 Let P lie in the exterior of a circle C. Let X be the midpoint of a chord formed by the intersection of C and a line through P. Find the locus of all such points X.

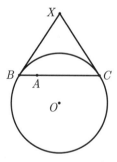

FIGURE 6.27.

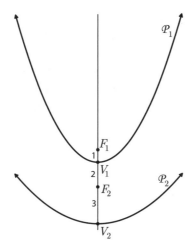

FIGURE **6.28.**

*S*6.6 Fix two points A and B in the plane. Describe the set of all points that are obtained by reflecting A about lines through B.

*S*6.7 Consider parabolas $\mathcal{P}_1$ and $\mathcal{P}_2$ with respective foci F_1 and F_2 and vertices V_1 and V_2. Suppose further that $F_1 - V_1 - F_2 - V_2$, with respective distances between these points of 1, 2, and 3, as shown in Figure 6.28. Show that the parabolas are similar by defining a homothety from one to the other. Give the center and coefficient of homothety. The reader is invited to explore the question in greater generality for $V_1 F_2 > 0$ and $F_2 V_2 > 1$.

*S*6.8 Fix a circle $\mathcal{C} = C(O, r)$ and a point K not on $\mathcal{C}$. Find the locus of all points X that are centers of circles through K and tangent to $\mathcal{C}$. Consider the cases where K is an interior and exterior point of $\mathcal{C}$ separately.

*S*6.9 Given two distinct points A and B, find the locus of points X for which $AX^2 - BX^2$ is constant.

*S*6.10 Let line l_1 intersect line l_2 at O, and let $k > 0$. Find the locus of all points X such that $|d(X, l_1) - d(X, l_2)| = k$.

*S*6.11 Let A and C be two fixed points on a circle, $\mathcal{C}$. Let point B move along the circle. Find the locus of all points X where X is the centroid of $\triangle ABC$.

*S*6.12 Let A and B be distinct points on a parabola $\mathcal{P}(F, l)$, with tangent lines AS and BS meeting at some point S. Let A' and B' be the feet of perpendiculars from A and B to l. Prove the following.

   (a) S is the circumcenter of $\triangle A'B'F$.
   (b) $\angle FAS \cong \angle FA'B' \cong \angle FSB$.
   (c) $\angle FBS \cong \angle FB'A' \cong \angle FSA$.
   (d) Triangles BFS, SFA, and $B'FA'$ are similar.
   (e) F lies on $\overline{AB}$ iff S lies on $\overline{A'B'}$ and, in this case, $m\angle ASB = 90°$.

*S*6.13 Any three tangent lines to a parabola form a triangle. Show that the circumcircle of this triangle passes through the focus of the parabola. Hint: Use Problem S6.12 twice.

S6.14 Assume that the hyperbola $\mathcal{H} = \mathcal{H}(F_1, F_2; d)$ divides the plane into three regions. Prove that the region of the plane bounded by $\mathcal{H}$ and not containing either focus is the exterior of $\mathcal{H}$ and the remaining two regions of the plane form the interior.

S6.15 Let a circle $\mathcal{C} = \mathcal{C}(O, r)$, and a point B inside of $\mathcal{C}$ be fixed. For an arbitrary point B on $\mathcal{C}$, construct the perpendicular bisector of $\overline{AC}$. Prove that the union of these bisectors is an ellipse and its exterior. (Notice the similarity between this problem and Example 13.)

7

Trigonometry

Knowledge is derived from the shadow. The shadow is derived from the gnomon. And the combination of right angle with numbers is what guides and rules the ten thousand things.

— Shang Gao[1]

In this chapter we present a basic discussion of trigonometry. We will briefly mention some of its earliest applications before moving to standard definitions and notation. While we suspect most readers have encountered trigonometry before, we still provide some elementary examples and problems for review. The ultimate goal, of course, is to apply trigonometric methods to solve problems in Euclidean plane geometry.

7.1   A Short History

The word trigonometry, coined in 1595 with Bartholomew Pitiscus's *Trigonometria*, comes from the Greek words *trigonon* and *metron*, meaning *triangle* and *measure*. The very earliest societies used geometry to determine the time of day, the day of the year, the heights of tall objects, etc., by observing shadows cast by vertical objects. Many of these geometric measurements involved the use of a shadow-casting pole called the *gnomon*, which originated with the Chaldean astronomers of Babylon as early as 1700 B.C. and was later brought to Greece. The simple notion of similar right triangles had great effect not just on geometry and mathematics as we know it now, but on related disciplines like surveying and astronomy.

Consider how one might measure a tall object like a tree using simple tools and similar triangles. First, measure a pole and place it vertically in the ground on a sunny day. Then simply wait until the length of the shadow cast by the pole is equal to the length of the pole, and measure the length of the shadow cast by the tree. (See Figure 7.1.) We assume that the rays from the sun are parallel and that the ground is flat. Thus, at any given moment, the triangle formed by the tree and its shadow is similar to the triangle formed by the pole and its shadow. As the latter triangle is isosceles, so is the former. Of course, with a moment's thought we realize that the same task can be accomplished

---

[1] From *Science and Civilization in China* by Joseph Needham, Vol. 3, Cambridge University Press, 1959, as cited in [57], page 33.

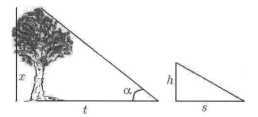

FIGURE 7.1.

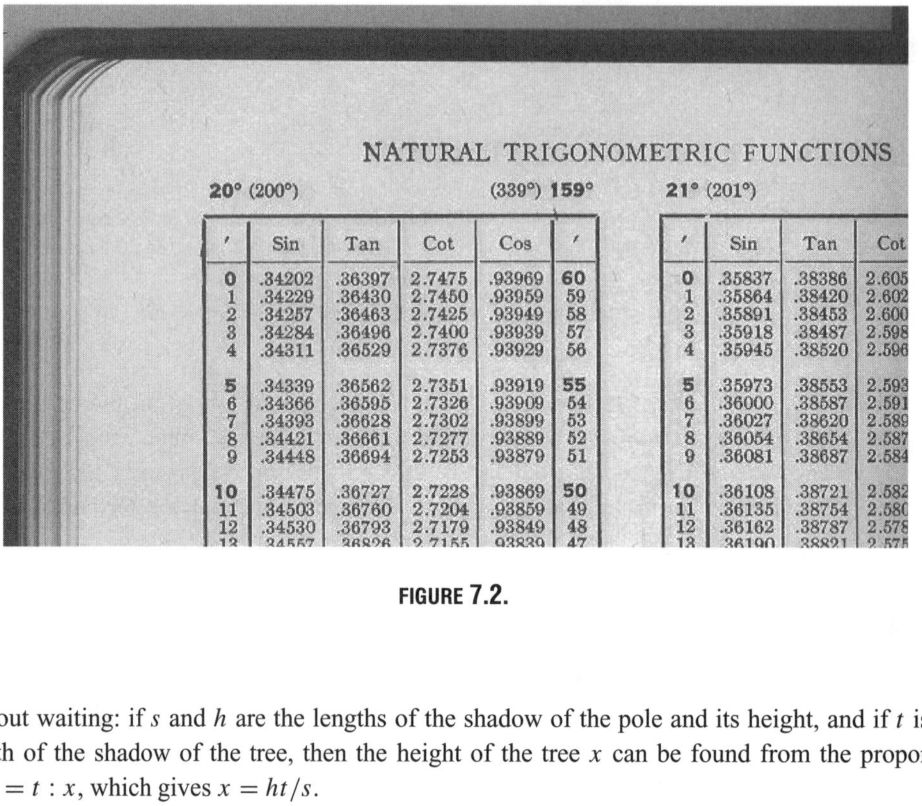

NATURAL TRIGONOMETRIC FUNCTIONS

20° (200°)   (339°) 159°   21° (201°)

′	Sin	Tan	Cot	Cos	′		′	Sin	Tan	Cot
0	.34202	.36397	2.7475	.93969	60		0	.35837	.38386	2.605
1	.34229	.36430	2.7450	.93959	59		1	.35864	.38420	2.602
2	.34257	.36463	2.7425	.93949	58		2	.35891	.38453	2.600
3	.34284	.36496	2.7400	.93939	57		3	.35918	.38487	2.598
4	.34311	.36529	2.7376	.93929	56		4	.35945	.38520	2.596
5	.34339	.36562	2.7351	.93919	55		5	.35973	.38553	2.593
6	.34366	.36595	2.7326	.93909	54		6	.36000	.38587	2.591
7	.34393	.36628	2.7302	.93899	53		7	.36027	.38620	2.589
8	.34421	.36661	2.7277	.93889	52		8	.36054	.38654	2.587
9	.34448	.36694	2.7253	.93879	51		9	.36081	.38687	2.584
10	.34475	.36727	2.7228	.93869	50		10	.36108	.38721	2.582
11	.34503	.36760	2.7204	.93859	49		11	.36135	.38754	2.580
12	.34530	.36793	2.7179	.93849	48		12	.36162	.38787	2.578
13	.34557	.36826	2.7155	.93839	47		13	.36190	.38821	2.575

FIGURE 7.2.

without waiting: if s and h are the lengths of the shadow of the pole and its height, and if t is the length of the shadow of the tree, then the height of the tree x can be found from the proportion $s : h = t : x$, which gives $x = ht/s$.

Thus, very early in history people realized that in a right triangle, the ratios of sides (s/h and t/x in Figure 7.1) did not depend on the size of the triangle, but only on the angles in the triangle. Eventually, these ratios were tabulated for different angles α, mostly by careful measurements. For example, for $\alpha = 20°$, the tables had the modern equivalent[2] (likely with less precision) of the number 0.36397 for the ratio of h/s and 0.34202 for the ratio of h to the length of the hypotenuse. See Figure 7.2 for a copy of a portion of a "modern" table, taken from the 12th edition of the C.R.C. Standard Mathematical Tables reference book, ©1959. Notice that .36397 appears as the approximation of the tangent of 20.0° in the first row and second column of the table.

---

[2] Surprisingly, it took quite some time for these tables to reach our modern tabular form containing ratios. The Greek tables consisted of lengths of chords in a circle and the Muslim/Indian tables contained lengths in geometric diagrams. Tables giving ratios of side lengths for a given angle may not have appeared until as late as the sixteenth century. ([62])

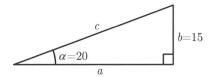

FIGURE 7.3.

FIGURE 7.4. Astrolabe.

Then if, as in Figure 7.3, we have a right triangle with a 20° angle, while the leg opposite to it has length $b = 15$ feet, we can find the lengths a and c of the second leg and the hypotenuse, respectively, by simple computations:

$$\frac{15}{a} = 0.36397 \Rightarrow a \approx 41.1 \text{ and } \frac{15}{c} = 0.34202 \Rightarrow c \approx 43.9.$$

The advantage of this method in practice was based on the fact that measuring angles was often much easier than measuring large lengths. Quite early in history people invented tools for measuring angles, and they perfected them through the centuries. Figure 7.4 contains a picture[3] of the front side of an *astrolabe*. The back of this ancient astronomical device was used for solving problems related to time and the position of the Sun and stars in the sky.

As geometers worked with these ratios, they noticed relationships between them and recognized that some ratios could be computed theoretically from others. They gave the ratios names, and their study began to form a "dialect" in the growing language of geometry. Moreover, what originally began as a study of right triangles assumed a much broader role as time passed. We now present the subject using modern terminology.

7.2 Definitions and Properties

Given two similar triangles, we have established that the ratios of corresponding sides are equal and determined by the angles in the triangle. For right triangles, these ratios were given names, which became the key terms in the language of trigonometry. For our purposes, we only discuss three of

---

[3] This image is provided courtesy of the Adler Planetarium and Astronomy Museum, Chicago, IL. The astrolabe is the Adler W-264, maker possibly Jean Fusoris, c. 1400, Paris, France.

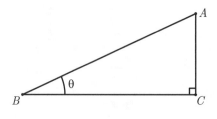

FIGURE 7.5.

these ratios: sine (sin), cosine (cos), and tangent (tan). Consider the right triangle ABC shown in Figure 7.5, where C is at the right angle and $m\angle ABC = \theta$. The sine, cosine, and tangent are defined as follows:

$$\sin\theta = \frac{AC}{AB}, \quad \cos\theta = \frac{BC}{AB}, \quad \tan\theta = \frac{AC}{BC}.$$

Suppose θ is measured in degrees. Obviously, $\sin\theta$ and $\cos\theta$ define functions with domain $(0°, 90°)$ and range $(0, 1)$, while $\tan\theta$ defines a function with domain $(0°, 90°)$ and range $(0, \infty)$.[4] Each of these non-zero ratios has a reciprocal (the cosecant, secant, and cotangent functions, respectively), but they are unnecessary and are only useful for avoiding writing the fractions $1/\sin\alpha$, $1/\cos\alpha$, and $1/\tan\alpha$.

Several relationships between these functions arise immediately. By the Pythagorean theorem, $AC^2 + BC^2 = AB^2$, which yields the identities in (7.1) below.

$$\sin^2\theta + \cos^2\theta = 1$$
$$\tan^2\theta + 1 = \frac{1}{\cos^2\theta} \tag{7.1}$$

Also, since the two acute angles in a triangle are complementary (sum to $90°$), by applying the definitions to either such angle in a right triangle we obtain the identities in (7.2).

$$\sin(90° - \theta) = \cos\theta$$
$$\cos(90° - \theta) = \sin\theta$$
$$\tan(90° - \theta) = 1/\tan\theta \tag{7.2}$$

By considering degenerate triangles, we can extend the domains of the sine and cosine functions to $[0°, 90°]$. Specifically, $\sin 0° = \cos 90° = 0$ and $\sin 90° = \cos 0° = 1$. Moreover, $\tan 0° = 0$ while $\tan 90°$ is not defined.

Knowing the identities in (7.2) makes it sufficient to construct tables for the sine and cosine functions for $0° \leq \theta \leq 45°$ only, or alternatively, to construct a table for just one of sine or cosine for $0° \leq \theta \leq 90°$. Of course, certain values can be obtained from the properties of $30° - 60° - 90°$ and $45° - 45° - 90°$ triangles; in particular $\sin 30° = \cos 60° = 1/2$, $\sin 45° = \cos 45° = \sqrt{2}/2$, $\tan 60° = 1/\tan 30° = \sqrt{3}$, and $\tan 45° = 1$. (See Problem 7.1.)

It turns out to be very useful to extend the above definitions of trigonometric functions to all angle measures in $(90°, 180°]$, in order to study obtuse triangles. With these extended definitions

---

[4] While these functions apply to the measure of an angle, we often conceptualize them as functions of the angle itself. For ease of notation, we will utilize both understandings. In Figure 7.5, for example, we permit ourselves to write any of $\sin B$, $\sin(\angle ABC)$, $\sin(m\angle ABC)$, or $\sin(m\angle B)$ for $\sin\theta$.

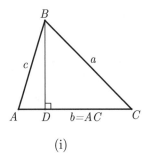
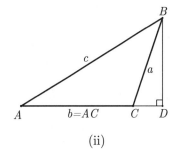

(i)   (ii)

FIGURE 7.6.

the geometric meaning of sine and cosine as ratios of side lengths of right triangles disappear. The definitions are given formally as follows: For $90° < \theta \leq 180°$,

$$\sin\theta := \sin(180° - \theta),$$
$$\cos\theta := -\cos(180° - \theta), \text{ and}$$
$$\tan\theta := -\tan(180° - \theta). \tag{7.3}$$

For example, $\sin 150° = \sin 30° = 1/2$ and $\cos 120° = -\cos 60° = -1/2$.

These extended definitions enable us to state and prove a remarkable result that supersedes the Pythagorean theorem in the sense that it applies to all triangles. The Pythagorean theorem is just a particular case of the result for right triangles.[5]

Theorem 7.1. *(Cosine Theorem) For every triangle ABC, $c^2 = a^2 + b^2 - 2ab\cos C$.*

Proof. Let $\overline{BD}$ be perpendicular to $\overleftrightarrow{AC}$, with D on $\overleftrightarrow{AC}$.

If $D = A$, then $b = a\cos C$, yielding

$$a^2 + b^2 - 2ab\cos C = a^2 + b^2 - 2b^2 = a^2 - b^2 = c^2,$$

where the last equality is due to the Pythagorean theorem. Hence, the statement is proven in this case.

If $D = C$, then $m\angle C = 90°$, so $\cos C = 0$ and the statement follows immediately from the Pythagorean theorem.

If $A - D - C$ (as in Figure 7.6(i)), then using the Pythagorean theorem twice and the fact that $CD = a\cos C$, we obtain

$$\begin{aligned}
c^2 &= BD^2 + AD^2 \\
&= BD^2 + (b - CD)^2 \\
&= (BD^2 + CD^2) + b^2 - 2b \cdot CD \\
&= a^2 + b^2 - 2ab\cos C.
\end{aligned}$$

---

[5] Despite the fact that the Pythagorean theorem is a particular case of the Cosine theorem the authors are not aware of any proof of the Cosine theorem that is independent of the Pythagorean theorem.

If $A - C - D$ (as in Figure 7.6(ii)), then $\angle C$ is obtuse, and we have

$$
\begin{aligned}
c^2 &= BD^2 + AD^2 \\
&= (a^2 - CD^2) + (CD + b)^2 \\
&= a^2 + b^2 + 2b \cdot CD.
\end{aligned}
$$

Notice that $CD = a\cos(m\angle BCD) = a\cos(180° - m\angle BCA) = -a\cos C$. Therefore, the last line simplifies to $c^2 = a^2 + b^2 - 2ab\cos C$, as desired.

Finally, if $D - A - C$, then $\angle C$ is acute, and we have

$$
\begin{aligned}
c^2 &= BD^2 + AD^2 \\
&= (a\sin C)^2 + (a\cos C - b)^2 \\
&= a^2(\sin^2 C + \cos^2 C) + b^2 - 2ab\cos C \\
&= a^2 + b^2 - 2ab\cos C.
\end{aligned}
$$

$\square$

Our ability to state and prove Theorem 7.1 for all triangles, including the obtuse ones, is based on our definition of $\cos\alpha$ as $-\cos(180° - \alpha)$ when $90° \le \alpha \le 180°$. This certainly makes a strong case for such a definition.

Another powerful theorem which can now be proved for all triangles is the Sine theorem.

Theorem 7.2. *(Sine Theorem) In a triangle ABC,*

$$
a/\sin A = b/\sin B = c/\sin C = 2R,
$$

where R is the radius of the circumcircle $\mathcal{C}$ of $\triangle ABC$.

Proof. If $\triangle ABC$ is a right triangle, then the hypotenuse is a diameter of $\mathcal{C}$ and the statement is obvious.

Suppose $\triangle ABC$ is not a right triangle. Let $\overline{CD}$ be a diameter of $\mathcal{C}$. (Clearly $D \ne A$ and $D \ne B$.) Then $\triangle ACD$ is a right triangle with the right angle at A.

Now we consider two cases, based on whether $\angle B$ is acute or obtuse. If $\angle B$ is acute, then $\angle B \cong \angle D$ as inscribed angles sharing the same arc. (See Figure 7.7.) Thus,

$$
\sin B = \sin D = \frac{b}{2R} \Rightarrow \frac{b}{\sin B} = 2R.
$$

If $\angle B$ is obtuse, then by Theorem 4.12,

$$
m\angle B = (1/2)(m(\widehat{ADC})) = (1/2)(360° - m(\widehat{ABC})) = 180° - m\angle D.
$$

Hence,

$$
\sin B = \sin(180° - m\angle D) = \sin D = \frac{b}{2R} \Rightarrow \frac{b}{\sin B} = 2R.
$$

Showing $a/\sin A = 2R$ is completely similar. In order to show that $c/\sin C = 2R$, simply consider diameter $\overline{AD}$ rather than $\overline{CD}$. $\square$

An alternate proof of the Sine theorem uses the fact that the area of $\triangle ABC$ can be expressed as

$$
\frac{1}{2}ab \cdot \sin C = \frac{1}{2}ac \cdot \sin B = \frac{1}{2}bc \cdot \sin A.
$$

We leave the details of this proof to Problem 7.10.

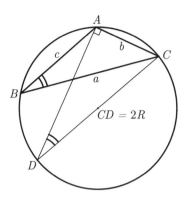

FIGURE 7.7.

Together, the Cosine theorem and the Sine theorem allow us to "solve a triangle" by determining its unknown angles and sides from known ones. Given the measure of an angle α, we can find (or at least estimate) $\sin\alpha$, $\cos\alpha$, or $\tan\alpha$ by using tables or a calculator. Conversely, knowing $\sin\alpha$, $\cos\alpha$, or $\tan\alpha$ allows us to determine possible values of α.

Example 15. Find the measures of all angles and the lengths of all sides in a triangle ABC where $AB = 10$, $BC = 8$, and $m\angle CAB = 40°$.

Solution: From the Sine theorem, $8/\sin 40° = 10/\sin C$; therefore,

$$\sin C = (5/4)\sin 40° \approx (1.25)(0.6428) \approx 0.8035.$$

At this point we notice there are two possible (supplementary) angles satisfying $\sin C \approx 0.8035$. If C is acute, $m\angle C \approx 53.46°$; if C is obtuse, $m\angle C \approx 126.54°$.

Suppose C is acute. Then we find that $m\angle B = 180° - m\angle A - m\angle C \approx 86.54°$. Now, since we know two sides and an angle, we could use the Cosine theorem to find $b = AC$. However, we also know all three angle measures, so we can use the Sine theorem (which involves fewer calculations). To wit,

$$\frac{10}{\sin C} = \frac{b}{\sin B} \text{ gives } b \approx \frac{10\sin(86.54°)}{\sin(53.46°)} \approx 12.42.$$

Similarly, when C is obtuse, we find that

$$m\angle B \approx 13.46° \text{ and } b \approx \frac{10\sin(13.46°)}{\sin(126.54°)} \approx 2.90.$$

$\square$

This example illustrates why we do not have an SSA law for congruent triangles, as there may exist two triangles with the specified dimensions.

There are other trigonometric identities besides those in (7.1), (7.2), and (7.3). In particular, it is convenient for many purposes to have angle addition and subtraction formulas for sine and cosine; namely, we seek expressions for $\sin(\alpha \pm \beta)$ and $\cos(\alpha \pm \beta)$ in terms of the sine and cosine of α and β. The angle addition and subtraction formulas were used by sixth century Hindu astronomers in creating a table of sines (derived using Ptolemy's theorem and lengths of specific chords in a unit circle). ([**18**], pp. 192 and 221.) At this point we restrict ourselves to nonobtuse angles, although the

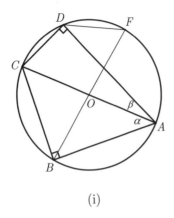
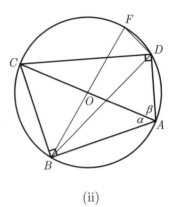

(i)   (ii)

FIGURE 7.8.

identities actually hold for all values of α and β. The extensions can be seen more readily when the trigonometric functions are defined for arbitrary angles, which we will do in Section 7.2.

Theorem 7.3. *(Angle Sum Formula for Sine) For nonobtuse angles α and β, $\alpha > \beta$,*

$$\sin(\alpha \pm \beta) = \sin\alpha\cos\beta \pm \sin\beta\cos\alpha.$$

Proof. If $\alpha = 90°$, then $\sin(90° + \beta) = \sin(180° - (90° - \beta)) = \sin(90° - \beta) = \cos\beta$. Since $\sin 90° = 1$ and $\cos 90° = 0$, the result follows. A similar calculation shows that the result holds when $\beta = 90°$.

We will prove the addition formula here for acute angles α and β, and leave the subtraction formula to the problem set. To this end, note that it is possible to inscribe a quadrilateral $ABCD$ in a circle of diameter 1, where $\overline{AC}$ is a diameter, $m\angle BAC = \alpha$, and $m\angle CAD = \beta$. Then both $\angle ADC$ and $\angle ABC$ are right angles. It follows that $AB = \cos\alpha$, $BC = \sin\alpha$, $CD = \sin\beta$, $AD = \cos\beta$, and $AC = 1$. We then claim that $BD = \sin(\alpha + \beta)$. We consider three cases, depending on the measure of $\angle BAD$.

Suppose first that $\alpha + \beta < 90°$ (as in Figure 7.8(i)), and consider $\triangle BDF$, where $\overline{BF}$ is a diameter of the circle. Clearly $m\angle BFD = \alpha + \beta$, since the angle intercepts the arc BD, just as angle BAD does, which has measure $\alpha + \beta$. Since $\overline{BF}$ is a diameter, $\angle BDF$ is a right angle, in which case $BD = \sin(\alpha + \beta)$. Applying Ptolemy's theorem (Theorem 4.23) to quadrilateral $ABCD$:

$$AD \cdot BC + AB \cdot CD = AC \cdot BD,$$

which, by substitution, yields the desired formula.

To handle the situation where $\alpha + \beta > 90°$ (as shown in Figure 7.8(ii)), we simply appeal to our definition of sine for obtuse angles. Note that $\angle BFD$ and $\angle BAD$ are supplementary, so $BD = \sin(\angle BFD) = \sin(\angle BAD) = \sin(\alpha + \beta)$, as in the first case. Applying Ptolemy's theorem as before gives the result.

We leave the case where $\alpha + \beta = 90°$ to Problem 7.8. $\qquad\qquad\qquad\qquad\qquad\square$

To prove the formula for $\sin(\alpha - \beta)$, one can consider two cases: $\alpha \geq \beta$ and $\alpha < \beta$. In the first case the proof is similar to the one above. Consider $ABCD$ with $\overline{AD}$ being a diameter, $AD = 1$,

$m\angle BAD = \alpha$, and $m\angle CAD = \beta$. Then $AB = \cos\alpha$, $CD = \sin\beta$, $AC = \cos\beta$, $BD = \cos\alpha$, and $BC = \sin(\alpha - \beta)$. (The expression for BC becomes obvious after introducing diameter BK and chord CK.) The second case can be reduced to the first one by using the properties $\sin(-x) = -\sin x$ and $\cos(-x) = \cos x$. (See also Problem 7.29.)

The proof above is based on Ptolemy's theorem for an inscribed quadrilateral. For an area proof similar to a famous proof of the Pythagorean theorem, see [48].

Theorem 7.4. *(Angle Sum Formula for Cosine) For nonobtuse angles α and β, $\alpha > \beta$,*

$$\cos(\alpha \pm \beta) = \cos\alpha\cos\beta \mp \sin\alpha\sin\beta.$$

Proof. The proof is left to Problem S7.9. $\qquad\qquad\qquad\qquad\qquad\qquad\qquad\qquad\qquad$ $\square$

7.3 Unit Circle Trigonometry

We now move beyond the early right triangle definitions of trigonometric functions to the modern ones utilizing the unit circle. We will extend the concept of an angle beyond that of its appearance in a triangle to the kinematic notion of an angle as a rotation. A motivation for this change in definition came from mechanics, where negative angles were introduced to distinguish the direction of rotation and angles of arbitrary size were utilized to describe the motion of a point around a circle. This conceptual shift on how we view angles will allow us to define the trigonometric functions on all real numbers, thereby extending the identities we have seen.

We assume here a familiarity with the rectangular coordinate system. For a review, the reader is invited to read the beginning pages of Chapter 8, where the Cartesian plane is introduced.

Consider the unit circle $\mathcal{C}$, having radius 1 and center at the origin O of the Cartesian plane. Define the point P_t on the unit circle to be the point obtained by rotating $P_0 : (1, 0)$ counterclockwise by angle t.[6] Then, we define the **cosine** of t and the **sine** of t to be the $x-$ and $y-$coordinates of P_t, respectively. (See Figure 7.9(i).) For $0° < t < 90°$, it is clear that these definitions match those from our original definition utilizing a right triangle with hypotenuse length 1. Furthermore, the symmetry evident in Figure 7.9(ii) shows that for $0° \leq t \leq 180°$, these definitions match those we gave in (7.3). Therefore, we retain the notation $\sin t$ and $\cos t$. Notice also that the Pythagorean theorem gives the identity $\sin^2 t + \cos^2 t = |\sin t|^2 + |\cos t|^2 = 1$.

Based on signs of the $x-$ and $y-$coordinates of P_t, the sine function will be positive in the first and second quadrants while the cosine function will be positive in the first and fourth quadrants. Because points P_t and P_{-t} are symmetric with respect to the x-axis, we have $\sin(-t) = -\sin t$ and $\cos(-t) = \cos t$. Furthermore, since $P_t = P_{t+360k}$, $\sin t = \sin(t + 360k)$ and $\cos t = \cos(t + 360k)$ for any integer k, which illustrates the periodic nature of the trigonometric functions.

As long as $\cos t \neq 0$, we define the **tangent** of t as $\tan t = \sin t / \cos t$. Note that for $0° < t < 90°$, this definition again coincides with the one coming from a right triangle. Like sine and cosine, the tangent of t can naturally be visualized as the coordinate of a point, in the following way. For fixed t, from similar triangles we see that $\tan t$ is the $y-$coordinate of the point of intersection of $\overleftrightarrow{OP_t}$ with the line $x = 1$. This demonstrates that the tangent function is positive when P_t lies in the first or third quadrant of the plane and is negative when P_t lies in the second or fourth quadrant.

---

[6] The switch from θ to t for the angle is due to the frequent use of the letter t to represent time.

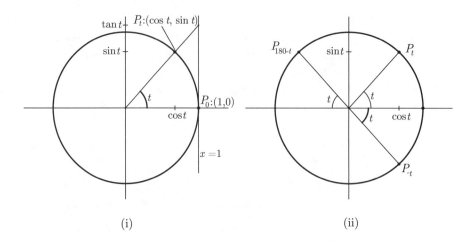

(i)   (ii)

FIGURE 7.9.

Invariably, one will need to solve equations (for x) of the form

$$\sin x = a, \quad \cos x = a, \quad \text{or} \quad \tan x = a,$$

where a is a given real number. Due to the periodic nature of the functions, if any of these equations has a solution, it has infinitely many solutions. It is clear that the first two equations may have solutions only if $|a| \le 1$. In order to refer to some "standard" solutions, we restrict values of x to a small interval of reals (depending on the equation) so that each equation has a unique solution inside the corresponding interval. From their unit circle definitions, it is apparent that sine increases from -1 to 1 on the interval $[-90°, 90°]$, that cosine decreases from 1 to -1 on the interval $[0°, 180°]$, and that tangent increases from $-\infty$ to $+\infty$ on the interval $(-90°, 90°)$. Since these functions are continuous on the corresponding intervals (which we take here for granted), the equations $\sin x = a$ and $\cos x = a$ have unique solutions in these intervals for all real a, $|a| \le 1$, and $\tan x = a$ has a unique solution in the interval $(-90°, 90°)$ for all real a.

The solution of $\sin x = a$ in $[-90°, 90°]$ is called **arcsine** of a, and is denoted by $\arcsin a$.
The solution of $\cos x = a$ in $[0°, 180°]$ is called **arccosine** of a, and is denoted by $\arccos a$.
The solution of $\tan x = a$ in $(-90°, 90°)$ is called **arctangent** of a, and is denoted by $\arctan a$.

7.4   Summary of Identities

We collect here the following identities that we have established.

$$\sin^2 \theta + \cos^2 \theta = 1 \qquad\qquad\qquad \tan^2 \theta + 1 = \frac{1}{\cos^2 \theta}$$

$$\sin(-\theta) = -\sin \theta \qquad\qquad\qquad\quad \cos(-\theta) = \cos \theta$$

$$\sin(180° \pm \theta) = \mp \sin \theta \qquad\qquad \cos(180° \pm \theta) = -\cos \theta$$

$$\sin(90° - \theta) = \cos \theta \qquad\qquad\quad \cos(90° - \theta) = \sin \theta$$

$$\sin(\alpha \pm \beta) = \sin \alpha \cos \beta \pm \sin \beta \cos \alpha \qquad \cos(\alpha \pm \beta) = \cos \alpha \cos \beta \mp \sin \alpha \sin \beta$$

$$\tan(180° \pm \theta) = \pm \tan \theta \qquad\qquad \tan(90° - \theta) = 1/\tan \theta$$

$$\tan(-\theta) = -\tan \theta$$

The following identity is a new one, obtained by dividing $\sin(\alpha \pm \beta)$ by $\cos(\alpha \pm \beta)$ and simplifying.

$$\tan(\alpha \pm \beta) = \frac{\tan \alpha \pm \tan \beta}{1 \mp \tan \alpha \tan \beta} \quad (\alpha, \beta, \alpha \pm \beta \neq 90° \cdot (2k+1) \text{ for any integer k}).$$

From the identities given so far and by using the properties of the trigonometric functions, one could compile a seemingly endless collection of identities, virtually all of which follow from $\sin^2 \theta + \cos^2 \theta = 1$ and the angle sum formulas. We don't intend to do this! However, the next section of this chapter demonstrates how geometric problems can sometimes be reduced to problems of trigonometry, where solving a trigonometric equation or proving a trigonometric identity provides a solution to the geometric problem. To facilitate this, we now derive and list just a few additional trigonometric identities that will be useful.

The so-called "double-angle formulas" for sine and cosine follow quickly from the respective angle sum formulas:

$$\sin 2\alpha = \sin(\alpha + \alpha) = \sin \alpha \cos \alpha + \sin \alpha \cos \alpha = 2 \sin \alpha \cos \alpha \tag{7.4}$$

$$\cos 2\alpha = \cos(\alpha + \alpha) = \cos \alpha \cos \alpha - \sin \alpha \sin \alpha = \cos^2 \alpha - \sin^2 \alpha \tag{7.5}$$

$$\tan 2\alpha = \frac{2 \tan \alpha}{1 - \tan^2 \alpha}. \tag{7.6}$$

Similar formulas for "triple angles" (and beyond) can be found recursively, as demonstrated below. Interestingly, both the sine triple-angle formula and the derivation of $\sin 18°$ in Example 19 arose in the construction of trigonometric tables in Islam and India. ([**62**])

$$\begin{aligned}
\sin 3\alpha &= \sin(2\alpha + \alpha) \\
&= \sin(2\alpha) \cos \alpha + \sin \alpha \cos(2\alpha) \\
&= 2 \sin \alpha \cos^2 \alpha + \sin \alpha(\cos^2 \alpha - \sin^2 \alpha) \\
&= 3 \sin \alpha \cos^2 \alpha - \sin^3 \alpha \\
&= 3 \sin \alpha(1 - \sin^2 \alpha) - \sin^3 \alpha \\
&= 3 \sin \alpha - 4 \sin^3 \alpha.
\end{aligned}$$

Likewise, we can derive the following identities (see Problem 7.6):

$$\cos 3\alpha = 4 \cos^3 \alpha - 3 \cos \alpha \quad \text{and} \quad \tan 3\alpha = \frac{3 \tan \alpha - \tan^3 \alpha}{1 - 3 \tan^2 \alpha}.$$

Example 16. Suppose $\sin \alpha = 1/3$, where $0° < \alpha < 90°$. Find $\cos \alpha$, $\sin 2\alpha$, and $\cos 3\alpha$.

Solution: Using the identity $\sin^2 \alpha + \cos^2 \alpha = 1$ and noting that $\cos \alpha > 0$ (since $0° < \alpha < 90°$), we find

$$\cos \alpha = \sqrt{1 - (1/3)^2} = 2\sqrt{2}/3.$$

Then, from the angle addition formulas,

$$\sin 2\alpha = 2 \sin \alpha \cos \alpha = (2)(1/3)(2\sqrt{2}/3) = \frac{4\sqrt{2}}{9} \quad \text{and}$$

$$\cos 3\alpha = 4 \cos^3 \alpha - 3 \cos \alpha = \frac{4 \cdot 16\sqrt{2}}{27} - 2\sqrt{2} = \frac{10\sqrt{2}}{27}.$$

$\square$

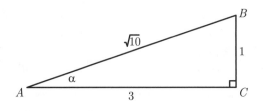

FIGURE **7.10.**

Example 17. Suppose $\tan \alpha = 1/3$. Find $\sin \alpha$ and $\cos \alpha$.

Solution: We give two solutions, one geometric and one algebraic. First, consider a right triangle ABC with leg lengths $AC = 3$ and $BC = 1$, as in Figure 7.10. By the Pythagorean theorem, $AB = \sqrt{10}$. If $0° < m\angle A = \alpha < 90°$, then $\sin \alpha = BC/AB = 1/\sqrt{10}$ and $\cos \alpha = AC/AB = 3/\sqrt{10}$. The other possibility is that $180° < \alpha < 270°$. For these values of α, the sine and cosine functions are both negative; in this case, $\sin \alpha = -1/\sqrt{10}$ and $\cos \alpha = -3/\sqrt{10}$.

   To obtain this solution algebraically, note that $\sin \alpha / \cos \alpha = 1/3$, so $\sin \alpha = (\cos \alpha)/3$. Substituting this into the identity $\sin^2 \alpha + \cos^2 \alpha = 1$ yields $10 \cos^2 \alpha = 9$, whence $\cos \alpha = \pm 3/\sqrt{10}$. The corresponding values for $\sin \alpha$ are $\pm 1/\sqrt{10}$, respectively.   □

Example 18. Suppose $\cos 2\alpha = 2/3$. Find $\tan \alpha$.

Solution: The identities $\sin^2 \alpha + \cos^2 \alpha = 1$ and $\cos 2\alpha = \cos^2 \alpha - \sin^2 \alpha$ yield

$$\cos 2\alpha = 2 \cos^2 \alpha - 1 = 1 - 2 \sin^2 \alpha.$$

From here we obtain two well-known identities, often called the "half-angle" formulas:

$$\sin^2 \alpha = \frac{1 - \cos 2\alpha}{2} \tag{7.7}$$

$$\cos^2 \alpha = \frac{1 + \cos 2\alpha}{2} \tag{7.8}$$

$$\tan^2 \alpha = \frac{1 - \cos 2\alpha}{1 + \cos 2\alpha}.$$

Substituting $\cos 2\alpha = 2/3$, we find $\tan^2 \alpha = 1/5$. As in the previous example, we have two solutions: $\tan \alpha = 1/\sqrt{5}$ and $\tan \alpha = -1/\sqrt{5}$.   □

Example 19. Show that the side length of a regular pentagon inscribed in a circle of radius 1 is $\frac{1}{2}\sqrt{10 - 2\sqrt{5}}$.

Solution: Let x be the length of each side of the pentagon; each angle of the pentagon has measure $108°$. Consider the triangle formed by one side of the pentagon and two radii. (See Figure 7.11.)

   From the Cosine theorem, $x^2 = 1^2 + 1^2 - 2 \cdot 1 \cdot 1 \cos 72°$. Then, noting that $\cos(3 \cdot 18°) = \cos 54° = \sin 36° = \sin(2 \cdot 18°)$ and using the triple angle formula for cosine and the double angle formula for sine, we obtain

$$4\cos^3 18° - 3\cos 18° = 2\sin 18° \cos 18° \iff$$
$$4\cos^2 18° - 3 = 2\sin 18° \iff$$
$$4\sin^2 18° + 2\sin 18° - 1 = 0 \iff$$
$$\sin 18° = \frac{-1 + \sqrt{5}}{4} \quad \text{(the negative root is discarded)}.$$

This gives

$$x^2 = 1^2 + 1^2 - 2 \cdot 1 \cdot 1 \cos 72° = 2(1 - \cos 72°) = 2(1 - \sin 18°) = 2\left(\frac{5 - \sqrt{5}}{4}\right),$$

so $x = \sqrt{10 - 2\sqrt{5}}/2 \approx 1.176$. $\qquad\qquad\qquad\qquad\qquad\qquad\qquad\qquad\qquad\qquad \square$

The angle sum formulas can also be manipulated to relate the sums of values of sines and cosines with the products of their (other) values. For example,

$$\sin(\alpha + \beta) + \sin(\alpha - \beta) = \sin\alpha\cos\beta + \sin\beta\cos\alpha + \sin\alpha\cos\beta - \sin\beta\cos\alpha$$
$$= 2\sin\alpha\cos\beta.$$

This formula, and others that can be obtained similarly, are listed below.

$$2\sin\alpha\cos\beta = \sin(\alpha + \beta) + \sin(\alpha - \beta) \tag{7.9}$$
$$2\sin\alpha\sin\beta = \cos(\alpha - \beta) - \cos(\alpha + \beta) \tag{7.10}$$
$$2\cos\alpha\cos\beta = \cos(\alpha + \beta) + \cos(\alpha - \beta). \tag{7.11}$$

If $\alpha + \beta = \theta$ and $\alpha - \beta = \gamma$, then

$$\alpha = \frac{\theta + \gamma}{2} \quad \text{and} \quad \beta = \frac{\theta - \gamma}{2}.$$

Substituting these expressions for θ and γ into (7.9), we get a useful equivalent version of (7.9). This formula and others like it are listed below.

$$\sin\theta + \sin\gamma = 2\sin\frac{\theta + \gamma}{2}\cos\frac{\theta - \gamma}{2} \tag{7.12}$$

$$\sin\theta - \sin\gamma = 2\cos\frac{\theta + \gamma}{2}\sin\frac{\theta - \gamma}{2} \tag{7.13}$$

$$\cos\theta + \cos\gamma = 2\cos\frac{\theta + \gamma}{2}\cos\frac{\theta - \gamma}{2} \tag{7.14}$$

$$\cos\theta - \cos\gamma = -2\sin\frac{\theta + \gamma}{2}\sin\frac{\theta - \gamma}{2} \tag{7.15}$$

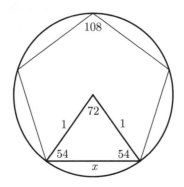

FIGURE 7.11.

Before we can use trigonometry to solve geometry problems, it will be helpful to become more fluent in manipulating trigonometric expressions. Here are several typical examples.

Example 20. Prove that

$$\frac{\sin 7\alpha - \sin 5\alpha}{\cos 7\alpha + \cos 5\alpha} = \tan \alpha.$$

Solution: Making use of (7.13) and (7.14), with $\theta = 7\alpha$ and $\gamma = 5\alpha$,

$$\frac{\sin 7\alpha - \sin 5\alpha}{\cos 7\alpha + \cos 5\alpha} = \frac{2 \cos 6\alpha \sin \alpha}{2 \cos 6\alpha \cos \alpha} = \frac{\sin \alpha}{\cos \alpha} = \tan \alpha.$$

$\square$

Example 21. Find all x in $[0°, 360°]$ such that $\cos^2 x = -\sin x$.

Solution: Letting $\phi = \sin x$ and using the identity $\sin^2 x + \cos^2 x = 1$, we arrive at $\phi^2 - \phi - 1 = 0$, which has solutions $\phi = (1 \pm \sqrt{5})/2$. But since $|\phi| \leq 1$, we need only consider $\sin x = (1 - \sqrt{5})/2 \approx -0.62$. There are two angles x in $[0°, 360°]$ satisfying this equation. If we wish to write the answer in terms of $\arcsin((1 - \sqrt{5})/2)$, we have $x = 360° + \arcsin((1 - \sqrt{5})/2)$ or $x = 180° - \arcsin((1 - \sqrt{5})/2)$. Using tables or a calculator, we obtain the corresponding approximations of $x \approx 321.8°$ and $x \approx 218.2°$. $\square$

Example 22. Prove that

$$\cos 20° \cos 40° \cos 60° \cos 80° = 1/16.$$

Solution: Using the fact that $\cos 60° = 1/2 = -\cos 120°$, and converting the product of cosines into the sum of cosines according to (7.11), we obtain:

$$\cos 20° \cos 40° \cos 60° \cos 80° = \frac{1}{2} \cos 20° \cos 40° \cos 80°$$

$$= \frac{1}{2} \cos 20° \frac{1}{2}(\cos 120° + \cos 40°)$$

$$= \frac{1}{4}\left(-\frac{1}{2} \cos 20° + \cos 20° \cos 40°\right)$$

$$= \frac{1}{4}\left(-\frac{1}{2} \cos 20° + \frac{1}{2}(\cos 20° + \cos 60°)\right)$$

$$= \frac{1}{4} \cdot \left(\frac{1}{2} \cos 60°\right) = \frac{1}{4} \cdot \frac{1}{4} = \frac{1}{16}$$

$\square$

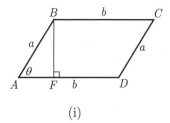

(i)

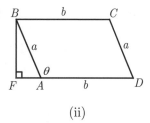

(ii)

FIGURE 7.12.

7.5 Applications

There is perhaps nothing which so occupies the middle position of mathematics as trigonometry.
— J.F. Herbart (1776–1841)[7].

Within the context of this book, trigonometry serves as a tool for solving Euclidean geometry problems. This section revisits some familiar facts and introduces some new problems, demonstrating situations in which trigonometry provides an alternate (and sometimes superior) method for obtaining a solution.

Example 23. Let $ABCD$ be a parallelogram with $AB = a$, $AD = b$, and $m\angle BAD = \theta$. Then $\text{Area}(ABCD) = ab \sin \theta$.

Solution: If $\theta = 90°$, then the parallelogram is actually a rectangle, with area $ab = ab \sin 90°$.
Let F be the base of the perpendicular from B to $\overleftrightarrow{AD}$. If $0° < \theta < 90°$, as in Figure 7.12(i), then $\sin \theta = BF/AB$, and it follows that the area of the parallelogram is

$$AD \cdot BF = AD \cdot AB \sin \theta = ab \sin \theta.$$

If $90° < \theta < 180°$, as in Figure 7.12(ii), then the same argument would yield an area of $ab \sin(180° - \theta) = ab \sin \theta.$ □

Since any triangle can be viewed as "half" of a parallelogram with the same base and height, we have the following immediate corollary.

Corollary 7.5. *The area of a triangle ABC is given by*

$$Area\,(\triangle ABC) = \frac{1}{2}ab \sin C = \frac{1}{2}bc \sin A = \frac{1}{2}ac \sin B.$$

Example 24. Show that a quadrilateral with diagonal lengths d_1 and d_2, with diagonals intersecting at an angle θ, has area $\frac{1}{2}d_1 d_2 \sin \theta$.

Solution: Suppose the point where the diagonals meet divides them so that $d_1 = a + b$ and $d_2 = c + d$, as shown in Figure 7.13.

---

[7] From the preface of [**36**].

The area A of the quadrilateral is then the sum of the areas of four triangles:

$$A = \frac{1}{2}bc\sin(180° - \theta) + \frac{1}{2}bd\sin\theta + \frac{1}{2}ad\sin(180° - \theta) + \frac{1}{2}ac\sin\theta$$

$$= \frac{1}{2}(bc + bd + ad + ac)\sin\theta$$

$$= \frac{1}{2}(a + b)(c + d)\sin\theta = \frac{d_1 d_2}{2}\sin\theta.$$

$\square$

From this example, we note that of all quadrilaterals with diagonal lengths d_1 and d_2, the ones with the greatest area will be those in which the diagonals meet at a right angle, since $\sin 90° = 1$.

Example 25. Consider a rectangle $ABCD$ sharing side $\overline{BC}$ with square $BFEC$, as shown in Figure 7.14. If $AB = 2BC$, find $m\angle BDC + m\angle FDE$. (One may also solve this problem by using complex numbers. See Problem 10.11.)

Solution: Let $\beta = m\angle BDC$ and $\alpha = m\angle FDE$, so $\tan\alpha = 1/3$ and $\tan\beta = 1/2$. Then,

$$\tan(\alpha + \beta) = \frac{\tan\alpha + \tan\beta}{1 - \tan\alpha\tan\beta} = \frac{1/3 + 1/2}{1 - 1/6} = 1.$$

Since $\tan(\alpha + \beta) = 1$ and $0° < \alpha + \beta < 180°$, we find that $m\angle BDC + m\angle FDE = \alpha + \beta = 45°$. $\square$

The following example appeared as Problem 4.28. Here we present a trigonometric solution.

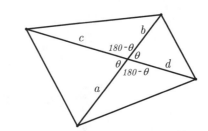

FIGURE 7.13.

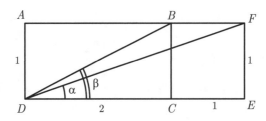

FIGURE 7.14.

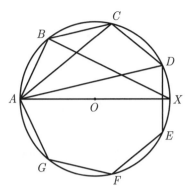

FIGURE 7.15.

Example 26. Let A, B, C, and D be four consecutive vertices of a regular 7-gon. Show that

$$\frac{1}{AB} = \frac{1}{AC} + \frac{1}{AD}.$$

Solution: Inscribe the regular 7-gon $ABCDEFG$ in a circle with center O, as shown in Figure 7.15.

Let $\overline{AX}$ be a diameter of the 7-gon, and assume that $OA = 1$. Observe that $m\angle AXB = (m\angle AOB)/2 = \pi/7$ (using radian measure). As $\angle ABX$ is a right angle and $AX = 2$, $AB = 2\sin(\pi/7)$. Similarly, $AC = 2\sin(2\pi/7)$ and $AD = 2\sin(3\pi/7)$.

Therefore, showing that $\dfrac{1}{AB} = \dfrac{1}{AC} + \dfrac{1}{AD}$ is equivalent to showing that

$$\frac{1}{2\sin\frac{\pi}{7}} = \frac{1}{2\sin\frac{2\pi}{7}} + \frac{1}{2\sin\frac{3\pi}{7}},$$

which, in turn, is equivalent to showing that

$$\sin\frac{2\pi}{7}\sin\frac{3\pi}{7} = \sin\frac{\pi}{7}\sin\frac{2\pi}{7} + \sin\frac{\pi}{7}\sin\frac{3\pi}{7}. \tag{7.16}$$

By identity (7.10),

$$\sin\frac{2\pi}{7}\sin\frac{3\pi}{7} = \frac{1}{2}\left[\cos\left(\frac{2\pi}{7} - \frac{3\pi}{7}\right) - \cos\left(\frac{2\pi}{7} + \frac{3\pi}{7}\right)\right] = \frac{1}{2}\left[\cos\left(-\frac{\pi}{7}\right) - \cos\frac{5\pi}{7}\right],$$

$$\sin\frac{\pi}{7}\sin\frac{2\pi}{7} = \frac{1}{2}\left[\cos\left(\frac{\pi}{7} - \frac{2\pi}{7}\right) - \cos\left(\frac{\pi}{7} + \frac{2\pi}{7}\right)\right] = \frac{1}{2}\left[\cos\left(-\frac{\pi}{7}\right) - \cos\frac{3\pi}{7}\right],$$

$$\sin\frac{\pi}{7}\sin\frac{3\pi}{7} = \frac{1}{2}\left[\cos\left(\frac{\pi}{7} - \frac{3\pi}{7}\right) - \cos\left(\frac{\pi}{7} + \frac{3\pi}{7}\right)\right] = \frac{1}{2}\left[\cos\left(-\frac{2\pi}{7}\right) - \cos\frac{4\pi}{7}\right].$$

Then (7.16) is equivalent to

$$\cos\left(-\frac{\pi}{7}\right) - \cos\frac{5\pi}{7} = \cos\left(-\frac{\pi}{7}\right) - \cos\frac{3\pi}{7} + \cos\left(-\frac{2\pi}{7}\right) - \cos\frac{4\pi}{7}.$$

Recalling that $\cos(-\theta) = \cos\theta$ and $\cos(\pi - \theta) = -\cos\theta$, the above equality is equivalent to

$$\cos\left(-\frac{\pi}{7}\right) + \cos\frac{2\pi}{7} = \cos\left(-\frac{\pi}{7}\right) - \cos\frac{3\pi}{7} + \cos\frac{2\pi}{7} + \cos\frac{3\pi}{7},$$

and the desired result is proven.   □

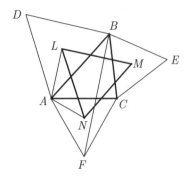

FIGURE 7.16.

Example 27. Consider $\triangle ABC$. In its exterior, construct an equilateral triangle ABD having $\overline{AB}$ as one side; construct equilateral triangles BCE and ACF similarly. (See Figure 7.16.) If L, M, and N are the centers of the equilateral triangles ABD, BCE, and ACF, respectively, then $\triangle LMN$ is equilateral.

Solution: Since triangles ADB, BEC, and CFA are equilateral, each of their interior angles has measure $60°$. The altitude, median, and angle bisector from any vertex of an equilateral triangle are concurrent.

Applying the Cosine theorem to $\triangle ALN$ yields

$$LN^2 = AL^2 + AN^2 - 2(AL)(AN)\cos(m\angle BAC + 60°). \qquad (7.17)$$

From Problem 3.2.10, $\overline{AL}$ has length equal to $2/3$ the length of the median from A in $\triangle ADB$. Since the median from A is also the altitude from A,

$$AL = \frac{2}{3}AB\cos 30° = \frac{AB}{\sqrt{3}}.$$

Similarly, $AN = AC/\sqrt{3}$. Substituting these values for AL and AN into (7.17), we get

$$LN^2 = \frac{AB^2}{3} + \frac{AC^2}{3} - 2\frac{(AB)(AC)}{3}\cos(m\angle BAC + 60°).$$

A similar expression can be obtained for NM^2:

$$NM^2 = \frac{BC^2}{3} + \frac{AC^2}{3} - 2\frac{(BC)(AC)}{3}\cos(m\angle BCA + 60°).$$

We wish to show that $LN = NM = LM$, which is equivalent to $LN^2 = NM^2 = LM^2$. Let us concentrate on $LN^2 = NM^2$, which holds if and only if

$$AB^2 + AC^2 - 2(AB)(AC)\cos(m\angle BAC + 60°)$$
$$= BC^2 + AC^2 - 2(BC)(AC)\cos(m\angle BCA + 60°),$$

which in turn holds if and only if

$$AB^2 + AF^2 - 2(AB)(AF)\cos(m\angle BAC + 60°)$$
$$= BC^2 + CF^2 - 2(BC)(CF)\cos(m\angle BCA + 60°).$$

But this last equality is true since both expressions are equal to BF^2, found by using the Cosine theorem on $\triangle ABF$ and $\triangle CBF$.

A similar argument shows that $LN^2 = LM^2$; thus, $\triangle LMN$ is equilateral, as claimed. $\square$

Our last example in this section presents a trigonometric version of Ceva's theorem (Theorem 3.17). It provides a necessary and sufficient condition for the concurrency of cevians based on the knowledge of how they divide the angles in a triangle.

Example 28. Let A_1, B_1, and C_1 be points chosen on the sides BC, AC, and AB of a triangle ABC, respectively. Then the segments AA_1, BB_1, and CC_1 are concurrent if and only if

$$\frac{\sin \angle ABB_1}{\sin \angle CBB_1} \cdot \frac{\sin \angle BCC_1}{\sin \angle ACC_1} \cdot \frac{\sin \angle CAA_1}{\sin \angle BAA_1} = 1.$$

Solution: Applying the Sine theorem to $\triangle ABB_1$ and $\triangle CBB_1$, we obtain

$$\frac{c}{\sin \angle AB_1B} = \frac{B_1A}{\sin \angle ABB_1} \quad \text{and} \quad \frac{a}{\sin \angle BB_1C} = \frac{CB_1}{\sin \angle CBB_1}.$$

Since the angles BB_1A and BB_1C are supplementary, their sines are equal. This implies

$$\frac{\sin \angle ABB_1}{\sin \angle CBB_1} = \frac{B_1A}{CB_1} \cdot \frac{a}{c}.$$

Similarly we obtain

$$\frac{\sin \angle BCC_1}{\sin \angle ACC_1} = \frac{C_1B}{AC_1} \cdot \frac{b}{a} \quad \text{and} \quad \frac{\sin \angle CAA_1}{\sin \angle BAA_1} = \frac{A_1C}{BA_1} \cdot \frac{c}{b}.$$

The result now follows from multiplying the last three equalities and applying Ceva's theorem. $\square$

7.6 Problems

> *Nothing in life is to be feared. It is only to be understood.*
> — Marie Skłodowska Curie (1867–1934).

As before, for a triangle ABC, we denote the length of the side opposite to A by a, the length of the side opposite B by b, etc. We also use A to denote the measure of the angle at A, and similarly for B and C.

7.1 Prove that $\sin 30° = \cos 60° = 1/2$, $\sin 60° = \cos 30° = \sqrt{3}/2$, and $\sin 45° = \cos 45° = \sqrt{2}/2$.

7.2 If $\tan \alpha = 3$, find $\sin 3\alpha$ and $\cos 3\alpha$ without using a calculator. Assume $0° < \alpha < 90°$.

7.3 Prove that in every triangle ABC, $\sin(A + B/2) = \sin(C + B/2)$.

7.4 Suppose $a \leq b \leq c$ in $\triangle ABC$. Prove that if $a^2 + b^2 > c^2$, then the triangle is acute, and if $a^2 + b^2 < c^2$, then it is obtuse.

7.5 Prove the identities $\sin 2\theta = \dfrac{2 \tan \theta}{1 + \tan^2 \theta}$ and $\cos 2\theta = \dfrac{1 - \tan^2 \theta}{1 + \tan^2 \theta}$.

7.6 Prove $\cos 3\alpha = 4\cos^3 \alpha - 3\cos \alpha$ and $\tan 3\alpha = \dfrac{3 \tan \alpha - \tan^3 \alpha}{1 - 3\tan^2 \alpha}$.

7.7 Find (approximate) the lengths of all sides and the measures of all angles in $\triangle ABC$, where $m\angle A = 75°$, $b = AC = 7$, and $c = AB = 5$.

7.8 Complete the proof of Theorem 7.3 by showing that for nonobtuse angles α and β, if $\alpha + \beta = 90°$, then $\sin\alpha\cos\beta + \sin\beta\cos\alpha = 1$.

7.9 Prove that if $\cos^2\alpha + \cos^2\beta = a$, then $\cos(\alpha+\beta)\cos(\alpha-\beta) = a - 1$.

7.10 Use the result of Corollary 7.5 to prove the Sine theorem: for any triangle ABC,

$$\frac{a}{\sin A} = \frac{b}{\sin B} = \frac{c}{\sin C}.$$

7.11 Consider a triangle with side lengths 4, 5, and 6. Prove that the measure of the smallest angle is half that of the measure of the largest angle.

7.12 Prove that $4\arctan\dfrac{1}{5} - \arctan\dfrac{1}{239} = \dfrac{\pi}{4}$.

7.13 Prove that $\dfrac{2\cos 40° - \cos 20°}{\sin 20°} = \sqrt{3}$.

7.14 Prove that for every α,

$$4\sin\alpha\sin(60° - \alpha)\sin(60° + \alpha) = \sin 3\alpha, \quad \text{and}$$
$$4\cos\alpha\cos(60° - \alpha)\cos(60° + \alpha) = \cos 3\alpha.$$

7.15 Prove that $\tan 55°\tan 65°\tan 75° = \tan 85°$.

7.16 Determine the length, x, of each side of the equilateral triangle shown in Figure 7.17.

7.17 If $\sin z + \cos z = x$ and $\sin^3 z + \cos^3 z = y$, then prove $y = (3x - x^3)/2$.

7.18 Let $\sin(\beta/2) \neq 0$. Prove that for all α,

$$\sum_{k=0}^{n}\sin(\alpha + k\beta) = \frac{\sin(\alpha + n\beta/2)\sin((n+1)\beta/2)}{\sin(\beta/2)}.$$

7.19 Let $\sin(\beta/2) \neq 0$. Prove that for all α,

$$\sum_{k=0}^{n}\cos(\alpha + k\beta) = \frac{\cos(\alpha + n\beta/2)\sin((n+1)\beta/2)}{\sin(\beta/2)}.$$

7.20 Prove the following identity:

$$\sum_{k=0}^{n-1}\sin\left(\alpha + \frac{2\pi}{n}k\right) = \sum_{k=0}^{n-1}\cos\left(\alpha + \frac{2\pi}{n}k\right) = 0.$$

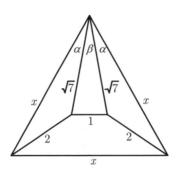

FIGURE 7.17.

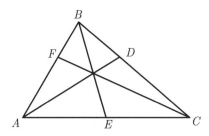

FIGURE 7.18.

7.21 Let P be the center of the square constructed on the hypotenuse $\overline{AC}$ of a right triangle ABC. Prove that $\overline{BP}$ bisects $\angle ABC$.

7.22 A hexagon inscribed in a circle has three consecutive sides of length a and three consecutive sides of length b. Determine the radius of the circle. Can you generalize the solution to an n-gon for even n?

7.23 Use trigonometry to prove Ceva's theorem:

Let points D, E, and F be chosen on the sides of $\triangle ABC$, such that $A - F - B$, $B - D - C$, and $C - E - A$. (See Figure 7.18.) Then $\overline{AD}$, $\overline{BE}$, and $\overline{CF}$ are concurrent if and only if
$$\frac{AE}{EC} \cdot \frac{CD}{DB} \cdot \frac{BF}{FA} = 1.$$

7.24 Given $\triangle ABC$, let $\overline{BB_1}$ be a segment joining B with a point B_1 on the side $\overline{AC}$. Then $\overline{BB_1}$ is a bisector of $\angle B$ if and only if $AB_1/B_1C = AB/BC = c/a$.

7.25 Prove that in any $\triangle ABC$,

$$\sin\frac{A}{2} \le \frac{a}{2\sqrt{bc}},$$

with equality if and only if $b = c$.

7.26 Label a triangle ABC so that the bisector of $\angle A$ meets $\overline{BC}$ at D and the bisector of $\angle C$ meets $\overline{AB}$ at E. Let O be the intersection of the bisectors, and suppose that $OD = OE$. Prove that either $m\angle B = 60°$ or that $\triangle ABC$ is isosceles (or both).

7.27 Consider an equilateral triangle with side lengths 7 and points C' and B' on $\overline{AB}$ and $\overline{AC}$, respectively, such that $AC' = CB' = 4$. Let P be the intersection of segments BB' and CC'. Find PC.

7.28 Consider a circle inscribed in a circular sector with central angle α radians. (See Figure 7.19.) Find the ratio of the area of the sector to the area of the inscribed circle.

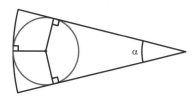

FIGURE 7.19.

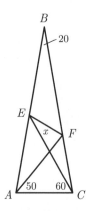

FIGURE 7.20.

7.29 Use Ptolemy's theorem to prove the following trigonometric identity for any two angles α and β, $0° < \beta \le \alpha < 90°$:

$$\sin(\alpha - \beta) = \sin\alpha\cos\beta - \sin\beta\cos\alpha.$$

7.30 Consider a triangle ABC with $m\angle ABC = 20°$ and $AB = AC$. Let $E \in \overline{AB}$ and $F \in \overline{BC}$ such that $m\angle FAC = 50°$ $m\angle ECA = 60°$, as shown in Figure 7.20. Find $x = m\angle CEF$. (Note: A calculator will actually be detrimental in solving this problem.)

7.31 Find the measures of the angles of $\triangle ABC$ if $\cos A + \cos B + \cos C = 3/2$.

7.32 Prove that the angle bisectors of a triangle are concurrent, using trigonometry.

7.33 Prove that among all triangles inscribed in a circle, the equilateral one has the greatest area.

7.34 Prove that among all triangles circumscribed around a circle, the equilateral one has the smallest area and the smallest perimeter.

7.7 Supplemental Problems

S7.1 Is there a triangle with two altitudes longer than 10 inches and with area smaller than 40 square inches?

S7.2 If $\sin\alpha = 2/5$, find $\tan 2\alpha$ and $\cos 2\alpha$ without using a calculator. Assume $0° < \alpha < 90°$.

S7.3 Prove the identity $\cos^4\theta - \sin^4\theta = \cos 2\theta$.

S7.4 Find (approximate) the lengths of all sides and the measures of all angles in $\triangle ABC$, where $m\angle A = 50°$, $m\angle B = 60°$, and $c = AB = 3$.

S7.5 For the triangle in Problem S7.4, find each of the following. Try not to use decimal approximations from the solution to that problem as intermediate steps in finding values for this problem.

  (a) The length of the angle bisector at A.
  (b) The length of the median at A.
  (c) The radius of the circumcircle of triangle ABC.

S7.6 You are grilling hamburger patties with your friend on a large grill. Each of the patties is perfectly round and they all have the same radius. In an effort to place as many patties on the grill as possible, you place them very close to each other. You notice that around one central patty, you are able to fit six more patties, where each is tangent to the central patty as well as to two others that are also tangent to the central patty.

(a) Your friend claims that the patties don't really fit perfectly as described, but that you smashed them a bit in order to make them tangent. What do you think?

(b) Your friend claims that if you keep the central hamburger patty the same size but make all of the others have half the radius of the central patty, then you will be able to arrange 12 patties around it so that each is tangent to the central patty as well as to two others. What do you think?

(c) Let O be a circle of radius 1 and let $n \geq 3$ be an integer. Suppose n circles of radius r lie externally tangent to O such that each is also externally tangent to two others. Find r.

(d) Repeat the previous part where each of the n circles lies internally tangent to O.

S7.7 Prove that if $x + y + z = \pi$, then

$$\sin x + \sin y + \sin z = 4 \sin \frac{x}{2} \sin \frac{y}{2} \sin \frac{z}{2}.$$

S7.8 Let $ABCD$ be a square, with $M \in \overline{CD}$, $AM = a$, $N \in \overline{BC}$ such that $\overline{AN}$ bisects $\angle BAM$. Find $BN + DM$.

S7.9 Prove the angle sum formula for cosine for nonobtuse angles α and β:

$$\cos(\alpha \pm \beta) = \cos \alpha \cos \beta \mp \sin \alpha \sin \beta.$$

S7.10 Out of all triangles inscribed in a given circle, which one has the greatest perimeter? Prove your answer.

S7.11 The distance between the centers of two discs of radius R in a plane is d, where $0 < d < 2R$. Determine the area of the intersection of the discs.

S7.12 In a triangle ABC, $1/b + 1/c = 1/l_a$, where l_a is the length of the triangle bisector of angle A. Prove that the measure of angle A is $120°$.

S7.13 Find the measures of angles of an isosceles triangle ABC ($AB = BC$) if the length of the altitude at B is twice the length of the bisector at A.

S7.14 In $\triangle ABC$, suppose $m\angle B = 18°$ and $m\angle C = 36°$. Show that $a - b = R$, where R is the circumradius of the triangle.

S7.15 Let $\overline{AD}$ and $\overline{CE}$ be the medians in a $\triangle ABC$, and let $m\angle BAD = m\angle BCE = 30°$. Prove that $\triangle ABC$ is equilateral.

S7.16 Consider a triangle ABC. Let A_1, B_1, and C_1 be points chosen on the sides BC, AC, and AB, respectively. Let A_2, B_2, and C_2 be points chosen on the sides B_1C_1, A_1C_1, and A_1B_1 of $\triangle A_1B_1C_1$, respectively. Suppose the lines AA_1, BB_1, and CC_1 are concurrent and the lines A_1A_2, B_1B_2, and C_1C_2 are concurrent. Prove that the lines AA_2, BB_2, and CC_2 are concurrent.[8]

S7.17 Let P be any point on the minor arc A_1A_{2n+1} of the circumcircle of a regular polygon $A_1A_2 \ldots A_{2n+1}$. Prove that

$$PA_1 + PA_3 + \cdots + PA_{2n+1} = PA_2 + PA_4 + \cdots + PA_{2n}.$$

---

[8] This problem was suggested by Sebastian Cioaba.

S7.18 Let $ABCD$ be a convex quadrilateral with $AB = a$, $BC = b$, $CD = c$, $DA = d$, and Area($ABCD$) $= S$, and let $p = (a + b + c + d)/2$ be its semiperimeter.

(a) Prove that $S = \sqrt{(p-a)(p-b)(p-c)(p-d)} - (abcd)\cos^2\left(\dfrac{B+D}{2}\right)$.

(b) Prove that $ABCD$ is inscribed in a circle if and only if
$$S = \sqrt{(p-a)(p-b)(p-c)(p-d)}.$$

(c) Prove that if $ABCD$ is both inscribed and circumscribed, then
$$S = \sqrt{abcd}.$$

(d) Compare the statements above with Heron's formula for the area of a triangle.

S7.19 Let d be the distance between the incenter and circumcenter of $\triangle ABC$. Let R and r be the inradius and circumradius, respectively. Prove that $d^2 = R^2 - 2Rr$.

8

Coordinatization

*In symbols one observes an advantage of discovery
which is greatest when they express the exact nature
of a thing briefly and, as it were, picture it; then
indeed the labor of thoughts is wonderfully dimin-
ished.*

— Gottfried Wilhelm Leibniz (1646–1716)[1]

In this chapter, we introduce and utilize coordinates as an analytic method for solving geometric problems. There is some debate among scholars as to who first used this approach,[2] but there is little dispute that the French mathematicians René Descartes (1596–1650) and Pierre Fermat (1601–1665) should be given the credit for (independently) formulating what we now call analytic geometry.

The analytic method is such a central, unifying part of mathematics that it is hard to conceive of its nonexistence. Essentially, this method intertwines algebra and geometry in such a way that a problem in one of these areas can be converted to a problem in the other, where a solution may be more easily obtained. The link between these two mathematical arenas is made by creating a one-to-one correspondence between numbers and points: each point on a line is represented by a real number, each point in the plane by an ordered pair of real numbers, and each point in 3-space by an ordered triple of real numbers. Once this correspondence is established and letters are used to represent the real numbers, figures (sets of points) are readily identified with equations or inequalities containing two or more variables. One can then study the algebraic properties of solutions to the equation $f(x, y) = 0$ in order to learn about the geometric properties of the figure in the plane whose point coordinates satisfy the equation, and vice-versa. In this way, the coordinate method serves as a "mathematical dictionary," used for translating between geometry and algebra.

8.1 The Real Number Line

The Ruler Postulate for Euclidean geometry (Postulate 2.3), first introduced in Section 2.4, allows us to claim the existence of a one-to-one correspondence between the set of real numbers and the set

---

[1] From *Calculus Gems*, by G. Simmons, MAA, 2007.

[2] The Egyptians and Romans used it for surveying and the Greeks for mapmaking. Nicole Oresme graphed one variable against another in the fourteenth century. ([**18**], pp. 346–347)

of points of an arbitrary line. The correspondence is usually defined by choosing two distinct points on the line, and then matching one of them with the number 0, and another with a positive number (often taken to be 1). Then every point of the line becomes matched with a unique real number, and vice versa, and the line is often referred to as the **real number line** or the **coordinate line**. The one-to-one correspondence is called the **coordinatization** of the line. For a given coordinatization, the number a corresponding to a point A is called the **coordinate of** A, and we denote this by writing $A : (a).$[3] By the Ruler Postulate, if $A : (a)$ and $B : (b)$, then the distance AB is equal to $|b - a|$.

Because the points of the line AB correspond precisely to the real numbers, we can find a point located at any desired distance from a fixed point. Furthermore, for any two distinct points, A and B, and any fixed positive constant, k, there exists a point C on $\overleftrightarrow{AB}$ so that $\frac{AC}{CB} = k$, as we now show. In the coordinatization of $\overleftrightarrow{AB}$, let $A : (a)$ and $B : (b)$. Assume that $a < b$. We wish to find a point $C : (c)$ on $\overleftrightarrow{AB}$ such that $\frac{AC}{CB} = \frac{|c-a|}{|c-b|} = k$. Observe the following:

$$\frac{|c - a|}{|c - b|} = k \Leftrightarrow \begin{cases} \frac{c-a}{-(c-b)} = k & \text{when } a < c < b \\ \frac{c-a}{c-b} = k & \text{when } c < a \text{ or } c > b \end{cases},$$

$$\Leftrightarrow \begin{cases} c - a = kb - kc & \text{when } a < c < b \\ c - a = kc - kb & \text{when } c < a \text{ or } c > b \end{cases},$$

$$\Leftrightarrow \begin{cases} c = \frac{kb+a}{k+1} & \text{when } a < c < b \\ c = \frac{kb-a}{k-1} & \text{when } c < a \text{ or } c > b, k \neq 1 \end{cases}.$$

Thus, unless $k = 1$, there are two points C dividing the segment $\overline{AB}$ into segments having ratio $k : 1$, one a point of internal division and the other a point of external division. (The point of external division could fall to the right of B or to the left of A, as determined by the value of k.) We summarize this result in the following proposition, accompanied by Figure 8.1.

Proposition 8.1. *For points* $A : (a)$, $B : (b)$, *and* $C : (c)$, $a < b$, *with* C *on* $\overleftrightarrow{AB}$ *and* $\frac{AC}{CB} = k \geq 0$,

$$c = \frac{kb - a}{k - 1} \text{ (when } c \leq a \text{ or } c > b, k \neq 1) \text{ or}$$

$$c = \frac{kb + a}{k + 1} \text{ (otherwise).}$$

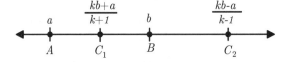

FIGURE 8.1. Internal and External divisions of $\overline{AB}$.

---

[3] This notation can be read as an independent clause (as in "A has the coordinate a") or as a noun (as in "A, with the coordinate a"), depending on the context.

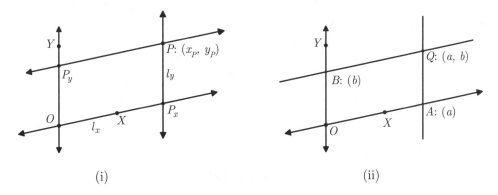

FIGURE 8.2. Coordinatization of the plane.

Of particular interest is a point C dividing $\overline{AB}$ exactly in half, that is, $\frac{AC}{CB} = k = 1$. In this case, $AC = CB$, C is the **midpoint** of $\overline{AB}$, and $c = \frac{a+b}{2}$.

8.2 Coordinates

To employ the full power of the coordinate method, we need to move beyond the one-dimensional real number line. Let Π be a plane, and let O, X, and Y be three non-collinear points of Π. Then, by Postulate 2.6, lines OX and OY are in Π. Each of these two lines can be coordinatized. We do this in such a way that O is at the point corresponding to 0 on each of the lines and both X and Y have positive coordinates. In addition, we assume that $\overrightarrow{OY}$ is obtained by rotating $\overrightarrow{OX}$ counterclockwise by an angle of positive measure less than $180°$.[4] Then we say that O, X, and Y define a **coordinate system** in Π, and we refer to it as the OXY-coordinate system. Point O is called the **origin**, and the coordinate lines OX and OY are called the *x*-**axis** and the *y*-**axis** of the coordinate system OXY, respectively.

Any such coordinate system can be used to coordinatize Π, as follows. Suppose that P is a point in Π. There is a unique line, l_y, through P that is parallel to $\overleftrightarrow{OY}$, and a unique line, l_x, through P that is parallel to $\overleftrightarrow{OX}$. Let $P_x = \overleftrightarrow{OX} \cap l_y$, and $P_y = \overleftrightarrow{OY} \cap l_x$. Points P_x and P_y are called the **projections** of P on the *x*- and *y*-axis, respectively. (See Figure 8.2(i).) Let x_P and y_P be the coordinates of points P_x and P_y on the *x*- and *y*-axis, respectively, i.e., $P_x : (x_P)$ and $P_y : (y_P)$. These numbers x_P and y_P are called the *x*- and *y*-**coordinates** of P in OXY, and we write $P : (x_P, y_P)$.[5]

From our discussion, it should be clear that once a coordinate system in the plane is chosen, each point of the plane can be assigned a unique ordered pair of real numbers – its coordinates. On the other hand, for every ordered pair of real numbers (a, b), there exists a unique point Q in the plane having coordinates (a, b) in the coordinate system. Indeed, point Q can be constructed as the intersection of two lines: one passing through point $A : (a)$ on the *x*-axis and parallel to the *y*-axis, and another passing through point $B : (b)$ on the *y*-axis and parallel to the *x*-axis. (See Figure 8.2(ii).) Therefore, as soon as a coordinate system in a plane is chosen, we obtain a one-to-one correspondence between the set of points of the plane and the elements of $\mathbb{R} \times \mathbb{R}$. This correspondence will justify using expressions like "consider the point $(3, 4)$," instead of "consider the point with coordinates $(3, 4)$."

---

[4] This is equivalent to saying that we fix an *orientation* in all our coordinate systems.

[5] As before, this notation can be read as an independent clause ("P has the coordinates (x_P, y_P)") or as a noun ("P, with the coordinates (x_P, y_P)"), depending on the context.

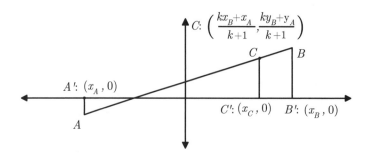

FIGURE 8.3. C divides $\overline{AB}$ internally in ratio k.

We now have the tools needed to generalize the formula given in Proposition 8.1. Specifically, we are interested in finding the coordinates of a point C that divides a line segment in the plane, $\overline{AB}$, in ratio $k : 1$. Let A and B be points in the plane, and assume that $x_A \le x_B$. We will determine first the coordinates (x_C, y_C) of a point C that divides $\overline{AB}$ internally in ratio k.

Consider the projections of points A, C, and B on the x-axis: $A' : (x_A, 0)$, $C' : (x_C, 0)$, and $B' : (x_B, 0)$, respectively, as shown in Figure 8.3. Since the lines $\overleftrightarrow{AA'}$, $\overleftrightarrow{CC'}$, and $\overleftrightarrow{BB'}$ are all parallel to the y-axis, they are parallel to each other. By Theorem 3.13, the segments on a pair of transversals that are created by a set of three or more parallel lines will be proportional. That is, $\frac{AC}{CB} = \frac{A'C'}{C'B'}$. If we want $\frac{AC}{CB} = k$, we simply need to ensure that $\frac{A'C'}{C'B'} = k$. In other words, finding x_C amounts to simply finding the coordinate of the appropriate point C' on the x-axis dividing $\overline{A'B'}$ internally in ratio k, a situation we have already encountered! We conclude that $x_C = (kx_B + x_A)/(k + 1)$.

Likewise, it can be shown that $y_C = (ky_B + y_A)/(k + 1)$. Thus, the coordinates of point C on $\overleftrightarrow{AB}$ dividing $\overline{AB}$ internally in ratio k ($k > 0$) are $(x_C, y_C) = \left(\frac{kx_B + x_A}{k+1}, \frac{ky_B + y_A}{k+1} \right)$. In particular, the point $C = \left(\frac{x_A + x_B}{2}, \frac{y_A + y_B}{2} \right)$ is the **midpoint** of $\overline{AB}$. Similarly, the coordinates of point C on $\overleftrightarrow{AB}$ dividing $\overline{AB}$ externally in ratio k ($k > 0, k \ne 1$) are $(x_C, y_C) = \left(\frac{kx_B - x_A}{k-1}, \frac{ky_B - y_A}{k-1} \right)$.

Theorem 8.2. *For points A, B, and C with C on $\overleftrightarrow{AB}$ and* $\dfrac{AC}{CB} = k \ge 0$,

$$(x_C, y_C) = \left(\frac{kx_B + x_A}{k + 1}, \frac{ky_B + y_A}{k + 1} \right) \ or$$

$$(x_C, y_C) = \left(\frac{kx_B - x_A}{k - 1}, \frac{ky_B - y_A}{k - 1} \right), \ k \ne 1.$$

To create the familiar coordinatization of the plane utilized by Descartes, we make the assumptions that the axes OX and OY are perpendicular and that the same scale is used on both lines. The axes divide the plane into four **quadrants**, with the location of a point within a quadrant or on an axis determined by the sign of its coordinates; this is illustrated in Figure 8.4, with x and y positive. The equality of scales on the x- and y-axes can be defined in the following way: the segment with the end points $(0, 0)$ and $(1, 0)$ is congruent to the segment with the end points $(0, 0)$ and $(0, 1)$. A

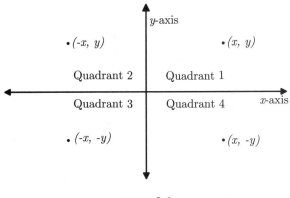

FIGURE 8.4.

plane with such a coordinatization is called the **Cartesian plane**, and the corresponding coordinate system is called the **Cartesian coordinate system**.

Given a coordinate line and two points $A : (a)$ and $B : (b)$ on it, we know that the length of $\overline{AB}$ is given by $AB = |a - b|$. Having a coordinate system in a plane and two points $A : (x_A, y_A)$ and $B : (x_B, y_B)$, we wish to likewise find AB. Here we do this for a Cartesian coordinate system only.

Consider the point $C : (x_B, y_A)$. Points A, B, and C are vertices of a right triangle with right angle at C (as shown in Figure 8.5). Using the Pythagorean theorem, $AC^2 + BC^2 = AB^2$. Since the points A and C agree in the y-coordinate, and the points B and C agree in the x-coordinate, $AC = |x_A - x_B|$ and $BC = |y_A - y_B|$. As $|a|^2 = a^2$ for any real number a, we obtain the following **distance formula in the plane**.

Proposition 8.3. *In a Cartesian coordinate system, the distance between two points A and B is given by*

$$AB = \sqrt{(x_A - x_B)^2 + (y_A - y_B)^2}.$$

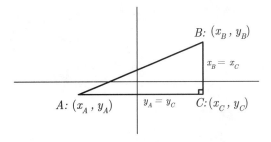

FIGURE 8.5.

8.3 Equations of a Line

Recall our earlier assertion that the coordinate method provides us with a mathematical dictionary, a claim referring to the correspondence between figures in the plane and algebraic relations in x and y. Our goal in this section is to find an algebraic description of a line in the plane. As discussed in Chapter 6, the set of all points satisfying a given property is called the **locus** for this property. When the property is expressed as an equation or inequality, the locus can be portrayed as a graph of that property. In this way, the graph is a visual representation of an algebraic relation.

If OXY is a coordinate system, the x-axis and all lines parallel to the x-axis are called **horizontal** lines; the y-axis and all lines parallel to it are **vertical** lines. It is easy to see that any horizontal line can be represented by an equation of the form $y = b$, as a point lies on the line if and only if it has coordinates of the form (x, b), where x can assume any real value. Similarly, any vertical line can be represented with an equation of the form $x = a$.

The reader is undoubtedly acquainted with the notion of the slope of a (non-vertical) line. Intuitively, the slope of a line specifies the direction in which the line rises and quantifies the "steepness" of the line. Let l be a non-vertical line. For any pair of distinct points $A : (x_A, y_A)$ and $B : (x_B, y_B)$ on l, we have $x_A \neq x_B$, so we can compute the number $(y_B - y_A)/(x_B - x_A)$. If we compute the same ratio for any other pair of distinct points A' and B' of l, the result will be the same, due to the similarity of the related triangles. (See Figure 8.6.) Therefore the ratio depends only on the line l, and it is called the **slope** of l. It is often denoted by m_l, or simply m. For $l = \overleftrightarrow{AB}$,

$$m_{\overleftrightarrow{AB}} = \frac{y_B - y_A}{x_B - x_A}.$$

Theorem 8.4. *Let l be a line having slope m, and let A be a point on l. A point P lies on l if and only if its coordinates satisfy the equation $y - y_A = m(x - x_A)$.*

Proof. Let $P : (x_P, y_P)$ be any point on l, distinct from A. Then $\overleftrightarrow{AP}$ is the same line as l, so $m = (y_P - y_A)/(x_P - x_A)$. Consequently, (x_P, y_P) satisfies $y - y_A = m(x - x_A)$, as desired.

Conversely, suppose $P : (x_P, y_P)$ with x_P and y_P satisfying the equation $y - y_A = m(x - x_A)$. Hence $y_P - y_A = m(x_P - x_A)$. Let P' be a point on l with the same x-coordinate as P, i.e., $x_{P'} = x_P$. Then, using the first part of the proof, $y_{P'} - y_A = m(x_{P'} - x_A) = m(x_P - x_A)$. Hence $y_P - y_A = y_{P'} - y_A$, so $y_P = y_{P'}$. This implies that $P = P'$. $\square$

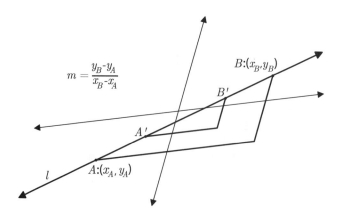

FIGURE 8.6. Slope of a line.

An equation of the form $y - y_A = m(x - x_A)$ is called the **point-slope form** for a line. The same line can also be described by equations presented in other forms. The point-slope form can be rewritten as $y = mx - mx_A + y_A$, or $y = mx + b$, where $b = y_A - mx_A$. The substitution $x = 0$ into any equation of the form $y = mx + b$ reveals that b is the y-coordinate of the point of intersection of the line with the y-axis; consequently, this form is known as the **slope-y-intercept form** of the equation for a line. Similarly, an equation of the form $y = mx + b$ can be rewritten as $\dfrac{x}{(-b/m)} + \dfrac{y}{b} = 1$, provided that neither m nor b is 0.

Each of these equations can be rewritten as $ax + by + c = 0$, which is often referred to as the **standard form**, or the general form, of the equation of the line. If $a = 0$ and $b \neq 0$, the graph of this equation is a horizontal line. If $a \neq 0$ and $b = 0$, it is a vertical line. If $a = b = 0$, then the graph of the equation is either the whole plane (if $c = 0$) or the empty set (if $c \neq 0$). The forms of linear equations are summarized in the table that follows.

Name of Form	Form
Standard	$ax + by + c = 0$ (at least one of a, b non-zero)
Point-Slope	$y - y_A = m(x - x_A)$ (for non-vertical lines)
Slope-Intercept	$y = mx + b$ (for non-vertical lines)
xy-Intercept	$\frac{x}{a} + \frac{y}{b} = 1$ $(a \neq 0, b \neq 0)$

One can easily observe the following.

> **Corollary 8.5.** *Lines are parallel if and only if they have the same slope (or both have no slope).*

Though many interesting geometric properties can be expressed in terms of coordinates in an arbitrary coordinate system OXY, most applications use Cartesian systems. Therefore,

> From now on, unless specified otherwise, a coordinate system in the plane is assumed to be Cartesian.

Suppose that l is a non-horizontal line in the coordinate plane. We define the **inclination** of l to be the smallest positive measure, θ, of an angle measured counterclockwise from the positive direction of the x-axis to l.[6] If l is parallel to the x-axis, then its inclination is defined to be $0°$. Observe that this definition restricts the value of θ so that $0° \leq \theta < 180°$. This gives a very useful relation: for any non-vertical line l,

$$m_l = \tan \theta.$$

(See Figure 8.7.) Note that l is vertical if and only if $\theta = 90°$, in which case neither the slope nor $\tan \theta$ exist.

This interpretation of the slope as tangent allows us to use trigonometry when the coordinate method is developed. In particular, we have the following useful statement.

---

[6] As usual, the angle is measured in degrees unless otherwise specified.

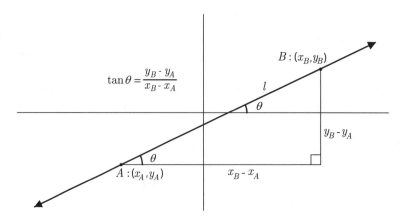

FIGURE 8.7.

Theorem 8.6. *Let α be the angle between intersecting non-vertical lines l_1 and l_2, having slopes m_1 and m_2, respectively. Then $\alpha = 90°$ if and only if $m_1 m_2 = -1$. Otherwise,*

$$\tan \alpha = \left| \frac{m_1 - m_2}{1 + m_1 m_2} \right|.$$

Proof. Let the inclinations of l_1 and l_2 be θ_1 and θ_2, respectively. (See Figure 8.8; let A, B, and C be the points of intersection shown in the figure.) Then $m_1 = \tan \theta_1$ and $m_2 = \tan \theta_2$. Since θ_1 is an exterior angle to $\triangle ABC$, $\theta_1 = \theta_2 + m\angle CAB$. Recall that the measure of an angle between two intersecting lines is the measure of the smaller angle formed by them, so $0° \leq \alpha \leq 90°$.

Suppose $\alpha \neq 90°$ and $m_1 m_2 \neq -1$. As $\alpha = m\angle CAB$ or $\alpha = 180° - m\angle CAB$, we always have $\tan \alpha = |\tan m\angle CAB|$. Thus, using the difference formula for the tangent function,

$$\tan \alpha = |\tan (\theta_1 - \theta_2)| = \left| \frac{\tan \theta_1 - \tan \theta_2}{1 + \tan \theta_1 \tan \theta_2} \right|.$$

Substitution yields the desired result.

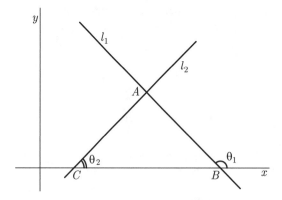

FIGURE 8.8.

Now, $\alpha = 90°$ if and only if $\theta_1 = \theta_2 + 90°$. As neither θ_1 nor θ_2 is a right angle, we have $\theta_2 \neq 0$. Therefore,

$$m_1 = \tan\theta_1 = \frac{\sin\theta_1}{\cos\theta_1} = \frac{\sin(\theta_2 + 90°)}{\cos(\theta_2 + 90°)} = \frac{\cos\theta_2}{-\sin\theta_2} = -\frac{1}{\tan\theta_2} = \frac{-1}{m_2}.$$

The implication, $m_1 m_2 = -1 \Rightarrow \alpha = 90°$, is left to the reader. $\qquad\square$

Example 29. Find the distance between the point $P : (-3, 4)$ and the line l given by $y - \frac{1}{2}x - 1 = 0$.

Solution: We first find the line through P that is perpendicular to l. As $m_l = 1/2$, this line (denoted $l_\perp$) will have slope $-1/m_l = -2$, so its equation is given by $y - 4 = -2(x + 3)$; equivalently, $y = -2x - 2$.

To find the coordinates of the point of intersection of these lines, we set $-2x - 2$ equal to $\frac{1}{2}x + 1$ and find $x = -\frac{6}{5}$. Thus, $y = -2\left(-\frac{6}{5}\right) - 2 = \frac{2}{5}$, and $(-6/5, 2/5)$ is the point of intersection.

It remains to find the distance between the points $(-3, 4)$ and $(-6/5, 2/5)$. Using the distance formula,

$$d\big((-3, 4), (-6/5, 2/5)\big) = \sqrt{\left(-3 - \left(-\frac{6}{5}\right)\right)^2 + \left(4 - \frac{2}{5}\right)^2}$$

$$= \sqrt{\left(-\frac{9}{5}\right)^2 + \left(\frac{18}{5}\right)^2}$$

$$= \sqrt{\left(\frac{9}{5}\right)^2 (1^2 + 2^2)} = \frac{9}{5}\sqrt{5}.$$

$\qquad\square$

The method of the preceding example can be used to find a general formula for the distance $d(P, l)$ between a point $P : (x_P, y_P)$ and a line, l, given by $ax + by + c = 0$. Notice first that if $a = 0$ and $b \neq 0$, then l is a horizontal line, equivalently expressed as $y = -c/b$. The line $l_\perp$ through P that is perpendicular to l is given by $x = x_P$ and the resulting distance from P to l is $|y_P - (-c/b)|$. Similarly, if $a \neq 0$ and $b = 0$, l is a vertical line, $x = -c/a$, and $d(P, l) = |x_P - (-c/a)|$.

If $a \neq 0$ and $b \neq 0$, then $m_l = -a/b$, and $l_\perp : y - y_P = (b/a)(x - x_P)$. Let $Q = l \cap l_\perp$. Finding its coordinates, we obtain

$$x_Q = \frac{b^2 x_P - ac - ab\, y_P}{a^2 + b^2} \quad \text{and} \quad y_Q = \frac{a^2 y_P - bc - ab\, x_P}{a^2 + b^2}.$$

Note that the expression for y_Q can be obtained from the one for x_Q by interchanging a with b and x with y, as one could anticipate. This implies

$$x_P - x_Q = \frac{a\,(a\,x_P + b\,y_P + c)}{a^2 + b^2} \quad \text{and} \quad y_P - y_Q = \frac{b\,(a\,x_P + b\,y_P + c)}{a^2 + b^2}.$$

Then,

$$d(P, l) = d(P, Q) = PQ = \sqrt{(x_P - x_Q)^2 + (y_P - y_Q)^2}$$

$$= \sqrt{\frac{a^2 (a\, x_P + b\, y_P + c)^2}{(a^2 + b^2)^2} + \frac{b^2 (a\, x_P + b\, y_P + c)^2}{(a^2 + b^2)^2}}$$

$$= \sqrt{\frac{(a^2 + b^2)(a\, x_P + b\, y_P + c)^2}{(a^2 + b^2)^2}}$$

$$= \sqrt{\frac{(a\, x_P + b\, y_P + c)^2}{a^2 + b^2}}$$

$$= \frac{|a\, x_P + b\, y_P + c|}{\sqrt{a^2 + b^2}}.$$

Note that though the formula is obtained under the assumption that neither a nor b is zero, it also gives the correct results if only one of them is zero. We state our finding in the following theorem.

Theorem 8.7. *Let P be a point, and let l be a line with equation $ax + by + c = 0$. Then the distance from P to l is given by*

$$d(P, l) = \frac{|a\, x_P + b\, y_P + c|}{\sqrt{a^2 + b^2}}.$$

Note that the formula confirms our answer for Example 29:

$$\frac{|a\, x_P + b\, y_P + c|}{\sqrt{a^2 + b^2}} = \frac{|(-1/2)(-3) + (1)(4) + (-1)|}{\sqrt{(-1/2)^2 + 1}} = \frac{9/2}{\sqrt{5/4}} = \frac{9}{\sqrt{5}}.$$

8.4 Applications

> *... in some deep sense the truth about how the world works resides in the equations ... to help our limited imaginations to visualize what is going on.*
> — Richard Feynman (1918–1988)[7]

The coordinatization of the plane and the accompanying algebraic representations of geometric figures can serve as powerful tools for solving problems of Euclidean geometry. Coordinatization of the plane requires choosing two intersecting lines to serve as axes and, sometimes, specifying the scale. The particular selection of the axes and the scale can impact (for better or worse) the ease with which a particular problem can be solved, as the examples and theorems in this section will demonstrate.

Theorem 8.8. *The medians of any triangle are concurrent and the point of concurrency divides each median into segments in ratio 2:1.*

---

[7] *Richard Feynman: A Life in Science*, John Gribbin and Mary Gribbin, Penguin Books, 1997, p. 27.

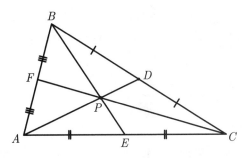

FIGURE 8.9.

Below we consider three solutions of this problem, all using the coordinate method. Each of them will illustrate advantages and difficulties of a particular approach, which is typical when the coordinate method is applied. Our first proof will be the most natural one, the second will represent a substantial improvement of the first, and the third proof will be based on a completely different approach.

Proof 1. Consider triangle ABC with medians $\overline{AD}$, $\overline{BE}$, and $\overline{CF}$. (See Figure 8.9.) Introduce an arbitrary coordinate system OXY in the plane such that no line defined by a median is vertical. (If one median is vertical in a given coordinate system, simply rotate the axes slightly so neither that median nor any others are vertical.) Then a point-slope form of $\overleftrightarrow{AD}$ is

$$y - y_A = \frac{y_D - y_A}{x_D - x_A}(x - x_A).$$

Being the midpoint of $\overline{BC}$, $D : (x_D, y_D) = \left(\dfrac{x_B + x_C}{2}, \dfrac{y_B + y_C}{2} \right)$. Substituting into the equation for $\overleftrightarrow{AD}$ and simplifying, we get

$$\overleftrightarrow{AD}: \quad y = \frac{y_B + y_C - 2y_A}{x_B + x_C - 2x_A}x + \frac{x_B y_A - x_A y_B + x_C y_A - x_A y_C}{x_B + x_C - 2x_A}. \tag{8.1}$$

In a similar way, we find an equation for $\overleftrightarrow{BE}$, but without actually redoing the computations from scratch. The following important (and obvious) observation makes the task trivial: in order to obtain an equation for $\overleftrightarrow{BE}$ by using (8.1), one simply needs to permute the letters A, B, and C in (8.1) so that A becomes B, B becomes C, and C becomes A. This permutation yields

$$\overleftrightarrow{BE}: \quad y = \frac{y_C + y_A - 2y_B}{x_C + x_A - 2x_B}x + \frac{x_C y_B - x_B y_C + x_A y_B - x_B y_A}{x_C + x_A - 2x_B}. \tag{8.2}$$

Similarly, if we permute the letters A, B, and C in (8.1) such that A becomes C, B becomes A, and C becomes B, we obtain an equation for $\overleftrightarrow{CF}$:

$$\overleftrightarrow{CF}: \quad y = \frac{y_A + y_B - 2y_C}{x_A + x_B - 2x_C}x + \frac{x_A y_C - x_C y_A + x_B y_C - x_C y_B}{x_A + x_B - 2x_C}. \tag{8.3}$$

Let $P = \overleftrightarrow{AD} \cap \overleftrightarrow{BE}$. Coordinates of P can be found by solving the system of equations (8.1) and (8.2), which is easier said than done! After tedious algebraic manipulations, which we omit, we finally arrive at

$$(x_P, y_P) = \left(\frac{x_A + x_B + x_C}{3}, \frac{y_A + y_B + y_C}{3} \right). \tag{8.4}$$

Applying our first permutation to all occurrences of indices A, B, and C in the system formed by (8.1) and (8.2), we obtain the system formed by equations (8.2) and (8.3), whose solutions give the coordinates of a point $P' = \overleftrightarrow{BE} \cap \overleftrightarrow{CF}$. It is clear that the coordinates of P' can be obtained from the coordinates of P by permuting the indices A, B, and C as before. Since the coordinates of P given by (8.4) **do not change** under any permutation of $\{A, B, C\}$, P and P' are the same point. Hence $\overleftrightarrow{AD}$ and $\overleftrightarrow{CF}$ intersect $\overleftrightarrow{BE}$ at the same point, which is another way to say that all three medians are concurrent. The statement about the medians being divided by P in proportion $2 : 1$ follows directly from (8.4) and Theorem 8.2. $\qquad\square$

Proof 2. Let us again use the point labeling as in Figure 8.9. A drawback of Proof 1 is that it requires us to determine coordinates of P given by (8.4) by solving the system formed by (8.1) and (8.2), and this requires a somewhat lengthy algebraic manipulation. The difficulty can be removed by choosing the coordinate system in a better way.

Introduce a coordinate system OXY, not necessarily Cartesian, such that E becomes the origin, $A : (a, 0)$, and $B : (0, 1)$. Then $a < 0$ and $C : (-a, 0)$. (See Figure 8.10.)

Then $D : (-a/2, 1/2)$, and a point-slope form of line AD is

$$\overleftrightarrow{AD} : \quad y = -\frac{1}{3a}(x - a) = -\frac{1}{3a}x + \frac{1}{3}.$$

(This equation is, of course, what one gets from (8.1) by substituting the coordinates of A, B, and C.) Similarly (or just by replacing a with $-a$ in the equation above) we obtain

$$\overleftrightarrow{CF} : \quad y = \frac{1}{3a}(x + a) = \frac{1}{3a}x + \frac{1}{3}.$$

Clearly, both $\overleftrightarrow{AD}$ and $\overleftrightarrow{CF}$ intersect $\overleftrightarrow{BE}$ at the same point $P : (0, 1/3)$ – their y-intercept. Again, the second part of the theorem follows from here immediately. $\qquad\square$

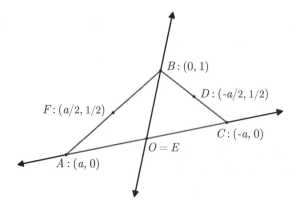

FIGURE 8.10.

We stress that use of a not-necessarily-Cartesian system in this solution has the advantage of reducing the number of independent coordinates. If we insist on a Cartesian system, and if no median of our triangle is also an altitude, then instead of one parameter a we are forced to introduce at least one more parameter. For example, keeping E as the origin and $A : (a, 0)$, we would have $B : (x_B, 1)$. Or, if we wish to preserve $A : (a, 0)$ and $B : (0, 1)$, then x_E or x_C would become a new parameter. A similar approach will still work, but the algebraic manipulations become a little harder.

Proof 3. We again use the point labeling of Figure 8.9. This proof will not use equations of lines, and is heavily motivated by the statement of the theorem concerning the $2 : 1$ ratios and Proposition 8.2.

Since D is the midpoint of $\overline{BC}$, D has coordinates $\left(\dfrac{x_B + x_C}{2}, \dfrac{y_B + y_C}{2} \right)$. Let P be the point on $\overline{AD}$ such that $AP/PD = 2/1$. Then by Theorem 8.2, P has coordinates

$$\left(\frac{x_A + 2x_D}{3}, \frac{y_A + 2y_D}{3} \right) = \left(\frac{x_A + x_B + x_C}{3}, \frac{y_A + y_B + y_C}{3} \right).$$

Likewise, since E is the midpoint of $\overline{AC}$, E has coordinates $\left(\dfrac{x_A + x_C}{2}, \dfrac{y_A + y_C}{2} \right)$. If P' is the point on $\overline{BE}$ such that $BP'/P'E = 2/1$, then P' has coordinates

$$\left(\frac{x_B + 2x_E}{3}, \frac{y_B + 2y_E}{3} \right) = \left(\frac{x_B + x_A + x_C}{3}, \frac{y_B + y_A + y_C}{3} \right).$$

Similarly, the point P'' on the segment CF such that $CP''/P''F = 2/1$ has coordinates

$$\left(\frac{x_C + x_A + x_B}{3}, \frac{y_C + y_A + y_B}{3} \right).$$

That is, P, P', and P'' are all the same! This establishes both claims of the theorem. $\qquad\square$

Note that in the last proof, the symmetry of the coordinates of P with respect to A, B, and C made it evident – even without further calculation – that P served the same role for all three medians, which made the proof even easier. The reader is invited to compare these solutions with those using other methods. This problem occurred as Problem 3.2.10 and it also follows from Ceva's Theorem (Theorem 3.17); in Chapter 12, we will revisit the statement once more as Theorem 12.10.

Our second example will use an approach similar to the one of Proof 2 above. It illustrates the advantages of a Cartesian coordinate system.

Theorem 8.9. *The altitudes of any triangle are concurrent.*

Proof. If our triangle is a right triangle, the proof is trivial: all altitudes meet at the vertex of the right angle. Therefore we assume that our triangle ABC has no right angle. Introduce a Cartesian coordinate system so that A and C lie on the x-axis and B lies on the positive y-axis, and choose the scale so that B has coordinates $(0, 1)$. As a result, the altitude from B will lie on the y-axis with foot at the origin. Label the remaining vertices $A : (x_A, 0)$ and $C : (x_C, 0)$. Since A and C are not at the origin, $x_A \neq 0$ and $x_C \neq 0$. (See Figure 8.11.)

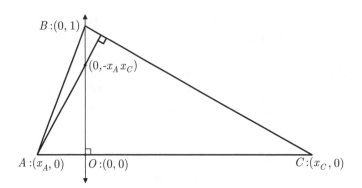

FIGURE 8.11.

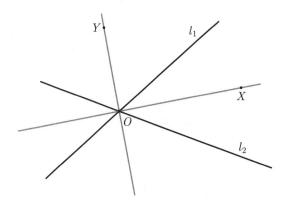

FIGURE 8.12.

We now find the equation for the altitude through A. The slope of $\overline{BC}$ is $\dfrac{0-1}{x_c-0} = \dfrac{-1}{x_c}$, so the altitude through A has slope x_c. The point-slope equation for the altitude is $y = x_c(x - x_A)$. Since the altitude from B corresponds to the vertical line $x = 0$, the point of intersection of the altitudes from A and from B has coordinates $(0, -x_A x_c)$.

Clearly the point of intersection of $\overleftrightarrow{BO}$ with the altitude through C will have coordinates $(0, -x_c x_A)$; indeed, in the argument above just replace C by A and A by C. The symmetry of the expression $x_c x_A$ with respect to A and C proves that the altitudes through A and C intersect $\overleftrightarrow{BO}$ at the same point! Hence, the three altitudes are concurrent. $\square$

The reader is invited to clarify how our proof used the assumption that the coordinate system was Cartesian. We also suggest comparing this solution with the one provided in the proof of Corollary 3.18.

We conclude this section with an application of the coordinate method to finding a locus.

Example 30.   Given a positive number k and two intersecting lines l_1 and l_2, find the locus of all points whose sum of distances to the lines is k.

Solution: We chose a coordinate system OXY in such a way that the coordinate axes OX and OY are each bisectors of the angles formed by l_1 and l_2. Clearly, the axes OX and OY are perpendicular and the obtained coordinate system is Cartesian. (See Figure 8.12.)

Then $l_1 : y = mx$ for some $m > 0$, and $l_2 : y = -mx$. A point (x, y) belongs to the locus if and only if

$$d((x, y), l_1) + d((x, y), l_2) = k.$$

Using Theorem 8.7, this is equivalent to

$$\frac{|mx - y|}{\sqrt{m^2 + 1}} + \frac{|-mx - y|}{\sqrt{(-m)^2 + 1}} = k.$$

To remove the clutter, let $K = k\sqrt{m^2 + 1}$. Then the equation can be rewritten as

$$|mx - y| + |mx + y| = K. \tag{8.5}$$

It is clear that if (x, y) satisfies this equation, then so do points $(-x, y)$, $(x, -y)$, and $(-x, -y)$. Therefore, the graph of the equation is symmetric with respect to both coordinate axes, and it is sufficient to construct it in the 1st quadrant only, i.e., we may assume that $x \geq 0$ and $y \geq 0$.

Since we are in the first quadrant, $mx + y \geq 0$, and (8.5) reduces to $|mx - y| + mx + y = k$. We consider the two cases where $mx - y \geq 0$ and $mx - y \leq 0$ separately.

Let us first illustrate the general argument with the specific example where $k = 1$ and $m = 2$, so that (8.5) gives $|2x - y| + |2x + y| = \sqrt{5}$. Then if $0 \leq y \leq 2x$, we have that $(2x - y) + (2x + y) = \sqrt{5}$, so $x = \sqrt{5}/4$. This means y will take on all values in the interval $[0, \sqrt{5}/2]$, so the graph is the "vertical" segment[8] with endpoints $(\sqrt{5}/4, 0)$ and $(\sqrt{5}/4, \sqrt{5}/2)$.

On the other hand, if $0 \leq 2x \leq y$, then $|2x - y| = -(2x - y)$, and we have $-(2x - y) + (2x + y) = \sqrt{5}$, so $y = \sqrt{5}/2$. In this case, the graph is the "horizontal" segment with endpoints $(0, \sqrt{5}/2)$ and $(\sqrt{5}/4, \sqrt{5}/2)$.

We now solve the general case in a similar fashion. If $0 \leq y \leq mx$, then

$$(8.5) \Leftrightarrow (mx - y) + (mx + y) = k\sqrt{m^2 + 1} \Leftrightarrow x = K/(2m).$$

The graph of this relation in the 1st quadrant is the vertical segment

$$\{(K/(2m), y) : 0 \leq y \leq K/2\}.$$

If $y > mx \geq 0$, then

$$(8.5) \Leftrightarrow -(mx - y) + (mx + y) = K \Leftrightarrow y = K/2.$$

The graph of this relation in the 1st quadrant is the horizontal segment

$$\{(x, K/2) : 0 \leq x \leq K/(2m)\}.$$

Therefore the locus is the rectangle shown in Figure 8.13. $\qquad\square$

8.5 Systems of Equations, Revisited

Given a pair of distinct curves in the plane, we may wish to determine their point(s) of intersection. Such a problem can be solved with a system of equations, using algebra. Often applications of a

---

[8] Recall that we have defined the vertical lines in a coordinate system to OXY to be those lines that are parallel to the y-axis, OY; the horizontal lines are those that are parallel to the x-axis, OX.

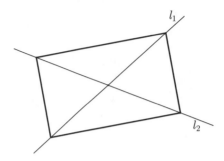

FIGURE **8.13.**

well-known fact or method are more exciting than the proof of the fact itself and, as a result, the "theoretical" part of the problem is not given sufficient attention. In this section, we show how insight into the theory of the very simple idea of equivalent systems of equations can be used to provide elegant solutions for a class of geometric problems for which students may typically choose a cumbersome algebraic approach. We begin with a simple example.

Example 31.   Solve the following system of equations:

$$2x + 3y = 11, \quad 4x - 2y = -2. \tag{8.6}$$

Solution: Multiplying both sides of the first equation by 2 and both sides of the second by 3, then adding the results and dividing by 16, we obtain $x = 1$ and, subsequently, $y = 3$. Therefore $\{(1, 3)\}$ is the solution set of the system.   □

How can one be sure that the new system obtained using the above method (namely, $2x + 3y = 11$, $x = 1$) is equivalent to (8.6), i.e., has the same set of solutions as (8.6)? Using a graphical interpretation of (8.6), the question can be restated as "why do lines whose equations are $4x - 2y = -2$ and $x = 1$ intersect the line whose equation is $2x + 3y = 11$ at the same point?"

The method we are using above is familiar to all algebra students. Despite its importance, very few books of algebra, precalculus, or even linear algebra prove that it is valid. The typical algebra student may have serious doubts that it requires any proof at all. Nevertheless, a thorough understanding of this method is important for understanding the solutions of later examples in this section. The following theorem legitimizes the method.

Theorem 8.10.   *Let $F_1(x, y)$ and $F_2(x, y)$ be two algebraic expressions. Let α and β be real numbers, $\beta \neq 0$. Then the following two systems are equivalent.*

$$F_1(x, y) = 0, \qquad F_2(x, y) = 0, \tag{8.7}$$

$$F_1(x, y) = 0, \quad \alpha F_1(x, y) + \beta F_2(x, y) = 0. \tag{8.8}$$

Proof. To prove that (8.7) and (8.8) are equivalent is to show that each solution of (8.7) is a solution of (8.8) and vice versa. If both solution sets are empty, the statement obviously holds. If one of the solution sets is not empty, then the following argument shows that neither is the other one, and they are equal.

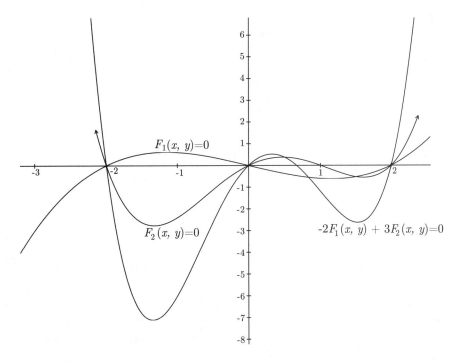

FIGURE 8.14.

(1) If (a, b) is a solution of (8.7), then $F_1(a, b) = 0$, $F_2(a, b) = 0$ and, as a result, $\alpha F_1(a, b) + \beta F_2(a, b) = 0 + 0 = 0$, which implies that (a, b) is also a solution of (8.8).

(2) If (a, b) satisfies (8.8) and $\beta \neq 0$ (by assumption), then the relations $F_1(a, b) = 0$ and $\alpha F_1(a, b) + \beta F_2(a, b) = 0$ imply $\alpha \cdot 0 + \beta F_2(a, b) = 0$, which is equivalent to $F_2(a, b) = 0$, and this completes the proof.

$\square$

In terms of the graphical interpretation of the equivalence of (8.7) and (8.8), the theorem says that the curves defined by the equations $F_2(x, y) = 0$ and $\alpha F_1(x, y) + \beta F_2(x, y) = 0$ intersect the graph of the curve with the equation $F_1(x, y) = 0$ at the same set of points. In Figure 8.14, we see that the curve of $F_1(x, y) = 0$ intersects the x-axis at $x = -2$, $x = 0$, and $x = 2$ while the curve of $F_2(x, y) = 0$ intersects the x-axis at $x = -2$, $x = 0$, $x = 1$, and $x = 2$. Thus, the system $F_1(x, y) = 0$, $F_2(x, y) = 0$ has solutions $(-2, 0)$, $(0, 0)$, and $(2, 0)$. From Theorem 8.10, then, the system $F_1(x, y) = 0$, $-2F_1(x, y) + 3F_2(x, y) = 0$ will also have solutions $(-2, 0)$, $(0, 0)$, and $(2, 0)$, as shown. Note that while $(2/3, 0)$ is a solution to $-2F_1(x, y) + 3F_2(x, y) = 0$, this is not a solution to the system.

We conclude this section with one more example. Several more impressive examples will be seen in the problem sets of future chapters, after we develop equations for different types of curves.

Example 32. Given an angle, $\angle CBD$, and a point A in its exterior, let $\overrightarrow{AR_i}$, $i = 1, 2, \ldots, n$, intersect $\overrightarrow{BC}$ at a point C_i and $\overrightarrow{BD}$ at a point D_i. Let $Q_{ij} = Q_{ji}$ be the intersection of the segments $C_i D_j$ and $C_j D_i$, $i, j \in \{1, 2, \ldots, n\}$, $i \neq j$. Prove that all points Q_{ij} are collinear.

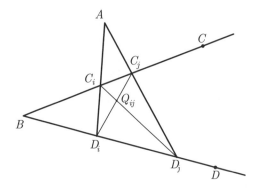

FIGURE 8.15.

Solution: Choose a coordinate system such that B is the origin, and lines BD and BC become the x-axis and the y-axis respectively. (The system is not necessarily Cartesian; see Figure 8.15.)

   Assume that A and $\overrightarrow{BC}$ lie on the same side of the line BD. Choose a scale such that point A has coordinates $(-1, a)$. Denote the slope of $\overleftrightarrow{AC_i}$ by m_i (which exists, since $\overleftrightarrow{AC_i}$ and $\overleftrightarrow{BC}$ are not parallel). Notice that $m_i \neq 0$, and $m_i \neq m_j$ for $i \neq j$. Then an equation of $\overleftrightarrow{AC_i}$ can be written in the form: $y - a = m_i(x + 1)$. Therefore, the points C_i and D_j have coordinates $(0, m_i + a)$ and $(-1 - a/m_j, 0)$, respectively, and the coordinates of the point Q_{ij} can be found by solving the following system of equations of the lines $\overleftrightarrow{C_i D_j}$ and $\overleftrightarrow{C_j D_i}$ with

$$F_1(x, y) = \frac{x}{-1 - a/m_j} + \frac{y}{m_i + a} - 1 = 0,$$

$$F_2(x, y) = \frac{x}{-1 - a/m_i} + \frac{y}{m_j + a} - 1 = 0.$$

Subtracting these equations – which amounts to letting $\alpha = 1$ and $\beta = -1$ in Theorem 8.10 – and simplifying the result, we obtain $y = ax$. Therefore, all points Q_{ij} lie on the line with the equation $y = ax$. $\qquad\square$

8.6   Equations of Circles

In Chapter 4, a **circle** was defined as the set of all points lying at a fixed distance (called the *radius*) from a specified point (the *center* of the circle). Consider any circle with center $O : (h, k)$ and radius r (denoted, as before, by $\mathcal{C}(O, r)$). Then a point $P : (x, y)$ is on $\mathcal{C}(O, r)$ if and only if $d(P, O) = r$. The formula for distance between two points yields the following equation for $\mathcal{C}(O, r)$:

$$(x - h)^2 + (y - k)^2 = r^2.$$

   Any circle can be represented with an equation of the above form, called the **standard form** for the equation of a circle. By expanding and simplifying, the standard form for the equation of a circle can be rewritten in what is called the **general form**:

$$x^2 + y^2 + Dx + Ey + F = 0. \tag{8.9}$$

Observe that the second-degree equation that describes a circle will have no xy-term. Will the graph of an equation of the form (8.9) always be a circle? Certainly not, since the solution set could be empty (such as for $x^2 + y^2 + 1 = 0$) or a single point (such as for $x^2 + y^2 = 0$).

However, if the equation $x^2 + y^2 + Dx + Ey + F = 0$ is satisfied by at least two distinct points, then its graph is indeed a circle. The center and radius can be determined by completing the square with respect to x and y, as we illustrate in the following example. Of course, the same is true for a general equation of the second degree of the form $Ax^2 + Cy^2 + Dx + Ey + F = 0$, if $A = C \neq 0$.

Example 33. Determine the graph of $4x^2 + 4y^2 - 4x + 24y - 27 = 0$.

Solution:

$$4x^2 + 4y^2 - 4x + 24y - 27 = 0 \Leftrightarrow$$
$$x^2 + y^2 - x + 6y = \frac{27}{4} \Leftrightarrow$$
$$\left(x^2 - x + \frac{1}{4}\right) + (y^2 + 6y + 9) = \frac{27}{4} + \frac{1}{4} + 9 \Leftrightarrow$$
$$\left(x - \frac{1}{2}\right)^2 + (y + 3)^2 = 16.$$

Thus, the equation $4x^2 + 4y^2 - 4x + 24y - 27 = 0$ describes a circle of radius 4 with center at the point $(1/2, -3)$. $\square$

Now we present several applications of the coordinate method to problems involving circles. We begin with a "coordinate" proof of a fundamental theorem, Theorem 4.2.

Example 34. Prove that any three non-collinear points determine a unique circle.

Solution: Let A, B, and C be three non-collinear points. In order to simplify our computations, coordinatize the plane so that $A : (0, 0)$, $C : (1, 0)$, and $B : (x_B, y_B)$ (with $y_B \neq 0$), as shown in Figure 8.16.

We wish to find a unique point $P : (x_P, y_P)$ so that $d(A, P) = d(B, P) = d(C, P)$. This condition is equivalent to the system

$$d(A, P) = d(B, P), \quad d(A, P) = d(C, P). \tag{8.10}$$

Using the distance formula we get:

$$d(A, P) = d(C, P) \Leftrightarrow \sqrt{x_P^2 + y_P^2} = \sqrt{(1 - x_P)^2 + y_P^2} \Leftrightarrow x_P = \frac{1}{2}.$$

(This makes sense; a point is equidistant from $A : (0, 0)$ and $C : (1, 0)$ if and only if it lies on the perpendicular bisector to $\overline{AC}$, which is the vertical line $x = 1/2$.)

Using the distance formula again, with $x_P = 1/2$ and $y_B \neq 0$, we get

$$d(A, P) = d(B, P) \Leftrightarrow \sqrt{\left(\frac{1}{2}\right)^2 + y_P^2} = \sqrt{\left(x_B - \frac{1}{2}\right)^2 + (y_B - y_P)^2} \Leftrightarrow$$
$$y_P = \frac{x_B^2 - x_B + y_B^2}{2y_B}.$$

Thus, system (8.10) is equivalent to

$$x_P = \frac{1}{2}, \quad y_P = \frac{x_B^2 - x_B + y_B^2}{2y_B}, \tag{8.11}$$

which means that the desired point P (the center of the circle) exists and is unique.

Our next two examples illustrate effective applications of Theorem 8.10.

Example 35. Find an equation of the line passing through the points of intersections of two circles C_1 and C_2 given respectively by the equations

$$x^2 + (y-3)^2 = 9 \quad \text{and} \quad (x-4)^2 + (y-2)^2 = 16. \tag{8.12}$$

Solution: First of all we observe that the circles do intersect, because the distance between the centers is $\sqrt{17}$ and the radii are 3 and 4. Therefore, system (8.12) has a solution. Replacing the second equation in (8.12) by its difference with the first one, we obtain system (8.13) which, according to Theorem 8.10, is equivalent to (8.12) (with $\alpha = 1$ and $\beta = -1$):

$$x^2 + (y-3)^2 = 9, \quad 4x - y = 2. \tag{8.13}$$

The line having equation $4x - y = 2$ is the one we are looking for, because it crosses C_1 at the same points as C_2. (See Figure 8.17.) $\square$

Notice that we solved the problem of Example 35 *without finding* the points of intersection of the circles. The latter could be done (if we wanted to) by solving the system (8.13). Many people solving this problem might not realize that obtaining the equation $4x - y = 2$ solves the problem.

Example 36. Let C_1, C_2, and C_3 be three circles in the plane, such that any two of them intersect at two points. Prove that the three common chords are concurrent. (See Figure 8.18.)

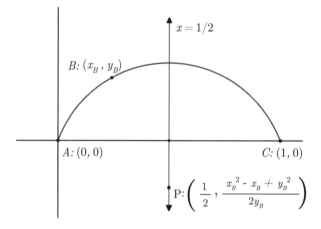

FIGURE 8.16.

Solution: Let $F_i(x, y)$ denote the equation of C_i in a Cartesian coordinate system; that is, $F_i(x, y) : (x - a_i)^2 + (y - b_i)^2 - (r_i)^2 = 0, i = 1, 2, 3$. Let L_{ij} be the line passing through the points of intersection of C_i and C_j. Using the same argument as in Example 35, we conclude that $F_i(x, y) - F_j(x, y) = (a_j - a_i)x + (b_j - b_i)y + c_{ij} = 0$ is an equation of L_{ij}, where c_{ij} is the constant term. The coordinates of the point of intersection of lines L_{12} and L_{13} can be found by solving the system

$$L_{12} : F_1(x, y) - F_2(x, y) = 0, \quad L_{13} : F_1(x, y) - F_3(x, y) = 0. \tag{8.14}$$

Replacing the second equation of (8.14) by its difference with the first one we obtain:

$$F_1(x, y) - F_2(x, y) = 0, \quad F_3(x, y) - F_2(x, y) = 0, \tag{8.15}$$

which, according to Theorem 8.10 (with $\alpha = 1, \ \beta = -1$) is equivalent to (8.14). But the second equation in (8.15) is an equation of the line $L_{32} (= L_{23})$! Therefore the lines L_{13} and L_{23} intersect the line L_{12} at the same point. $\qquad\square$

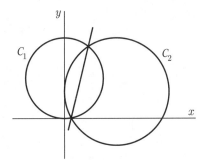

FIGURE 8.17.

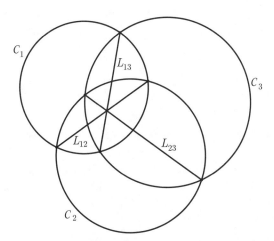

FIGURE 8.18.

Our last example is an application of the coordinate method to finding a locus.

Example 37. Let points A and B and a positive number k be given. Find the locus of all points C such that $CA/CB = k$.

Solution: Introduce a Cartesian coordinate system such that $A : (0, 0)$ and $B : (1, 0)$. A point $C : (x, y)$ is in the locus if and only if

$$\frac{AC}{BC} = k \quad \Leftrightarrow \quad \frac{\sqrt{x^2 + y^2}}{\sqrt{(x-1)^2 + y^2}} = k \quad \Leftrightarrow \quad \frac{x^2 + y^2}{(x-1)^2 + y^2} = k^2. \qquad (8.16)$$

For $k = 1$, (8.16) is equivalent to $x = 1/2$, so the locus is the perpendicular bisector of $\overline{AB}$. For $0 < k \neq 1$, (8.16) is equivalent to each of the following.

$$x^2(k-1) - 2k^2 x + y^2(k^2 - 1) + k^2 = 0,$$

$$x^2 - \left(\frac{2k^2}{k^2 - 1}\right)x + \left(\frac{k^2}{k^2 - 1}\right)^2 + y^2 = \frac{k^4}{(k^2 - 1)^2} - \frac{k^2}{(k^2 - 1)},$$

$$\left(x - \frac{k^2}{k^2 - 1}\right)^2 + y^2 = \left(\frac{k}{k^2 - 1}\right)^2.$$

Hence the locus is the circle centered at $(k^2/(k^2 - 1), 0)$ and of radius $k/|k^2 - 1|$. It is often referred to as "the circle of Apollonius." $\qquad\qquad\square$

Compare the solution above to the one for Problem 6.13. Isn't this one *much* easier?

8.7 Problems

You just keep pushing. I made every mistake that could be made. But I just kept pushing.
— René Descartes (1596–1650)

Try to solve the following problems by using the coordinate method. Then see whether you can solve them in other way(s), and compare the advantages of different approaches.

8.1 Prove that the diagonals of a parallelogram bisect each other.

8.2 Prove that the diagonals of a rhombus are perpendicular bisectors of each other.

8.3 Prove that in a right triangle ABC, where $\angle C$ is the right angle, the median at C has length equal to half the length of side AB.

8.4 Prove that the midpoints of the sides of any convex quadrilateral are vertices of a parallelogram.

8.5 Consider the triangle with vertices $A : (6, 0)$, $B : (2, 4)$, and $C : (1, -1)$.

   (a) Find the length of the median at C.

   (b) Find the length of the altitude from C.

8.6 Given a positive number k and two intersecting lines l_1 and l_2, find the locus of all points whose difference of distances to the lines is k or $-k$.

8.7 The midpoints of the sides $\overline{AB}$ and $\overline{CD}$ and of sides $\overline{BC}$ and $\overline{DE}$ of a convex pentagon (5-gon) $ABCDE$ are joined by two segments. The midpoints H and K of these segments are joined. Prove that $\overline{HK} \parallel \overline{AE}$ and $HK = \frac{1}{4}AE$.

8.8 Prove that the three perpendicular bisectors to the sides of a triangle are concurrent.

8.9 Let $ABCD$ be a parallelogram, and let $F \in \overline{AD}$ be such that $AF = \frac{1}{5}AD$. Let E be the intersection of $\overline{BF}$ and $\overline{AC}$. Prove that $AE = \frac{1}{6}AC$.

8.10 Let $ABCD$ be a trapezoid with $\overline{BC} \parallel \overline{AD}$, $AD = a$, $BC = b$, $a > b$, and let O be the intersection of the diagonals $\overline{AC}$ and $\overline{BD}$. A line through O that is parallel to the bases intersects lateral sides AB and CD at points E and F, respectively.

(a) Show that $EO = OF$.
(b) Find EF.
(c) Prove that the midpoints of bases AD and BC, point O, and the point of intersection of sides AB and CD are all collinear.

8.11 Let $ABCD$ be a trapezoid, and let P and Q be the midpoints of the bases $\overline{BC}$ and $\overline{AD}$, respectively. Take a point M on $\overleftrightarrow{AC}$ such that M is outside of the trapezoid. Let $\overleftrightarrow{MP}$ and $\overleftrightarrow{MQ}$ intersect the lateral sides, $\overline{AB}$ and $\overline{CD}$, at points H and K, respectively. Prove that $\overline{HK}$ is parallel to the bases.

8.12 Prove that the line joining the point of intersection of the extensions of the nonparallel sides of a trapezoid to the point of intersection of its diagonals bisects each base of the trapezoid.

8.13 Given an equilateral triangle, $\triangle ABC$, prove that the sum of three distances from any point in the interior or on the boundary of the triangle to the sides is constant.

8.14 Prove that the area of a triangle whose vertices are (x_A, y_A), (x_B, y_B), and (x_C, y_C) equals

$$\frac{1}{2}|(x_A y_B + x_B y_C + x_C y_A - x_B y_A - x_C y_B - x_A y_C)|$$

$$= \frac{1}{2}|(x_A - x_C)(y_B - y_C) - (x_B - x_C)(y_A - y_C)|.$$

8.15 Prove that if $\triangle ABC$ lies in a plane, where points A and B are fixed, then the center of the nine-point circle[9] for $\triangle ABC$ must lie on or in the exterior of a circle centered at the midpoint of $\overline{AB}$ and whose radius is equal to $AB/4$. Characterize those triangles for which the radius of the nine-point circle is exactly $AB/4$.

8.16 (Ceva's Theorem) Consider a triangle with vertices, A, B, and C, and points D, E, and F such that $A - F - B$, $B - D - C$, and $C - E - A$. Then $\overline{AD}$, $\overline{BE}$, and $\overline{CF}$ are concurrent if and only if

$$\frac{AF}{FB} \cdot \frac{BD}{DC} \cdot \frac{CE}{EA} = 1.$$

---

[9] The nine-point circle is a remarkable circle that can be constructed for any triangle. It contains the following nine points: the midpoint of each side of the triangle, the foot of each altitude, and the midpoint of the segment of each altitude from its vertex to the orthocenter.

8.8 Supplemental Problems

*S*8.1 Prove that the diagonals of a rectangle are congruent.

*S*8.2 Each vertex of a quadrilateral is joined to the centroid of the triangle formed by the other three vertices. Prove that the four segments are concurrent. (This problem also appears as Problem S3.3.15.)

*S*8.3 Let $M_1, \ldots, M_6$ be the midpoints of consecutive sides in an arbitrary hexagon (6-gon). Prove that the centroids of $\triangle M_1 M_3 M_5$ and $\triangle M_2 M_4 M_6$ coincide.

*S*8.4 Given $\triangle ABC$, let $A' \in \overline{BC}$ and $B' \in \overline{AC}$, and let N be the point of intersection of $\overline{AA'}$ and $\overline{BB'}$. Find $AB'/B'C$, if $AN = 4NA'$ and $BN = 3NB'$.

*S*8.5 Let $ABCD$ and $A_1 B_1 C_1 D_1$ be parallelograms with $A_1 B_1 C_1 D_1$ inscribed in $ABCD$ in such a way that vertices A_1, B_1, C_1 and D_1 lie on $\overline{AB}, \overline{BC}, \overline{CD}$, and $\overline{DA}$, respectively. Prove that the centers of these parallelograms coincide.

---

*S*8.6 Points A and B are moving along two rays that initiate at point O such that

$$\frac{1}{OB} - \frac{1}{OA} = 1.$$

Prove that all lines AB pass through a fixed point.

*S*8.7 Let $ABCD$ be a square, and let E be the midpoint of $\overline{AB}$. Let points $F \in \overline{BC}$ and $G \in \overline{CD}$ be chosen in such a way that $\overline{EF} \parallel \overline{AG}$. Prove that $\overline{FG}$ is tangent to the incircle of the square.

*S*8.8 Given two circles in the plane such that each circle is in the exterior of another, let A_1 and B_1 be the points of intersection of the first circle with two tangents from its center to the second circle; likewise, let A_2 and B_2 be the points of intersection of the second circle with two tangents from its center to the first circle. Prove that $A_1 B_1 = A_2 B_2$. (This problem also appeared as Problem S4.15.)

*S*8.9 Fix a diameter in a circle, and let K be a point on the diameter. Let $\overline{AB}$ be a chord of the circle that passes through K and forms a $45°$ angle with the diameter. Prove that the sum $AK^2 + KB^2$ does not depend on the position of K on the diameter. (This problem also appeared as Problem S4.11.)

*S*8.10 Two pairs of perpendicular lines in a plane intersect in such a way that four right triangles are formed. Prove that the midpoints of the hypotenuses of these triangles are vertices of a rectangle.

*S*8.11 Given two circles, one in the exterior of another, find the set of all points X in the plane such that the lengths of tangent segments from X to the circles are equal.

*S*8.12 Two points A_1 and A_2 are moving with equal constant velocity v along two intersecting lines l_1 and l_2, respectively. Prove that there exists a fixed point P in the plane such that at any moment of time the distances $A_1 P$ and $A_2 P$ are equal.

*S*8.13 (Menelaus' Theorem[10]) Consider $\triangle ABC$. Suppose $C_1 \in \overline{AB}$, $A_1 \in \overline{BC}$, and $B_1 \in \overleftrightarrow{AB}$ such that $A - C - B_1$. Then points A_1, B_1, and C_1 are collinear if and only if

$$\frac{AC_1}{C_1 B} \cdot \frac{BA_1}{A_1 C} \cdot \frac{CB_1}{B_1 A} = 1.$$

---

[10] Menelaus of Alexandria (c. 70–140 A.D.)

$S8.14$ (Newton's Theorem) A quadrilateral is circumscribed around a circle. Prove that the center of the circle coincides with the midpoint of the segment joining the midpoints of the diagonals of the quadrilateral.

$S8.15$ (Pappus' Theorem[11])

(a) Suppose points A_1, B_1, and C_1 are on line l_1 with $A_1 - B_1 - C_1$, and points A_2, B_2, and C_2 are on line l_2 with $A_2 - B_2 - C_2$. Let $C_3 = A_1B_2 \cap A_2B_1$, $B_3 = A_1C_2 \cap A_2C_1$, and $A_3 = B_1C_2 \cap B_2C_1$. Then points A_3, B_3, and C_3 are collinear.

(b) Prove the same result if the six points are on a circle with

$$A_1 - B_1 - C_1 - A_2 - B_2 - C_2.$$

$S8.16$ (Bow-tie Theorem) Let $\overline{AB}$ be a chord in a circle, and C be the midpoint of $\overline{AB}$. Let $\overline{KL}$ and $\overline{MN}$ be two chords of the circle which pass through C. Assume that K and M are on one side of line AB. Let Q and P be the points of intersection of chords KN and ML with the chord AB, respectively. Prove that $CP = CQ$.

---

[11] Pappus of Alexandria (c. 290–c. 350 B.C.)

9

Conics

One cannot escape the feeling that ... mathematical formulas have an independent existence and an intelligence of their own, that they are wiser than we are, wiser even than their discoverers, that we get more out of them than was originally put into them.
— Heinrich Hertz[1] (1857–1894)

Compared to the synthetic approach presented in Chapter 6, analytic geometry often provides a superior method for studies of curves. A **conic** is a figure obtained by the intersection of a circular cone with a plane. It can be proven—though we do not do it in this text, which is concerned with plane geometry—that any conic is one of the following figures: an ellipse, a hyperbola, a parabola, a pair of intersecting lines, a line, or a point. The first three of these figures, namely ellipse, hyperbola, and parabola, are called **non-degenerate conics**, and the other three are called **degenerate conics**. The goal of this chapter is to find equations of conics and illustrate some ways in which these equations can be used. In all cases, a Cartesian coordinate system will be chosen in such a way that the obtained equation of a curve is relatively simple.

9.1 Parabolas

A **parabola** has been previously described as the set of all points that are equidistant from a given fixed point F (called the *focus*) and a given fixed line l (called the *directrix*). We assume that the focus does not lie on the directrix. (See Figure 9.1.) The *axis* of a parabola is the line passing through the focus and perpendicular to the directrix; the *vertex* is the point on the axis halfway between the focus and the directrix.

Let OXY be a Cartesian coordinate system such that the y-axis coincides with the axis of the parabola, the vertex of the parabola is at the origin, and the focus is $F : (0, p)$. Then the directrix l is

---

[1] Quoted by ET Bell in *Men of Mathematics*, 1937.

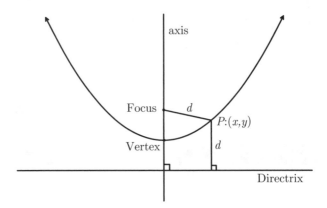

FIGURE 9.1.

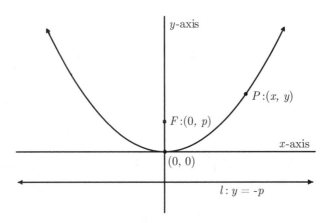

FIGURE 9.2.

parallel to the x-axis and $l : y = -p$. Let $P : (x, y)$ be any point on the parabola. (See Figure 9.2.) Then P lies on the parabola if and only if

$$
\begin{aligned}
d(P, F) &= d(P, l) & \Leftrightarrow \\
\sqrt{x^2 + (y - p)^2} &= |y - (-p)| & \Leftrightarrow \text{ (as both sides are nonnegative)} \\
x^2 + (y - p)^2 &= (y + p)^2 & \Leftrightarrow \\
x^2 &= 4py.
\end{aligned}
$$

The equation $x^2 = 4py$ is often referred to as the **standard form** of the equation of the parabola with vertex at $(0, 0)$, focus $F : (0, p)$, and directrix given by $y = -p$. If the scale is chosen in such a way that $4p = 1$, then the equation becomes just $y = x^2$. What could be simpler?! More generally, when the vertex is at (h, k) and the focus is at $(h, k + p)$, the standard form of the equation of the parabola can be transformed easily into

$$
(x - h)^2 = 4p(y - k). \tag{9.1}
$$

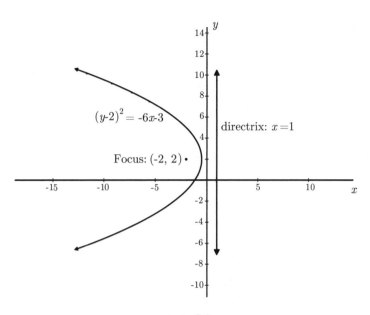

FIGURE 9.3.

As $p \neq 0$, this equation can always be rewritten in the form

$$y = ax^2 + bx + c, \tag{9.2}$$

where $a, b, c \in \mathbb{R}$ and $a \neq 0$. Conversely, every equation of the form (9.2), can be rewritten in the form (9.1) by completing the square with respect to x. (See also Problem 9.3.)

Example 38.   Describe the curve whose equation is $y^2 + 6x - 4y + 7 = 0$.

Solution: The equation is of the form (9.2) with x and y interchanged. As points (a, b) and (b, a) are symmetric with respect to line $y = x$, we conclude that the curve is a parabola with directrix parallel to the x-axis. In order to find the vertex, the focus, and the directrix, we complete the square with respect to y, and then represent -6 in the form $4p$. This gives

$$y^2 + 6x - 4y + 7 = 0 \qquad\qquad \Leftrightarrow$$
$$(y - 2)^2 = -6x - 3 \qquad\qquad \Leftrightarrow$$
$$(y - 2)^2 = 4\left(-\frac{3}{2}\right)\left(x + \frac{1}{2}\right).$$

Therefore the curve is a parabola having

(1) vertex at $(-1/2, 2)$,
(2) focus at $(-1/2 + -3/2, 2) = (-2, 2)$, and
(3) directrix $x = -1/2 - (-3/2) = 1$. (See Figure 9.3.)   □

Example 39.   In how many points can a line and a parabola intersect?

  The answer to this question may seem obvious to the reader, and if so, try to prove your claim! The intuitive answer (of 0, 1, or 2 points) undoubtedly comes from the "mental picture" we have in our minds of what a parabola looks like, but where did that picture come from? Likely not from the definition of a parabola, so a proof is needed.

Solution: Let l denote the line and $\mathcal{P}$ denote the parabola. Choose a Cartesian coordinate system such that the equation of $\mathcal{P}$ is $y = x^2$. If l is vertical in this coordinate system, then its equation is $l : x = a, a \in \mathbb{R}$, and l intersects $\mathcal{P}$ only at the point (a, a^2).

If l is not vertical, then its equation can be written as $l : y = mx + b$. In order to find the coordinates of the intersection points, we solve the system formed by equations $y = x^2$ and $y = mx + b$. Substituting for y in the first equation, we obtain a quadratic equation in x: $x^2 - mx - b = 0$. Let $D = m^2 + 4b$; this is called the *discriminant* of the equation. If $D < 0$, there are are no points of intersection, and, if $D > 0$, there are two distinct points of intersection. If $D = 0$, then, just as we might say that "a quadratic equation has two equal solutions," we say that l intersects $\mathcal{P}$ at two equal points. It can be explained that in this case l is tangent to $\mathcal{P}$.   □

The ideas above allow us to find an equation of the tangent line to $\mathcal{P}$ at an arbitrary point (a, a^2) without the use of calculus (more precisely, without using derivatives). These ideas were used in the seventeenth century for finding tangent lines to various curves, and led to far-reaching generalizations, which have become part of modern mathematics. Here is the argument for the parabola.

As a vertical line is never tangent to $\mathcal{P}$, the equation of a tangent line can be written as $y - a^2 = m(x - a)$ or $y = mx + (a^2 - ma)$. Then

$$D = 0 \Leftrightarrow m^2 + 4(a^2 - ma) = 0 \Leftrightarrow (m - 2a)^2 = 0 \Leftrightarrow m = 2a.$$

Hence the tangent line has an equation $y = 2ax - a^2$.

Example 40. Find an equation for a parabola passing through the points $A : (-1, 2)$, $B : (-2, -3)$, and $C : (3, -1)$. Is such a parabola unique?

Solution: Suppose that the parabola passing through the specified points has a vertical axis. Then it can be described by an equation of the form $y = ax^2 + bx + c$. Substituting in the coordinates of the three given points, we obtain a system of three equations in three unknowns:

$$a - b + c = 2,$$
$$4a - 2b + c = -3,$$
$$9a + 3b + c = -1.$$

Solving this system of equations, we obtain $a = -23/20$, $b = 31/20$, $c = 47/10$, so an equation for a parabola passing through the points $(-1, 2)$, $(-2, -3)$, and $(3, -1)$ is given by

$$y = \frac{-23}{20}x^2 + \frac{31}{20}x + \frac{47}{10}.$$

Note that if one assumes the parabola has a horizontal line of symmetry, then it is defined by the equation $x = ay^2 + by + c$, and an almost identical argument will yield the equation

$$x = \frac{-23}{30}y^2 - \frac{17}{30}y + \frac{16}{5}.$$

This shows that there are at least two parabolas passing through the given three points. (See Figure 9.4.) It can be shown that there are infinitely many parabolas passing through these three points whose axes are neither horizontal nor vertical.   □

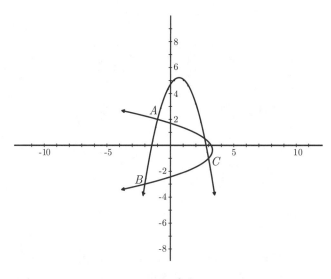

FIGURE 9.4.

Example 41. Use the coordinate method to prove the reflection property of parabolas (as stated in Corollary 6.4).

Solution: Choose a Cartesian coordinate system such that the equation of the parabola is $y = x^2$. Then its focus F is at $(0, 1/4)$. Let $A = (a, a^2)$ be an arbitrary point on the parabola. If $a = 0$, then the point is the vertex of the parabola and the statement is obvious.

If $a \neq 0$, then, as shown in Example 39, the slope of the tangent line is $m_1 = 2a$. The line through A parallel to the axis of the parabola has equation $x = a$, and the slope of $\overleftrightarrow{FA}$ is $m_2 = (a^2 - 1/4)/(a - 0) = a - 1/(4a)$. It is sufficient to prove that lines $x = a$ and $\overleftrightarrow{FA}$ form equal angles with the tangent line. If α is the measure of the inclination of the tangent, then the measure of the angle β between $x = a$ and the tangent line is $|\pi/2 - \alpha|$. Hence, $\tan \beta = 1/|2a|$. Using Theorem 8.6, we find that the tangent of the angle γ between the tangent line and $\overleftrightarrow{FA}$ is

$$\tan \gamma = \left| \frac{m_1 - m_2}{1 + m_1 m_2} \right| = \left| \frac{2a - (a - \frac{1}{4a})}{1 + 2a(a - \frac{1}{4a})} \right| = \frac{1}{2|a|}.$$

As $\tan \beta = \tan \gamma$, and both angles are acute, we conclude that $\beta = \gamma$, and the proof is finished. $\qquad\square$

9.2 Ellipses

In this section we wish to find equations of ellipses, proceeding straight from the definition of an ellipse. Then we illustrate how having these equations facilitates studying the properties of ellipses.

We recall that an **ellipse** consists of all points the sum of whose distances from two fixed points (the *foci*) is constant. Let F_1 and F_2 be the foci; then $F_1 F_2 = 2c \geq 0$ and, for every point P on the ellipse, $P F_1 + P F_2 = 2a > 0$. (See Figure 9.5.) Let us denote the ellipse by $\mathcal{E}$. By the triangle inequality, $a \geq c$. If $a = c$, then $\mathcal{E}$ is just $\overline{F_1 F_2}$. For the rest of this section, we assume $a > c \geq 0$.

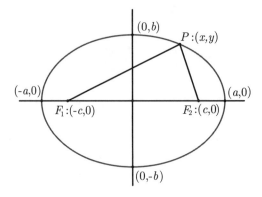

FIGURE 9.5.

In order to derive an equation for an ellipse, we introduce a Cartesian coordinate system such that the origin bisects $\overline{F_1 F_2}$ and the x-axis coincides with $\overleftrightarrow{F_1 F_2}$. Then $F_1 : (-c, 0)$ and $F_2 : (c, 0)$, and let $P : (x, y)$. Using the distance formula we obtain:

$$P \in \mathcal{E} \;\Leftrightarrow\; d(P, F_1) + d(P, F_2) = 2a \qquad \Leftrightarrow \tag{9.3}$$

$$\sqrt{(x + c)^2 + y^2} + \sqrt{(x - c)^2 + y^2} = 2a \;\Leftrightarrow \tag{9.4}$$

$$\sqrt{(x + c)^2 + y^2} = 2a - \sqrt{(x - c)^2 + y^2} \;\Rightarrow \tag{9.5}$$

$$(x + c)^2 + y^2 = 4a^2 - 4a\sqrt{(x - c)^2 + y^2} + (x - c)^2 + y^2 \;\Leftrightarrow \tag{9.6}$$

$$a\sqrt{(x - c)^2 + y^2} = a^2 - cx \;\Rightarrow \tag{9.7}$$

$$(a^2 - c^2)x^2 + a^2 y^2 = a^2(a^2 - c^2). \tag{9.8}$$

We now introduce another parameter, $b = \sqrt{a^2 - c^2}$. Since $a > c \geq 0$, then $b > 0$. Now the last equation is equivalent to

$$\frac{x^2}{a^2} + \frac{y^2}{b^2} = 1. \tag{9.9}$$

With care, one can show that (9.4) $\Leftrightarrow$ (9.9).[2]

---

[2] In deriving equation (9.9) from (9.4), we twice squared both sides of some intermediate equations. Therefore, (9.4) $\Rightarrow$ (9.9). Let us explain that (9.9) $\Rightarrow$ (9.4), which will mean that during our transformations we have not introduced extraneous solutions to (9.4).

Indeed, (9.9) $\Leftrightarrow$ (9.8) which can be rewritten as $a^2((x - c)^2 + y^2) = (a^2 - cx)^2$. The last equation is equivalent to

$$a\sqrt{(x - c)^2 + y^2} = |a^2 - cx|. \tag{9.10}$$

Also (9.9) implies that $|x| \leq a$. Together with $a > c \geq 0$, we get $|cx| \leq ca \leq a^2$. Hence $|a^2 - cx| = a^2 - cx$, and we obtain (9.9) $\Rightarrow$ (9.7) $\Leftrightarrow$ (9.6). As (9.5) $\Leftrightarrow$ (9.4), it is sufficient to show that (9.6) $\Rightarrow$ (9.5).

Clearly, (9.6) $\Leftrightarrow$ $\sqrt{(x + c)^2 + y^2} = |2a - \sqrt{(x - c)^2 + y^2}|$. Equation (9.9) implies that $y^2 \leq b^2 = a^2 - c^2$, and, as we mentioned above, $|xc| \leq a^2$. Since

$$(x - c)^2 + y^2 = x^2 - 2xc + c^2 + y^2 \leq a^2 + 2|xc| + c^2 + (a^2 - c^2) \leq a^2 + 2a^2 + a^2 = 4a^2,$$

we get $2a - \sqrt{(x - c)^2 + y^2} \geq 2a - 2a = 0$, and (9.6) $\Rightarrow$ (9.5). Finally, we can conclude that (9.9) $\Rightarrow$ (9.4). This proves that (9.4) $\Leftrightarrow$ (9.9).

Equation (9.9) is often referred to as the **standard equation** of the ellipse $\mathcal{E}$, and it can be used to derive all of the properties of ellipses which we mentioned in Chapter 6 and many more. It is clear, for example, that a point (x, y) satisfies (9.9) if and only if any one of $(-x, y)$, $(x, -y)$, or $(-x, -y)$ satisfies (9.9). This implies that $\mathcal{E}$ is symmetric with respect to the coordinate axes and the origin. Finding its points of intersection with the coordinate axes, we obtain the vertices of $\mathcal{E}$: $(-a, 0)$, $(a, 0)$, $(0, -b)$, and $(0, b)$.

A **chord** of an ellipse is the segment joining two of its points. The chord containing both foci is called the **major axis** of the ellipse, and the chord joining the other two vertices, which is perpendicular to the major axis, is called the **minor axis**. The major and minor axes bisect each other, and their intersection is called the **center** of the ellipse. The length of the major axis is $2a$, and of the minor is $2b$. As $a \geq \sqrt{a^2 - c^2} = b$, the major axis is always at least as long as the minor. If their lengths are equal, then $c = 0$, $F_1 = F_2$, and $\mathcal{E}$ is a circle. The numbers a and b are often called the **semi-axes lengths** of $\mathcal{E}$. The foci of $\mathcal{E}$ are at $(\pm c, 0)$, where $c = \sqrt{a^2 - b^2}$.

Observe that (9.9) implies that $|x| \leq a$ and $|y| \leq b$. Therefore $\mathcal{E}$ lies inside the rectangle bounded by lines $x = \pm a$ and $y \pm b$. It is easy to see that the *interior* of the ellipse is formed by all points (x, y) satisfying

$$x^2/a^2 + y^2/b^2 < 1,$$

and the *exterior* by all points (x, y) satisfying

$$x^2/a^2 + y^2/b^2 > 1.$$

In order to show this, simply use the inequality signs in lines (9.3) – (9.9).

More generally, an ellipse with center (h, k) and semi-axes lengths a and b has an equation

$$\frac{(x - h)^2}{a^2} + \frac{(y - k)^2}{b^2} = 1. \tag{9.11}$$

If $a > b > 0$, then the major axis of the ellipse is horizontal, and points $(\pm c + h, k)$, with $c = \sqrt{a^2 - b^2}$, are its foci. If $b > a > 0$, then the major axis of the ellipse is vertical, and points $(h, \pm c + k)$, with $c = \sqrt{b^2 - a^2}$, are its foci.

We leave several other properties of ellipses as problems, and devote the end of this section to four examples.

Example 42.   Show that the graph of the equation $9x^2 + 4y = -4 - y^2 + 18x$ is an ellipse. Find its center, semi-axes lengths, vertices, and foci.

Solution: The equation is equivalent to $9x^2 - 18x + y^2 + 4y + 4 = 0$. Completing the squares with respect to x and y, we obtain $9(x - 1)^2 + (y + 2)^2 = 9$, which is equivalent to

$$\frac{(x - 1)^2}{1^2} + \frac{(y + 2)^2}{3^2} = 1.$$

Therefore the curve is an ellipse with center $(1, -2)$, semi-axes lengths 1 and 3, and vertices at $(1 \pm 1, -2)$ and $(1, -2 \pm 3)$, i.e., at $(0, -2)$, $(2, -2)$, $(1, 1)$, and $(1, -5)$. As $a = 1$ and $b = 3$, $c = \sqrt{3^2 - 1^2} = 2\sqrt{2}$, and the foci are at $(1, -2 \pm 2\sqrt{2})$. (See Figure 9.6.) $\square$

Suppose a line passes through an interior point of an ellipse. Perhaps it seems obvious that such a line must intersect the ellipse at two distinct points. (Recall that we explored a similar question with the parabola.) But how could we prove it using the definition of interior? Below we present an intuitively appealing argument based on the Intermediate Value theorem.

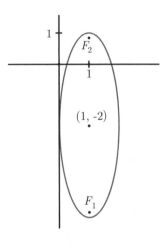

FIGURE 9.6.

Another proof, which also allows us to find the intersection points, is presented in our solution to Problem 9.20.

Example 43. Prove that a line passing through an interior point of an ellipse intersects the ellipse at two distinct points.

Solution: Let $\mathcal{E} = \mathcal{E}(F_1, F_2; s)$ denote the ellipse, and let P be an interior point. Let l be a line through P and let $A \in l$. Then $AF_1 + AF_2$ is a continuous function f of A.[3] As P is an interior point, $f(P) < s$. Moving A along l far enough in each direction from P, we eventually achieve $f(A) > s$. By the Intermediate Value theorem, there should be points A_1 and $A_2 \in l$ such that $A_1 - P - A_2$ and $f(A_1) = f(A_2) = s$. Hence l intersects $\mathcal{E}$ in *at least* two distinct points.

Suppose a coordinate system is introduced in such a way that the equation of $\mathcal{E}$ is of the form (9.9). Then finding the points of intersection of l and $\mathcal{E}$ amounts to solving a system formed by (9.9) and an equation for l, which is of the form $y = mx + b$ or $x = c$. Substituting $mx + b$ for y or c for x in (9.9) leads to a quadratic equation with respect to x or y, respectively. Therefore, the system has *at most* two distinct solution, which implies that the intersection contains *at most* two distinct points. Thus, l intersects $\mathcal{E}$ at exactly two distinct points. $\square$

Example 44. Let $\mathcal{E}_1$ and $\mathcal{E}_2$ be two ellipses whose major axes are perpendicular. Show that if the ellipses intersect at four points, then the four points lie on a circle. (See Figure 9.7.)

Solution: In a Cartesian coordinate system with axes parallel to those of the two ellipses, the coordinates of the points of intersection are solutions of the system:

$$F_1(x, y) = \frac{(x - h_1)^2}{a_1{}^2} + \frac{(y - k_1)^2}{b_1{}^2} - 1 = 0,$$

$$F_2(x, y) = \frac{(x - h_2)^2}{a_2{}^2} + \frac{(y - k_2)^2}{b_2{}^2} - 1 = 0. \qquad (9.12)$$

---

[3] If one is uncomfortable with f being a continuous function of a point, it is easy to reduce the discussion to a more familiar setting. Introduce a coordinate system in such a way that l is not vertical. As l is fixed, the y-coordinate of any point A on l is a linear function of its x-coordinate. Therefore, as F_1 and F_2 are fixed, the distance formula implies that $AF_1 + AF_2$ is a continuous function of x only.

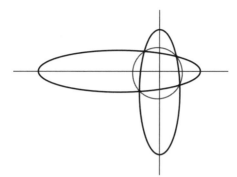

FIGURE 9.7.

In order to show that the points of intersection lie on a circle, we replace the second equation in system (9.12) by the linear combination $\alpha F_1(x, y) + \beta F_2(x, y) = 0$, where α and β are chosen so that the equation $\alpha F_1(x, y) + \beta F_2(x, y) = 0$ is an equation of a circle, as we show next. Then, the result will follow immediately from Theorem 8.10.

Assuming the graph of $Ax^2 + By^2 + Cx + Dy + F = 0$ is not empty or a single point, it will be a circle if and only if $A = B \neq 0$; thus α and β can be found by equating the coefficients of x^2 and y^2 in $\alpha F_1(x, y) + \beta F_2(x, y)$. Hence,

$$\frac{\alpha}{a_1{}^2} + \frac{\beta}{a_2{}^2} = \frac{\alpha}{b_1{}^2} + \frac{\beta}{b_2{}^2}. \tag{9.13}$$

If $|a_1| = |b_1|$, then the ellipse $\mathcal{E}_1$ is a circle and the statement is proven. If $|a_1| \neq |b_1|$, set $\beta = 1$ in (9.13) and solve (9.13) for α. This completes the proof. $\square$

Example 45. Suppose a straight rod moves on a plane such that its endpoints slide along the sides of a given right angle. Fix a point on the rod. What is the trajectory of this point?

Solution: Introduce a Cartesian coordinate system in the plane such that the right angle with its interior forms the 1st quadrant. Choose the scale in such a way that the length of the rod is 1. Suppose the fixed point on the rod divides it into two pieces of length a and b, with $a + b = 1$. (See Figure 9.8.)

Let $\overline{AB}$ represent a position of the rod, let $C : (x, y)$ be the position of the point, and let $m\angle ABO = \theta$. Then $a \cos \theta = x$, and $b \sin \theta = y$. As $\sin^2 \theta + \cos^2 \theta = 1$, we obtain

$$\frac{x^2}{a^2} + \frac{y^2}{b^2} = 1. \tag{9.14}$$

Hence the point moves along an arc of the ellipse given by (9.14). The endpoints of the arc are two vertices of the ellipse: $(a, 0)$ and $(0, b)$. $\square$

9.3 Hyperbolas

Finally, we consider hyperbolas in the coordinate plane. In order to avoid repeating many of the arguments and ideas presented for the parabola and ellipse, we will try to be brief, stressing the

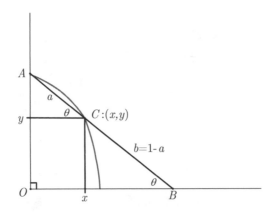

FIGURE **9.8.**

properties of hyperbolas that distinguish them from other conics. As previously defined, a **hyperbola** $\mathcal{H} = \mathcal{H}(F_1, F_2; d)$ consists of all points P such that $|PF_1 - PF_2| = d > 0$. (See Figure 9.9.) Distinct points F_1 and F_2 are the *foci* of $\mathcal{H}$. The line determined by the foci intersects the hyperbola at two points, the *vertices* of the hyperbola. The segment joining any two points of a hyperbola is called a **chord**.

Choose a Cartesian coordinate system such that $\overleftrightarrow{F_1F_2}$ is the x-axis and the y-axis bisects $\overline{F_1F_2}$. Let $F_1F_2 = 2c > 0$ and $d = 2a$. Then $F_1 : (-c, 0)$ and $F_2 : (c, 0)$. From the triangle inequality we obtain $0 < a < c$.

An argument similar to the one which led us to the standard equation of an ellipse (9.9) gives that $P : (x, y)$ is on $\mathcal{H}$ if and only if

$$\frac{x^2}{a^2} - \frac{y^2}{b^2} = 1, \tag{9.15}$$

where $b = \sqrt{c^2 - a^2}$. Equation (9.15) is called the **standard equation** of a hyperbola. It immediately implies that $\mathcal{H}$ is symmetric with respect to both coordinate axes (the *axes* of the hyperbola), and the origin is its center of symmetry (the **center** of the hyperbola). A hyperbola crosses only one of its axes at its vertices, which have coordinates $(-a, 0)$ and $(a, 0)$. The chord of the hyperbola joining the vertices is called its **major axis**. The segment joining points $(0, -b)$ and $(0, b)$ is called the

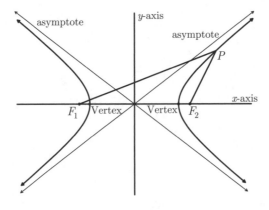

FIGURE **9.9.**

minor axis of the hyperbola. Note that the minor axis is not a chord of the hyperbola. The hyperbola is not a connected curve, but rather consists of two congruent pieces, its **branches**.

Expressing y through x, we get

$$y = \pm b \sqrt{\frac{x^2}{a^2} - 1}.$$

As a and b are fixed, for large x the value of y is close to $(b/a)x$ or $-(b/a)x$. Therefore, for large $|x|$, a point (x, y) of $\mathcal{H}$ is close to one of the two lines $y = \pm(b/a)x$, called the **asymptotes** of $\mathcal{H}$.

If a hyperbola has its center at (h, k) and the axis through the foci is horizontal, then

$$\frac{(x - h)^2}{a^2} - \frac{(y - k)^2}{b^2} = 1,$$

or, if the axis through the foci is vertical,

$$\frac{(y - k)^2}{a^2} - \frac{(x - h)^2}{b^2} = 1.$$

If $a = b$, then the slopes of the asymptotes are ± 1, and they are perpendicular. In this case the hyperbola is called a **rectangular** hyperbola.

We conclude this section with several examples.

Example 46. Consider a family of parallel chords of a branch of a hyperbola. Prove that the midpoints of the chords are collinear.

Proof. Let us denote the branch of the hyperbola by $\mathcal{H}$, and choose a Cartesian coordinate system in such a way that

$$\mathcal{H} : \frac{x^2}{a^2} - \frac{y^2}{b^2} = 1, \ x > 0.$$

If all chords are vertical, then their midpoints lie on the x-axis, and the proof is left to the reader. Suppose each chord determines a line with same slope m. Then $m \neq 0$, as the branch has no horizontal chords. Every such line l has an equation of the form $y = mx + t$. Finding the x-coordinates of the intersection of l with $\mathcal{H}$ amounts to solving the quadratic (in x) equation

$$x^2/a^2 - (mx + t)^2/b^2 = 1 \ \Leftrightarrow \ (b^2 - a^2m^2)x^2 - 2a^2mtx - (a^2t^2 + a^2b^2) = 0. \quad (9.16)$$

If $b^2 - a^2m^2 = 0$ (i.e., $m = \pm b/a$) then (9.16) has only one solution. Of course! In this case the lines are parallel to one of the asymptotes of $\mathcal{H}$, and there are no chords of $\mathcal{H}$ with this slope. This is an example of an instance when careful algebra points to a geometrically significant fact. So we assume that $m \neq \pm b/a$.

If x_1 and x_2 are solutions of (9.16), then the x-coordinate of the midpoint of the chord is given by

$$x_M = (x_1 + x_2)/2 = a^2mt/(b^2 - a^2m^2) \quad (9.17)$$

by Vièta's theorem.[4] Note that we found x_M without finding the x-coordinates of the endpoint of the chord. Similarly, rewriting $y = mx + t$ as $x = (y - t)/m$ (permitted since $m \neq 0$) and substituting into $x^2/a^2 - y^2/b^2 = 1$ leads to a quadratic equation in y:

$$(y/m - t/m)^2/a^2 - y^2/b^2 = 1 \ \Leftrightarrow \ (b^2 - a^2m^2)y^2 - 2tb^2y + (t^2 - a^2m^2)b^2 = 0. \quad (9.18)$$

---

[4] A special case of Vièta's theorem, applied to quadratics, says that the sum of the roots of $ax^2 + bx + c$ is $-b/a$ and the product of the roots is c/a.

If y_1 and y_2 are solutions of (9.18), then using Vièta's theorem,

$$y_M = (y_1 + y_2)/2 = b^2 t/(b^2 - a^2 m^2). \tag{9.19}$$

Comparing x_M and y_M, given by (9.17) and (9.19), respectively, we see that

$$y_M = \frac{b^2}{a^2 m} x_M.$$

Hence the midpoints of the chords with slope $m \neq 0$ and $m \neq \pm b/a$ lie on the line

$$y = \frac{b^2}{a^2 m} x. \qquad \square$$

Example 47. Given two points A and B, find the loci of all points P such that $m\angle PAB = 2m\angle PBA$.

Solution: We introduce a Cartesian coordinate system such that A is at the origin and $B : (1, 0)$. Let $P : (x, y)$, and assume that $m\angle PBA = \alpha$ and $m\angle PAB = 2\alpha$. As 2α is the measure of the inclination angle of line PA, $\tan 2\alpha$ is equal to the slope of line PA, giving $\tan 2\alpha = y/x$ for $x \neq 0$. For our problem, $x = 0$ corresponds to $\alpha = \pi/4$, and, obviously, there exists a point in the locus corresponding to this angle.

As $\pi - \alpha$ is the measure of the inclination angle of $\overleftrightarrow{PB}$, $\tan \alpha$ is opposite of the slope of $\overleftrightarrow{PB}$: $\tan \alpha = -(y - 0)/(x - 1) = -y/(x - 1)$ for $x \neq 1$. It is clear that no point of the line $x = 1$ is in the locus. Using the formula $\tan 2\alpha = 2 \tan \alpha/(1 - \tan^2 \alpha)$, we obtain

$$\frac{y}{x} = 2 \left(\frac{-y/(x - 1)}{1 - y^2/(x - 1)^2} \right).$$

If $y = 0$, then P is an interior point of $\overline{AB}$ and $\alpha = 0$. If $P = A$ or B, then the measure of one of the angles is not defined. Assuming $y \neq 0$ and $x \neq 0, 1$, the last equation is reduced to $3x^2 - 4x + 1 - y^2 = 0$, which is equivalent to

$$\frac{(x - 2/3)^2}{(1/3)^2} - \frac{y^2}{(1/\sqrt{3})^2} = 1. \tag{9.20}$$

This is a standard form of an equation of a hyperbola with center $(2/3, 0)$, $a = 1/3$, and $b = 1/\sqrt{3}$. Hence its vertex is at $(1/3, 0)$. This is interesting, since it is not obvious from the beginning that points of the locus close to $\overleftrightarrow{AB}$ are close to the point C on $\overline{AB}$ such that $AC/CB = 1 : 2$.[5] As $a^2 + b^2 = 4/9$, $c = \pm 2/3$, and the foci of the hyperbola are at $(0, 0)$ and $(4/3, 0)$. As $b/a = \sqrt{3}$, the asymptotes have inclination angles $\pi/3$ and $2\pi/3$. Hence, the locus is the interior of $\overline{AB}$ together with one branch of a hyperbola. In Figure 9.10, the locus and asymptotes are shown. The figure also shows the right branch of the hyperbola, which would be part of the locus if points A and B were interchanged. $\qquad \square$

Readers no doubt already believe that the graph of the equation $y = 1/x$, is a hyperbola. The following example provides an explanation as to why this curve is indeed a hyperbola.

---

[5] This also could be seen from the Sine theorem for a triangle and the fact that for small x, $\sin x \approx x$.

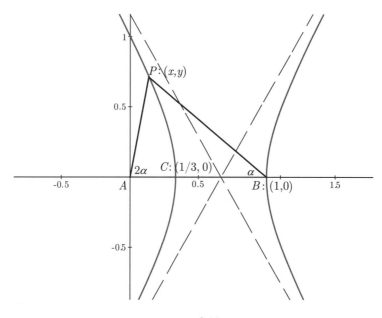

FIGURE 9.10.

Example 48. Consider the curve with equation $xy = 1$ in a Cartesian coordinate system. Prove that this curve is a rectangular hyperbola.

Solution: Plotting the graph, or remembering how it should look from our previous mathematical experience, we begin to suspect that it is a hyperbola, with asymptotes $y = 0$ and $x = 0$, axes $y = \pm x$, and vertices at $(-1, -1)$ and $(1, 1)$. (See Figure 9.11.)

Moreover, if the curve is a hyperbola, then the distance from the center to a vertex, which is $\sqrt{2}$, will be the value of a in the standard form of the equation of a hyperbola. As the asymptotes are perpendicular, $b = a$ and $c^2 = a^2 + b^2 = 2 + 2 = 4$. So $c = 2$, which implies that points $F_1 : (-\sqrt{2}, -\sqrt{2})$ and $F_2 : (\sqrt{2}, \sqrt{2})$ are the foci. Therefore, in order to prove that our curve is indeed a hyperbola, it is sufficient to check that for these F_1 and F_2 and an arbitrary point P on the curve, $|PF_1 - PF_2| = 2\sqrt{2}$.

For any point $P : (x, y)$ on the curve, $x \neq 0$ and $y = 1/x$. Therefore,

$$|PF_1 - PF_2| = \left| \sqrt{(-\sqrt{2} - x)^2 + (-\sqrt{2} - 1/x)^2} - \sqrt{(\sqrt{2} - x)^2 + (\sqrt{2} - 1/x)^2} \right|$$

$$= \left| \sqrt{4 + 2\sqrt{2}(x + 1/x) + (x^2 + 1/x^2)} \right.$$

$$\left. - \sqrt{4 - 2\sqrt{2}(x + 1/x) + (x^2 + 1/x^2)} \right|.$$

Let $u = x + 1/x$. Then $x^2 + 2 + 1/x^2 = u^2$, and

$$|PF_1 - PF_2| = \left| |u + \sqrt{2}| - |u - \sqrt{2}| \right|.$$

As $u^2 \geq 2$, $|u| \geq \sqrt{2}$. This allows us to simplify the expression above as follows.

For $u \leq -\sqrt{2}$, $|PF_1 - PF_2| = |(-u - \sqrt{2}) - (-(u - \sqrt{2}))| = |-2\sqrt{2}| = 2\sqrt{2}$.

For $u \geq \sqrt{2}$, $|PF_1 - PF_2| = |(u + \sqrt{2}) - (u - \sqrt{2})| = |2\sqrt{2}| = 2\sqrt{2}$. □

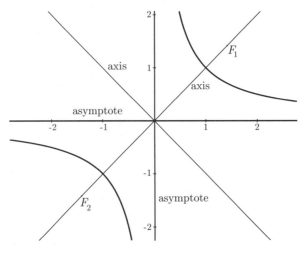

FIGURE 9.11.

The last example raises a natural question: how much can the equation of a conic deviate from its standard form, and how does one recognize a conic from an equation? We will discuss this question in Section 9.6.

9.4 A Unified Definition of Conics as Loci

Though our definitions of an ellipse and a hyperbola evidence some similarity, the definition of a parabola seems to have little in common with them. However, it turns out that the three curves can be defined in a unified way. Let e be a non-negative real number, F a point, and l a line not containing F. Consider the set S of all points P in the plane such that the ratio of the distances from P to F and from P to l is e; equivalently, consider all points P such that $PF = e \cdot d(P, l)$. Then the following theorem holds.

> **Theorem 9.1.** *If $0 < e < 1$, then S is an ellipse. If $e = 1$, then S is a parabola. If $e > 1$, then S is a hyperbola.*

It is not hard to show that every ellipse, parabola, and hyperbola may be formed this way for a suitable e, F, and l. The number e is called the **eccentricity**, the point F is called the **focus**, and the line l is the **directrix** of the curve. Clearly, for a parabola, this definition coincides with the original. It can be shown that F also coincides with one of the "old" foci of an ellipse and hyperbola.[6] The proof of the equivalence of the definitions can be found in many books, and Problem 9.17 encourages the reader to find one.

> In what follows in this chapter, unless noted otherwise, whenever we mention one or more coordinate systems in the plane, we assume that all of them are Cartesian with congruent unit segments. The orientation in all coordinate systems will be the same.

---

[6] It is easy to check that when the lengths of the semi-axes of an ellipse become close to each other, the distance between the focus and the directrix becomes large and the eccentricity tends to zero. That is why we define the eccentricity of a circle to be zero and its directrix to be "at infinity."

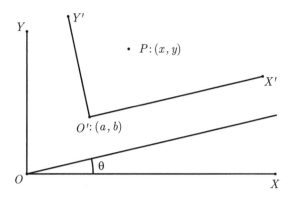

FIGURE 9.12.

9.5   Change of Coordinates

Things should be as simple as possible, but not simpler.
— Albert Einstein (1879–1955)

The Fahrenheit and Celsius scales, which we use to measure temperature, can be thought of as two coordinate lines.[7] If F is the value of the temperature in Fahrenheit, and C is the same temperature in Celsius, then the well-known formula $C = (5/9)(F - 32)$ allows us to convert one to the other, and the formula $F = (9/5)C + 32$ converts in the opposite direction. For example, $212°F = 100°C$.

Here is a bit more difficult situation. Suppose OXY and $O'X'Y'$ are two coordinate systems and P is a point in the plane such that $P : (x, y)$ in OXY. What are the coordinates (x', y') of P in in $O'X'Y'$?

In order to answer this question, we need to know more about the mutual position of the coordinate axes. There are different ways to describe the position. For example, if we say that $O' : (a, b)$ in OXY, and that $\overrightarrow{OX}$ has to be rotated (counterclockwise) around O by an angle of measure $\theta \in [0°, 360°)$ to become directed as $\overrightarrow{O'X'}$, then the position of $O'X'Y'$ relative to OXY becomes completely defined. (See Figure 9.12.) Under this setup, the problem can be restated more precisely:

Given a, b, θ, and $P : (x, y)$ in OXY, find the coordinates (x', y') of P in $O'X'Y'$.

We solve this problem in several steps.

If $\theta = 0$—i.e., the directions of the corresponding axes are the same—then,

$$x' = x - a \quad \text{and} \quad y' = y - b, \tag{9.21}$$

which is illustrated in Figure 9.13. These formulae can be rewritten in terms of x and y, and the resulting equations facilitate the change of coordinates from $O'X'Y'$ to OXY. Let us refer to all such formulae as **translation formulae**.

Now we suppose that $O = O'$. Since the orientation of the two coordinate systems is the same, $OX'Y'$ can be obtained by rotating OXY around O by angle θ to make them coincide.

Suppose $m\angle POX = \alpha$, as in Figure 9.14. Then $m\angle POX' = \alpha - \theta$, $x = OP\cos\alpha$, $y = OP\sin\alpha$, $x' = OP\cos(\alpha - \theta)$, and $y' = OP\sin(\alpha - \theta)$. Expanding the cosine and sine of the

---

[7] More precisely, they are segments of the lines, with the endpoints determined by modern physics.

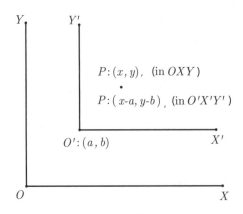

FIGURE 9.13.

difference of two angles, we obtain

$$x' = \cos\theta\, x + \sin\theta\, y, \tag{9.22}$$

$$y' = -\sin\theta\, x + \cos\theta\, y. \tag{9.23}$$

For $\theta = \pi/2$, the equations simplify to $x' = y$, $y' = -x$.

Solving (9.22) for x and y in terms of x' and y', which, clearly, is equivalent to replacing θ with $-\theta$, we arrive at

$$x = \cos\theta\, x' - \sin\theta\, y',$$

$$y = \sin\theta\, x' + \cos\theta\, y'.$$

We refer to all these formulae as **rotation formulae**.

In deriving the rotation formulae we relied heavily on Figure 9.14. It is natural to ask how the formulae depend on the position of P. It turns out that they do not, and are completely general. A careful proof of this is equivalent to a careful proof of the fact that the addition formula for cosine holds for arbitrary angles. Not wanting to bore the reader, and following a great tradition, we do not check it either. See Chapter 7 for a related discussion.

Given coordinate systems OXY and $O'X'Y'$, imagine a third coordinate system $O'X^*Y^*$ such that the directions of the axes $O'X^*$ and $O'Y^*$ are the same as of OX and OY, respectively. Let

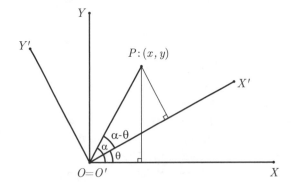

FIGURE 9.14.

$P : (x^*, y^*)$ be a point in $O'X^*Y^*$. Then, combining the results above, we have

$$x^* = x - a \qquad x' = \cos\theta\, x^* + \sin\theta\, y^*$$
$$y^* = y - b \qquad \text{and} \qquad y' = -\sin\theta\, x^* + \cos\theta\, y^*.$$

This leads to a complete answer to our original question, which we summarize in the following theorem.

Theorem 9.2. (Change of Coordinates: same point, different coordinate systems.) *Let OXY and $O'X'Y'$ be two Cartesian coordinate systems with congruent unit segments. Let $O' : (a, b)$ in OXY, and suppose $\overrightarrow{OX}$ has to be rotated counterclockwise around O by an angle of measure θ to become directed as $\overrightarrow{O'X'}$. If $P : (x, y)$ in OXY and $P : (x', y')$ in $O'X'Y'$, then*

$$x' = \cos\theta\,(x - a) + \sin\theta\,(y - b) \qquad\qquad x = \cos\theta\, x' - \sin\theta\, y' + a$$
$$y' = -\sin\theta\,(x - a) + \cos\theta\,(y - b) \qquad \text{and} \qquad y = \sin\theta\, x' + \cos\theta\, y' + b.$$

Other useful results, which are very close to the ones we just obtained, are stated in the following theorem. Here we fix a coordinate system, and move a point. The proof is immediate and is omitted.

Theorem 9.3. (Change of Coordinates: same coordinate system, different points.) *Let OXY be a Cartesian coordinate system. Suppose a point P is translated a units along the x-axis and b units along the y-axis to a point P^*, and then P^* is rotated to a point P' by an angle of measure θ around the origin O. Let $P : (x, y)$, $P^* : (x^*, y^*)$, and $P' : (x', y')$ in OXY. Then*

$$x^* = x + a \qquad x' = \cos\theta\,(x + a) - \sin\theta\,(y + b)$$
$$y^* = y + b \qquad \text{and} \qquad y' = \sin\theta\,(x + a) + \cos\theta\,(y + b).$$

The reader may feel a little lost among all these formulae. They are supposed to be understood, not memorized. One looks them up when they are needed, or they can be derived easily. In the latter case, it is always a good idea to check the results by taking, say, $P : (2, 1)$, $\theta = 9\pi/4$, $a = 3$, and $b = 4$. (You should find $P' : (0, 5\sqrt{2})$.)

The usefulness of the theorems above is clear: for certain coordinate systems they provide simple ways of expressing the coordinates in one of them in terms of the coordinates in another. But the theorems can also be used to change equations of figures. Let ϕ be the graph of an equation $F(x, y) = 0$ in coordinate system OXY. Rewriting the equation by using change of coordinates formulae, one can obtain an equation $F'(x', y') = 0$ in another coordinate system $O'X'Y'$. Often this equation may be simpler or more familiar – for example, it could be the standard form of the equation of ϕ.

Example 48 can now be viewed as recognizing the curve given by $xy - 1 = 0$ in OXY as the curve given by the equation

$$\frac{x'^2}{(\sqrt{2})^2} - \frac{y'^2}{(\sqrt{2})^2} = 1$$

in another coordinate system, $OX'Y'$. This new coordinate system is obtained by rotating OXY by $\pi/4$ around the origin. The corresponding change of coordinates formulae are

$$x = \cos\frac{\pi}{4}x' - \sin\frac{\pi}{4}y' \;\;=\; \frac{\sqrt{2}}{2}x' - \frac{\sqrt{2}}{2}y',$$

$$y = \sin\frac{\pi}{4}x' + \cos\frac{\pi}{4}y' \;\;=\; \frac{\sqrt{2}}{2}x' + \frac{\sqrt{2}}{2}y'.$$

Let's check what we got:

$$xy = 1 \;\Leftrightarrow\; \left(\frac{\sqrt{2}}{2}x' - \frac{\sqrt{2}}{2}y\right)\left(\frac{\sqrt{2}}{2}x' + \frac{\sqrt{2}}{2}y'\right) = 1$$

$$\Leftrightarrow\; \frac{x'^2}{(\sqrt{2})^2} - \frac{y'^2}{(\sqrt{2})^2} = 1. \quad \text{Yes!}$$

Similarly, Example 38 can now be viewed as representing the curve given by $y^2 + 6x - 4y + 7 = 0$ in OXY as the curve given by the equation

$$y^{*2} = 4(-3/2)x^*,$$

in $O^*X^*Y^*$, which has axes with the same directions as the corresponding ones in OXY, and $O^* : (-1/2, 2)$ in OXY. This equation reminds us of the standard form of the equation of a parabola, with x and y switched. This switching corresponds to rewriting $y^{*2} = 4(-3/2)x^*$ in the coordinate system $O^*X'Y'$ as $x'^2 = 4(3/2)y'$, which is precisely the standard form of the equation of a parabola. The system $O^*X'Y'$ is obtained from $O^*X^*Y^*$ by a rotation around O^* by $\pi/2$. The change of coordinates formulae between $O^*X'Y'$ and OXY are

$$x = \cos\frac{\pi}{2}x' - \sin\frac{\pi}{2}y' + \left(-\frac{1}{2}\right) = -y' - \frac{1}{2},$$

$$y = \sin\frac{\pi}{2}x' + \cos\frac{\pi}{2}y' + 2 = x' + 2.$$

Let's check what we got:

$$y^2 + 6x - 4y + 7 = 0 \;\Leftrightarrow\; (x'+2)^2 + 6(-y'-1/2) - 4(x'+2) + 7 = 0$$

$$\Leftrightarrow x'^2 = 6y' \;\Leftrightarrow\; x'^2 = 4(3/2)y'. \text{ Yes!}$$

9.6 The Graph of $Ax^2 + Bxy + Cy^2 + Fx + Gy + H = 0$

Consider a polynomial equation of degree at most two with respect to x and y:

$$Ax^2 + Bxy + Cy^2 + Fx + Gy + H = 0, \tag{9.24}$$

where A, B, C, F, G, and H are real numbers. What is the graph of this equation? Is it always one of the figures that we have already discussed, or could it be a new one?

Why do we ask these questions? One reason is that while thinking about equations of the figures we have been studying, we observe that many of them are particular cases of (9.24).[8]

---

[8] Other reasons are that the answer to this question was a great achievement of mathematics, that we know the answer and have developed everything needed in order to present it, and that this piece of knowledge is useful. But the main reason is that we like this subject so much that we cannot wait to share it with the reader!

- For $A = B = C = 0$, we get $Fx + Gy + H = 0$, and the graph is a line (if $F \neq 0$ or $G \neq 0$), the empty set, (if $F = G = 0$ and $H \neq 0$), or the whole plane (if $F = G = H = 0$).

- For $A = 1$, $C = -1$, $B = F = G = H = 0$, we get $x^2 - y^2 = (x - y)(x + y) = 0$, so the graph is the union of two intersecting lines $x + y = 0$ and $x - y = 0$.

- For $A \neq 0$, but $B = C = G = 0$, we obtain $Ax^2 + Fx + H = 0$, a quadratic equation for x. So the graph in OXY is the union of two distinct vertical lines (if roots are distinct), one vertical line (or two equal vertical lines, if roots are equal), or the empty set (if the quadratic has no real roots).

- For $A = 1$, $B = F = G = 0$, $C = 1$, $H = -1$, we get $x^2 + y^2 = 1$, and the graph is a unit circle. For $A = 1/4$, $B = F = G = 0$, $C = 1$, $H = -1$ we get $x^2/4 + y^2 = 1$, an ellipse. For $A = 1$, $B = F = G = 0$, $C = -1$, $H = -1$, we get $x^2 - y^2 = 1 = 0$, a hyperbola. For $A = 1$, $B = C = F = H = 0$, $G = -1$, we get $x^2 = y$, a parabola.

- We have also seen that the graph of $9x^2 + 4y - 4 + y^2 - 18x = 0$ is an ellipse (Example 42) and the graph of $xy - 1 = 0$ is a hyperbola (Example 48).

- The graph of (9.24) can also be a point: $x^2 + y^2 = 0$.

It seems that in each instance of (9.24) that we have encountered, the graph was a figure we have discussed. Can it be anything else? The answer is "NO," and we are going to prove this. We will also understand how one can recognize quickly the graph type of (9.24) by considering select coefficients.

Let us begin with simple observations. We have seen that if $A = B = C = 0$, then the graph is a line, the empty set, or the whole plane. So, we restrict ourselves to the cases when at least one of the coefficients A, B, or C is not 0.

Case 1. Suppose $A \neq 0$, but $B = 0$. Then (9.24) is of the form $Ax^2 + Cy^2 + Fx + Gy + H = 0$.

If $C \neq 0$, after completing the square with respect to x and with respect to y, we will obtain an equation of the form $Ax'^2 + Cy'^2 + H' = 0$, where $x' = x - F/(2A)$ and $y' = y - G/(2C)$. So in the coordinate system $O'X'Y'$ that corresponds to this change of coordinates, the graph of this equation is an ellipse, a hyperbola, a pair of distinct lines, a point, or the empty set; we leave the proof to Problem 9.2.

If $C = 0$, then (9.24) is of the form $Ax^2 + Fx + Gy + H = 0$. If $G \neq 0$, the equation can be rewritten in the form $y = A'x^2 + F'x + H'$, $A' \neq 0$, which can be easily recognized as a parabola. (See Problem 9.3.) If $G = 0$, then the equation is of the form $A'x^2 + F'x + H' = 0$, and its graph was discussed earlier in this section.

Case 2. Suppose $C \neq 0$, but $B = 0$. The analysis is similar to Case 1.

Case 3. Suppose $B \neq 0$. We wish to show that in this case, there exists another coordinate system $OX'Y'$, where (9.24) is represented by an equation in x', y' of the same form but with the coefficient at $x'y'$ being zero. This will reduce the problem to Case 1 or Case 2, which we already understand.

Let us search for a system $OX'Y'$ that is obtained from OXY by rotating it by an angle of measure θ around O. Then, by Theorem 9.2,

$$x = \cos\theta\, x' - \sin\theta\, y',$$

$$y = \sin\theta\, x' + \cos\theta\, y'.$$

Substituting these expressions for x and y into (9.24), we obtain

$$A'x'^2 + B'x'y' + C'y'^2 + F'x' + G'y' + H' = 0, \tag{9.25}$$

where

$$B' = B(\cos^2 \theta - \sin^2 \theta) + (2C - 2A)\sin\theta\cos\theta$$
$$= B\cos 2\theta + (C - A)\sin 2\theta.$$

Our goal is to show that for some θ, $B' = 0$. If $\sin 2\theta = 0$, then $B' \neq 0$, as $B \neq 0$ and $|\cos 2\theta| = 1$. Therefore $B' = 0$ is equivalent to $\cot 2\theta = (A - C)/B$. As the range of the cotangent function is $\mathbb{R}$, this equation has a solution. This ends the proof of Case 3.

Provided that the details left for Problems 9.2 and 9.3 are completed, we have proven part (1) of the following theorem. The proof of part (2) is left to Problem 9.4.

Theorem 9.4. *Consider an equation of the form (9.24).*

(1) *The graph of (9.24) in a Cartesian system OXY is one of the following figures: a conic (an ellipse, a hyperbola, or a parabola), a pair of lines, a point, the whole plane, or the empty set. In each case, the standard forms of equations are attained in coordinate systems obtained from OXY by translations and rotations only.*

(2) *For an arbitrary Cartesian system OXY, an equation of each figure mentioned in part (1) can be written in the form (9.24).*

The discussion preceding the theorem allows us to identify which of these possibilities occurs. We finish this section with a beautiful result that speeds the identification process up considerably. It relates to a number,

$$D = B^2 - 4AC,$$

which is called the **discriminant** of (9.24). Though the discriminants for two equivalent equations of type (9.24) may differ, such as for $x^2 + y^2 - 1 = 0$ and $-2x^2 - 2y^2 + 2 = 0$, its sign does not.

Theorem 9.5. *Let OXY be a Cartesian coordinate system. Let ϕ be the graph of (9.24). Then,*

(1) *if ϕ is an ellipse, then $D < 0$,*

(2) *if ϕ is a parabola, then $D = 0$, and*

(3) *if ϕ is a hyperbola, then $D > 0$.*

Before proceeding with the proof, we make several remarks.

First of all, Theorem 9.5 holds if (9.24) is the standard equation of an ellipse, hyperbola, or parabola.

- For an ellipse we have $A = 1/a^2$, $B = F = G = 0$, $C = 1/b^2$, and $H = -1$. Then $D = 0^2 - 4(1/a^2)(1/b^2) = -4/(a^2 b^2) < 0$.
- For a parabola, $A = 1$, $B = C = F = H = 0$, and $G = -1$, so $D = 0^2 - 1 \cdot 0 = 0$.
- For a hyperbola, $A = 1/a^2$, $B = F = G = 0$, $C = -1/b^2$, and $H = -1$, so $D = 0^2 - 4(1/a^2)(-1/b^2) = 4/(a^2 b^2) > 0$.

Secondly, no statement of Theorem 9.5 is sufficient. For example, for $x^2 - y^2 = 0$, $D = 4 > 0$, but the graph is a pair of intersecting lines $y = \pm x$, rather than a hyperbola. For $x^2 - 1 = 0$, $D = 0$, but the graph is a pair of two parallel lines $x = 1$ and $x = -1$, rather than a parabola. For $x^2 + y^2 + 1 = 0$, $D = -4$, but the graph is the empty set, rather than an ellipse.

Proof. The remarkable thing about the discriminant of (9.24) is that it does not change when the coordinates change according to the translation or the rotation formulae. We say that the discriminant is *invariant* with respect to such changes. A proof can be obtained by the following straightforward verifications.

For translations, if $x = x' + a$, $y = y' + b$, then the following are equivalent:

$$0 = Ax^2 + Bxy + Cy^2 + Fx + Gy + H, \tag{9.24}$$
$$0 = A(x' + a)^2 + B(x' + a)(y' + b) + C(y' + b)^2 + F(x' + a) + G(y' + b) + H,$$
$$0 = A'x'^2 + B'x'y' + C'y'^2 + \ldots \tag{9.26}$$

where $A' = A$, $B' = B$, and $C' = C$. Therefore the discriminants of (9.24) and (9.26) are equal.

Now we check for invariance of the discriminant under a rotation. If $x = \cos\theta\, x' - \sin\theta\, y'$ and $y = \sin\theta\, x' + \cos\theta\, y'$, then (9.24) is equivalent to

$$0 = A(\cos\theta\, x' - \sin\theta\, y')^2 + B(\cos\theta\, x' - \sin\theta\, y')(\sin\theta\, x' + \cos\theta\, y')$$
$$+ C(\sin\theta\, x' + \cos\theta\, y')^2 + \ldots$$
$$\Leftrightarrow \quad A'x'^2 + B'x'y' + C'y'^2 + \ldots = 0,$$

where

$$A' = \frac{1}{2}\big[(A + C) + B\sin 2\theta + (A - C)\cos 2\theta\big],$$

$$B' = B\cos 2\theta - (A - C)\sin 2\theta, \text{ and}$$

$$C' = \frac{1}{2}\big[(A + C) - B\sin 2\theta - (A - C)\cos 2\theta\big].$$

Simplifying $D' = B'^2 - 4A'C'$, we find that it is equal to D. (The details are left to Problem 9.11.) Therefore the discriminants of (9.24) and (9.26) are equal.

We have understood that $O'X'Y'$ could always be obtained from OXY by a sequence of a translation and a rotation. As each of these changes of coordinates transforms an equation of type (9.24) into an equation of the same type, and with the same discriminant, the value of the discriminant does not depend on the coordinate system. As the theorem holds for the standard form of the equation of each curve (which corresponds to some of these coordinate systems), it must hold for every equation of type (9.24). $\square$

Looking more closely at the particular instances of (9.24), it is possible to state necessary and sufficient conditions on the coefficients of (9.24) to define each of the figures listed in Theorem 9.4(1), but we will not do it here. The complete analysis can be found in many texts on analytic geometry or linear algebra, e.g., see [**40**]. Instead we present examples which illustrate that one can identify the figure by using the facts we have already established along with various simple ideas.

Example 49. Describe the graph, ϕ, of the equation $x^2 + xy - 2y^2 + 2x + 4y = 0$.

Solution 1: Looking at the discriminant of the equation, we get $D = 1^2 - 4 \cdot 1 \cdot (-2) = 9 > 0$. Hence, by Theorem 9.5, ϕ is neither an ellipse nor a parabola. If $x = 0$, then $y = 0$ or $y = 2$. If $y = 0$, then $x = 0$ or $x = -2$. Hence we have three non-collinear points of ϕ: $(0, 0)$, $(-2, 0)$, and $(0, 2)$. Therefore, by Theorem 9.4 (1), ϕ can only be a hyperbola or a pair of intersecting lines. In the latter case there exists a line through the origin that coincides with one of them. Therefore, we study the intersection of ϕ with an arbitrary line $y = mx$. We have

$$x^2 + x(mx) - 2(mx)^2 + 2x + 4(mx) = 0 \iff (1 + m - 2m^2)x^2 + (2 + 4m)x = 0 \iff$$
$$(x = 0) \text{ or } (1 - m)(1 + 2m)x = -2(1 + 2m).$$

Hence, if $1 + 2m = 0$, the line $x + 2y = 0$ intersects ϕ at infinitely many points. This proves that ϕ is a union of two lines, with $x + 2y = 0$ being one of them. It passes through the origin.

If we wish to find the second line, we either divide $x^2 + xy - 2y^2 + 2x + 4y$ by $x + 2y$, or find an equation of a line through $(-2, 0)$ and $(0, 2)$. Each way leads to the line $x - y + 2 = 0$. Hence, ϕ is the union of points of two lines: $x + 2y = 0$ and $x - y + 2 = 0$.

Solution 2: One should never forget about trying to complete the square:

$$\begin{aligned}
x^2 + xy - 2y^2 + 2x + 4y &= x^2 + (y + 2)x - 2y^2 + 4y \\
&= \left(x + \frac{y+2}{2}\right)^2 - \frac{(y+2)^2}{4} - 2y^2 + 4y \\
&= \left(x + \frac{y+2}{2}\right)^2 - \left(\frac{9}{4}y^2 - 3y + 1\right) \\
&= \left(x + \frac{y+2}{2}\right)^2 - \left(\frac{3}{2}y - 1\right)^2 \\
&= \left(x + \frac{y+2}{2} + \frac{3}{2}y - 1\right)\left(x + \frac{y+2}{2} - \frac{3}{2}y + 1\right) \\
&= (x + 2y)(x - y + 2) = 0.
\end{aligned}$$

Again ϕ is the union of points of two lines: $x + 2y = 0$ and $x - y + 2 = 0$. $\qquad\square$

Example 50. Describe the graph of the equation $5x^2 + 2xy + 2y^2 + 14x + 4y + 10 = 0$.

Solution 1: Let ϕ denote the graph. Looking at the discriminant of the equation, we find $D = 2^2 - 4 \cdot 5 \cdot 2 = -36 < 0$. Hence, by Theorem 9.5, ϕ is neither a hyperbola nor a parabola. It is certainly not the whole plane, as $(0, 0)$ does not belong to the graph. We also observe that ϕ has no y-intercept.

Let us now try to intersect ϕ with an arbitrary non-vertical line through the origin, $y = mx$. Substituting mx for y, we obtain:

$$(5 + 2m + 2m^2)x^2 + (14 + 4m)x + 10 = 0. \tag{9.27}$$

The discriminant of this quadratic equation is

$$(14 + 4m)^2 - 4 \cdot (5 + 2m + 2m^2) \cdot 10 = -4(1 - 4m)^2.$$

This means that a line $y = mx$ intersects ϕ if and only if $m = 1/4$, as this is the only value of m for which the discriminant $-4(1 - 4m)^2$ is non-negative. For $m = 1/4$, (9.27) gives $(45/8)x^2 + 15x + 10 = 0$, which is equivalent to $(3x + 4)^2 = 0$, or $x = -4/3$. The corresponding y-coordinate is $y = (1/4)(-4/3) = -1/3$. So, the only line through the origin that meets ϕ is the line $y = (1/4)x$, and the intersection is the unique point $(-4/3, -1/3)$. Hence, $\phi = \{(-4/3, -1/3)\}$.

Solution 2: Completing the square with respect to x, we obtain:

$$5x^2 + 2xy + 2y^2 + 14x + 4y + 10 = 5\left(x + \frac{y + 7}{5}\right)^2 + \frac{(3y + 1)^2}{5} = 0.$$

Hence the graph consists of only one point, $(-4/3, -1/3)$. $\qquad\square$

9.7 Problems

The value of a problem is not so much coming up with the answer as in the ideas and attempted ideas it forces on the would-be solver.
— I. N. Herstein (1923–1988)

Try to solve the following problems by using the coordinate method. Then see whether you can solve them in other way(s), and compare the advantages of different approaches.

9.1 Find (without using calculus) the slope of the tangent line to the curve given by the equation

  (a) $y = 4 - x^2$ at the point $(3, -5)$;
  (b) $x^2 + y^2/9 = 1$ at point $(1/3, 2\sqrt{2})$;
  (c) $x^2 - y^2 = 1$ at the point $(2, \sqrt{3})$.

9.2 Let A, C, and H be real numbers, $A \neq 0$, $C \neq 0$. Prove that the graph of the equation $Ax^2 + Cy^2 + H = 0$ is either an ellipse, a hyperbola, a pair of two distinct intersecting lines, a point, or the empty set. Find the conditions on the coefficients, A, C, and H that define each of these cases.

9.3 Let A, F, and H be real numbers, $A \neq 0$. Prove that the graph of the equation $y = Ax^2 + Fx + H$ is always a parabola.

9.4 Prove that in every Cartesian coordinate system, the equation of a conic, a pair of distinct lines, a pair of parallel lines, a point, the whole plane, or the empty set, can be written in the form $Ax^2 + Bxy + Cy^2 + Fx + Gy + H = 0$.

9.5 The cable of a suspension bridge assumes the shape of a parabola if the weight of the suspended roadbed (together with that of the cable) is uniformly distributed horizontally. Suppose that the towers of a bridge are 240 feet apart and 60 feet high and that the lowest point of the cable is 20 feet above the roadway; find the two vertical distances from the roadway to the cable at a distance of 60 feet from a tower.

9.6 Consider all right triangles with a common hypotenuse. Find the locus of the vertices of their right angles.

9.7 Determine the type of conic that is the graph of each of the following equations.

(a) $6xy - 8y^2 - 12x + 26y + 11 = 0$
(b) $x^2 + 2xy + y^2 - 8x + 4 = 0$
(c) $9x^2 - 4xy + 6y^2 + 6x - 8y + 2 = 0$

9.8 Let $\overline{AB}$ be a chord through the center of an ellipse described via the equation $x^2/a^2 + y^2/b^2 = 1$. Let P be any point on the ellipse for which the slopes of lines AP and BP are defined. Prove that the product of these slopes is $-b^2/a^2$.

Note that this problem is a generalization of the fact that a triangle inscribed in a circle, where one of the sides of the triangle is a diameter of the circle, will be a right triangle, since in a circle, $a = b$ and $\angle APB$ will be a right angle.

9.9 Prove without using calculus that an equation of the line tangent at the point (x_0, y_0) to

(a) an ellipse $\mathcal{E}$ given by $x^2/a^2 + y^2/b^2 = 1$ can be written as

$$(x_0/a^2)x + (y_0/b^2)y = 1, \text{ and}$$

(b) a hyperbola $\mathcal{H}$ given by $x^2/a^2 - y^2/b^2 = 1$ can be written as

$$(x_0/a^2)x - (y_0/b^2)y = 1.$$

9.10 Consider a ladder of length c standing vertically against a building, and let Z be a point on the ladder one-third of the distance from the top to the bottom. Find the locus of Z as the base is pulled away from the building. Compare to Problem 6.2.

9.11 In our proof of Theorem 9.5, the details of showing that $D' = D$ when the change of variables is performed via rotation formulae were omitted. Supply the details.

9.12 Consider a family of parallel chords of an ellipse. Prove that their midpoints are collinear. Does the result hold for a parabola?

9.13 Let $\mathcal{P}$ be the parabola given by $y^2 = 4ax$. Prove that all chords of $\mathcal{P}$ that subtend a right angle at the vertex of $\mathcal{P}$ are concurrent.

9.14 Can a plane be covered with

(a) finitely many parabolas and their interiors?
(b) finitely many hyperbolas and their interiors?

9.15 Let $\mathcal{P}_1$ and $\mathcal{P}_2$ be two parabolas whose axes are perpendicular. Show that if the parabolas intersect at four points, then the four points lie on a circle.

9.16 Prove that the union of an ellipse with its interior is a convex figure, i.e., show that if A and B are two points from the union, then the union also contains all points of $\overline{AB}$.

9.17 Prove Theorem 9.1.

9.18 Find the locus of the midpoints of all chords of a parabola that pass through its focus.

9.19 Suppose an ellipse touches all sides of a rhombus. Prove that the vertices of the rhombus lie on the axes of the ellipse.

9.20 By using the ideas of Example 39, prove that a line passing through an interior point of an ellipse intersects the ellipse at two distinct points.

9.21 Is it possible to find six points not on an ellipse, so that the six conics through subsets of any five of them are all distinct ellipses? What if we have hyperbolas instead of ellipses? Or parabolas?

9.8 Supplemental Problems

*S*9.1 (a) Find an equation for an ellipse passing through the points $(-1, 2), (-2, -3)$, and $(3, -1)$. Is such an ellipse unique?

(b) Find an equation for an ellipse passing through the points $(-1, 2), (-2, -3), (3, -1)$, and $(5, 5)$. Is such an ellipse unique?

(c) Suppose we are given four points such that no three of them are collinear. Is there always an ellipse passing through them?

*S*9.2 Find an equation of the circle that is tangent to the line $3x - 4y = 2$ at the point $(2, 1)$ and passes through the origin.

*S*9.3 The ellipse $\mathcal{E} : x^2/6 + y^2/3 = 1$ is tangent to all four sides of a square. Find equations of the four lines defined by the sides of a square.

*S*9.4 Let l be a tangent to an ellipse $\mathcal{E}$.

(a) Prove that the product of distances from the foci of $\mathcal{E}$ to l does not depend on l.

(b) Does the property in (a) hold for a hyperbola?

*S*9.5 Prove that for a given hyperbola, the product of the two distances from each of its points to the asymptotes is constant.

*S*9.6 Suppose two parabolas with vertices V_1 and V_2 share the same focus F, and let the rays FV_1 and FV_2 have opposite directions. Prove that the parabolas are orthogonal, i.e., their tangents at an intersection point are perpendicular.

*S*9.7 Find an equation of the common tangents to an ellipse $\mathcal{E} : x^2/45 + y^2/20 = 1$ and a parabola $\mathcal{P} : 3y^2 = 20x$.

*S*9.8 Identify the type of conic corresponding to the graphs of each of the following equations.

(a) $x^2 - 2xy + y^2 + 4x - 5 = 0$

(b) $x^2 + 4xy + y^2 - 3 = 0$

(c) $4x^2 - 4xy + y^2 + 4x - 2y + 1 = 0$

---

*S*9.9 What is the greatest number of intersection points of two parabolas? Of a parabola and an ellipse? Prove your answer.

*S*9.10 (a) Is there a parabola passing through points $(0, 0), (1, 0)$, and $(0, 1)$? Answer the same question for an ellipse and for a hyperbola.

(b) Is there a parabola passing through points $(0, 0), (1, 0), (0, 1)$, and $(1, 1)$? Answer the same question for an ellipse and for a hyperbola.

*S*9.11 Let l and m be perpendicular lines, and let a and b be two positive numbers. Find the locus of centers of all circles that cut on l a segment of length a and on m a segment of length b.

*S*9.12 Let $\mathcal{E}$ be an ellipse, and let $\overline{PA}$ and $\overline{PB}$ be tangent to $\mathcal{E}$ at points A and B. Find the locus of all P such that $m\angle APB = 90°$.

*S*9.13 Prove that an ellipse and a hyperbola that share common foci are orthogonal, i.e., their tangents at an intersection point are perpendicular.

*S*9.14 Consider two Cartesian coordinate systems OXY and $O'X'Y'$, where the second is obtained from the first by using translation and rotation only. Suppose a figure Φ is presented in OXY

as a graph of

$$Ax^2 + Bxy + Cy^2 + Fx + Gy + H = 0,$$

and in $O'X'Y'$ as a graph of

$$A'x'^2 + B'x'y' + C'y'^2 + F'x' + G'y' + H' = 0.$$

Prove that $A + B = A' + B'$.

Comment. The value of $A + B$, like $D = B^2 - 4AC$, is another invariant of conics with respect to coordinate changes consisting of translations and rotations.

S9.15 Let $A_i : (x_i, y_i)$, $i = 1, 2, \ldots, 5$ be five distinct points in a Cartesian plane such that $A_1 : (0, 0)$, $A_2 : (1, 0)$, and $A_3 : (0, 1)$, and no three of A_i are collinear. Prove that there exists a unique non-degenerate conic passing through all of them.

S9.16 Prove that if a parabola is tangent to all four sides of a quadrilateral, then its axis is parallel to the line passing through the midpoints of the diagonal of the quadrilateral.

S9.17 Is there an infinite set of points on a plane such that they are not all collinear and the distances between any two of them are integers?

10

Complex Numbers

*I have found another kind of cubic radicals
which is very different from the others.*
— R. Bombelli[1] (1526–1572)

The notion of a number is as old as mathematics itself, and its development is inseparable from that of mathematics. Usually, each newly-introduced set of numbers contained the previous set, or, as we often say, extended it. In this way, the set $\{1, 2, \ldots\}$ of natural numbers has been extended by joining to it zero, the negative integers, rational numbers (fractions), irrational numbers, complex numbers, etc. The actual extensions happened not in the order suggested by the previous sentence, and the history of the process is fascinating. The main motivation for each extension came from mathematics itself: having a greater set of numbers allowed mathematicians to express themselves with better precision and fewer words, i.e., with greater ease. Numbers form one of the most important parts of mathematical "language," and in this regard, their development is very similar to the development of live languages, where the vocabulary increases mostly for convenience, rather than out of necessity.

Often, a motivation for inventing new numbers is presented by mathematicians in a simplistic way as the desire to have solutions of a certain equation. Using modern notation, one can say that the numbers 0, -3, $5/7$, $\sqrt{2}$, and i appeared when one wanted to solve equations like $x + 5 = 5$, $x + 3 = 0$, $7x - 5 = 0$, $x^2 - 2 = 0$, and $x^2 + 1 = 0$, respectively. Newly introduced numbers have always been treated with suspicion (hence the terms *irrational, transcendental, imaginary*), and their acceptance by many took centuries. Some numbers were related to "real life" more obviously than others: zero represented absence of a quantity, negative numbers were easily interpreted as debts, $1/3$ stood as one part of something divided into three equal parts, the symbol $\sqrt{2}$ represented the length of the diagonal of a unit square, and π denoted the ratio of the length of the circumference to the diameter of a circle. These numbers, and many others, called real numbers, and the set of all reals is denoted by $\mathbb{R}$. An easily perceived correspondence between real numbers and points of a straight line made them even more "real" and led to the coordinate method, which we discussed in Chapter 8. We are certain that the reader is well acquainted with real numbers and has no doubt about their usefulness.

---

[1] From Bombelli's *L'algebra Opera*, Bologna,1572; as presented in [37].

10.1   Introduction

Let us now discuss complex numbers. Complex numbers are usually introduced in algebra or precalculus courses, and we hope that the reader has previously encountered them. What is presented below can be thought of as a quick introduction to the subject or as a short review. We concentrate mostly on the relationship between complex numbers and geometry. Complex numbers arise naturally in a quest to find solutions to certain polynomial equations with real coefficients of degree at least two, like

$$x^2 + 1 = 0 \quad \text{or} \quad x^3 - x + 10 = 0,$$

and, actually, cubic equations played a much greater role in their birth than did quadratics.[2]

One way to define complex numbers is the following. A **complex number** is a formal expression $a + bi$, where a and b are real numbers and i is merely a symbol. Let $\mathbb{C}$ denote the set of all complex numbers.[3] At this stage bi has nothing to do with any multiplication, nor does the sign $+$ have anything to do with addition. We say that $a + bi = c + di$ if $a = c$ and $b = d$. We define the addition and multiplication of complex numbers by the following rules:

$$(a + bi) + (c + di) := (a + c) + (b + d)i, \quad \text{and}$$
$$(a + bi)(c + di) := (ac - bd) + (ad + bc)i.$$

For example, we have $(1 + 3i) + (2 + (-1)i) = 3 + 2i$, and $(1 + 3i)(2 + (-1)i) = 5 + 5i$.

Note that in the above definition of addition in $\mathbb{C}$, the same symbol $+$ denotes three different things: between the parentheses in $(a + bi) + (c + di)$ it denotes the just-introduced addition of complex numbers, inside the parentheses $(a + b)$ and $(c + d)$ it denotes usual addition of real numbers, and in the expression $x + iy$ it has no meaning at all (for now). Similarly, we have several types of multiplication: between complex numbers, between the letters in the expression like ac (the usual multiplication of reals), and in expressions like yi (which has no meaning for now). The definition of addition of complex numbers is essentially the same as for linear polynomials, where a, b, c, and d denote the coefficients and i denotes the indeterminate. The definition of multiplication of complex numbers may look rather strange; it also can be thought of as multiplication of linear polynomials of i, but with $ii = i^2$ replaced by -1.

Let us not dwell on the motivation for introducing complex numbers and their operations in the way we did. Whatever they are, one can easily check that these operations possess all the nice properties of addition and multiplication of real numbers, such as commutativity, associativity, and the distributive law.

Moreover, the role of the unique identity element for addition is played by $0 + 0i$, and for multiplication, by $1 + 0i$. The unique additive inverse for $z = a + bi$ is $(-a) + (-b)i$, denoted by $-z$. It is easy to check that $-z = (-1 + 0i)z$.

It is also easy to show that the unique multiplicative inverse of $z = a + bi$ (where $z \neq 0 + 0i$), is the number

$$\frac{a}{a^2 + b^2} + \frac{-b}{a^2 + b^2}\, i,$$

which is denoted by $1/z$ or z^{-1}. Note that the condition that $a + bi$ is not zero in the set of complex numbers is equivalent to the fact that at least one of the real numbers a or b is not equal to zero in the

---

[2] The history and applications of complex numbers are presented well in many places. See, for example, [17], and more recent books [1], [26], or [37], or [42].

[3] C.F. Gauss was apparently the first to refer to this set as "complex" numbers in 1831; prior to that date, they were commonly called "imaginary numbers" or "impossible numbers."

set of real numbers, which is equivalent to $a^2 + b^2 \neq 0$. So the multiplicative inverse of any non-zero z exists.[4] For example, if $z = 1 + 3i$, then $-z = (-1) + (-3)i$, and $z^{-1} = 1/10 + ((-3)/10)i$.

Once we have additive and multiplicative inverses, the subtraction and division of complex numbers z_1 and z_2 can be defined in the usual way:

$$z_1 - z_2 := z_1 + (-z_2), \quad \text{and} \quad z_1/z_2 := z_1 z_2^{-1} \text{ if } z_2 \neq 0 + 0i.$$

For example,

$$(1 + 3i) - (2 + (-1)i) = (-1) + 4i$$

and

$$(1 + 3i)/(2 + (-1)i) = (1 + 3i)(2/5 + (1/5)i) = (-1/5) + (7/5)i.$$

After some computing with complex numbers, it soon becomes clear that the difference between computing with numbers of the form $x + 0i$ and real numbers x is just notational. Indeed,

$$(a + 0i) + (b + 0i) = (a + b) + (0 + 0)i = (a + b) + 0i \text{ and}$$
$$(a + 0i)(b + 0i) = (ab - 0 \cdot 0i) + (a \cdot 0 + b \cdot 0i) = (ab - 0) + (0 + 0)i = (ab) + 0i.$$

This allows us to simplify notation for some complex numbers, and also to assign meaning to $+$ in $a + bi$ and to multiplication in bi, as we discuss below.

We begin with just writing x instead of $x + 0i$, still remembering that x in this discourse is a complex number. Continuing this for a while, we do even more: we *identify* a complex number $x + 0i$ with the real number x, which makes $\mathbb{R}$ a subset of $\mathbb{C}$.[5] Other simplified notations are bi for $0 + bi$ and i for $1i$.

Since

$$bi = (b + 0i)(0 + 1i), \quad \text{and} \quad a + bi = (a + 0i) + (b + 0i)(0 + 1i),$$

the $+$ sign in $a + bi$ and the multiplication in bi acquire clear meanings as addition and multiplication of complex numbers written by using the simplified notation. Finally, $(-b)i$ is the same as $(-1)(bi)$, or $-(bi)$. The latter is usually written as $-bi$. Thus, $a - bi = a + (-b)i$.

Note that

$$i^2 = (0 + 1i)(0 + 1i) = -1 + 0i = -1.$$

That is why $i^2 = -1$. Here is the first surprise: the square of a complex number can be a negative real number! This suggests calling i a square root of -1. Note that $-i$ is another square root of -1:

$$(-i)^2 = (-i)(-i) = ((-1)i)((-1)i) = (-1)^2 i^2 = 1(-1) = -1.$$

Euler is credited with introducing the notation i in 1777. It came from the first letter of the Latin word *imaginarius*. Often one writes $i = \sqrt{-1}$, but we will try to avoid this for the following reason. The symbol $\sqrt{}$, and more generally, $\sqrt[n]{}$, when applied to nonnegative real numbers, always denotes the nonnegative square or the n-th root, respectively. As we will soon explain, <u>any</u> nonzero complex number possesses n distinct n-th roots. On the other hand, notions of positive or negative complex numbers and their ordering are not introduced.[6]

---

[4] If the discussion in this paragraph is too fast, keep reading. You can verify or clarify all of these statements later.

[5] Those familiar with the notion of ring isomorphism realize that what we have is a subring of $\mathbb{C}$ which is isomorphic to $\mathbb{R}$.

[6] See, e.g., [26], Section 1.5, for the reasons why not.

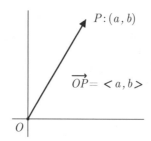

FIGURE 10.1. The Complex Plane

We call a and b the **real** and the **imaginary** part of $z = a + bi$, and write $a = \operatorname{Re} z$ and $b = \operatorname{Im} z$, respectively.

10.2   Geometric Interpretation

> *This miracle of analysis, this marvel of the world*
> *of ideas, an almost amphibian object between Being*
> *and Non-being that we call the imaginary number.*
> — Gottfried Wilhelm von Leibniz (1646–1716)

Though several famous mathematicians used complex numbers freely in the 17th and 18th centuries, they remained the "miracle of analysis" and were treated by many with distrust until the 19th century, when they were interpreted geometrically. While the idea of interpreting complex numbers as vectors[7] or points seems natural nowadays, it first appeared in print in the work of G. Wessel in 1799, and in 1806 in the work of J.-R. Argand, who did not know about Wessel's publication.

The geometric interpretation is this. Fix a Cartesian coordinate system in the plane. Then a complex number $z = a + bi$ corresponds to a point P with coordinates (a, b), and we write $P : (z)$. Or we may assume that z corresponds to a vector $\overrightarrow{OP} = \langle a, b \rangle$. When the points of a plane are interpreted as complex numbers in this way, we refer to the plane as the **complex plane**.[8] (See Figure 10.1.)

The points of the x-axis correspond to real numbers, and those of the y-axis correspond to numbers of the form bi. That is why these axes are often referred to as the **real axis** and the **imaginary axis**, respectively. The origin corresponds to the (complex) number $0 = 0 + 0i$.

The addition of complex numbers corresponds to the addition of vectors. (For a discussion of vector addition, see Chapter 11.) This is immediate from the definitions. Both are based on the Parallelogram Rule, as shown in Figure 10.2(i).

Likewise, subtraction of complex numbers can be represented graphically as subtraction of vectors, as illustrated in Figure 10.2(ii). It is clear, then, that the distance between two points z_1 and z_2 may be represented as $d(z_1, z_2) = |z_1 - z_2|$. For example, the solutions of the equation $|z - 2i| = 3$ form a circle of radius 3 centered at the (complex) point $2i$.

The desire for the product of two vectors in the plane to be a vector was motivated by the goals of bringing geometry closer to algebra and reducing geometric problems to direct computations. After all, real numbers could be thought of as vectors which could be both added and multiplied, and real numbers were useful.

---

[7] Vectors will be defined in Chapter 11. At this point, the reader may think of a vector as a directed line segment.

[8] The term complex *line* would be much better in order to keep the analogy with the reals, but we will follow the tradition.

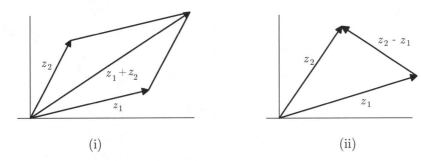

(i)   (ii)

FIGURE 10.2. Addition and Subtraction of Complex Numbers

But how could one multiply vectors in the plane? Note that when $a + bi$ is multiplied by i, the result is $-b + ai$. So, if we interpret $a + bi$ as the vector $\langle a, b \rangle$ (or as the point (a, b)) then $-b + ai$ is interpreted as the vector $\langle -b, a \rangle$ (or as the point $(-b, a)$) obtained by the rotation of $\langle a, b \rangle$ by 90° around the origin counterclockwise! This is the first glimpse into the geometric effect of multiplication of complex numbers. Let us uncover the whole story.

The **magnitude**, or the **modulus**, of $z = a + bi$, denoted $|z|$, is defined as $\sqrt{a^2 + b^2}$ and is the distance from the origin to the point (a, b) or the length of the vector $\langle a, b \rangle$. For any $z \neq 0$, the line containing z and 0 makes an angle t with the real axis (measured counterclockwise from its positive direction) for some $t \in (0, 2\pi)$. (See Figure 10.3.) Every angle that differs from t by an integer multiple of 2π – i.e., every angle of the form $t + 2\pi n$ where n is an integer – is called an **argument** of z, and the set of all arguments of z is denoted by $\arg z$.

The number $|z|$ and the set $\arg z$ are uniquely determined by z, and vice versa. It is clear that $\operatorname{Re} z = a = |z| \cos t$, and $\operatorname{Im} z = b = |z| \sin t$. Denoting $|z|$ by r, we have

$$z = r(\cos t + \sin t \, i).$$

This form of writing of z is called the **trigonometric form** of z or the **polar form** of z. For $z = 0$, the argument – and, hence, the polar form – is not defined. Note that the magnitude of the second factor, $\cos t + \sin t \, i$, is always 1:

$$|\cos t + \sin t \, i| = \sqrt{\cos^2 t + \sin^2 t} = 1.$$

If $z = a + bi$, where $a, b \in \mathbb{R}$, we say that z is represented in the **standard form**. If $0 \neq z = a + bi = r(\cos t + \sin t \, i), 0 \leq t < 2\pi$, then $r = \sqrt{a^2 + b^2}$. Furthermore, if $a \neq 0$, then $t = \arctan(b/a)$ if $b/a \geq 0$ and $t = 2\pi + \arctan(b/a)$ if $b/a < 0$.

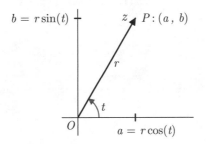

FIGURE 10.3. $|z| = r = \sqrt{a^2 + b^2}$

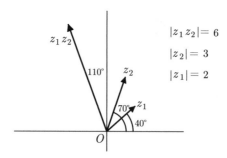

FIGURE 10.4. Multiplication of Complex Numbers

The geometric interpretation of multiplication in $\mathbb{C}$ becomes apparent if one utilizes the trigonometric form. Indeed, if $z_1 = r_1(\cos t_1 + \sin t_1\, i)$ and $z_2 = r_2(\cos t_2 + \sin t_2\, i)$, then

$$
\begin{aligned}
z_3 = z_1 \cdot z_2 &= r_1(\cos t_1 + \sin t_1\, i) \cdot r_2(\cos t_2 + \sin t_2\, i) \\
&= r_1 r_2 ([\cos t_1\, \cos t_2 - \sin t_1\, \sin t_2] + [\cos t_1\, \sin t_2 + \sin t_1\, \cos t_2]\, i) \\
&= r_1 r_2 \left(\cos(t_1 + t_2) + \sin(t_1 + t_2)\, i\right).
\end{aligned}
$$

Since the complex number $\cos\theta + \sin\theta\, i$ has magnitude 1 for any value of θ, we see that the magnitude of $z_1 z_2$ is found by multiplying the magnitudes of the two factors. If $\arg z_1 + \arg z_2$ denotes the set $\{t_1 + t_2 : t_1 \in \arg z_1,\ t_2 \in \arg z_2\}$, then we have

$$
|z_1 z_2| = |z_1||z_2|, \qquad \arg(z_1 z_2) = \arg z_1 + \arg z_2. \tag{10.1}
$$

(See Figure 10.4.) Put differently, multiplying any complex number z_1 by a complex number z_2 amounts to changing the length of the vector corresponding to z_1 by a factor of $|z_2|$, and rotating the vector by an angle from $\arg z_2$.

Using induction, one can easily generalize (10.1) to $n \geq 2$ factors:

$$
|z_1 z_2 \cdots z_n| = |z_1||z_2| \cdots |z_n|,
$$

$$
\arg(z_1 z_2 \cdots z_n) = \arg z_1 + \arg z_2 + \cdots + \arg z_n.
$$

When all numbers are equal, we obtain the following very useful corollary, known as De Moivre's theorem.[9] Let $n \arg z$ denote the sum of n sets, each equal to $\arg z$. It is easy to see that $n \arg z = \{nt : t \in \arg z\}$.

Theorem 10.1. (De Moivre's Theorem). *Let n be a positive integer, and let*

$$
z = r(\cos t + \sin t\, i).
$$

Then

$$
z^n = r^n(\cos nt + \sin nt\, i),
$$

or, equivalently, $|z^n| = |z|^n$, and $\arg z^n = n \arg z$.

---

[9] Abraham De Moivre (1667–1754) was a French-born mathematician who spent most of his professional life in England. The theorem was apparently known to I. Newton as early as 1676.

Example 51.   Find $(1 - i)^7$.

Solution 1. We will present the answer in both polar and standard forms. First note that the trigonometric form of $z = 1 - i$ is given by $\sqrt{2}\,(\cos(7\pi/4) + \sin(7\pi/4)\,i)$. Applying De Moivre's theorem, we obtain

$$(1 - i)^7 = (\sqrt{2})^7 \left(\cos\left(49 \cdot \frac{\pi}{4}\right) + \sin\left(49 \cdot \frac{\pi}{4}\right) i \right).$$

Reducing the angle modulo 2π gives $\pi/4$, since

$$49 \cdot \frac{\pi}{4} - 6 \cdot 2\pi = \frac{\pi}{4}.$$

Hence $(1 - i)^7 = 8\sqrt{2}(\cos(\pi/4) + \sin(\pi/4)\,i)$ is the polar form of the answer. Finally,

$$(1 - i)^7 = 8\sqrt{2}(\cos(\pi/4) + \sin(\pi/4)\,i) = 8\sqrt{2}(\sqrt{2}/2 + (\sqrt{2}/2)\,i) = 8 + 8i,$$

so $8 + 8i$ is the standard form of the answer.

Solution 2. Those who are familiar with the binomial formula could use it to first find the result in the standard form:

$$(1 - i)^7 = \sum_{k=0}^{7} (-1)^k \binom{7}{k} 1^{7-k}\, i^k = \sum_{k=0}^{7} (-1)^k \binom{7}{k} i^k$$

$$= \left(\binom{7}{0} - \binom{7}{2} + \binom{7}{4} - \binom{7}{6} \right) + \left(-\binom{7}{1} + \binom{7}{3} - \binom{7}{5} + \binom{7}{7} \right) i$$

$$= (1 - 21 + 35 - 7) + (-7 + 35 - 21 + 1)\, i = 8 + 8i.$$

The polar form of the answer follows. As $|(1 - i)^7| = |8 + 8i| = \sqrt{8^2 + 8^2} = 8\sqrt{2}$, letting t be an argument of $(1 - i)^7$, we obtain $(1 - i)^7 = 8\sqrt{2}\,(\cos t + \sin t\,i) = 8 + 8i$. Hence $\cos t = \sin t = 8/(8\sqrt{2}) = 1/\sqrt{2}$. This gives $t = \pi/4 + 2\pi n$, for any integer n. Taking $n = 0$ gives $(1 - i)^7 = 8\sqrt{2}\,(\cos \pi/4 + \sin \pi/4\,i)$, which is the polar form of $(1 - i)^7$. $\square$

Example 52.   Find all complex numbers z such that $z^6 = 1$.

Solution: Let $z = r(\cos t + \sin t\,i)$, where r and t are unknown. Then $z^6 = r^6(\cos 6t + \sin 6t\,i)$. Writing $1 = 1 \cdot (\cos 0 + \sin 0\,i)$, we get

$$z^6 = 1 \quad \Leftrightarrow \quad r^6(\cos 6t + \sin 6t\,i) = 1 \cdot (\cos 0 + \sin 0\,i).$$

Since two nonzero complex numbers are equal if and only if their magnitudes are equal and any pair of their arguments differ by a multiple of 2π, we obtain $r^6 = 1$, and $6t - 0 = 2\pi k$, for some integer k. As r is a positive real number, $r = 1$, and $t = (2\pi/6)k = (\pi/3)k$. Since it is sufficient to determine all t in $[0, 2\pi)$, we can assume $n = 0, 1, \ldots, 5$. This gives the following six distinct solutions of $z^6 = 1$:

$$z_k = \cos \frac{\pi}{3}k + \sin \frac{\pi}{3}k\, i, \quad k = 0, 1, 2, 3, 4, 5.$$

As the values of the sine and cosine of integer multiples of $\pi/3$ can be easily written in radicals, the solutions can be rewritten in the standard form as:

$$z_0 = 1, \quad z_1 = \frac{1}{2} + \frac{\sqrt{3}}{2}\,i, \quad z_2 = -\frac{1}{2} + \frac{\sqrt{3}}{2}\,i,$$

$$z_3 = -1, \quad z_4 = -\frac{1}{2} - \frac{\sqrt{3}}{2}\,i, \quad z_5 = \frac{1}{2} - \frac{\sqrt{3}}{2}\,i.$$

$\square$

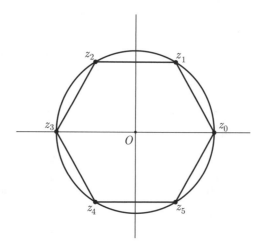

FIGURE 10.5. The 6th Roots of Unity

The solutions of the equation $z^n = 1$, where n is a positive integer, are called the nth **roots of unity**. We just demonstrated that there are exactly six distinct sixth roots of unity. Similarly, one can prove that there are exactly n distinct nth roots of unity, and they are given by the following formula:

$$z_k = \cos\frac{2\pi}{n}k + \sin\frac{2\pi}{n}k\,i, \quad \text{where} \;\; k = 0, 1, \ldots, n-1.$$

When viewed geometrically, the nth roots of unity fall at the vertices of a regular n-gon inscribed in the unit circle. For example, Figure 10.5 illustrates the 6th roots of unity.

More generally, there are exactly n distinct n^{th} roots of an arbitrary nonzero complex number c.

Theorem 10.2. *The equation $z^n = c$, $c \neq 0$, has n distinct roots. Let $c = r(\cos t + \sin t\,i)$ be a polar form of c. Then these roots can be represented by*

$$z_k = r^{1/n}\left(\cos\left(\frac{t}{n} + \frac{2\pi}{n}k\right) + \sin\left(\frac{t}{n} + \frac{2\pi}{n}k\right)i\right),$$

where $k = 0, 1, \ldots, n-1$. Geometrically, they represent vertices of a regular n-gon inscribed in a circle of radius $r^{1/n}$ centered at the origin.

The proof of the theorem is left to Problem 10.8.

For $z = a + bi$, the complex number $a - bi$ is called the **conjugate** of z and is denoted by $\bar{z}$. Note that $z\bar{z} = (a + bi)(a - bi) = (a^2 + b^2) + (-ab + ba)i = a^2 + b^2 = |z|^2$ is the square of the magnitude of z. The following useful properties of conjugates and magnitudes are easily verified.

Proposition 10.3. *For any complex numbers z, z_1, and z_2,*

(1) $\overline{z_1 + z_2} = \overline{z_1} + \overline{z_2}$;

(2) $\overline{z_1 z_2} = \overline{z_1}\,\overline{z_2}$;

(3) $\overline{\left(\dfrac{z_1}{z_2}\right)} = \dfrac{\overline{z_1}}{\overline{z_2}}, z_2 \neq 0;$

(4) $z^{-1} = \dfrac{1}{z} = \dfrac{\overline{z}}{z\overline{z}} = \dfrac{\overline{z}}{|z|^2}, z \neq 0;$

(5) For $z \neq 0, t \in \arg z \Leftrightarrow -t \in \arg \overline{z} \Leftrightarrow -t \in \arg z^{-1};$

(6) $\arg\left(\dfrac{z_1}{z_2}\right) = \arg(z_1) - \arg(z_2);$

(7) $|z_1 z_2| = |z_1||z_2|;$

(8) $\left|\dfrac{z_1}{z_2}\right| = \dfrac{|z_1|}{|z_2|}, z_2 \neq 0;$

(9) *The Triangle Inequality:* $|z_1 + z_2| \le |z_1| + |z_2|$, *with equality if and only if one of the numbers is zero or* $\arg z_1 = \arg z_2$.

We leave the proof of Proposition 10.3 to Problem 10.6. Clearly, one can generalize properties (1), (2), (7), and (9) to any $n \ge 2$ complex numbers.

We assume that the reader is familiar with polynomials, and, in particular, with the fact that the number a is a root of the polynomial p, $\deg p \ge 1$, if and only if p is divisible by $x - a$, i.e.,

$$p(a) = 0 \quad \Leftrightarrow \quad p(x) = (x - a)q(x), \quad \deg q(x) = \deg p(x) - 1.$$

Perhaps the most important fact about complex numbers is the following theorem, first proved rigorously by Gauss. All known proofs of the theorem are quite difficult, although the reader can easily find one in the literature.

Theorem 10.4. *Let* $p(x) = a_n x^n + a_{n-1}x^{n-1} + \cdots + a_1 x + a_0$ *be a polynomial of degree* $n \ge 1$ *with complex coefficients. Then there exists a complex number z such that $p(z) = 0$. Equivalently, there exist complex numbers $z_1, z_2, \ldots, z_n$, not necessarily distinct, such that*

$$p(x) = a_n(x - z_1)(x - z_2)\cdots(x - z_n).$$

The second statement of the preceding theorem can be rephrased as: every polynomial of degree $n \ge 1$ with complex coefficients has exactly n complex roots.

10.3 Applications

The shortest path between two truths in the real domain passes through the complex domain.
— Jacques Hadamard (1865–1963)

One type of application of complex numbers to plane geometry is based on the fact that, with respect to addition and multiplication by real numbers, the complex numbers do not differ from vectors.

As we know, a point P with coordinates (a, b) corresponds to the complex number $z = a + bi$. Or, as we will see in Chapter 11, if O is the origin, then $\overrightarrow{OP}$ corresponds to $z = a + bi$. Therefore, if a problem is solved by using vectors and without using the dot product of vectors, its solution can be rewritten *verbatim* by using complex numbers. For example, if z_1 and z_2 are complex numbers corresponding to points P_1 and P_2 in the plane, then the number $\frac{1}{2}(z_1 + z_2)$ corresponds to the midpoint of the segment $P_1 P_2$, and the centroid of a triangle with vertices corresponding to z_1, z_2, and z_3 is $(z_1 + z_2 + z_3)/3$. We will not stress this type of application, as there is no benefit to using complex numbers instead of vectors in this case.

Complex numbers are particularly useful in solving problems where one can take advantage of being able to multiply and divide them. They can also sometimes be used to prove trigonometric facts, whether these appear in solutions of geometric problems or on their own. While complex numbers can be a very effective tool for solving certain problems, they may be bad for solving others. Since they permeate modern mathematics, one is advised to keep them in the toolbox. In this section we present only several impressive applications of the method. More examples can be found in the problems and in the references (see [1], [26], [39], [42], [53], and [70]).

Example 53. Suppose points $A_1, A_2, \ldots, A_n$ lie on a circle $\mathcal{C}(O, r)$ and divide the circle into n congruent arcs. Prove that

$$\sum_{i=1}^{n} \overrightarrow{OA_i} = \vec{0}.$$

Proof. Let the center of the circle correspond to the origin and let $A_1 : (z)$, where z is a positive real number. Assume that the points are numbered counterclockwise around the circle. Let $\zeta = \cos\frac{2\pi}{n} + \sin\frac{2\pi}{n} i$ be an nth root of unity. Then $A_k : (z\zeta^{k-1})$, for $k = 2, 3, \ldots, n$, and the sum $\sum_{i=1}^{n} \overrightarrow{OA_i}$ corresponds to the following sum of complex numbers:

$$z + z\zeta + z\zeta^2 + \cdots + z\zeta^{n-1}.$$

Using the formula for the sum of the geometric series with the first term r and ratio $\zeta \neq 1$, along with the equality $\zeta^n = 1$, we obtain

$$z(1 + \zeta + \zeta^2 + \ldots + \zeta^{n-1}) = \frac{z(1 - \zeta^n)}{1 - \zeta} = \frac{z(1 - 1)}{1 - \zeta} = 0.$$

Thus, $\sum_{i=1}^{n} \overrightarrow{OA_i} = \vec{0}.$ □

The appearance of a geometric progression is the key to this solution. The previous example will appear again as Example 66 in Chapter 11, in which a very different geometric solution will be provided.

As an immediate corollary we obtain the following identities for trigonometric sums: for every angle α and every integer $n \geq 2$,

$$\sum_{k=0}^{n-1} \cos\left(\alpha + \frac{2\pi}{n}k\right) = \sum_{k=0}^{n-1} \sin\left(\alpha + \frac{2\pi}{n}k\right) = 0.$$

This follows from denoting an argument of z by α and equating to zero the real and imaginary parts of the sum $z + z\zeta + z\zeta^2 + \cdots + z\zeta^{n-1}$.

Our next example is a generalization of Ptolemy's theorem from Chapter 4, due to Euler. We say that several points are **cocyclic** if they all lie on a circle or are collinear.

> **Example 54.** Let A, B, C, and D be distinct points in a plane. Prove that
> $$AB \cdot CD + BC \cdot DA \geq AC \cdot BD,$$
> with equality if and only if the four points are cocyclic in the order A-B-C-D.

Proof. Let z_1, z_2, z_3, and z_4 be complex numbers corresponding to the points A, B, C, and D, respectively. Then the inequality we are proving can be rewritten as:

$$|z_1 - z_2| \cdot |z_3 - z_4| + |z_2 - z_3| \cdot |z_4 - z_1| \geq |z_1 - z_3| \cdot |z_2 - z_4| \quad \Leftrightarrow$$

$$|(z_1 - z_2) \cdot (z_3 - z_4)| + |(z_2 - z_3) \cdot (z_4 - z_1)| \geq |(z_1 - z_3) \cdot (z_2 - z_4)|.$$

By expanding the products, one verifies that

$$(z_1 - z_2)(z_3 - z_4) + (z_2 - z_3)(z_1 - z_4) = (z_1 - z_3)(z_2 - z_4).$$

As $z_1 - z_4 = -(z_4 - z_1)$, we achieve inequality by applying the triangle inequality to the complex numbers $(z_1 - z_2)(z_3 - z_4)$ and $(z_2 - z_3)(z_1 - z_4)$![10]

Now we prove the part concerning the equality. One may use the sufficient condition for a quadrilateral to be inscribed in a circle, namely that the sum of measures of the interior angles at points A_1 and A_3 is π. We know that the equality in the triangle inequality happens if and only if one of the numbers is zero or their arguments are equal. As the A_i are distinct, so are all z_i; therefore neither $(z_1 - z_2)(z_3 - z_4)$ nor $(z_2 - z_3)(z_1 - z_4)$ is 0. Hence $\arg((z_1 - z_2)(z_3 - z_4)) = \arg((z_2 - z_3)(z_1 - z_4))$, which is equivalent to their ratio

$$\frac{(z_1 - z_2)(z_3 - z_4)}{(z_2 - z_3)(z_1 - z_4)}$$

having argument $2\pi n$, for some integer n. Let us transform the ratio as:

$$\frac{(z_1 - z_2)(z_3 - z_4)}{(z_2 - z_3)(z_1 - z_4)} = \frac{z_1 - z_2}{z_1 - z_4} \cdot (-1) \cdot \frac{z_4 - z_3}{z_2 - z_3}.$$

(Refer to Figure 10.6.) The values of the arguments of

$$(z_1 - z_2)/(z_1 - z_4) = (z_2 - z_1)/(z_4 - z_1) \text{ and } (z_4 - z_3)/(z_2 - z_3)$$

in $[0, 2\pi)$ are $m\angle A_4 A_1 A_2$ and $m\angle A_2 A_3 A_4$, respectively. As π is an argument of -1, we obtain

$$m\angle A_4 A_1 A_2 + \pi + m\angle A_2 A_3 A_4 = 2\pi n,$$

for some integer n. As the sum $m\angle A_4 A_1 A_2 + \pi + m\angle A_2 A_3 A_4$ is in $[\pi, 3\pi]$, we have $n = 1$, and this gives

$$m\angle A_4 A_1 A_2 + m\angle A_2 A_3 A_4 = \pi,$$

which completes the proof. Note that collinearity corresponds to the "circle of infinite radius." $\quad\square$

---

[10] If you are a mathematics major and not impressed with this proof, switch your major.

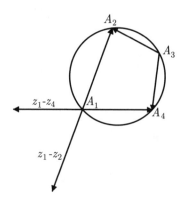

FIGURE 10.6.

Corollary 10.5 gives another simple criteria for four points being cocyclic. It will be used in Chapter 13.

Corollary 10.5. *Distinct complex numbers a, b, c, and d represent four cocyclic points if and only if the ratio*

$$\frac{a-c}{a-d} \div \frac{b-c}{b-d}$$

is a real number.

Proof. The proof is essentially contained in the proof of Example 54. If the distinct numbers a, b, c, and d represent cocyclic points, then it is clear from the proof of Example 54 that the given ratio is a real number. If $k = \dfrac{a-c}{a-d} \div \dfrac{b-c}{b-d}$ is a positive real number, then a, b, c, and d correspond to z_1, z_2, z_3, and z_4, respectively, in the proof above (so the numbers correspond to cocyclic points). If k is a negative real number, then interchanging the roles of b and c will yield a ratio of $1-k$, which is a positive real number, [11] so once again the numbers correspond to cocyclic points. $\square$

The following example illustrates the relation between rotations and multiplication of complex numbers.

Example 55. On each side of a triangle ABC, erect an equilateral triangle lying exterior to the original triangle, as shown in Figure 10.7. Then the centers of the three equilateral triangles are themselves vertices of an equilateral triangle.[12]

---

[11] If $k = \dfrac{a-c}{a-d} \div \dfrac{b-c}{b-d}$, it can actually be shown that any permutation of the four numbers, a, b, c, and d, in the ratio will always yield one of the values k, $1/k$, $1-k$, $1/(1-k)$, $k/(k-1)$, or $(k-1)/k$. Thus, if the ratio is real for one of the orderings of these numbers, it will be real for all orderings. (See Problem S10.8.) By the way, the preceding argument implies that the six expressions above define a set of six functions of k that is closed under composition.

[12] This statement is often referred to as Napoléon's Theorem, though it seems there is no evidence that Napoléon Bonaparte had anything to do with it. We suggest to stop this nonsense.

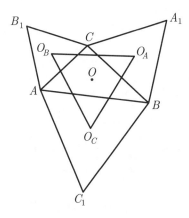

FIGURE **10.7.**

Proof. In what follows, the notation of a point and a number corresponding to it will be the same. Suppose that the centroid of the triangle coincides with the origin, and let A, B, and C be three complex numbers that correspond to the vertices of the triangle (taken in the counterclockwise direction). Then by the Triangle Rule (see Chapter 11),

$$A + B + C = 0.$$

Let ω be a complex number of unit modulus with an argument of $-60°$. The reason for introducing ω is that multiplication by ω rotates every complex number by $60°$ in the clockwise direction. Let ABC_1, BCA_1, and CAB_1 be the equilateral triangles constructed on the sides of triangle ABC. We then have

$$A_1 = B + \omega(C - B),$$
$$B_1 = C + \omega(A - C),$$
$$C_1 = A + \omega(B - A).$$

The arithmetic average of B, C, and A_1 yields the centroid O_A of $\triangle BCA_1$. Similarly, we compute the centers O_B and O_C of triangles CAB_1 and ABC_1, respectively:

$$O_A = (2B + C + \omega(C - B))/3,$$
$$O_B = (2C + A + \omega(A - C))/3,$$
$$O_C = (2A + B + \omega(B - A))/3.$$

We must show that the three points O_A, O_B, and O_C are vertices of an equilateral triangle. First note that the centroid of $\triangle O_A O_B O_C$ lies at the origin, because the sum of the three numbers is 0 (just as with the original triangle). Indeed,

$$O_A + O_B + O_C = (A + B + C) + \frac{\omega}{3} \cdot ((C - B) + (A - C) + (B - A)) = A + B + C = 0.$$

Now, if $\triangle O_A O_B O_C$ is to be an equilateral triangle centered at the origin, it must be possible to move from any vertex to some other through a rotation of $120°$ clockwise, i.e., by multiplying the corresponding number by ω^2. To this end, let us show that $\omega^2 \cdot O_A = O_C$. Note that $\omega^3 = -1$. Hence,

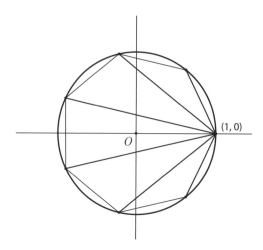

FIGURE 10.8. Example 56, with $n = 7$

$\omega^3 + 1 = (\omega + 1)(\omega^2 - \omega + 1) = 0$. As $\omega \neq -1$, we conclude that $\omega^2 - \omega + 1 = 0$, or $\omega^2 = \omega - 1$. Therefore we have:

$$\begin{aligned}
\omega^2 \cdot O_A &= \omega^2 \cdot (2B + C + \omega(C - B))/3 \\
&= ((\omega - 1)(2B + C) + (\omega^3)(C - B))/3 \\
&= ((\omega - 1)(2B + C) + (B - C))/3 \\
&= (\omega(B + B + C) - B - 2C))/3 \\
&= (\omega(B - A) - B + 2(A + B))/3 \quad (\text{as} \quad A + B + C = 0) \\
&= (2A + B + \omega(B - A))/3 \\
&= O_C.
\end{aligned}$$

The proofs that vertices O_B and O_C are rotated into O_A and O_B, respectively, are absolutely similar and are left to the reader. ◻

The proof above demonstrates the power of using complex numbers when a geometric problem can be interpreted as one involving rotations. Our next example shows how a relation between (complex) roots of a polynomial and its coefficients can be used in geometry.

Example 56. Let there be $n > 2$ equally spaced points on a circle of radius 1 (the vertices of a regular n-gon). Pick any point and draw the $n - 1$ chords from this point to the other $n - 1$ points. Prove that the product of the lengths of these $n - 1$ chords is n.

Proof. We will use complex numbers to represent the points. Place the unit circle with its center at the origin, and let axis OX pass through one of the points. Then the vertices of the n-gon correspond to the n roots of unity, $z_0, \ldots, z_{n-1}$, which are the roots of the equation $x^n - 1 = 0$. As

$$x^n - 1 = (x - 1)(x^{n-1} + x^{n-2} + \cdots + x + 1),$$

short of $z_0 = 1$, the rest of the roots satisfy $p(x) := x^{n-1} + x^{n-2} + \cdots + x + 1 = 0$. Using $z_0 = 1$ as the common vertex of the chords, we obtain that the length of the chord drawn to vertex z_k is

$|z_k - 1|$, and the problem is reduced to proving that

$$|z_1 - 1| \cdot |z_2 - 1| \cdot \cdots \cdot |z_{n-1} - 1| = n.$$

Since $|ab| = |a||b|$ for any complex numbers a and b, it is natural to try to find the magnitude of

$$A := (z_1 - 1) \cdot (z_2 - 1) \cdot \cdots \cdot (z_{n-1} - 1).$$

Our first key observation is that the numbers $z_k - 1$, for $k = 1, \ldots, n - 1$, are exactly the roots of the polynomial

$$g(x) := p(x + 1) = (x + 1)^{n-1} + \cdots + (x + 1)^2 + (x + 1) + 1,$$

since $g(z_k - 1) = p(z_k) = 0$. Hence,

$$g(x) = (x - (z_1 - 1))(x - (z_2 - 1)) \cdots (x - (z_{n-1} - 1)),$$

and the product $(-1)^{n-1} A$ is the constant term of $g(x)$! This is our second key observation. Finally, we observe that the constant term of $g(x)$ can also be computed as

$$g(0) = 1 + 1 + \cdots + 1 = n.$$

Thus, $(-1)^{n-1} A = n$, and therefore $|A| = n$. $\qquad\square$

Since $|z_k - 1| = 2 \sin \frac{\pi}{n} k$ (see Problem 10.10), as an immediate corollary we obtain the following interesting trigonometric identity:

$$\prod_{k=1}^{n-1} \sin\left(\frac{k}{n}\pi\right) = \frac{n}{2^{n-1}}.$$

Can you suggest a better proof of this identity? We cannot.

Our last example is a recent gem and uses complex numbers to prove Heron's formula for the area of a triangle (see [16]).

Example 57. Let $s = (a + b + c)/2$ be the semiperimeter of $\triangle ABC$. Then

$$Area\,(\triangle ABC) = \sqrt{s(s - a)(s - b)(s - c)}.$$

Proof. Let I be the center of the inscribed circle of $\triangle ABC$, with lengths and angles as shown in Figure 10.9. Observe that $a = y + z$, $b = x + z$, and $c = x + y$. Clearly $2\alpha + 2\beta + 2\gamma = 2\pi$, so $\alpha + \beta + \gamma = \pi$. Then

$$r + xi = u(\cos \gamma + \sin \gamma\, i),$$
$$r + yi = v(\cos \beta + \sin \beta\, i), \text{ and}$$
$$r + zi = w(\cos \alpha + \sin \alpha\, i),$$

for real numbers $u = IA$, $v = IB$, and $w = IC$. This gives

$$(r + xi)(r + yi)(r + zi) = uvw(\cos(\alpha + \beta + \gamma) + \sin(\alpha + \beta + \gamma)i)$$
$$= uvw(\cos \pi + \sin \pi\, i) = -uvw.$$

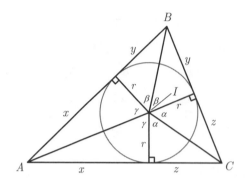

FIGURE **10.9.**

Since $-uvw$ is a real number,

$$
\begin{aligned}
0 &= \operatorname{Im}\left[(r + xi)(r + yi)(r + zi)\right] \\
&= \operatorname{Im}\left[(r^3 - ryz - rxz - rxy) + (r^2(x + y + z) - xyz)i\right] \\
&= r^2(x + y + z) - xyz.
\end{aligned}
$$

Noting that $x + y + z = 1/(2a + 2b + 2c) = s$, we obtain the following expression for r:

$$
r = \sqrt{\frac{xyz}{x + y + z}} = \sqrt{\frac{(s - a)(s - b)(s - c)}{s}}.
$$

Since the area of $\triangle ABC$ is the sum of the areas of $\triangle AIB$, $\triangle BIC$, and $\triangle CIA$, we have

$$
\operatorname{Area}(\triangle ABC) = \frac{ra}{2} + \frac{rb}{2} + \frac{rc}{2} = rs = \sqrt{s(s - a)(s - b)(s - c)},
$$

and the proof is complete. $\square$

10.4 Problems

> *Each problem that I solved became a rule which served afterwards to solve other problems.*
> — René Descartes (1596–1650)[13]

10.1 Compute and present the results in standard form.

   (a) $\dfrac{(1 - 3i)(2 + i)}{i - 3} + \dfrac{1}{i^{24}}$

   (b) $(1 + i)^{13}$

   (c) $1 + i + i^2 + \cdots + i^{2007}$

10.2 In a complex plane, let $A : (2 + 3i)$ and $B : (4 - i)$. Find the complex number corresponding to the point obtained from the midpoint of $\overline{AB}$ rotated around A by $120°$.

10.3 Describe the sets of points $z \in \mathbb{C}$ satisfying the following conditions.

   (a) $|z| = 3$

   (b) $|z + 2 - 3i| > 2$

   (c) $|z - 2| + |z - 1 + 3i| = 5$

---

[13] Le Discours de la Méthode (1637)

     (d) $|z - 2| + |z - 1 + 3i| = 3$

     (e) $|z - 2i| - |z - 2 + 3i| = 1$

     (f) $\operatorname{Re} z \geq -1$ and $\operatorname{Im} z \leq 3$.

10.4 Given three vertices of a parallelogram, $1 + 2i$, $3 - i$, and $2 + i$, find the fourth vertex. How many solutions does the problem have?

10.5 Prove that

     (a) $\bar{z} = z$ if and only if $\operatorname{Im} z = 0$, i.e., z is a real number.

     (b) $\bar{z} = -z$ if and only if $\operatorname{Re} z = 0$, i.e., $z = bi$, $b \in \mathbb{R}$. Such complex numbers are often called **purely imaginary**.

     (c) $\bar{z} = 1/z$ if and only if $|z| = 1$.

10.6 Prove all statements of Proposition 10.3.

10.7 Prove that if both x and y are a sum of squares of two integers, then so is xy. For example: if $x = 13 = 2^2 + 3^2$ and $y = 61 = 5^2 + 6^2$, then $xy = 13 \cdot 61 = 793 = 8^2 + 27^2$.

10.8 Prove Theorem 10.2.

10.9 Express $\sin 3\alpha$ in terms of $\sin \alpha$, and express $\cos 6\alpha$ in terms of $\cos \alpha$.

10.10 Justify the claim made after Example 56 that $|z_k - 1| = 2 \sin \dfrac{\pi}{n} k$.

10.11 Use the product $(2 + i)(3 + i)$ to prove the identity

$$\arctan \frac{1}{2} + \arctan \frac{1}{3} = \frac{\pi}{4}.$$

     Generalize the method to show that for any real positive numbers p and q,

$$\arctan \frac{1}{p + q} + \arctan \frac{q}{p^2 + pq + 1} = \arctan \frac{1}{p}.$$

10.12 Prove that the centroids of $\triangle ABC$ and $\triangle O_A O_B O_C$ in Example 55 coincide.

10.13 Prove that for $0 < \alpha < 2\pi$,

$$\sum_{k=0}^{n-1} \cos k\alpha = \frac{\cos \frac{(n-1)\alpha}{2} \cdot \sin \frac{n\alpha}{2}}{\sin \frac{\alpha}{2}},$$

$$\sum_{k=0}^{n-1} \sin k\alpha = \frac{\sin \frac{(n-1)\alpha}{2} \cdot \sin \frac{n\alpha}{2}}{\sin \frac{\alpha}{2}}.$$

10.14 Compute the sum of squares of lengths of all sides and diagonals of a regular n-gon inscribed in a circle of radius R.

10.15 Let $A_1 A_2 \ldots A_n$ be a regular polygon inscribed in a circle $\mathcal{C}(O; R)$. Let X be a point on a ray OA_1 such that $OX = a$. Compute the product of all distances XA_k, $k = 1, \ldots, n$.

10.16 Use the product $(5 + i)^4(-239 + i)$ to show that

$$4 \arctan \frac{1}{5} - \arctan \frac{1}{239} = \frac{\pi}{4}.$$

10.17 Let A, B, and C be complex numbers, and let $\triangle ABC$ denote the corresponding triangle in the complex plane with vertices ordered clockwise. Let ζ be the complex number of unit modulus and argument $60°$. Prove that $\triangle ABC$ is equilateral if and only if $A\zeta - B\zeta^2 - C = 0$.

10.18 Let z_1 and $z_2 \in \mathbb{C}$ represent two distinct points in the complex plane.

(a) Prove that

$$\left\{ z \in \mathbb{C} : z - z_1 = \frac{z_2 - z_1}{\overline{z_2} - \overline{z_1}} (\overline{z} - \overline{z_1}) \right\}$$

represents the set of all points of the line passing through z_1 and z_2. Therefore

$$z - z_1 = \frac{z_2 - z_1}{\overline{z_2} - \overline{z_1}} (\overline{z} - \overline{z_1}) \tag{10.2}$$

is called an **equation of the line in the complex plane** passing through z_1 and z_2.

(b) The ratio $\kappa = \dfrac{z_2 - z_1}{\overline{z_2} - \overline{z_1}}$ is called the **complex slope** of the line. Check that

$$\overline{\kappa} = \kappa^{-1} \quad \text{and} \quad |\kappa| = 1.$$

(c) Let κ_1 and κ_2 be complex slopes of two lines in the complex plane. Prove that the lines are

(i) parallel if and only if $\kappa_1 = \kappa_2$;
(ii) perpendicular if and only if $\kappa_1 = -\kappa_2$.

10.19 Let a, b, c, and d be complex numbers corresponding to distinct points A, B, C, and D in the complex plane. Prove that $\overline{AD}$ and $\overline{BC}$ are perpendicular if and only if $(d - a)/(c - b)$ is purely imaginary.

10.20 Consider $\triangle ABC$, and let H and O represent its orthocenter and circumcenter, respectively. Suppose a Cartesian coordinate system is introduced in such a way that O is at the origin. Let $a, b, c,$ and $h \in \mathbb{C}$ represent complex numbers corresponding to A, B, C, and H, respectively. Prove that $h = a + b + c$. Then use this fact to show that if G is the centroid of $\triangle ABC$, then $G \in \overline{OH}$ and $OG : GH = 1 : 2$.

10.21 On each side of a convex quadrilateral, draw a square externally. Show that the two segments joining the centers of the opposite squares are perpendicular to each other and are of the same length.

10.22 On each side of a parallelogram, draw a square externally. Show that the centers of the squares form the vertices of a square.

10.5 Supplemental Problems

$S10.1$ Prove the distributive law for complex numbers: $(a + b)c = ac + bc$, for every a, b, and $c \in \mathbb{C}$.

$S10.2$ Express $\tan 5\alpha$ in terms of $\tan \alpha$.

$S10.3$ Use equations of lines in the complex plane to find the point of intersection of the line l passing through the points $A : (2 - i)$ and $B : (1 + 3i)$ and the line m perpendicular to l and passing through the point $C : (i)$. (See Problem 10.18 for an equation of a line in the complex plane.)

$S10.4$ In a complex plane, let $A : (1 - i)$ and $B : (4i)$. Find the complex number corresponding to the point B' obtained by rotating B around the midpoint of the segment AB by $60°$.

$S10.5$ Let $a = \cos(2\pi/9) + i \sin(2\pi/9)$, so $a^9 = 1$. Prove that the complex slope of the line through points $K : (a^5)$ and $L : (a^8)$ is $-a^4$.

$S10.6$ Find all real numbers a such that the point $D : (a + 2i)$ lies on the circle defined by points $A : (1 + i)$, $B : (2 - i)$, and $C : (2 + 3i)$.

$S10.7$ Let $A_i : (z_i)$, $i = 1, 2, 3$, be three points on the unit circle centered at the origin in a complex plane. Prove that $A_1 A_2 = A_1 A_3$ if and only if $z_1^2 = z_2 z_3$.

$S10.8$ This problem justifies that the condition for cocyclicity of four points given in Corollary 10.5 doesn't depend on the order of the points (as should be the case). Let a, b, c, and d be distinct complex numbers, and let

$$k = \frac{a - c}{a - d} \div \frac{b - c}{b - d}.$$

(a) Prove that interchanging any two of a, b, c, d in $\dfrac{a - c}{a - d} \div \dfrac{b - c}{b - d}$ results in one of the four values k, $1/k$, $1 - k$, or $k/(k - 1)$. (Make use of a computer, if you wish.)

(b) Prove that any permutation of a, b, c, d in $\dfrac{a - c}{a - d} \div \dfrac{b - c}{b - d}$ results in one of the six values k, $1/k$, $1 - k$, $1/(1 - k)$, $k/(k - 1)$, or $(k - 1)/k$.

(c) Prove that if k is a real number, then any permutation of a, b, c, d in $\dfrac{a - c}{a - d} \div \dfrac{b - c}{b - d}$ results in a real value of the ratio and there exists a permutation that results in a positive value of the ratio.

Note that since a, b, c, and d are all distinct, $k \neq 0$ and $k \neq 1$.

$S10.9$ (a) Prove that for any complex number $D \neq 0$, the equation $z^2 = D$ has exactly two distinct solutions.

(b) Prove that for any $a, b, c \in \mathbb{C}$, $a \neq 0$, the quadratic equation $az^2 + bz + c = 0$ has exactly two solutions (which may not be distinct).

$S10.10$ Use complex numbers to show that the altitudes of a triangle are concurrent.

$S10.11$ Let O be the center of a circle. Consider three equilateral triangles $O A_1 B_1$, $O A_2 B_2$, and $O A_3 B_3$, such that all points A_k and B_k lie on the circle, and the order $O A_k B_k$ defines the same orientation in all triangles. Let M_1 be the midpoint of $\overline{B_1 A_2}$, M_2 the midpoint of $\overline{B_2 A_3}$, and M_3 the midpoint of $\overline{B_3 A_1}$. Prove that $\triangle M_1 M_2 M_3$ is equilateral.

$S10.12$ Prove that the equation of any line in a complex plane can be written in the form

$$Az + \overline{A}\overline{z} + C = 0,$$

where $A \neq 0$ and C is a real number. Also prove the converse statement.

$S10.13$ Prove that the distance from a point $P : (p)$ to a line $l : Az + \overline{A}\overline{z} + C = 0$, where $A \neq 0$ and C is a real number, can be computed as

$$d(P, l) = \frac{|Ap + \overline{A}\overline{p} + C|}{2|A|}.$$

$S10.14$ Two pirates decide to hide a stolen treasure on a desert island, which has a well (W), a birch tree (B), and a pine tree (P). To bury the treasure, one of the pirates starts at W and walks to B and after reaching turns right $90°$ and walks the same distance as WB reaching the point Q. The second pirate starts at W and walks to P, after which he turns left $90°$ degrees and walks the same distance as WP reaching the point R. Then they bury the treasure halfway between Q and R. Some months later the two pirates return to dig up the treasure only to discover that the well was gone. Can they find the treasure?

$S10.15$ Prove that for every integer $n \geq 2$,

$$\prod_{k=1}^{n-1} \sin \frac{\pi}{2n} k = \frac{\sqrt{n}}{2^{n-1}}.$$

$S10.16$ Consider a triangle ABC inscribed in a unit circle with center O; let $a = BC, b = AC$, and $c = AB$. Let G be the centroid and H be the orthocenter of the triangle.

(a) Prove that $OG^2 = 1 - \frac{1}{9}(a^2 + b^2 + c^2)$.

(b) Express OH^2 in terms of a, b, and c, where H is the orthocenter of $\triangle ABC$.

$S10.17$ Let G be the centroid of a triangle ABC, where $m\angle A \neq 90°$, and let $\overline{BB_1}$ and $\overline{CC_1}$ represent the altitudes of the triangle. Prove that G is on the line $B_1 C_1$ if and only if $b^2 + c^2 = 3a^2$.

$S10.18$ Let $A_1 A_2 \ldots A_7$ be a regular heptagon (7-gon). Consider a triangle formed by lines $A_1 A_2$, $A_3 A_5$, and $A_4 A_7$. Prove that the circumcircle of this triangle passes through A_6.

11

Vectors

11.1   Introduction

Lord Kelvin would be surprised to see the fundamental role that vectors played in mathematics after
his death. Understanding the notion of a vector and possessing facility with vector computations
are fundamental for modern mathematics. For our purposes in this book, vectors provide another
tool for solving problems in plane geometry. Our use of vectors in plane geometry will also lay a
foundation for their (even more impressive) use in spatial geometry, but such usage is beyond our
goals here. As will become apparent, vectors are closely related to other topics presented in this
book – the coordinate method, trigonometry, complex numbers, and transformations.

In mathematics and physics, some quantities need just one real number to describe them; the mass
of an object, the length of a line segment, the area of a plane figure, time, and temperature are all
examples of such quantities, often called **scalars**. Other quantities require two or more parameters
in order to be understood; for example, force and velocity are each characterized by both magnitude
and direction. Such quantities are called **vectors**, often depicted as directed line segments (arrows).
The length of the arrow indicates the **magnitude** of the vector and the "direction" of the arrow
corresponds to the **direction** of the vector.[2]

Vectors are not figures in a plane, which we defined in this book as subsets of points. The same
vector has infinitely many different representations in geometric diagrams as a directed segment: the
length and direction of the arrow depicting the vector are fixed, but the placement within the plane is
not. For many applications, the location of this placement does not matter, and we want vectors with
identical lengths and directions to be considered "equal." This brings us to a more formal definition

---

[1] Taken from `http://en.wikiquote.org/wiki/Lord_Kelvin`. Letter to G. F. FitzGerald (1896) as quoted in *A
History of Vector Analysis: The Evolution of the Idea of a Vectorial System* (1994) by Michael J. Crowe, p. 120

[2] While the definition of *direction* can be made more rigorous by describing it in terms of an angle, that isn't necessary
for our purpose.

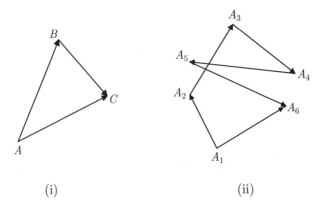

(i)   (ii)

FIGURE 11.1. Triangle Rule and Polygon Rule.

of a vector. Often we will represent a vector as a directed segment beginning at the origin. If such a vector terminates at $P : (a_1, a_2)$, meaning that the arrow points to P, we denote it as $\mathbf{a} = \vec{a} = \langle a_1, a_2 \rangle$ and call it the **position vector** of point P. We agree that any directed segment from a point $A : (x, y)$ to point $B : (x + a_1, y + a_2)$ will be a representative of the vector $\overrightarrow{AB} = \vec{a} = \langle a_1, a_2 \rangle$.[3] Then A is called the **initial point** of $\overrightarrow{AB}$ and B is called the **terminal point**. Thus, the position vector is a particular representation for the entire class of vectors having the same length and direction. The numbers a_1 and a_2 are the **components** of the vector $\vec{a}$. We say that two vectors are **equal** if and only if their corresponding components are equal.

Since the magnitude of a vector is simply the distance from the initial point to the terminal point of any of the arrows representing the vector, by the distance formula the magnitude of a vector $\overrightarrow{AB} = \vec{a} = \langle a_1, a_2 \rangle$ is

$$AB = |\vec{a}| = \sqrt{a_1{}^2 + a_2{}^2}.$$

We assume that the zero vector, $\vec{0} = \langle 0, 0 \rangle$, is the only vector without a direction. Clearly, it is also the only vector with magnitude zero.

Example 58. Vectors $\vec{a} = \langle 3, -1 \rangle$ and $\overrightarrow{AB}$, with $A : (2, 5)$ and $B : (5, 4)$, are equal and have magnitude $\sqrt{10}$.

11.2 Computations

Physics provides both the motivation for and the main source of applications of vectors outside of mathematics. Consider, for example, the combined effect of two forces, each having its own magnitude and direction. Each force has a vector representation and, by a law of physics, the vector corresponding to the resulting force can be found as their **sum**, defined in the following way: if $\vec{a} = \langle a_1, a_2 \rangle$ and $\vec{b} = \langle b_1, b_2 \rangle$, then

$$\vec{a} + \vec{b} = \langle a_1 + b_1, a_2 + b_2 \rangle.$$

Vector addition can be illustrated geometrically in at least two ways. Figure 11.1(i) demonstrates the "Triangle Rule" of vector addition, in which the terminal point, B, of vector $\overrightarrow{AB}$ is coincident

---

[3] While it is unfortunate that the notation $\overrightarrow{AB}$ is used both to represent a ray and a vector, the meaning will usually be clear from the context.

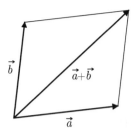

FIGURE 11.2. Vector Addition.

with the initial point of $\overrightarrow{BC}$; the sum of these two vectors is the vector $\overrightarrow{AC}$, which forms the third side of the triangle. Notice that one consequence of the Triangle Rule is that if the sides of any triangle are represented as vectors, $\overrightarrow{AB}$, $\overrightarrow{BC}$, and $\overrightarrow{CA}$, all placed from the terminal point of one to the initial point of another, then the sum of the vectors is the zero vector, $\vec{0}$. The Triangle Rule can also be generalized to the "Polygon Rule" for addition of more than two vectors, as shown in Figure 11.1(ii).

In Figure 11.2 the "Parallelogram Rule" is illustrated. The two vectors $\vec{a}$ and $\vec{b}$ have the same initial point and form the sides of the parallelogram; the vector sum $\vec{a} + \vec{b}$ corresponds to the diagonal of the parallelogram.

The reader might think that it would now be natural to discuss the multiplication of two vectors. While there are several such ways to define multiplication (and we saw one of them in Chapter 10 where we multiplied complex numbers), we instead proceed in another direction. Multiplication of a vector by a scalar (that is, by a real number) is not only possible but essential. We envision multiplication of vector $\vec{a} = \langle a_1, a_2 \rangle$ by scalar $k > 0$ as creating the vector $k\vec{a}$ having the same direction as $\vec{a}$ and length k times the length of $\vec{a}$; if $k < 0$, the direction of the scaled vector is opposite that of the original. (See Figure 11.3.) Thus, we define

$$k\vec{a} = k\langle a_1, a_2 \rangle = \langle ka_1, ka_2 \rangle.$$

With this definition, the length of $k\vec{a}$ is $|k|$ times that of $\vec{a}$:

$$|k\vec{a}| = \sqrt{(ka_1)^2 + (ka_2)^2} = \sqrt{k^2(a_1{}^2 + a_2{}^2)} = \sqrt{k^2}\sqrt{a_1{}^2 + a_2{}^2} = |k||\vec{a}|.$$

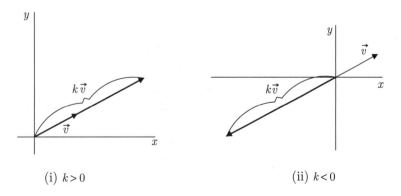

(i) $k > 0$ (ii) $k < 0$

FIGURE 11.3. Scalar Multiplication.

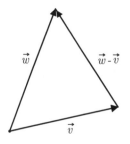

FIGURE **11.4.**

Two nonzero vectors $\vec{v}$ and $\vec{w}$ are **parallel**, or **collinear**, if $\vec{v} = k\vec{w}$ for some scalar $k \neq 0$. The unique vector $\vec{u}$ such that $\vec{u} + \vec{v} = \vec{0}$ is called the **opposite** of $\vec{v}$ and is denoted by $-\vec{v}$. Clearly, $-\vec{v} = (-1)\vec{v}$, and the vector $-\vec{v}$ has the same length as $\vec{v}$ but points in the opposite direction.

The following theorem, which relates the collinearity of two vectors to their coordinates, is very useful.

Theorem 11.1. *The following conditions are equivalent for two nonzero vectors,* $\vec{a} = \langle a_1, a_2 \rangle$ *and* $\vec{b} = \langle b_1, b_2 \rangle$.

(1) $\vec{a}$ *and* $\vec{b}$ *are collinear.*

(2) $a_1/a_2 = b_1/b_2$ *or* $a_2 = b_2 = 0$.

(3) $a_1 b_2 - a_2 b_1 = 0$.

Proof. The proof is straightforward and is left to the reader. $\square$

As seen in Figure 11.1(i), the length of the sum of two vectors will be less than the sum of their lengths, unless the vectors point in the same direction. This result is stated below, and it follows immediately from the triangle inequality we have already seen (by letting the terminal points of $\vec{v}$, $\vec{w}$, and $\vec{v} + \vec{w}$ represent the vertices of a triangle, as in the figure, where $\vec{v} = \overrightarrow{AB}$ and $\vec{w} = \overrightarrow{BC}$).

Theorem 11.2. *(Triangle Inequality for Vectors) For vectors* $\vec{v}$ *and* $\vec{w}$,

$$|\vec{v} + \vec{w}| \leq |\vec{v}| + |\vec{w}|,$$

with equality if and only if $\vec{v} = k\vec{w}$ *for some scalar* k.

Using scalar multiplication and vector addition, we define the **difference** of two vectors as $\vec{w} - \vec{v} = \vec{w} + (-\vec{v})$. Illustrated geometrically using the Triangle Rule, the difference $\vec{w} - \vec{v}$ is the vector whose initial point is the terminal point of $\vec{v}$ and whose terminal point is the terminal point of $\vec{w}$, when the vectors $\vec{v}$ and $\vec{w}$ are placed so that they have a shared initial point, as in Figure 11.4.

Vectors have many properties that are readily verifiable geometrically or algebraically. We list them for completeness; the proofs can be easily supplied by the reader or found in many standard calculus or linear algebra texts.

> **Theorem 11.3.** *Let $\vec{u}$, $\vec{v}$, and $\vec{w}$ be vectors, and let c and d be scalars. Then*
>
> (1) $\vec{u} + \vec{v} = \vec{v} + \vec{u}$
> (2) $\vec{u} + \vec{0} = \vec{u}$
> (3) $\vec{u} + (\vec{v} + \vec{w}) = (\vec{u} + \vec{v}) + \vec{w}$
> (4) $\vec{u} + (-\vec{u}) = \vec{0}$
> (5) $c(\vec{u} + \vec{v}) = c\vec{u} + c\vec{v}$
> (6) $(c + d)\vec{u} = c\vec{u} + d\vec{u}$
> (7) $(cd)\vec{u} = c(d\vec{u})$
> (8) $1\vec{u} = \vec{u}$.

> **Example 59.** Let $\vec{a} = \langle 3, -4 \rangle$ and $\vec{b} = \langle 2, 6 \rangle$. Find $|\vec{a}|$ and the vectors $\vec{a} + \vec{b}, \vec{a} - \vec{b}, 4\vec{a}$, and $3\vec{a} + 2\vec{b}$.
>
> *Solution:*
>
> $$\begin{aligned}
> |\vec{a}| &= \sqrt{3^2 + (-4)^2} = 5, \\
> \vec{a} + \vec{b} &= \langle 3 + 2, -4 + 6 \rangle = \langle 5, 2 \rangle, \\
> \vec{a} - \vec{b} &= \langle 3 - 2, -4 - 6 \rangle = \langle 1, -10 \rangle, \\
> 4\vec{a} &= \langle 4 \cdot 3, 4(-4) \rangle = \langle 12, -16 \rangle, \\
> 3\vec{a} + 2\vec{b} &= 3\langle 3, -4 \rangle + 2\langle 2, 6 \rangle = \langle 9 + 4, -12 + 12 \rangle = \langle 13, 0 \rangle.
> \end{aligned}$$
>
> $\square$

A **linear combination** of vectors $\vec{a}$ and $\vec{b}$ is any vector of the form $c\vec{a} + d\vec{b}$, where $c, d \in \mathbb{R}$. All vectors in Example 59 are examples of linear combinations of vectors $\vec{a}$ and $\vec{b}$. The set of all (real) linear combinations of $\vec{a}$ and $\vec{b}$, $\{c\vec{a} + d\vec{b} : c, d \in \mathbb{R}\}$, is called the **span** of $\vec{a}$ and $\vec{b}$. In many cases this span will equal $\mathbb{R} \times \mathbb{R}$, and when this is so, we say that $\vec{a}$ and $\vec{b}$, as a set, form a **basis** for $\mathbb{R} \times \mathbb{R}$.

Often we find it beneficial to express one vector as a linear combination of two other vectors. The next two theorems give two ways to do this: the first one holds for nonparallel vectors and the second is a vector form of the result we have seen about the coordinates of a point dividing a given segment in given ratio (see Theorem 8.2).

> **Theorem 11.4.** *If $\vec{a}$ and $\vec{b}$ are two nonparallel nonzero vectors, then for every $\vec{c}$, there exists a unique pair of real numbers x and y such that $\vec{c} = x\vec{a} + y\vec{b}$.*

Proof. Suppose $\vec{a} = \langle a_1, a_2 \rangle$, $\vec{b} = \langle b_1, b_2 \rangle$, and $\vec{c} = \langle c_1, c_2 \rangle$. We wish to find x and y such that $\vec{c} = x\vec{a} + y\vec{b}$, which in component form is equivalent to $c_1 = xa_1 + yb_1$ and $c_2 = xa_2 + yb_2$. This is a system of two linear equations with two unknowns, with a unique solution of

$$x = \frac{c_2 b_1 - c_1 b_2}{a_2 b_1 - a_1 b_2} \quad \text{and} \quad y = \frac{c_2 a_1 - c_1 a_2}{a_1 b_2 - a_2 b_1}.$$

Note that since $\vec{a}$ and $\vec{b}$ are not parallel, $a_1 b_2 \neq a_2 b_1$, so the desired values of x and y do exist. $\square$

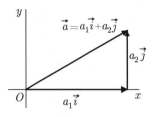

FIGURE 11.5.

Theorem 11.5. *Let C lie on segment AB such that $AC/CB = k$. Then for any point O,*

$$\overrightarrow{OC} = \frac{1}{k+1}\,\overrightarrow{OA} + \frac{k}{k+1}\,\overrightarrow{OB}.$$

Proof. Given the location of C on $\overline{AB}$, it is clear that $AC/AB = k/(k+1)$. Therefore, $\overrightarrow{AC} = k/(k+1)\,\overrightarrow{AB}$. For any point O,

$$\begin{aligned}
\overrightarrow{OC} &= \overrightarrow{OA} + \overrightarrow{AC} \\
&= \overrightarrow{OA} + \frac{k}{k+1}(\overrightarrow{OB} - \overrightarrow{OA}) \\
&= \frac{1}{k+1}\,\overrightarrow{OA} + \frac{k}{k+1}\,\overrightarrow{OB}\,.
\end{aligned}$$

$\square$

A vector having length 1 is called a **unit vector**. Two unit vectors,

$$\vec{i} = \langle 1, 0 \rangle \quad \text{and} \quad \vec{j} = \langle 0, 1 \rangle,$$

are of special importance. They form a commonly used basis of $\mathbb{R} \times \mathbb{R}$, called the **standard basis**. In this basis, the components of a vector $\vec{a} = \langle a_1, a_2 \rangle$ become the coefficients in the corresponding linear combination of $\vec{i}$ and $\vec{j}$, as shown in Figure 11.5. For example $\langle 7, 4 \rangle = 7\vec{i} + 4\vec{j}$. For any nonzero vector $\vec{a} = \langle a_1, a_2 \rangle$, the vector $\vec{u} = \vec{a}/|\vec{a}|$ is clearly a unit vector in the direction of $\vec{a}$.

Example 60. Find the unit vector, $\vec{u}$, having the same direction as $\vec{a} = \langle 3, -7 \rangle$.

Solution:

$$\vec{u} = \frac{\vec{a}}{|\vec{a}|} = \frac{\langle 3, -7 \rangle}{\sqrt{3^2 + (-7)^2}} = \left\langle \frac{3}{\sqrt{58}}, \frac{-7}{\sqrt{58}} \right\rangle.$$

$\square$

Our next example provides an application to physics. When several forces act on an object, one may sum these forces (as vectors) to find the resulting force on the object (called the **resultant force**).

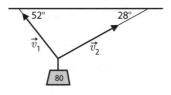

FIGURE **11.6.**

Example 61. An 80-lb weight hangs from two wires as shown in Figure 11.6. Find the vectors $\vec{v}_1$ and $\vec{v}_2$ corresponding to the tensions in these two wires. Also give the magnitude of these tensions (in pounds).

Solution: Our goal is to express $\vec{v}_1$ and $\vec{v}_2$ in terms of their horizontal and vertical components, $\vec{i}$ and $\vec{j}$. From the figure, we have

$$\vec{v}_1 = -|\vec{v}_1|\cos 52^\circ \vec{i} + |\vec{v}_1|\sin 52^\circ \vec{j},$$
$$\vec{v}_2 = |\vec{v}_2|\cos 28^\circ \vec{i} + |\vec{v}_2|\sin 28^\circ \vec{j}.$$

Now, the resultant vector $\vec{v}_1 + \vec{v}_2$ is completely counterbalanced by the 80-lb weight, whose force is all exerted downward, represented by the vector $-80\vec{j}$. Therefore, $\vec{v}_1 + \vec{v}_2 = 80\vec{j}$. Equating the $\vec{i}$ and $\vec{j}$ components of the two sides of this equation, we obtain

$$-|\vec{v}_1|\cos 52^\circ + |\vec{v}_2|\cos 28^\circ = 0,$$
$$|\vec{v}_1|\sin 52^\circ + |\vec{v}_2|\sin 28^\circ = 80.$$

Solving the first equation for $|\vec{v}_1|$ and substituting into the second yields

$$|\vec{v}_2|\frac{\cos 28^\circ}{\cos 52^\circ}\sin 52^\circ + |\vec{v}_2|\sin 28^\circ = 80.$$

Therefore,

$$|\vec{v}_2| = \frac{80}{(\cos 28^\circ)(\tan 52^\circ) + \sin 28^\circ} \approx 50.01 \text{ lbs.}$$

From here, $|\vec{v}_1| = |\vec{v}_2|\cos 28^\circ / \cos 52^\circ \approx 71.73$ lbs. Substituting the values for $|\vec{v}_1|$ and $|\vec{v}_2|$ into our first two equations, we find our tension vectors. Notice that the sum of the horizontal components is 0 and the sum of the vertical components is 80, as expected:

$$\vec{v}_1 \approx -44.16\vec{i} + 56.52\vec{j} \ \text{ and } \ \vec{v}_2 \approx 44.16\vec{i} + 23.48\vec{j}.$$

$\square$

11.3   Applications to Geometry

We now turn to using vectors to solve problems in Euclidean geometry. In many cases the problems themselves will make no mention of vectors, but their usage will provide a very nice solution. Several of the problems will be ones we have seen before.

Example 62. Let $A : (a_1, a_2)$, $B : (b_1, b_2)$, $C : (c_1, c_2)$, and $D : (d_1, d_2)$ be any four distinct points. Let E and F be the midpoints of $\overline{AB}$ and $\overline{CD}$, respectively. Show that $EF \leq \frac{1}{2}(BC + AD)$, with equality if and only if $\overrightarrow{AD}$ and $\overrightarrow{BC}$ have the same direction.

Proof. We begin by finding an expression for $\overrightarrow{EF}$. To this end,

$$\overrightarrow{EF} = \langle (c_1 + d_1)/2 - (a_1 + b_1)/2, (c_2 + d_2)/2 - (a_2 + b_2)/2 \rangle$$
$$= \frac{1}{2} \langle c_1 - b_1, c_2 - b_2 \rangle + \frac{1}{2} \langle d_1 - a_1, d_2 - a_2 \rangle$$
$$= \frac{1}{2} \left(\overrightarrow{BC} + \overrightarrow{AD} \right).$$

From here, note that $EF = |\overrightarrow{EF}| = \frac{1}{2}|\overrightarrow{BC} + \overrightarrow{AD}|$. The result now follows from the triangle inequality for vectors. Equality will occur if and only if $\overrightarrow{AD}$ and $\overrightarrow{BC}$ have the same direction. $\quad\square$

The property that three points are collinear can be expressed in vector terminology, as shown in the next example. The reader is invited to compare this result with that of Theorem 3.26(2).

Example 63. Given points A, B, and C, show that for every point O there is a scalar k such that $\overrightarrow{OC} = k\,\overrightarrow{OA} + (1-k)\overrightarrow{OB}$ if and only if the points A, B, and C are collinear.

Proof. Let O be a point in the plane and suppose that $\overrightarrow{OC} = k\,\overrightarrow{OA} + (1-k)\overrightarrow{OB}$ for some scalar k. Then

$$\overrightarrow{AC} = \overrightarrow{OC} - \overrightarrow{OA} = k\overrightarrow{OA} + (1-k)\overrightarrow{OB} - \overrightarrow{OA}$$
$$= (k-1)\overrightarrow{OA} + (1-k)\overrightarrow{OB}$$
$$= (1-k)\overrightarrow{AB}.$$

This shows that $\overrightarrow{AC}$ is a scalar multiple of $\overrightarrow{AB}$, which proves that A, B, and C are collinear.

Now suppose that A, B, and C are collinear. Then $\overrightarrow{AC} = p\overrightarrow{AB}$ for some scalar p, and we can reverse the steps in the above proof (by letting $1 - k = p$) to see that $\overrightarrow{OC} = (1-p)\overrightarrow{OA} + p\overrightarrow{OB}$. $\quad\square$

Example 64. In parallelogram $OABC$, let M and N be the midpoints of sides OA and AB, respectively. Find the ratio $x = OP : ON$, where P is the intersection of segments ON and CM. (See Figure 11.7.)

Solution: Let $\vec{a} = \overrightarrow{OA}$ and $\vec{c} = \overrightarrow{OC}$. Then

$$\overrightarrow{OP} = x\overrightarrow{ON} = x\left(\vec{a} + \frac{1}{2}\vec{c}\right) = x\vec{a} + \frac{x}{2}\vec{c}.$$

The collinearity of O, P, and N justifies the first equality; the Triangle Rule for vector addition justifies the second.

Applying the result of Example 63 to the collinear points C, P, and M, there must be a constant k such that

$$\overrightarrow{OP} = k\overrightarrow{OM} + (1-k)\overrightarrow{OC} = k\left(\frac{1}{2}\vec{a}\right) + (1-k)\vec{c} = \frac{k}{2}\vec{a} + (1-k)\vec{c}.$$

Noting by Theorem 11.4 that the scalars at $\vec{a}$ and $\vec{c}$ must be unique in the linear combination for $\overrightarrow{OP}$, we have the equations $x = k/2$ and $x/2 = 1 - k$. Therefore, $2x = 1 - x/2$, so $x = 2/5$. $\quad\square$

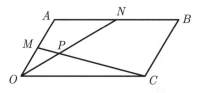

FIGURE 11.7.

Example 65. Given a set of points $A_1 : (x_1, y_1)$, $A_2 : (x_2, y_2), \ldots, A_n : (x_n, y_n)$, prove there exists a unique point G such that $\overrightarrow{GA_1} + \overrightarrow{GA_2} + \cdots + \overrightarrow{GA_n} = \vec{0}$.

Proof. Let O be an arbitrary point in the plane, and let G be a point with the desired property. (We will demonstrate that such a point exists by constructing it.) Then $\overrightarrow{OA_i} = \overrightarrow{OG} + \overrightarrow{GA_i}$ for each i. Therefore

$$\vec{0} = \overrightarrow{GA_1} + \overrightarrow{GA_2} + \cdots + \overrightarrow{GA_n} = \sum_{i=1}^{n}(\overrightarrow{OA_i} - \overrightarrow{OG}) = \left(\sum_{i=1}^{n} \overrightarrow{OA_i}\right) - n\overrightarrow{OG}.$$

This gives

$$\overrightarrow{OG} = \frac{1}{n}\sum_{i=1}^{n} \overrightarrow{OA_i}.$$

Since each of the steps above is reversible, for each point O we have determined a unique G with the desired property. Does G depend on the choice of O? No, it does not. For suppose O' is another point, giving a point G' with the desired property. We want to show that $G = G'$. Repeating the argument above, we obtain $\overrightarrow{O'G'} = \frac{1}{n}\sum_{i=1}^{n} \overrightarrow{O'A_i}$. Writing each vector $O'A_i$ as the sum $\overrightarrow{O'O} + \overrightarrow{OA_i}$, we obtain

$$\overrightarrow{O'G'} = \frac{1}{n}\sum_{i=1}^{n} \overrightarrow{O'A_i} = \frac{1}{n}\sum_{i=1}^{n}(\overrightarrow{O'O} + \overrightarrow{OA_i}) = \overrightarrow{O'O} + \overrightarrow{OG} = \overrightarrow{O'G}.$$

Therefore, $\overrightarrow{O'G'} = \overrightarrow{O'G}$, so $G = G'$. $\qquad \square$

When O is the origin, $\overrightarrow{OA_i} = \langle x_i, y_i \rangle$ for each i, so

$$G = (g_1, g_2) = \left(\frac{1}{n}\sum x_i, \frac{1}{n}\sum y_i\right).$$

The point G found in the previous example is called the **centroid** of the set of points $A_1, A_2, \ldots, A_n$. The notion of the centroid of a set of points is motivated by the concept of the center of mass of n point-masses. Suppose that mass m_i is at point A_i for $1 \leq i \leq n$. Then the center of these n point-masses can be defined as the point G such that

$$\sum_{i=1}^{n} m_i \overrightarrow{GA_i} = \vec{0}.$$

Repeating the argument above, one can obtain that for an arbitrary point O,

$$\overrightarrow{OG} = \sum_{i=1}^{n} \frac{m_i}{m} \overrightarrow{OA_i},$$

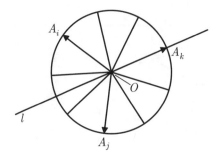

FIGURE 11.8.

where $m = m_1 + m_2 + \cdots + m_n \neq 0$.[4] When all masses are equal, the center of point masses coincides with the centroid of the set of points.

Example 66. Suppose points $A_1, A_2, \ldots, A_n$ lie on a circle $\mathcal{C}$ with center O, and divide the circle into n congruent arcs. Prove that

$$\sum_{i=1}^{n} \overrightarrow{OA_i} = \vec{0}.$$

Proof. Can the sum be anything other than $\vec{0}$? Many people think that if this question is asked, the answer "No" is so obvious that no explanation is really needed. Some may just say, "Obvious due to symmetry." This is probably one of those rare cases when we can accept such a "proof." On the other hand, it is not hard to present a more conventional argument, which follows here.

If n is even, then for each vector $\overrightarrow{OA_i}$, there is a vector $\overrightarrow{OA_j} = -\overrightarrow{OA_i}$, $i \neq j$, so that $\overrightarrow{OA_i} + \overrightarrow{OA_j} = \vec{0}$. The result follows.

Suppose n is odd, and for sake of contradiction assume $\sum_{i=1}^{n} \overrightarrow{OA_i} \neq \vec{0}$. Then the figure formed by the n points $A_1, A_2, \ldots, A_n$ has n axes of symmetry; let l be one of them. One of the n vectors (identified as $\overrightarrow{OA_k}$ in Figure 11.8) is parallel to l. The other $n - 1$ vectors are in pairs which are symmetric with respect to l. In Figure 11.8, this is illustrated with the vectors $\overrightarrow{OA_i}$ and $\overrightarrow{OA_j}$. The sum of each of these pairs is a vector parallel to l. Consequently, $\sum_{i=1}^{n} \overrightarrow{OA_i}$ is the sum of $1 + (n-1)/2$ vectors that are each parallel to the designated line of symmetry, l. Therefore, $\sum_{i=1}^{n} \overrightarrow{OA_i}$ is itself a vector parallel to l. Now select a different line of symmetry for the figure, l'. Repeating the above argument, we conclude that $\sum_{i=1}^{n} \overrightarrow{OA_i}$ is a vector parallel to l'. Since l' is not parallel to l, this leads to a contradiction. We conclude that $\sum_{i=1}^{n} \overrightarrow{OA_i} = \vec{0}$. □

In the next section we will discuss more fully how angles can be utilized in the study of vectors. For now, we simply define what we mean by the angle between two vectors, in order to consider the next example.

---

[4] The condition $m = m_1 + \cdots + m_n \neq 0$ reflects the fact that in some applications it is convenient to have negative values for m_i.

The angle between two nonzero vectors $\vec{u} = \overrightarrow{OA}$ and $\vec{v} = \overrightarrow{OB}$ is the angle AOB such that $0 \leq m\angle AOB \leq 180°$. If two nonzero vectors do not share an initial point, we place them so that the initial points coincide, and the angle between them is defined as the angle between the vectors in such a placement. Clearly, if the angle between $\vec{u}$ and $\vec{v}$ is θ, then the angle between $\vec{u}$ and $-\vec{v}$ is $180° - \theta$.

The statement of the next example is equivalent to the one in Example 66, but is phrased in trigonometric terms. We will use it several times ahead in this book.

Example 67.   Show that $\sum_{i=1}^{n} \cos\left(\alpha + \frac{2\pi}{n}i\right) = \sum_{i=1}^{n} \sin\left(\alpha + \frac{2\pi}{n}i\right) = 0$ for any angle α (measured in radians) and any integer $n \geq 2$.

Proof. Consider n points $A_1, A_2, \ldots, A_n$ on the unit circle centered at the origin, O, with point A_i having coordinates $(\cos\left(\alpha + \frac{2\pi}{n}i\right), \sin\left(\alpha + \frac{2\pi}{n}i\right))$. Then for any i, $\overrightarrow{OA_i} = \langle\cos\left(\alpha + \frac{2\pi}{n}i\right), \sin\left(\alpha + \frac{2\pi}{n}i\right)\rangle$. By the result of Example 66, the sum of these n vectors $\overrightarrow{OA_i}$ is $\vec{0} = \langle 0, 0 \rangle$. Therefore, the corresponding sums of each of the coordinates must be 0. $\square$

Example 68.   Suppose that $\sin x + \sin y + \sin z = \cos x + \cos y + \cos z = 0$. Compute

$$\sin 2000x + \sin 2000y + \sin 2000z.$$

Proof.   Consider the three points on the unit circle centered at the origin obtained by rotating the point $(1, 0)$ in the positive direction by angles x, y, and z, respectively. The corresponding position vectors (i.e., the vectors obtained by joining the origin with each of these points) are $\vec{a} = \langle\cos x, \sin x\rangle$, $\vec{b} = \langle\cos y, \sin y\rangle$, and $\vec{c} = \langle\cos z, \sin z\rangle$. (See Figure 11.9.) We are given that the sum of the three x-coordinates and the sum of the three y-coordinates is zero; therefore, $\vec{a} + \vec{b} + \vec{c} = \vec{0}$. Since $\vec{a}$, $\vec{b}$, and $\vec{c}$ are unit vectors, by the Triangle Rule, we can view each of these vectors as sides of an equilateral triangle. Thus, for example, the angle between $\vec{a}$ and $\vec{a} + \vec{b} = -\vec{c}$ is $\pi/3$, which proves that the angle between $\vec{a}$ and $\vec{c}$ is $\pi - \pi/3 = 2\pi/3$. Similarly, the angle between any two of the three vectors is $2\pi/3$.

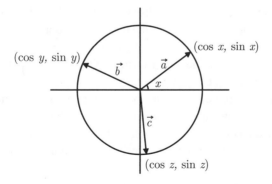

FIGURE 11.9.

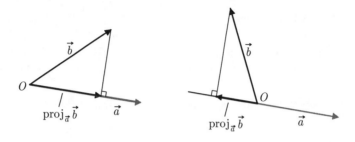

FIGURE 11.10.

Therefore, $y = x + 2\pi/3 + 2\pi k$ and $z = x - 2\pi/3 + 2\pi n$ for some integers k and n. (Note that k or n might be negative, depending on the order the points occur on the unit circle.) Thus, $2000y = 2000x + 4\pi/3 + 2\pi k_1$ and $2000z = 2000x + 2\pi/3 + 2\pi n_1$ for some integers k_1 and n_1. We conclude that the position vectors of the points of the unit circle corresponding to the angles $2000x$, $2000y$, and $2000z$ form angles of $2\pi/3$ with one another. Consequently, their sum is the zero vector. This implies that

$$\sin 2000x + \sin 2000y + \sin 2000z = 0;$$

it also implies that

$$\cos 2000x + \cos 2000y + \cos 2000z = 0.$$

$\square$

11.4   The Dot Product

Consider two nonzero vectors, $\vec{a}$ and $\vec{b}$, that are not parallel. Despite having different directions, we can visualize some "portion" of $\vec{b}$ as being in the direction of $\vec{a}$, in the following way. Position $\vec{a}$ and $\vec{b}$ so that they share a common initial point, O, and construct the perpendicular from the terminal point of $\vec{b}$ to the line defined by $\vec{a}$. (Two representations of such pairs of vectors and the corresponding constructed perpendiculars are shown in Figure 11.10.) The vector with initial point at O and terminal point at the foot of the perpendicular is the **vector projection of $\vec{b}$ onto $\vec{a}$**, denoted $\text{proj}_{\vec{a}}\vec{b}$.

The vector projection of one vector onto another has a natural application: computing the work done on an object by a given force over a given distance. Because it has both magnitude and direction, force is a quantity best described as a vector. If the force is constant and in the same direction as the movement of the object, then the work done can be found by simply multiplying the magnitude of the force by the distance the object is moved.[5] However, if the direction of the force is different from the direction of the movement of the object, then the magnitude of the force vector must be replaced by the magnitude of the vector projection of the force vector onto the direction vector. Suppose that a force vector, $\vec{F}$, forms the hypotenuse of a right triangle, as shown in Figure 11.11. Let θ be the angle between $\vec{F}$ and the direction vector, $\vec{D}$. Then the magnitude of $\text{proj}_{\vec{D}}\vec{F}$ is $|\vec{F}|\cos\theta$. For any θ, $0 \le \theta \le 180$, the number $|\vec{F}|\cos\theta$ is called the **scalar projection of $\vec{F}$ onto $\vec{D}$**, or the **component**

---

[5] This is the familiar formula, Work = Force × Distance; in this equation, "Force" and "Distance" must be understood as indicating vector magnitudes.

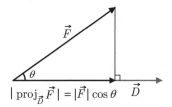

$$| \operatorname{proj}_{\vec{D}} \vec{F} | = |\vec{F}| \cos \theta \qquad \vec{D}$$

FIGURE 11.11.

of $\vec{F}$ along $\vec{D}$. The equation for the work done by a force $\vec{F}$ in the direction of $\vec{D}$ is thus given by $W = |\vec{D}||\vec{F}| \cos \theta$.[6]

The following definition of the dot product of two vectors is motivated by the computation of work done by a force, as well as by other examples which we have not mentioned. If θ is the measure of the angle between nonzero vectors $\vec{a}$ and $\vec{b}$, then the **dot product** of $\vec{a}$ and $\vec{b}$ is defined as

$$\vec{a} \cdot \vec{b} = |\vec{a}||\vec{b}| \cos \theta.\text{[7]}$$

The following theorem provides us with a simple way of computing the dot product in terms of components of the vectors.

Theorem 11.6. *If $\vec{a} = \langle a_1, a_2 \rangle$ and $\vec{b} = \langle b_1, b_2 \rangle$, then*

$$\vec{a} \cdot \vec{b} = a_1 b_1 + a_2 b_2.$$

Proof. Let θ be the angle between $\vec{a}$ and $\vec{b}$, and let O, A, and B be as shown in Figure 11.12. Then, applying the Cosine theorem to triangle OAB, we obtain

$$AB^2 = OA^2 + OB^2 - 2(OA)(OB) \cos \theta,$$

which has the vector form

$$\begin{aligned} |\vec{b} - \vec{a}|^2 &= |\vec{b}|^2 - 2\vec{a} \cdot \vec{b} + |\vec{a}|^2 \\ &= (a_1^2 + a_2^2) + (b_1^2 + b_2^2) - 2\vec{a} \cdot \vec{b}. \end{aligned}$$

Now, $|\vec{b} - \vec{a}|^2 = (b_1 - a_1)^2 + (b_2 - a_2)^2 = a_1^2 + a_2^2 + b_1^2 + b_2^2 - 2(a_1 b_1 + a_2 b_2)$. Equating the right-hand sides of our last two equations gives that $\vec{a} \cdot \vec{b} = a_1 b_1 + a_2 b_2$, as desired. $\qquad \square$

Example 69. Find $\vec{a} \cdot \vec{b}$ if

(1) $|\vec{a}| = 2$, $|\vec{b}| = 3$, and angle between $\vec{a}$ and $\vec{b}$ is $120°$;
(2) $\vec{a} = \langle -2, 2\sqrt{3} \rangle$ and $\vec{b} = \langle 1, \sqrt{3} \rangle$.

Solution:

(1) From the definition of dot product, $\vec{a} \cdot \vec{b} = |\vec{a}| |\vec{b}| \cos(120°) = (2)(3)(-1/2) = -3$.
(2) From Theorem 11.6, $\vec{a} \cdot \vec{b} = (-2)(1) + (2\sqrt{3})(\sqrt{3}) = -2 + 6 = 4$. $\qquad \square$

---

[6] In Figure 11.11, we depict an acute right triangle. If instead, θ is obtuse, the same formula for the work will hold, but the work will be negative.

[7] If one of $\vec{a}$ or $\vec{b}$ is the zero vector, we define $\vec{a} \cdot \vec{b}$ as the number zero.

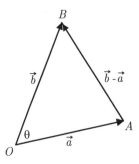

FIGURE 11.12.

Notice that unlike the results of vector addition, vector subtraction, and multiplication of a vector by a scalar, the dot product of two vectors yields not another vector, but rather a scalar. For this reason, the dot product is often referred to as the **scalar product** of $\vec{a}$ and $\vec{b}$.

The dot product provides a convenient expression for the projection of one nonzero vector $\vec{a}$ onto another nonzero vector $\vec{b}$:

$$\text{proj}_{\vec{b}}\vec{a} = \left(\frac{\vec{a} \cdot \vec{b}}{|\vec{b}|^2}\right)\vec{b}.$$

Theorem 11.7 summarizes some of the handy and easily proven properties of the dot product, many of which are analogs of properties of the real numbers. Its proof is relegated to the exercises.

Theorem 11.7. *(Properties of the Dot Product)*
If $\vec{a}$, $\vec{b}$, and $\vec{c}$ are vectors in the plane and k is a scalar, then

(1) $\vec{a} \cdot \vec{b} = \vec{b} \cdot \vec{a}$

(2) $\vec{a} \cdot (\vec{b} + \vec{c}) = \vec{a} \cdot \vec{b} + \vec{a} \cdot \vec{c}$

(3) $k(\vec{a} \cdot \vec{b}) = k\vec{a} \cdot \vec{b} = \vec{a} \cdot k\vec{b}$

(4) $\vec{0} \cdot \vec{a} = 0$

(5) $\vec{a} \cdot \vec{a} = |\vec{a}|^2$

(6) *Nonzero vectors $\vec{a}$ and $\vec{b}$ are perpendicular if and only if $\vec{a} \cdot \vec{b} = 0$. (Two vectors with zero dot product are said to be* **orthogonal***.)*

Theorem 11.8. *Let $\vec{v}_1$, $\vec{v}_2$, ..., $\vec{v}_n$ be vectors. The following statements are equivalent:*

(1) $\sum_{i=1}^{n} \vec{v}_i = \vec{0}$.

(2) *There are two nonparallel vectors $\vec{a}$ and $\vec{b}$ such that the sum of the vector projections of all $\vec{v}_i$ on each of these two vectors is $\vec{0}$.*

(3) *There are two nonparallel vectors $\vec{a}$ and $\vec{b}$ such that the sum of the scalar projections of all $\vec{v}_i$ on each of these two vectors is 0.*

Proof. Denote $\vec{v}_i$ as $\langle v_{i,1}, v_{i,2} \rangle$ for $i = 1, 2, \ldots n$.

(1) $\Rightarrow$ (2). Assume that $\sum_{i=1}^{n} \vec{v}_i = \vec{0} = \langle 0, 0 \rangle$, in which case $\sum_{i=1}^{n} v_{i,1} = \sum_{i=1}^{n} v_{i,2} = 0$.

Then the sum of the vector projections of $\vec{v}_1, \vec{v}_2, \ldots, \vec{v}_n$ onto any given vector $\vec{b}$ (and therefore onto each one of any two vectors $\vec{a}$ and $\vec{b}$) is given by

$$\sum_{i=1}^{n} \text{proj}_{\vec{b}} \vec{v}_i = \sum_{i=1}^{n} \left(\frac{\vec{v}_i \cdot \vec{b}}{|\vec{b}|^2} \right) \vec{b}$$

$$= \frac{1}{|\vec{b}|^2} \sum_{i=1}^{n} \langle v_{i,1} b_1{}^2 + v_{i,2} b_1 b_2 , \; v_{i,1} b_1 b_2 + v_{i,2} b_2{}^2 \rangle$$

$$= \left\langle \frac{b_1{}^2}{|\vec{b}|^2} \sum_{i=1}^{n} v_{i,1} + \frac{b_1 b_2}{|\vec{b}|^2} \sum_{i=1}^{n} v_{i,2} , \; \frac{b_1 b_2}{|\vec{b}|^2} \sum_{i=1}^{n} v_{i,1} + \frac{b_2{}^2}{|\vec{b}|^2} \sum_{i=1}^{n} v_{i,2} \right\rangle$$

$$= \langle 0 + 0, 0 + 0 \rangle = \vec{0}.$$

(2) $\Rightarrow$ (3). Assume there are two nonparallel vectors $\vec{a}$ and $\vec{b} = \langle b_1, b_2 \rangle$ for which the sum of the vector projections of $\vec{v}_1, \vec{v}_2, \ldots, \vec{v}_n$ onto each is $\vec{0}$. Taking the dot product of this sum with $\vec{b}$, we obtain

$$\sum_{i=1}^{n} \left(\frac{\vec{v}_i \cdot \vec{b}}{|\vec{b}|^2} \right) \vec{b} \cdot \vec{b} = \vec{0} \cdot \vec{b}.$$

Since $\vec{b} \cdot \vec{b} = |\vec{b}|^2$ and $\vec{0} \cdot \vec{b} = 0$, we see that $\sum_{i=1}^{n} \vec{v}_i \cdot \vec{b} = 0$ which shows that the sum of the scalar projections of $\vec{v}_1, \vec{v}_2, \ldots, \vec{v}_n$ onto $\vec{b}$ is 0. Similarly, the sum of the scalar projections onto $\vec{a}$ is 0.

(3) $\Rightarrow$ (1). Assume there are nonparallel vectors $\vec{a} = \langle a_1, a_2 \rangle$ and $\vec{b} = \langle b_1, b_2 \rangle$ such that

$$\sum_{i=1}^{n} \frac{\vec{v}_i \cdot \vec{a}}{|\vec{a}|} = \sum_{i=1}^{n} \frac{\vec{v}_i \cdot \vec{b}}{|\vec{b}|} = 0.$$

For convenience, let

$$x = \sum_{i=1}^{n} v_{i,1} \quad \text{and} \quad y = \sum_{i=1}^{n} v_{i,2}.$$

Then,

$$\sum_{i=1}^{n} \left(a_1 v_{i,1} + a_2 v_{i,2} \right) = 0,$$

so $a_1 x + a_2 y = 0$; similarly, $b_1 x + b_2 y = 0$. However, since $\langle a_1, a_2 \rangle$ and $\langle b_1, b_2 \rangle$ are not parallel, then $a_1 b_2 \neq a_2 b_1$ by Theorem 11.1 and the only solution to this system is $x = y = 0$. Hence,

$$\sum \vec{v}_i = \langle x, y \rangle = \langle 0, 0 \rangle = \vec{0},$$

which is what we needed to prove. $\qquad\qquad\qquad\qquad\qquad\qquad\qquad\qquad\qquad\qquad$ $\square$

The astute reader may note that we could add the following statements to those of Theorem 11.8. We omit the proof.

(4) For any nonzero $\vec{b}$, the sum of the vector projections of nonzero vectors $\vec{v}_1, \vec{v}_2, \ldots, \vec{v}_n$ onto $\vec{b}$ is $\vec{0}$.

(5) For any nonzero $\vec{b}$, the sum of the scalar projections of nonzero vectors $\vec{v}_1, \vec{v}_2, \ldots, \vec{v}_n$ onto $\vec{b}$ is $\vec{0}$.

Theorem 11.9. *(Cauchy-Schwartz Inequality)*
Given any pair of vectors, $\vec{a}$ and $\vec{b}$, $|\vec{a} \cdot \vec{b}| \leq |\vec{a}||\vec{b}|$.

Proof. This important result follows immediately from our definition of dot product, since $0 \leq |\cos\theta| \leq 1$, where θ is the angle between $\vec{a}$ and $\vec{b}$. If one of the vectors is the zero vector, the inequality (which is actually an equality) is trivial. $\qquad\square$

Suppose that $\vec{a} = \langle a_1, a_2 \rangle \neq \vec{0}$ and $\vec{a}$ makes an angle α with the positive direction of the x-axis. Then, using Theorem 11.6,

$$\vec{a} \cdot \vec{i} = (a_1)(1) + (a_2)(0) = |\vec{a}||\vec{i}| \cos\alpha.$$

Therefore $a_1 = |\vec{a}| \cos\alpha$, so $\cos\alpha = \dfrac{a_1}{|\vec{a}|}$.

Similarly, if β is the angle that $\vec{a}$ makes with the positive direction of the y-axis, $\cos\beta = \dfrac{a_2}{|\vec{a}|}$.

These cosine values are called the **direction cosines** of $\vec{a}$. Note that $\vec{a}$ can be expressed as $\vec{a} = |\vec{a}|\langle \cos\alpha, \cos\beta \rangle$. Thus,

$$\frac{\vec{a}}{|\vec{a}|} = \langle \cos\alpha, \cos\beta \rangle;$$

that is, the direction cosines are the components of the unit vector in the direction of $\vec{a}$.

Example 70. Consider two vectors $\vec{a} = \langle 2, 3 \rangle$ and $\vec{b} = \langle -1, 4 \rangle$.

(1) Find the lengths of $\vec{v} = 2\vec{a} - 3\vec{b}$ and $\vec{w} = \vec{a} + \vec{b}$, and determine the cosine of the angle between $\vec{v}$ and $\vec{w}$.

(2) Find a scalar k such that $\vec{a} + k\vec{b}$ is orthogonal to $\vec{v} = 2\vec{a} - 3\vec{b}$.

Solution:

(1) $\vec{v} = 2\vec{a} - 3\vec{b} = \langle 7, -6 \rangle$, and $\vec{w} = \vec{a} + \vec{b} = \langle 1, 7 \rangle$. Therefore,

$$|\vec{v}| = \sqrt{(7)^2 + (-6)^2} = \sqrt{85} \text{ and } |\vec{w}| = \sqrt{(1)^2 + (7)^2} = \sqrt{50}.$$

Let γ be the angle between $\vec{v}$ and $\vec{w}$. By definition, $\vec{v} \cdot \vec{w} = |\vec{v}||\vec{w}| \cos\gamma$, so

$$\cos\gamma = \frac{\vec{v} \cdot \vec{w}}{|\vec{v}||\vec{w}|} = \frac{7 - 42}{(\sqrt{50})(\sqrt{85})} = \frac{-35}{5\sqrt{170}} = \frac{-7}{\sqrt{170}}.$$

(2) We wish to find k such that $(2\vec{a} - 3\vec{b}) \cdot (\vec{a} + k\vec{b}) = 0$. This is equivalent to

$$\langle 7, -6 \rangle \cdot \langle 2 - k, 3 + 4k \rangle = 0,$$
$$14 - 7k - 18 - 24k = 0,$$
$$k = -\frac{4}{31}.$$

$\qquad\square$

Example 71. Consider two vectors $\vec{a}$ and $\vec{b}$ such that $|\vec{a}| = 2$, $|\vec{b}| = 3$, and the angle between them is $\theta = 120°$.

(1) Find the lengths of $\vec{v} = 2\vec{a} - 3\vec{b}$ and $\vec{w} = \vec{a} + \vec{b}$, and determine the cosine of the angle between $\vec{v}$ and $\vec{w}$.

(2) Find a scalar k such that $\vec{a} + k\vec{b}$ is orthogonal to $\vec{v} = 2\vec{a} - 3\vec{b}$.

Solution:

(1) Making use of the properties of the dot product, we find the length of $\vec{v}$ as follows:

$$|\vec{v}| = |2\vec{a} - 3\vec{b}| = \sqrt{(2\vec{a} - 3\vec{b}) \cdot (2\vec{a} - 3\vec{b})}$$
$$= \sqrt{(2\vec{a} \cdot 2\vec{a}) - 2(2\vec{a} \cdot 3\vec{b}) + (3\vec{b} \cdot 3\vec{b})}$$
$$= \sqrt{|2\vec{a}|^2 + |3\vec{b}|^2 - 12(\vec{a} \cdot \vec{b})}$$
$$= \sqrt{4|\vec{a}|^2 + 9|\vec{b}|^2 - 12|\vec{a}||\vec{b}|\cos\theta}$$
$$= \sqrt{16 + 81 + 36} = \sqrt{133}.$$

Similar computations show that $|\vec{w}| = |\vec{a} + \vec{b}| = \sqrt{7}$.

Let γ be the angle between $\vec{v}$ and $\vec{w}$. By definition, $\vec{v} \cdot \vec{w} = |\vec{v}||\vec{w}|\cos\gamma$. Thus,

$$(2\vec{a} - 3\vec{b}) \cdot (\vec{a} + \vec{b}) = \sqrt{133}\sqrt{7}\cos\gamma$$
$$2\vec{a} \cdot \vec{a} + 2\vec{a} \cdot \vec{b} - 3\vec{a} \cdot \vec{b} - 3\vec{b} \cdot \vec{b} = \sqrt{931}\cos\gamma$$
$$2\vec{a} \cdot \vec{a} - \vec{a} \cdot \vec{b} - 3\vec{b} \cdot \vec{b} = 7\sqrt{19}\cos\gamma$$
$$2|\vec{a}|^2 - 3|\vec{b}|^2 - |\vec{a}||\vec{b}|\cos\theta = 7\sqrt{19}\cos\gamma$$
$$2(4) - 3(9) - (2)(3)(-1/2) = 7\sqrt{19}\cos\gamma$$
$$-\frac{16}{7\sqrt{19}} = \cos\gamma.$$

(2) We wish to find k such that $(2\vec{a} - 3\vec{b}) \cdot (\vec{a} + k\vec{b}) = 0$.

$$(2\vec{a} - 3\vec{b}) \cdot (\vec{a} + k\vec{b}) = 0$$
$$2|\vec{a}|^2 - 3k|\vec{b}|^2 + (2k - 3)\vec{a} \cdot \vec{b} = 0$$
$$2(4) - 3k(9) + (2k - 3)(2)(3)(-1/2) = 0$$
$$8 - 27k - 6k + 9 = 0$$
$$k = \frac{17}{33}.$$

□

11.5 Applications of the Dot Product

The dot product often becomes a useful tool when one attempts to solve geometric problems by using vectors. The first two examples in this section establish results that have previously been proven using other methods. As you read through the solutions, consider whether or not the vector solution compares favorably with other solutions.

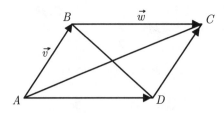

FIGURE 11.13.

Example 72. Let $ABCD$ be a parallelogram. Use vector dot products to show that

(1) $AC^2 + BD^2 = 2(AB^2 + BC^2)$,

(2) $ABCD$ is a rectangle if and only if $AC = BD$, and

(3) $ABCD$ is a rhombus if and only if $\overline{AC}$ is perpendicular to $\overline{BD}$.

Proof. Let $ABCD$ be a parallelogram, and assume $m\angle DAB \leq 90°$. Let $\vec{v} = \overrightarrow{AB} = \overrightarrow{DC}$; let $\vec{w} = \overrightarrow{BC} = \overrightarrow{AD}$. Then $\overrightarrow{AC} = \vec{v} + \vec{w}$ and $\overrightarrow{BD} = \vec{w} - \vec{v}$. (See Figure 11.13.)

(1) By substitution,

$$
\begin{aligned}
AC^2 + BD^2 &= |\vec{v} + \vec{w}|^2 + |\vec{w} - \vec{v}|^2 \\
&= (\vec{v} + \vec{w}) \cdot (\vec{v} + \vec{w}) + (\vec{w} - \vec{v}) \cdot (\vec{w} - \vec{v}) \\
&= |\vec{v}|^2 + 2\vec{v} \cdot \vec{w} + |\vec{w}|^2 + |\vec{w}|^2 - 2\vec{v} \cdot \vec{w} + |\vec{v}|^2 \\
&= 2(|\vec{v}|^2 + |\vec{w}|^2) \\
&= 2(AB^2 + BC^2).
\end{aligned}
$$

(2) The following statements are all equivalent, which demonstrates that $ABCD$ is a rectangle if and only $AC = BD$.

$$
\begin{aligned}
AC &= BD \\
|\vec{v} + \vec{w}|^2 &= |\vec{w} - \vec{v}|^2 \\
|\vec{v}|^2 + |\vec{w}|^2 + 2\vec{v} \cdot \vec{w} &= |\vec{v}|^2 + |\vec{w}|^2 - 2\vec{v} \cdot \vec{w} \\
2\vec{v} \cdot \vec{w} &= -2\vec{v} \cdot \vec{w} \\
\vec{v} \cdot \vec{w} &= 0
\end{aligned}
$$

(3) The following statements are all equivalent, which demonstrates that $ABCD$ is a rhombus if and only $\overline{AC} \perp \overline{BD}$.

$$
\begin{aligned}
|\vec{v}| &= |\vec{w}| \\
|\vec{w}|^2 - |\vec{v}|^2 &= 0 \\
\vec{v} \cdot \vec{w} + |\vec{w}|^2 - |\vec{v}|^2 - \vec{v} \cdot \vec{w} &= 0 \\
(\vec{v} + \vec{w}) \cdot (\vec{w} - \vec{v}) &= 0 \\
(\vec{v} + \vec{w}) &\perp (\vec{w} - \vec{v}) \\
\overline{AC} &\perp \overline{BD}
\end{aligned}
$$

$\square$

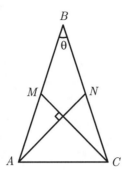

FIGURE 11.14.

Example 73. Let ABC be an isosceles triangle with $AB = BC$. Suppose that the medians, CM and AN, to sides AB and CD, respectively, are perpendicular. (See Figure 11.14.) Find $m\angle ABC$.

Solution: Let $\overrightarrow{AB} = \vec{v}$, $\overrightarrow{CB} = \vec{w}$, and $\theta = m\angle ABC$; note that $|\vec{v}| = |\vec{w}|$. Then $\overrightarrow{AN} = \overrightarrow{AB} - \overrightarrow{NB} = \vec{v} - \frac{1}{2}\vec{w}$, and $\overrightarrow{CM} = \overrightarrow{CB} - \overrightarrow{MB} = \vec{w} - \frac{1}{2}\vec{v}$. Since $\overrightarrow{AN} \perp \overrightarrow{CM}$, we obtain the following equivalent equations.

$$\left(\vec{v} - \frac{1}{2}\vec{w}\right) \cdot \left(\vec{w} - \frac{1}{2}\vec{v}\right) = 0$$

$$\vec{v} \cdot \vec{w} - \frac{1}{2}\vec{v} \cdot \vec{v} - \frac{1}{2}\vec{w} \cdot \vec{w} + \frac{1}{4}\vec{v} \cdot \vec{w} = 0$$

$$\frac{5}{4}|\vec{v}||\vec{w}|\cos\theta - \frac{1}{2}|\vec{v}|^2 - \frac{1}{2}|\vec{w}|^2 = 0$$

$$\frac{5}{4}|\vec{v}|^2 \cos\theta = |\vec{v}|^2$$

$$\cos\theta = \frac{4}{5}$$

$$\theta \approx 36.9°$$

$\square$

Example 74. Find the measures of all angles in the triangle with vertices $A : (4, 5)$, $B : (-2, 3)$, and $C : (1, -7)$.

Solution: Let $\vec{u} = \overrightarrow{BA}$, $\vec{v} = \overrightarrow{BC}$, and $\vec{w} = \overrightarrow{AC}$. Then $\vec{u} = \langle 6, 2\rangle$, $\vec{v} = \langle 3, -10\rangle$, and $\vec{w} = \langle -3, -12\rangle$. From Theorem 11.6, $\vec{u} \cdot \vec{v} = (6)(3) + (2)(-10) = -2$. On the other hand, $\vec{u} \cdot \vec{v} = |\vec{u}||\vec{v}|\cos(\angle ABC) = \sqrt{40}\sqrt{109}\cos(\angle ABC)$. Therefore, $\cos(\angle ABC) = -2/\sqrt{4360} \approx -0.030$, so $m\angle ABC \approx 91.7°$. Similarly, $111 = \vec{v} \cdot \vec{w} = \sqrt{109}\sqrt{153}\cos(\angle BCA)$; thus $\cos(\angle BCA) \approx 0.860$, so $m\angle BCA \approx 30.7°$. In the same way, $\cos(\angle BAC) = \dfrac{\vec{u} \cdot \vec{w}}{|\vec{u}||\vec{w}|} \approx 0.537$, so $m\angle BAC \approx 57.5°$.

$\square$

Example 75. Suppose points $A_1, A_2, \ldots, A_n$ lie on the circle $\mathcal{C}(O, 1)$ and divide the circle into n congruent arcs.

(1) Find $\sum_{i=1}^{n} A_1 A_i{}^2$.
(2) Find the sum of the squares of the lengths of all segments $A_i A_j$ for $1 \le i < j \le n$.

Solution:

(1) For each i, let $\vec{v}_i = \overrightarrow{OA_i}$. Then $|\vec{v}_i| = 1$, and $\overrightarrow{A_1 A_i} = \vec{v}_i - \vec{v}_1$. Furthermore, from Example 66, $\sum_{i=1}^{n} \vec{v}_i = \vec{0}$. Therefore,

$$\sum_{i=1}^{n} |\overrightarrow{A_1 A_i}|^2 = \sum_{i=1}^{n} (\vec{v}_i - \vec{v}_1)^2 = \sum_{i=1}^{n} (|\vec{v}_i|^2 - 2\vec{v}_i \cdot \vec{v}_1 + |\vec{v}_1|^2)$$

$$= \sum_{i=1}^{n} (2 - 2\vec{v}_i \cdot \vec{v}_1) = 2n - 2\vec{v}_1 \cdot \sum_{i=1}^{n} \vec{v}_i$$

$$= 2n - 2\vec{v}_1 \cdot \vec{0} = 2n - 0 = 2n.$$

Thus, $\sum_{i=1}^{n} A_1 A_i{}^2 = \sum_{i=1}^{n} |\overrightarrow{A_1 A_i}|^2 = 2n$. In addition, it is clear from the symmetry of the regular n-gon $A_1 A_2 \ldots A_n$ that replacing the vertex A_1 in the sum with any other vertex leads to the same result.

(2) By part (1), for any fixed i, $\sum_{1 \le j \le n} |\overrightarrow{A_i A_j}|^2 = 2n$. Hence,

$$\sum_{1 \le i < j \le n} |A_i A_j|^2 = \frac{1}{2} \sum_{1 \le i \le n} \left(\sum_{1 \le j \le n} |A_i A_j|^2 \right)$$

$$= \frac{1}{2} \sum_{1 \le i \le n} (2n) = \frac{1}{2}(2n)(n) = n^2.$$

The first equality above is due to the facts that every addend from the first sum appears in the second sum exactly two times and $|A_i A_j| = 0$ for all i. $\qquad \square$

11.6 Problems

A pessimist sees the difficulty in every opportunity; an optimist sees the opportunity in every difficulty.
— Winston Churchill (1874–1965)

11.1 Prove that for any set of four points, A, B, C, and D,
$$\overrightarrow{AB} \cdot \overrightarrow{CD} + \overrightarrow{BC} \cdot \overrightarrow{AD} + \overrightarrow{CA} \cdot \overrightarrow{BD} = \vec{0}.$$

11.2 Prove that the segment joining the midpoints of two sides of a triangle is parallel to and has half the length of the third side.

11.3 Prove that the midpoints of the sides of a quadrilateral form a parallelogram.

11.4 If $\vec{c} = |\vec{b}|\vec{a} + |\vec{a}|\vec{b}$, where $\vec{a}$, $\vec{b}$, and $\vec{c}$ are all nonzero vectors, prove that $\vec{c}$ bisects the angle between $\vec{a}$ and $\vec{b}$.

11.5 Let $M_1, M_2, \ldots, M_6$ be the midpoints of the sides of a convex 6-gon, $A_1 A_2 \ldots A_6$. Prove that there exists a triangle whose sides are parallel to and congruent to the segments $M_1 M_2$, $M_3 M_4$, and $M_5 M_6$.

11.6 Prove the dot product properties given in Theorem 11.7.

11.7 A line in the plane is uniquely determined when a point on the line and the direction of the line are given. The notion of position vectors allows us a useful vector representation for equations of lines and segments: if $l = \overleftrightarrow{AB}$ is a line in the plane, one can think of the vector representation of l as the set L of all position vectors $\overrightarrow{OX}$, where O corresponds to the origin and X lies on l.

   (a) Prove that $L = \{\overrightarrow{OA} + k\overrightarrow{AB} : k \in \mathbb{R}\}$.
   (b) Prove that $\{\overrightarrow{OA} + k\overrightarrow{AB} : k \in [0, 1]\}$ corresponds to the position vectors of the points on $\overline{AB}$.

11.8 Prove that a quadrilateral is a parallelogram if and only if the sum of the squares of its side lengths equals the sum of the squares of its diagonal lengths.

11.9 Prove that the medians of a triangle ABC intersect at a point M which divides their length in ratio 2:1 (starting from the vertex).

11.10 For a triangle ABC, prove that $\overrightarrow{MA} + \overrightarrow{MB} + \overrightarrow{MC} = \vec{0}$ if and only if M is the point of intersection of the medians of the triangle.

11.11 An immediate consequence of Problem 11.10 is that, in any triangle, the three medians are the sides of another triangle. Is the same true of the three angle bisectors as vectors?

11.12 For a quadrilateral $ABCD$ having side lengths a, b, c, and d, prove that Area$(ABCD) \leq (a + c)(b + d)/4$ with equality if and only if $ABCD$ is a rectangle. (See Problem 1.12.)

11.13 Suppose that in $\triangle ABC$, $a^2 + b^2 = 5c^2$. Prove that the medians to the sides AC and BC are perpendicular.

11.14 Let A, B, C, and D be four points such that $\overline{AB} \perp \overline{CD}$ and $\overline{BC} \perp \overline{AD}$.

   (a) Prove that $\overline{AC} \perp \overline{BD}$.
   (b) Prove also that in this case, $AB^2 + CD^2 = AC^2 + BD^2$. Does this equality imply that $\overline{AB} \perp \overline{CD}$?

11.15 Prove that the three lines defined by the altitudes of a triangle are concurrent.

11.16 Let $A_1 A_2 \ldots A_n$ be a regular n-gon inscribed in a circle, $\mathcal{C}(O, R)$. Then, for any point X such that $OX = d$, prove that $A_1 X^2 + A_2 X^2 + \cdots + A_n X^2 = n(R^2 + d^2)$.

11.17 Given a triangle ABC, find a point X such that the sum $AX^2 + BX^2 + CX^2$ is the smallest.

11.18 Express the distance between the midpoints of the diagonals of a quadrilateral in terms of the lengths of its four sides and two diagonals.

11.19 Let O be an arbitrary point inside $\triangle ABC$. Prove that

$$(\text{Area}(\triangle BOC))\,\overrightarrow{OA} + (\text{Area}(\triangle AOC))\,\overrightarrow{OB} + (\text{Area}(\triangle AOB))\,\overrightarrow{OC} = \vec{0}.$$

11.7 Supplemental Problems

$S11.1$ Let $|\vec{a}| = 1$ and $|\vec{b}| = 3$, and let the measure of the angle between $\vec{a}$ and $\vec{b}$ be $45°$.

   (a) Find the measure of the angle between $2\vec{a} + \vec{b}$ and $\vec{b} - \vec{a}$.
   (b) Let $\vec{u}$ be a unit vector parallel to the bisector of the angle formed by $\vec{a} + \vec{b}$ and $\vec{b} - \vec{a}$. Express $\vec{u}$ as a linear combination of $\vec{a}$ and $\vec{b}$.

*S*11.2 Let $A : (1, 2)$, $B : (2, 4)$, and $C : (5, 0)$. Let M be the midpoint of the median of $\triangle ABC$ at A, and let E be the point of intersection of line CM with the side AB. Find CM/ME.

*S*11.3 Prove that the segment joining the midpoints of the non-parallel sides of a trapezoid is parallel to the bases and has length equal to one-half the sum of their lengths.

*S*11.4 Consider a parallelogram $ABCD$. Let $E \in \overline{BC}$ and $F \in \overline{CD}$ such that $BE/EC = 2$ and $CF/FD = 3$. Let H be the point of intersection of segments AF and DE. Find AH/HF.

*S*11.5 Find $\vec{a} \cdot b + \vec{b} \cdot c + \vec{c} \cdot a$, where $\vec{a}$, $\vec{b}$ and $\vec{c}$ are three unit vectors such that $\vec{a} + \vec{b} + \vec{c} = \vec{0}$.

*S*11.6 Let E and F be the midpoints of the sides AB and CD of a quadrilateral $ABCD$. Prove that the midpoints of segments AF, BF, CE, and DE are vertices of a parallelogram.

---

*S*11.7 Suppose the diagonals of a quadrilateral Q are perpendicular. Prove that any quadrilateral with sides congruent to the sides of Q also has perpendicular diagonals.

*S*11.8 Given triangle ABC, let $\overline{AD}$ be the bisector of angle A. Prove that

$$AD = \frac{2bc \cos (A/2)}{b + c}.$$

*S*11.9 Consider a 2×2 matrix $\mathbf{A}$ (a rectangular array): $\mathbf{A} = \begin{bmatrix} a & b \\ c & d \end{bmatrix}$.

Suppose two row vectors of A, $\vec{r}_1 = \langle a, b \rangle$ and $\vec{r}_2 = \langle c, d \rangle$, and two column vectors of A, $\vec{c}_1 = \langle a, c \rangle$ and $\vec{c}_2 = \langle b, d \rangle$, are all unit vectors. Prove that $\vec{r}_1$ and $\vec{r}_2$ are orthogonal if and only if $\vec{c}_1$ and $\vec{c}_2$ are orthogonal.

*S*11.10 Two points A and B are given on opposite sides of a strip defined by two parallel lines. The strip represents a river, and the points represent cities on the opposite sides of the river. Determine the shortest possible road between A and B, assuming that the land parts of the road are straight line segments and the bridge is perpendicular to the banks.[8]

*S*11.11 Prove that the area of a triangle ABC is equal to

$$\frac{1}{2}\sqrt{AB^2 AC^2 - (\overrightarrow{AB} \cdot \overrightarrow{AC})^2}.$$

*S*11.12 Points A, B, and C are moving with an equal constant angular speed and in the same direction (say counterclockwise) around three circles, one point on each circle. Prove that the point of intersection of medians of $\triangle ABC$ is also moving around a circle.

---

*S*11.13 Given triangle ABC, let $\vec{u}$, $\vec{v}$ and $\vec{w}$ be unit vectors perpendicular to its sides AB, BC, and CA, respectively, and directed towards the exterior of the triangle. Prove that

$$AB\vec{u} + BC\vec{v} + CA\vec{w} = \vec{0}.$$

*S*11.14 Consider ten vectors with the property that the magnitude of the sum of any nine of them is smaller than the magnitude of the sum of all of them. Prove that there exists a vector such that the scalar projection of each of the ten vectors on it is positive.

*S*11.15 Prove that for any $n \geq 2$, and any real numbers $a_1, a_2, \ldots a_n$, not all zero, and any real numbers $b_1, b_2, \ldots b_n$,

$$\left| \sum_{i=1}^{n} a_i b_i \right| \leq \left(\sum_{i=1}^{n} a_i^2 \right)^{1/2} \left(\sum_{i=1}^{n} b_i^2 \right)^{1/2},$$

---

[8] This problem was taken from the website www.cut-the-knot.org

where the equality is achieved if and only if there exists a number λ such that $b_i = \lambda a_i$ for all $i = 1, 2, \ldots, n$.

Comment. This is a generalization of the Cauchy-Schwartz inequality to higher dimensions. Since we have not considered vectors in higher dimensions, your solution must differ from the $n = 2$ case.

12

Affine Transformations

Chaotic features of the World erase
And you will see its Beauty.
— Alexander A. Block (1880–1921)[1]

12.1   Introduction

Suppose we are struggling with a geometric problem concerning an arbitrary triangle or an arbitrary parallelogram. How often we would wish for the triangle to be an equilateral or $45° - 90° - 45°$ triangle, or for the parallelogram to be a square! The solution is so easy in these cases. But we know that these would be just very particular instances of the problem. Solving them will make us feel better, but not much better. Well, the good news is that for *some* problems, solving just a particular instance turns out to be sufficient to claim that the problem is solved in complete generality! In this chapter we learn how to recognize some of these problems, and we justify such an approach.

We start by reviewing some familiar concepts. Let A and B be sets. A **function** or **mapping** f from A to B, denoted $f : A \to B$, is a set of ordered pairs (a, b), where $a \in A$ and $b \in B$, with the following property: for every $a \in A$ there exists a unique $b \in B$ such that $(a, b) \in f$. The fact that $(a, b) \in f$ is usually denoted by $f(a) = b$, and we say that f maps a to b. Another way to denote that f maps a to b is $f : a \mapsto b$; if it is clear which function is being discussed, we will often just write $a \mapsto b$. We also say that b is the **image** of a (in f), and that a is a **preimage** of b (in f). The set A is called the **domain** of f and the set B is the **codomain** of f. The set $f(A) = \{f(a) : a \in A\}$ is a subset of B, called the **range** of f.

A function $f : A \to B$ is **surjective** (or **onto**) if $f(A) = B$; that is, f is surjective if every element of B is the image of at least one element of A. A function $f : A \to B$ is **injective** (or **one-to-one**) if each element in the range of f is the image of *exactly* one element of A; that is, f is injective if $f(x) = f(y)$ implies $x = y$. A function $f : A \to B$ is **bijective** if it is both surjective and injective.

---

[1] Translated from the Russian by Vera Zubareva.

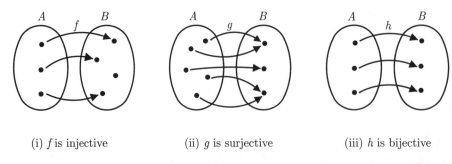

(i) f is injective   (ii) g is surjective   (iii) h is bijective

FIGURE 12.1.

If $f : A \to B$ and $g : B \to C$ are functions, then the **composition** of f and g, denoted $g \circ f$, is a function from A to C such that $(g \circ f)(a) = g(f(a))$ for any $a \in A$. The proof of Theorem 12.1 is left to the reader and can be found in many texts.

Theorem 12.1. *A composition of two bijections is a bijection.*

If $f : A \to B$, then $f^{-1} : B \to A$ is the **inverse** of f if $(f^{-1} \circ f)(a) = a$ for any $a \in A$ and $(f \circ f^{-1})(b) = b$ for any $b \in B$. A function f has an inverse if and only if f is a bijection.

Let $\mathbb{E}^2$ denote the Euclidean plane. Introducing a coordinate system[2] OXY on $\mathbb{E}^2$, we can identify every point P with the ordered pair of its coordinates (x_P, y_P); alternatively, P can be identified with its position vector, $\overrightarrow{OP} = \langle x_P, y_P \rangle$. The collection of all such vectors form a vector space,[3] namely $\mathbb{R}^2$. If $\vec{x}$ represents the vector with initial point at the origin and terminal point at (x_P, y_P), then $\overrightarrow{OP}$, $\langle x_P, y_P \rangle$, and $\mathbf{x}$ can also be used to denote $\vec{x}$.

A **transformation** of a set is a bijection of the set to itself. It is easy to see that any transformation $f : \mathbb{E}^2 \to \mathbb{E}^2$ corresponds to a bijection $\tilde{f} : \mathbb{R}^2 \to \mathbb{R}^2$, in that $\tilde{f}(\langle x_P, y_P \rangle) = \langle x_{P'}, y_{P'} \rangle$ whenever $f(P) = P'$. Since f and $\tilde{f}$ uniquely define one another within a fixed coordinate system, we will also refer to $\tilde{f}$ as a transformation of the plane, and we will write f to denote either a mapping of $\mathbb{E}^2$ to $\mathbb{E}^2$ or a mapping of $\mathbb{R}^2$ to $\mathbb{R}^2$. It will be clear from the context which of the two mappings f represents.

Just as any point P in OXY corresponds to a unique vector $\overrightarrow{OP}$, each figure φ in $\mathbb{E}^2$ uniquely corresponds to a set of vectors $\overrightarrow{OP}$ of $\mathbb{R}^2$, where $P \in \varphi$. We say that this set of vectors is a **figure in $\mathbb{R}^2$**, and we denote it again by φ. The set $f(\varphi)$ is defined as $\{f(P) : P \in \varphi \subseteq \mathbb{E}^2\}$, or $\{f(\overrightarrow{OP}) : \overrightarrow{OP} \in \varphi \subseteq \mathbb{R}^2\}$. It is not hard to make the relationship between point spaces and vector spaces more precise, but we will not do it here.[4] In fact, we freely interchange the representations of point and vector, (x, y) and $\langle x, y \rangle$, when they are domain elements of a function f.

Transformations of the plane and their application to solving geometry problems form the focus of this chapter. The transformations we study will be of two types, illustrated by the following examples:

$$f(\langle x, y \rangle) = \langle 2x - 3y, x + y \rangle \quad \text{and} \quad g(\langle x, y \rangle) = \langle 2x - 3y + 1, x + y - 4 \rangle.$$

---

[2] Recall that OXY denotes a coordinate system (not necessarily Cartesian) with axes $\overleftrightarrow{OX}$ and $\overleftrightarrow{OY}$.

[3] Students who have studied some linear algebra may recall that a vector space is a collection of objects on which an "addition" operation may be performed in such a way that nice properties like commutativity and the existence of additive inverses hold, but a precise definition of vector space is not necessary in order to continue reading.

[4] See, for example, [**34**], [**50**], or [**65**] for rigorous expositions.

At this point it is not obvious that f and g are bijections, but this will be verified later in the chapter. To get a more concrete sense of what f and g do, consider how they "transform" the vectors $\langle 0, 0 \rangle$, $\langle 0, 1 \rangle$, $\langle 1, 0 \rangle$, and $\langle 1, 1 \rangle$.

$\vec{x}$	$f(\vec{x})$	$g(\vec{x})$
$\langle 0, 0 \rangle$	$\langle 0, 0 \rangle$	$\langle 1, -4 \rangle$
$\langle 0, 1 \rangle$	$\langle -3, 1 \rangle$	$\langle -2, -3 \rangle$
$\langle 1, 0 \rangle$	$\langle 2, 1 \rangle$	$\langle 3, -3 \rangle$
$\langle 1, 1 \rangle$	$\langle -1, 2 \rangle$	$\langle 0, -2 \rangle$

Notice that the origin, $\vec{0}$, is fixed under f, while $g(\langle 0, 0 \rangle) = \langle 1, -4 \rangle$. Notice also that $f(\langle 0, 1 \rangle + \langle 1, 0 \rangle) = f(\langle 0, 1 \rangle) + f(\langle 1, 0 \rangle)$; again, this is not true of g. These properties of f are indicative of the **linearity** of that mapping. A function $T : \mathbb{R}^2 \to \mathbb{R}^2$ is called **linear** if $T(\vec{x} + \vec{y}) = T(\vec{x}) + T(\vec{y})$ for any vectors $\vec{x}$ and $\vec{y}$, and $T(k\vec{x}) = kT(\vec{x})$ for any vector $\vec{x}$ and scalar k. The reader can verify that these properties hold for f but not for g.

As will be shown later in this chapter, both f and g map a line segment to a line segment. Therefore, knowing where f and g map the points corresponding to the vectors $\langle 0, 0 \rangle$, $\langle 0, 1 \rangle$, $\langle 1, 1 \rangle$, and $\langle 1, 0 \rangle$ is sufficient for determining the image of the unit square, S, having vertices at these four points. Figure 12.2 shows S together with $f(S)$ and $g(S)$. Notice that both $f(S)$ and $g(S)$ are parallelograms; Theorem 12.7 will prove that this is not a coincidence.

12.2 Matrices

Transformations of $\mathbb{E}^2$ or $\mathbb{R}^2$ are often studied via another type of mathematical object, the matrix. Though the benefits of using the language of matrices are not striking when we study $\mathbb{E}^2$, matrices

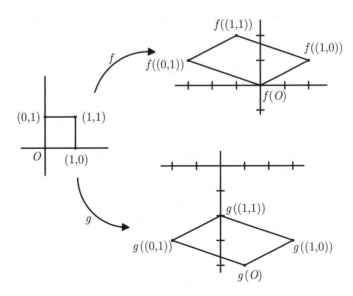

FIGURE 12.2.

turn out to be very convenient when generalizing geometric notions of the plane to spaces of higher dimensions.[5]

An $m \times n$ **matrix A** is a rectangular array of real numbers,

$$
\mathbf{A} = \begin{bmatrix}
a_{11} & a_{12} & \ldots & a_{1n} \\
a_{21} & a_{22} & \ldots & a_{2n} \\
\cdot & \cdot & & \cdot \\
\cdot & \cdot & & \cdot \\
\cdot & \cdot & & \cdot \\
a_{m1} & a_{m2} & \ldots & a_{mn}
\end{bmatrix}.
$$

The entry in the ith row and the jth column is denoted a_{ij}, and we often write $\mathbf{A} = [a_{ij}]$. Two matrices $\mathbf{A} = [a_{ij}]$ and $\mathbf{A}' = [a'_{ij}]$ are called **equal** if they have an equal number of rows, an equal number of columns, and $a_{ij} = a'_{ij}$ for all i and j. When the matrix is $n \times n$, so that there are an equal number of rows and columns, the matrix is called a **square** matrix. Notice that a vector $\vec{v} = \langle v_1, v_2 \rangle$ can be thought of as the 1×2 matrix $[v_1 \; v_2]$, called a "row vector." It can also be thought of as a "column vector" by writing $\vec{v}$ as the 2×1 matrix $\begin{bmatrix} v_1 \\ v_2 \end{bmatrix}$.

If $\mathbf{A} = [a_{ij}]$ and $\mathbf{B} = [b_{ij}]$ are both $m \times n$ matrices, then the **sum A + B** is the $m \times n$ matrix $\mathbf{C} = [c_{ij}]$ in which $c_{ij} = a_{ij} + b_{ij}$. If $\mathbf{A} = [a_{ij}]$ is an $m \times n$ matrix and $c \in \mathbb{R}$, then the **scalar multiple** of $\mathbf{A}$ by c is the $m \times n$ matrix $c\mathbf{A} = [ca_{ij}]$. (That is, $c\mathbf{A}$ is obtained by multiplying each entry of $\mathbf{A}$ by c.)

The **product AB** of two matrices is defined when $\mathbf{A} = [a_{ij}]$ is an $m \times n$ matrix and $\mathbf{B} = [b_{ij}]$ is an $n \times p$ matrix. Then $\mathbf{AB} = [c_{ij}]$, where $c_{ij} = \sum_{k=1}^{n} a_{ik}b_{kj}$. For example, if $\mathbf{A}$ is a 2×2 matrix, and $\mathbf{B}$ is a 2×1 matrix, then

$$
\mathbf{AB} = \begin{bmatrix} a_{11} & a_{12} \\ a_{21} & a_{22} \end{bmatrix} \begin{bmatrix} b_{11} \\ b_{21} \end{bmatrix} = \begin{bmatrix} a_{11}b_{11} + a_{12}b_{21} \\ a_{21}b_{11} + a_{22}b_{21} \end{bmatrix}.
$$

We say that here we multiply $\mathbf{A}$ by a (column) vector. Notice that $\mathbf{BA}$ is not defined in this case.

If $\mathbf{A}$ and $\mathbf{B}$ are both 2×2 matrices,

$$
\mathbf{AB} = \begin{bmatrix} a_{11} & a_{12} \\ a_{21} & a_{22} \end{bmatrix} \begin{bmatrix} b_{11} & b_{12} \\ b_{21} & b_{22} \end{bmatrix} = \begin{bmatrix} a_{11}b_{11} + a_{12}b_{21} & a_{11}b_{12} + a_{12}b_{22} \\ a_{21}b_{11} + a_{22}b_{21} & a_{21}b_{12} + a_{22}b_{22} \end{bmatrix}.
$$

Although $\mathbf{BA}$ is defined in this case, *in general* $\mathbf{BA}$ is not equal to $\mathbf{AB}$. So matrix multiplication is not commutative. These two instances of matrix multiplication (when $\mathbf{A}$ is a 2×2 matrix and $\mathbf{B}$ is a 2×1 or a 2×2 matrix) are the only ones we will need in this book. In what follows, no matter whether $\vec{x}$ is a 1×2 vector or 2×1 vector, when it is used in the expression $\mathbf{A}\vec{x}$, it is always understood as a column vector, i.e., as a 2×1 matrix.

Theorem 12.2 summarizes some of the most useful properties of matrix operations. Its proof can easily be produced by the reader (part (4) is the most difficult) or may be found in a standard linear algebra text.

---

[5] Here, when we say "language," we mean the objects, their notation, operations on the objects, and properties of those operations – similar to the "languages" of trigonometry, algebra, logic, and calculus.

> **Theorem 12.2.**
>
> (1) *If* **A** *and* **B** *are* $m \times n$ *matrices, then* $\mathbf{A} + \mathbf{B} = \mathbf{B} + \mathbf{A}$.
>
> (2) *If* **A**, **B**, *and* **C** *are* $m \times n$ *matrices, then* $\mathbf{A} + (\mathbf{B} + \mathbf{C}) = (\mathbf{A} + \mathbf{B}) + \mathbf{C}$.
>
> (3) *Given an* $m \times n$ *matrix* **A**, *there exists a unique* $m \times n$ *matrix* **B** *such that* $\mathbf{A} + \mathbf{B} =$ **B** + **A** *is the* **zero matrix** *(that is, the matrix with 0 in every entry).*
>
> (4) *If* **A** *is an* $m \times n$ *matrix,* **B** *is an* $n \times p$ *matrix, and* **C** *is a* $p \times q$ *matrix, then* $\mathbf{A}(\mathbf{B}\mathbf{C}) = (\mathbf{A}\mathbf{B})\mathbf{C}$.
>
> (5) *If* **A** *and* **B** *are* $m \times n$ *matrices,* **C** *is an* $n \times p$ *matrix, and* **D** *is a* $q \times m$ *matrix, then* $(\mathbf{A} + \mathbf{B})\mathbf{C} = \mathbf{A}\mathbf{C} + \mathbf{B}\mathbf{C}$ *and* $\mathbf{D}(\mathbf{A} + \mathbf{B}) = \mathbf{D}\mathbf{A} + \mathbf{D}\mathbf{B}$.
>
> (6) *If* $r, s \in \mathbb{R}$, **A** *is an* $m \times n$ *matrix, and* **B** *is an* $n \times p$ *matrix, then*
>
>   *(a)* $r(s\mathbf{A}) = (rs)\mathbf{A} = s(r\mathbf{A})$, *and*
>   *(b)* $\mathbf{A}(r\mathbf{B}) = r(\mathbf{A}\mathbf{B})$.
>
> (7) *If* $r, s \in \mathbb{R}$, *and* **A** *and* **B** *are* $m \times n$ *matrices, then*
>
>   *(a)* $(r + s)\mathbf{A} = r\mathbf{A} + s\mathbf{A}$, *and*
>   *(b)* $r(\mathbf{A} + \mathbf{B}) = r\mathbf{A} + r\mathbf{B}$.

Using the notation of matrices, we can represent the functions

$$f(\langle x, y \rangle) = \langle 2x - 3y, x + y \rangle \quad \text{and} \quad g(\langle x, y \rangle) = \langle 2x - 3y + 1, x + y - 4 \rangle$$

using matrix multiplication as follows. First, let $\vec{x} = \begin{bmatrix} x \\ y \end{bmatrix}$, and let

$$\mathbf{A} = \begin{bmatrix} 2 & -3 \\ 1 & 1 \end{bmatrix}.$$

Then

$$f(\vec{x}) = \mathbf{A}\vec{x} = \begin{bmatrix} 2 & -3 \\ 1 & 1 \end{bmatrix} \begin{bmatrix} x \\ y \end{bmatrix}.$$

One way to think about the matrix **A** corresponding to the transformation f is that the columns of **A** specify the images of the vectors $\vec{i} = \langle 1, 0 \rangle$ and $\vec{j} = \langle 0, 1 \rangle$. Using matrix multiplication, we see that $\mathbf{A}\vec{i} = \begin{bmatrix} 2 & -3 \\ 1 & 1 \end{bmatrix} \begin{bmatrix} 1 \\ 0 \end{bmatrix} = \begin{bmatrix} 2 \\ 1 \end{bmatrix}$, and $\mathbf{A}\vec{j} = \begin{bmatrix} 2 & -3 \\ 1 & 1 \end{bmatrix} \begin{bmatrix} 0 \\ 1 \end{bmatrix} = \begin{bmatrix} -3 \\ 1 \end{bmatrix}$, as illustrated in Figure 12.3.

If we let $\vec{b} = \begin{bmatrix} 1 \\ -4 \end{bmatrix}$, then the same 2×2 matrix **A** gives

$$g(\vec{x}) = \mathbf{A}\vec{x} + \vec{b} = \begin{bmatrix} 2 & -3 \\ 1 & 1 \end{bmatrix} \begin{bmatrix} x \\ y \end{bmatrix} + \begin{bmatrix} 1 \\ -4 \end{bmatrix}$$

$$= \begin{bmatrix} 2x - 3y \\ x + y \end{bmatrix} + \begin{bmatrix} 1 \\ -4 \end{bmatrix} = \begin{bmatrix} 2x - 3y + 1 \\ x + y - 4 \end{bmatrix}.$$

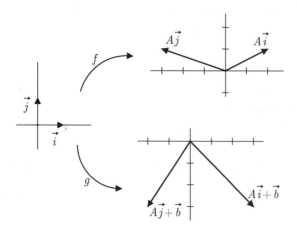

FIGURE 12.3.

Now, $\mathbf{A}\vec{i} + \vec{b} = \begin{bmatrix} 2 & -3 \\ 1 & 1 \end{bmatrix} \begin{bmatrix} 1 \\ 0 \end{bmatrix} + \begin{bmatrix} 1 \\ -4 \end{bmatrix} = \begin{bmatrix} 3 \\ -3 \end{bmatrix}$, and

$\mathbf{A}\vec{j} + \vec{b} = \begin{bmatrix} 2 & -3 \\ 1 & 1 \end{bmatrix} \begin{bmatrix} 0 \\ 1 \end{bmatrix} + \begin{bmatrix} 1 \\ -4 \end{bmatrix} = \begin{bmatrix} -2 \\ -3 \end{bmatrix}$, again illustrated in Figure 12.3.

Notice that using column form for vectors allows us to write the elements of the domain of f and g on the right side of the matrix representing the function, just as the variable is on the right when using the notation $f(x)$. If we compose two functions, f and g, where $f(\vec{x}) = \mathbf{A}\vec{x}$ and $g(\vec{x}) = \mathbf{B}\vec{x}$, then $(g \circ f)(\vec{x}) = g(f(\vec{x})) = \mathbf{B}(\mathbf{A}\vec{x}) = (\mathbf{BA})\vec{x}$. Hence the matrix that corresponds to the composition $g \circ f$ is $\mathbf{BA}$.[6]

The 2×2 **identity matrix**, $\mathbf{I}_2 = \begin{bmatrix} 1 & 0 \\ 0 & 1 \end{bmatrix}$, has special significance. It is easy to check that $\mathbf{I}_2$ is the only matrix with the property that if $\mathbf{A}$ is any 2×2 matrix, $\mathbf{AI}_2 = \mathbf{I}_2\mathbf{A} = \mathbf{A}$, and $\mathbf{I}_2\vec{x} = \vec{x}$ for each $\vec{x}$ in $\mathbb{R}^2$. Clearly $\mathbf{I}_2$ is a matrix analog of the number 1.[7]

Furthermore, for *some*[8] square matrices $\mathbf{A}$, there exists a matrix $\mathbf{B}$ such that $\mathbf{AB} = \mathbf{BA} = \mathbf{I}_2$. It is easy to show that if $\mathbf{B}$ exists, then it is unique. Such a matrix $\mathbf{A}$ is called **invertible** or **nonsingular**, and the corresponding matrix $\mathbf{B}$ (more often denoted $\mathbf{A}^{-1}$) is called the **inverse** of $\mathbf{A}$. For example, the matrix $\mathbf{A} = \begin{bmatrix} 2 & -3 \\ 1 & 1 \end{bmatrix}$ is invertible, with $\mathbf{A}^{-1} = \begin{bmatrix} 1/5 & 3/5 \\ -1/5 & 2/5 \end{bmatrix}$, because

$$\begin{bmatrix} 2 & -3 \\ 1 & 1 \end{bmatrix} \begin{bmatrix} 1/5 & 3/5 \\ -1/5 & 2/5 \end{bmatrix} = \begin{bmatrix} 1/5 & 3/5 \\ -1/5 & 2/5 \end{bmatrix} \begin{bmatrix} 2 & -3 \\ 1 & 1 \end{bmatrix} = \begin{bmatrix} 1 & 0 \\ 0 & 1 \end{bmatrix}.$$

As $\mathbf{AA}^{-1} = \mathbf{A}^{-1}\mathbf{A} = \mathbf{I}_2$, the matrix $\mathbf{A}^{-1}$ is also invertible and $\mathbf{A}$ is its inverse.

---

[6] The order of matrices in this multiplication matches the order of the corresponding functions in the notation $g \circ f$, but the order in which the two functions are composed does not match the order in which they are written. For this reason, some authors prefer to replace the notation $f(\vec{x})$ with $(\vec{x})f$. Then $\vec{x}$ can be thought of as a row vector, and we write $(\vec{x})f = \vec{x}\mathbf{A}$. For f and g as in our case, this would make $\vec{x}\mathbf{AB}$ correspond to $(\vec{x}f)g$. While this notation may be less familiar, at least the orders match! One cannot have it all. . . .

[7] These statements can be made in greater generality. The $n \times n$ identity matrix, $I_n = [c_{ij}]$, is the matrix having $c_{ij} = 1$ if $i = j$ and $c_{ij} = 0$ otherwise. Then, if $\mathbf{A}$ is any $m \times n$ matrix, $\mathbf{AI}_n = \mathbf{I}_m\mathbf{A} = \mathbf{A}$.

[8] Actually, for *most* of them, but we will not discuss the meaning of "most" at this point.

Let **A** be an invertible matrix, let $\vec{b}$ be a vector, and let $f : \mathbb{R}^2 \to \mathbb{R}^2$ be defined via $\vec{x} \mapsto \mathbf{A}\vec{x} + \vec{b}$. For any vector $\vec{y}$, the following are all equivalent.

$$f(\vec{x}) = \vec{y}$$
$$\mathbf{A}\vec{x} + \vec{b} = \vec{y}$$
$$\mathbf{A}\vec{x} = \vec{y} - \vec{b}$$
$$\mathbf{A}^{-1}(\mathbf{A}\vec{x}) = \mathbf{A}^{-1}(\vec{y} - \vec{b})$$
$$(\mathbf{A}^{-1}\mathbf{A})\vec{x} = \mathbf{A}^{-1}\vec{y} - \mathbf{A}^{-1}\vec{b}$$
$$\mathbf{I}_2\vec{x} = \mathbf{A}^{-1}\vec{y} - \mathbf{A}^{-1}\vec{b}$$
$$\vec{x} = \mathbf{A}^{-1}(\vec{y} - \vec{b})$$

We conclude that f^{-1} exists and can be given by $f^{-1}(\vec{x}) = \mathbf{A}^{-1}(\vec{x} - \vec{b})$. (One can also easily check that for every vector $\vec{x}$, $(f^{-1} \circ f)(\vec{x}) = \vec{x}$ and $(f \circ f^{-1})(\vec{x}) = \vec{x}$.) Therefore, both f and f^{-1} are bijections on $\mathbb{R}^2$, also called **transformations of the plane**.

A transformation f of the plane of the form $f(\vec{x}) = \mathbf{A}\vec{x} + \vec{b}$ where **A** is an invertible matrix is called an **affine transformation** of the plane. Since $\mathbf{A}^{-1}$ is invertible if and only if **A** is, we have just proven the following.

> **Theorem 12.3.** *An affine transformation of the plane has an inverse that is also an affine transformation of the plane.*

Obviously, it will be useful to know whether a given matrix has an inverse. Fortunately, there is a nice computational tool available for this. The **determinant** of a 2×2 matrix $\mathbf{A} = \begin{bmatrix} a & b \\ c & d \end{bmatrix}$ is the number $ad - bc$, denoted $\det \mathbf{A}$. The primary significance of the determinant follows from Theorem 12.4.

> **Theorem 12.4.** *Let* **A** *and* **B** *be* 2×2 *matrices. Then*
>
> (1) **A** *is invertible if and only if* $\det \mathbf{A} \neq 0$.
> (2) *If* $\det \mathbf{A} \neq 0$, *then* $\mathbf{A}^{-1} = \dfrac{1}{\det \mathbf{A}} \begin{bmatrix} d & -b \\ -c & a \end{bmatrix}$.
> (3) $\det(\mathbf{AB}) = (\det \mathbf{A})(\det \mathbf{B})$.

Proof. (3) Suppose that $\mathbf{A} = \begin{bmatrix} a & b \\ c & d \end{bmatrix}$ and $\mathbf{B} = \begin{bmatrix} a' & b' \\ c' & d' \end{bmatrix}$. Then

$$\mathbf{AB} = \begin{bmatrix} aa' + bc' & ab' + bd' \\ ca' + dc' & cb' + dd' \end{bmatrix}.$$

Consequently,

$$\det(\mathbf{AB}) = (aa' + bc')(cb' + dd') - (ab' + bd')(ca' + dc')$$
$$= aa'dd' + bb'cc' - ab'dc' - ba'cd' = (ad - bc)(a'd' - c'b') = (\det \mathbf{A})(\det \mathbf{B}).$$

(2) We demonstrate that $\mathbf{A}^{-1} = \dfrac{1}{\det \mathbf{A}} \begin{bmatrix} d & -b \\ -c & a \end{bmatrix}$ by matrix multiplication:

$$\mathbf{A} \begin{bmatrix} d & -b \\ -c & a \end{bmatrix} = \begin{bmatrix} ad - bc & -ab + ba \\ cd - dc & -cb + da \end{bmatrix} = \begin{bmatrix} ad - bc & 0 \\ 0 & ad - bc \end{bmatrix} = (ad - bc)\mathbf{I}_2.$$

By part (6) of Theorem 12.2, $\mathbf{A} \cdot \left(\frac{1}{\det \mathbf{A}}\right) \begin{bmatrix} d & -b \\ -c & a \end{bmatrix} = \mathbf{I}_2$. It can similarly be demonstrated that $\mathbf{A}^{-1}\mathbf{A} = \mathbf{I}_2$.

(1) Part (2) above shows that if $\det \mathbf{A} \neq 0$, then $\mathbf{A}$ has an inverse.

Suppose that $\det \mathbf{A} = 0$. If $\mathbf{A}^{-1}$ exists, then $\mathbf{A}\mathbf{A}^{-1} = \mathbf{I}_2$, and by Part (3) of this theorem, $(\det \mathbf{A})(\det \mathbf{A}^{-1}) = \det \mathbf{I}_2$. Since $\det \mathbf{I}_2 = 1 \cdot 1 - 0 \cdot 0 = 1$, this gives $0 \cdot \det \mathbf{A}^{-1} = 1$, a contradiction. $\square$

Corollary 12.5. *A composition of affine transformations is an affine transformation.*

Proof. Let $f(\vec{x}) = \mathbf{A}\vec{x} + \vec{a}$ and $g(\vec{x}) = \mathbf{B}\vec{x} + \vec{b}$ be affine transformations. Then $(g \circ f)(\vec{x}) = g(f(\vec{x})) = \mathbf{B}(\mathbf{A}\vec{x} + \vec{a}) + \vec{b} = (\mathbf{B}\mathbf{A})\vec{x} + (\mathbf{B}\vec{a} + \vec{b})$. Since $\mathbf{A}$ and $\mathbf{B}$ are invertible matrices, $\mathbf{B}\mathbf{A}$ is invertible. This can be seen in several ways.

Note that

$$(\mathbf{A}^{-1}\mathbf{B}^{-1})(\mathbf{B}\mathbf{A}) = \mathbf{A}^{-1}(\mathbf{B}^{-1}\mathbf{B})\mathbf{A} = \mathbf{A}^{-1}(\mathbf{I}_2)\mathbf{A} = \mathbf{A}^{-1}\mathbf{A} = \mathbf{I}_2,$$

and similarly, $(\mathbf{B}\mathbf{A})(\mathbf{A}^{-1}\mathbf{B}^{-1}) = \mathbf{I}_2$. Thus,

$$(\mathbf{B}\mathbf{A})^{-1} = \mathbf{A}^{-1}\mathbf{B}^{-1}.$$

Therefore $\mathbf{B}\mathbf{A}$ is invertible, and we conclude that $g \circ f$ is an affine transformation.

Alternatively, by Theorem 12.4(1), since $\mathbf{A}$ and $\mathbf{B}$ are invertible, $\det \mathbf{A}$ and $\det \mathbf{B}$ are both nonzero. Hence, by Theorem 12.4(3), $\det(\mathbf{B}\mathbf{A}) = (\det \mathbf{B})(\det \mathbf{A}) \neq 0$. Therefore, by Theorem 12.4(1), $\mathbf{B}\mathbf{A}$ is invertible, and we again conclude that $g \circ f$ is an affine transformation. $\square$

The following simple theorem, whose proof is left to the reader, relates the determinant to collinearity of vectors.

Theorem 12.6. *Let $\mathbf{A}$ be a 2×2 matrix. Then the following statements are equivalent.*

(1) $\det \mathbf{A} = 0$.

(2) *The row vectors of $\mathbf{A}$ are collinear.*

(3) *The column vectors of $\mathbf{A}$ are collinear.*

Homotheties, in which the vector $\vec{x}$ is mapped to the vector $k\vec{x}$ where $k \neq 0$ (see Section 3.2.7), provide examples of one type of affine transformation. Two other kinds of affine transformations are of particular interest: translations and rotations.

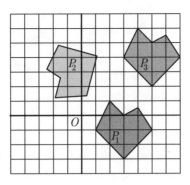

FIGURE 12.4.

A **translation** is an affine transformation of the form

$$f(\vec{x}) = \vec{x} + \vec{b} = \mathbf{I}_2\vec{x} + \vec{b}.$$

A translation can be pictured as "sliding" all points of the plane in the direction given by $\vec{b}$, by the distance $|\vec{b}|$.

A **rotation** is an affine transformation of the form

$$f(\vec{x}) = R_0^\theta(\vec{x}),$$

where $R_0^\theta = \begin{bmatrix} \cos\theta & -\sin\theta \\ \sin\theta & \cos\theta \end{bmatrix}$. Under a rotation, the vector $\overrightarrow{OP}$ is mapped to the vector $\overrightarrow{OP'}$, where O is the origin, $m(\angle POP') = \theta$, and $|\overrightarrow{OP}| = |\overrightarrow{OP'}|$. This transformation can be pictured by imagining sticking a pin at the origin to fix that point, and then rotating the entire plane counterclockwise by the angle θ.

In Figure 12.4, the original figure, P_1, is mapped to P_2 via rotation by an angle of $120°$, and mapped to P_3 via translation by the vector $\langle 2, 5 \rangle$. The effect of translations and rotations on conic sections will be explored in Section 12.5.

12.3   Properties

Some things never change.

— Various[9]

One of the essential aspects of affine transformations is that certain geometric properties are preserved, or *invariant*, under any affine transformation. If a geometric figure φ possesses a property that is invariant under affine transformations, then the image, $f(\varphi)$, under any affine transformation f will also have that property. Theorem 12.7 establishes the invariance of key properties under affine transformations. Note that the proof regularly uses the linearity of the function $\vec{x} \mapsto A\vec{x}$, i.e., the facts that $\mathbf{A}(t\vec{u}) = t(\mathbf{A}\vec{u})$ and $\mathbf{A}(\vec{u} + \vec{w}) = \mathbf{A}\vec{u} + \mathbf{A}\vec{w}$, where $\mathbf{A}$ is a 2×2 matrix and t is a scalar. Remembering that vectors can be thought of as 2×1 matrices, these facts follow from parts (6)(b) and (5), respectively, of Theorem 12.2.

---

[9] In the context of this section the phrase was used in the title of [**32**].

> **Theorem 12.7.** *Let $f(\vec{x}) = \mathbf{A}\vec{x} + \vec{b}$ be an affine transformation. Then f*
>
> (1) *maps a line to a line,*
>
> (2) *maps a line segment to a line segment,*
>
> (3) *preserves the property of parallelism among lines and line segments,*
>
> (4) *maps an n-gon to an n-gon,*
>
> (5) *maps a parallelogram to a parallelogram,*
>
> (6) *preserves the ratio of lengths of two parallel segments, and*
>
> (7) *preserves the ratio of areas of two figures.*

Proof.

(1) Let l be a line, and let $l : \vec{p} + t\vec{u}$, $t \in \mathbb{R}$, be an equation of l in vector form (as specified in Problem 11.7). Then, for every $t \in \mathbb{R}$,

$$f(\vec{p} + t\vec{u}) = \mathbf{A}(\vec{p} + t\vec{u}) + \vec{b} = (\mathbf{A}\vec{p} + \vec{b}) + t(\mathbf{A}\vec{u}) = \vec{p}_1 + t\vec{u}_1,$$

where $\vec{p}_1 = \mathbf{A}\vec{p} + \vec{b}$ and $\vec{u}_1 = \mathbf{A}\vec{u}$. Hence $f(l) = l_1$, where $l_1 : \vec{p}_1 + t\vec{u}_1$, $t \in \mathbb{R}$, is again a line.

(2) The proof is the same as that for (1), with t restricted to $[0, 1]$.

(3) Suppose that $l : \vec{p} + t\vec{u}$ and $m : \vec{q} + t\vec{v}$, $t \in \mathbb{R}$, are parallel lines. Then $\vec{v} = k\vec{u}$ for some $k \in \mathbb{R}$. Therefore,

$$f(\vec{p} + t\vec{u}) = \mathbf{A}(\vec{p} + t\vec{u}) + \vec{b} = (\mathbf{A}\vec{p} + \vec{b}) + t(\mathbf{A}\vec{u}) = \vec{p}_1 + t\vec{u}_1 \text{ and}$$
$$f(\vec{q} + t\vec{v}) = f(\vec{q} + t(k\vec{u})) = \mathbf{A}(\vec{q} + t(k\vec{u})) + \vec{b}$$
$$= (\mathbf{A}\vec{q} + \vec{b}) + t(\mathbf{A}k\vec{u}) = \vec{q}_1 + t(k\vec{u}_1).$$

That is, l and m are mapped to lines l_1 and m_1 that are parallel.

It is clear that for two line segments or a line and a line segment the proof is absolutely analogous.

(4) We prove this by strong induction on n. For the base case, when $n = 3$, consider a triangle T. Then T and its interior can be represented in vector form as $T : \vec{u} + s\vec{v} + t\vec{w}$, where $s, t \in [0, 1]$, $s + t \leq 1$, and the vectors $\vec{v}$ and $\vec{w}$ are not collinear. Then

$$f(T) = f(\vec{u} + s\vec{v} + t\vec{w}) = \mathbf{A}(\vec{u} + s\vec{v} + t\vec{w}) + \vec{b}$$
$$= (\mathbf{A}\vec{u} + \vec{b}) + s(\mathbf{A}\vec{v}) + t(\mathbf{A}\vec{w})$$
$$= \vec{u}_1 + s\vec{v}_1 + t\vec{w}_1,$$

where $s, t \in [0, 1]$, $s + t \leq 1$. By (3), $\vec{v}_1 = \mathbf{A}\vec{v}$ and $\vec{w}_1 = \mathbf{A}\vec{w}$ are not parallel. Thus, T is mapped to a triangle T_1, which completes the proof of the base case.

Now suppose that f maps each n-gon to an n-gon for all n, $3 \leq n \leq k$, and let $\mathcal{P}$ be a polygon with $k + 1$ sides. In the solution to Problem 3.2.30, we saw that every polygon with at least 4 sides has a diagonal contained completely in its interior. Let $\overline{AB}$ be such a diagonal in $\mathcal{P}$. This diagonal divides $\mathcal{P}$ into two polygons, $\mathcal{P}_1$ and $\mathcal{P}_2$, containing t and $k + 3 - t$ sides, respectively, for some t, $3 \leq t \leq k$. By the inductive hypothesis, $f(\mathcal{P}_1)$ and $f(\mathcal{P}_2)$ will be t-sided and $(k + 3 - t)$-sided polygons, respectively. Since each of these polygons will have the segment from $f(A)$ to $f(B)$ as a diagonal, the union of $\mathcal{P}_1$ and $\mathcal{P}_2$ will form a polygon with $k + 1$ sides, which concludes the proof.

(5) The proof that a parallelogram is mapped to a parallelogram is analogous to the proof that triangles get mapped to triangles in (4), by simply dropping the condition that $s + t \leq 1$.

(6) Consider parallel line segments, S_1 and S_2, given in vector form as $S_i : \vec{p_i} + t\vec{u_i}$, $t \in [0, 1]$. Because they are parallel, $\vec{u_2} = k\vec{u_1}$ for some $k \in \mathbb{R}$. As $|\vec{u_i}|$ is the length of S_i, the ratio of lengths of S_2 and S_1 is $|k|$. From parts (1) and (2), S_i is mapped into a segment of length $|\mathbf{A}\vec{u_i}|$. Since $\mathbf{A}\vec{u_2} = \mathbf{A}(k\vec{u_1}) = k(\mathbf{A}\vec{u_1})$, $|\mathbf{A}\vec{u_2}| = |k||\mathbf{A}\vec{u_1}|$, which shows that the ratio of lengths of $f(S_2)$ and $f(S_1)$ is also $|k|$.

(7) We postpone discussion of the proof of this property until the end of this section.

$\square$

Theorems 12.7 and 12.8 (to be proven below) are the vehicles by which we will be able to accomplish the goals promised at the beginning of the chapter – proving a geometric fact in complete generality simply by proving that it is true for a specific case.

Theorem 12.8. *(Fundamental Theorem of Affine Transformations) Given two ordered sets of three non-collinear points each, there exists a unique affine transformation f mapping one set onto the other.*

Proof. We first show that the special (ordered) triple of vectors,

$$\left\{ \vec{0} = \begin{bmatrix} 0 \\ 0 \end{bmatrix}, \vec{i} = \begin{bmatrix} 1 \\ 0 \end{bmatrix}, \vec{j} = \begin{bmatrix} 0 \\ 1 \end{bmatrix} \right\},$$

can be mapped by an appropriate affine transformation to an arbitrary (ordered) triple of vectors,

$$\left\{ \vec{p} = \begin{bmatrix} p_1 \\ p_2 \end{bmatrix}, \vec{q} = \begin{bmatrix} q_1 \\ q_2 \end{bmatrix}, \vec{r} = \begin{bmatrix} r_1 \\ r_2 \end{bmatrix} \right\},$$

which corresponds to three non-collinear points. Let

$$\mathbf{A} = \begin{bmatrix} q_1 - p_1 & r_1 - p_1 \\ q_2 - p_2 & r_2 - p_2 \end{bmatrix} \quad \text{and} \quad \vec{b} = \vec{p} = \begin{bmatrix} p_1 \\ p_2 \end{bmatrix}.$$

One can immediately verify that

$$\mathbf{A}\vec{0} + \vec{b} = \vec{p}, \quad \mathbf{A}\vec{i} + \vec{b} = \vec{q}, \quad \text{and} \quad \mathbf{A}\vec{j} + \vec{b} = \vec{r}.$$

Note that the columns of $\mathbf{A}$ correspond to the vectors $\vec{q} - \vec{p}$ and $\vec{r} - \vec{p}$. Since the points (p_1, p_2), (q_1, q_2), and (r_1, r_2) are non-collinear, the vectors $\vec{q} - \vec{p}$ and $\vec{r} - \vec{p}$ are non-parallel vectors. Hence, by Theorem 12.6, the determinant of $\mathbf{A}$ is nonzero. Thus, by Theorem 12.4, $\mathbf{A}$ is invertible, and $f(\vec{x}) = \mathbf{A}\vec{x} + \vec{b}$ is an affine transformation by definition.

Let $\{\vec{p}, \vec{q}, \vec{r}\}$ and $\{\vec{p'}, \vec{q'}, \vec{r'}\}$ be two ordered triples of position vectors representing two arbitrary triples of non-collinear points. Using the result we have just proven, there exist affine transformations f and g mapping the special triple $\{\vec{0}, \vec{i}, \vec{j}\}$ to $\{\vec{p}, \vec{q}, \vec{r}\}$ and to $\{\vec{p'}, \vec{q'}, \vec{r'}\}$, respectively. Then $g \circ f^{-1}$ is an affine transformation that maps $\{\vec{p}, \vec{q}, \vec{r}\}$ to $\{\vec{p'}, \vec{q'}, \vec{r'}\}$. The uniqueness of this transformation is left to Problem 12.1. $\square$

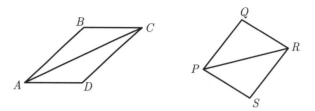

FIGURE **12.5.**

Corollary 12.9.

(1) *Given any two triangles, there exists an affine transformation mapping one to the other.*

(2) *Given any two parallelograms, there exists an affine transformation mapping one to the other.*

Proof.

(1) By Theorem 12.8, the three vertices of one triangle can be mapped to the three vertices of any other triangle. Then use Theorem 12.7.

(2) Consider parallelograms $ABCD$ and $PQRS$, with diagonals $\overline{AC}$ and $\overline{PR}$, as shown in Figure 12.5.

By (1), there is an affine transformation, f, mapping $\triangle ABC$ to $\triangle PQR$, with $f(A) = P$, $f(B) = Q$, and $f(C) = R$. Furthermore, by Theorem 12.7(3), the images of lines AD and CD, namely $\overleftrightarrow{PS}$ and $\overleftrightarrow{RS}$, must be parallel to lines QR and QP, respectively. So, $f(D) = S$. $\square$

Since, by Corollary 12.9, any triangle can be mapped to any other triangle, we say that all triangles are **affine equivalent**; likewise for all parallelograms. We conclude that, in particular, any triangle can be mapped by an affine transformation to an equilateral triangle or to a $45° - 90° - 45°$ triangle, and every parallelogram can be mapped to a square.[10]

We now are prepared to discuss the general idea of a proof of property (7) of Theorem 12.7. First, impose upon the plane a grid of congruent squares. (See Figure 12.6(i).) The first four properties of Theorem 12.7 imply that an affine transformation f will map this grid of squares into a grid of parallelograms, and property (6) implies that these parallelograms are all congruent to each other. (See Figure 12.6(ii).)

Let φ_1 and φ_2 be two figures in the plane, with images $f(\varphi_1)$ and $f(\varphi_2)$, respectively, under the map. If the grid of squares is sufficiently fine, then the ratio of the number of squares in the interior of φ_1 to the number of squares in the interior of φ_2 will differ by arbitrarily little from the ratio Area(φ_1)/Area(φ_2). (Indeed, Area(φ_1)/Area(φ_2) is often defined as the limit of the ratio of the number of squares in φ_1 to the number of squares in φ_2 as the side of the square in the grid decreases indefinitely.[11]) Similarly, the ratio of the number of parallelograms in the interior of $f(\varphi_1)$

---

[10] Affine equivalent figures differ in shape, but not too much. This probably prompted Euler to introduce the term *'affinatas'* to identify transformations of the type $x' = x/m$, $y' = y/n$ in his *Introductio in analysin infinitorum* in 1748. The meanings of the word "affinity" include: a resemblance, or an inherent similarity between things.

[11] A proof of the existence of this limit requires rigorous calculus concepts, which are not assumed for this book.

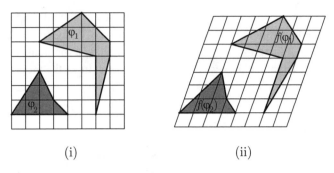

FIGURE **12.6.**

to the number of parallelograms in the interior of $f(\varphi_2)$ will differ by arbitrarily little from the ratio Area $f(\varphi_1)$/Area $f(\varphi_2)$.

An equivalent way of stating property (7) of Theorem 12.7 is this: for every affine transformation f, there exists a positive real number k such that the area of every figure is altered by a factor of k, i.e., Area$(f(\varphi)) = k\cdot$ Area(φ). In order to find k, we may concentrate on the change of area of the unit square defined by vectors $\vec{i}$ and $\vec{j}$. As previously noted, if $\mathbf{A} = \begin{bmatrix} a & b \\ c & d \end{bmatrix}$ is the 2×2 matrix corresponding to an affine transformation f, the first column of $\mathbf{A}$ is $\vec{v} = f(\vec{i})$ and the second column is $\vec{w} = f(\vec{j})$. Under f, the unit square with sides given by $\vec{i}$ and $\vec{j}$ is mapped to a parallelogram with sides defined by $\vec{v} = \langle a, c \rangle$ and $\vec{w} = \langle b, d \rangle$. The area of the parallelogram can be found by subtracting the areas of two pairs of congruent triangles from the area of a rectangle. This is pictured in Figure 12.7 for the case when $a > b > 0$, and $d > c > 0$.

Therefore, the area of the parallelogram is

$$(a + b)(c + d) - 2\left(\frac{1}{2}(a + b)c\right) - 2\left(\frac{1}{2}b(c + d)\right) = ad - bc = \det\mathbf{A}.$$

By similar arguments one can show that essentially the same result holds if we remove the conditions imposed on a, b, c, and d. More precisely, the unit square defined by $\vec{i}$ and $\vec{j}$ is always mapped to a parallelogram having area equal to $|\det(\mathbf{A})|$. From this we conclude that the area of any figure is altered by a factor equalling the absolute value of the determinant of $\mathbf{A}$ under the transformation f.

Restating some parts of Theorem 12.7 in terms of invariants, we can say that certain properties of a figure, such as being a line, a segment, or a triangle, are invariant under affine transformations, as are ratios of lengths of parallel segments and ratios of areas of figures. The list can be continued. For example, the property of a segment being a median in a triangle, the property of a set of lines being concurrent, the property of a point being the centroid of a triangle, and the property of a quadrilateral being a trapezoid are all invariant under affine transformations.

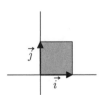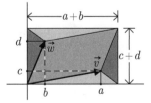

FIGURE **12.7.**

On the other hand, there are many properties that are *not* invariant under affine transformations: the ratio of lengths of non-parallel segments, the property of lines being perpendicular, the property of a triangle being isosceles, the property of a quadrilateral being a rhombus, the property of a ray being the bisector of an angle, the property of a figure being a circle, the property of a point being the center of the in-circle of a triangle, etc.

12.4   Applications

> *A mathematician is a person who can find analogies between theorems; a better mathematician is one who can see analogies between proofs and the best mathematician can notice analogies between theories. One can imagine that the ultimate mathematician is one who can see analogies between analogies.*
> — Stefan Banach (1892–1945)

We begin with a theorem that we have seen before, but with a new proof that illustrates well the ideas of this chapter.

Theorem 12.10.   *The three medians of a triangle are concurrent.*

Proof. Given a triangle ABC, by Corollary 12.9 there is an affine transformation, f, mapping $\triangle ABC$ to an equilateral triangle, $\triangle DEF$. By Theorem 12.7(2), f maps each side of $\triangle ABC$ to a side of $\triangle DEF$; we may assume that $\overline{AB}$ maps to $\overline{DE}$. Let C' be the midpoint of $\overline{AB}$, so that $AC' : C'B = 1 : 1$. By property (6) of Theorem 12.7, $f(C') = F'$ is the midpoint of $\overline{DE}$. Consequently, f maps the medians of $\triangle ABC$ to the medians of $\triangle DEF$.

Proving that the medians of $\triangle DEF$ are concurrent is easier than the general case, due to the many "symmetries" of an equilateral triangle. For example, in an equilateral triangle, the medians are also the perpendicular bisectors and the angle bisectors. These properties can be used to show that the three segments are concurrent, which will prove that the property holds for $\triangle ABC$ as well, and thus for all triangles. We leave the details to the reader. $\square$

Note that we can also conclude that the point of concurrency of the medians (the centroid) divides each median in a ratio 2:1, starting from the vertex of the triangle. Triangles DGF' and FGD', as shown in Figure 12.8, are congruent $30° - 60° - 90°$ triangles. By properties of $30° - 60° - 90°$ triangles, $F'G : GD = 1 : 2$. By equating the lengths of congruent sides of the two triangles, $GD = GF$, so $F'G : GF = 1 : 2$. Because ratios of parallel segments are preserved under affine transformations, this ratio must also hold in an arbitrary triangle.

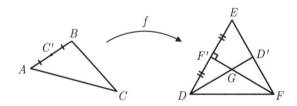

FIGURE 12.8.

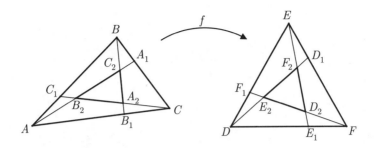

FIGURE **12.9.**

Theorem 12.11. *Let f be an affine transformation and let $\mathcal{P}$ be a polygon. Then f maps the centroid of $\mathcal{P}$ to the centroid of $f(\mathcal{P})$.*

Proof. The discussion prior to the statement of the theorem establishes the result in the case where P is a triangle. The proof for the general case is left to Problem 12.5. $\square$

Our proof of Theorem 12.10 used the method of affine transformations to re-prove a fact we have previously established. We know from earlier chapters that the three angle bisectors of a triangle and the three altitudes of a triangle are also concurrent. However, the method employed above does not work to prove the concurrence of these latter trios; when a triangle is mapped via an affine transformation onto an equilateral triangle, the property of a segment being an angle bisector or an altitude is not necessarily preserved. The mapping of medians to medians is a consequence of the invariance of ratios of parallel line segments, a property that is not relevant to angle bisectors or altitudes.

Example 76. Let A_1, B_1, and C_1 be points on the sides $\overline{BC}$, $\overline{CA}$, and $\overline{AB}$, respectively, of $\triangle ABC$, such that
$$\frac{BA_1}{A_1C} = \frac{CB_1}{B_1A} = \frac{AC_1}{C_1B} = \frac{1}{2}.$$
Let A_2, B_2, and C_2 be the points of intersections of the segments BB_1 and CC_1, CC_1 and AA_1, and AA_1 and BB_1, respectively. (See Figure 76). Prove that
$$\frac{\text{Area}(\triangle A_2B_2C_2)}{\text{Area}(\triangle ABC)} = \frac{1}{7}.$$

Solution: As in the previous example, we use an affine transformation, f, that maps $\triangle ABC$ to an equilateral triangle, $\triangle DEF$. The points $D_1 = f(A_1)$, $E_1 = f(B_1)$, and $F_1 = f(C_1)$ will divide the sides of $\triangle DEF$ in the same 1:2 ratio. Therefore, $\overline{DF_1}$, $\overline{ED_1}$, and $\overline{FE_1}$ will all have the same length. Let us assume that this length is 1.

Let D_2, E_2, and F_2 be the points of intersections of the segments EE_1 and FF_1, FF_1 and DD_1, and DD_1 and EE_1, respectively. Rotating $\triangle DEF$ clockwise by $120°$ around its center, we see that $D_1 \mapsto E_1 \mapsto F_1 \mapsto D_1$. This implies that $\overline{DD_1} \mapsto \overline{EE_1} \mapsto \overline{FF_1} \mapsto \overline{DD_1}$, and therefore $D_2 \mapsto E_2 \mapsto F_2 \mapsto D_2$. This proves that $\triangle D_2E_2F_2$ is equilateral.

Using the Cosine theorem for $\triangle DF_1F$, we get

$$FF_1 = \sqrt{1^2 + 3^2 - 2 \cdot 1 \cdot 3 \cdot \cos(\pi/3)} = \sqrt{7}.$$

Now, $\triangle DE_2F_1 \sim \triangle DED_1$, since they have two pairs of congruent angles. Thus,

$$\frac{E_2F_1}{F_1D} = \frac{ED_1}{D_1D} \implies \frac{E_2F_1}{1} = \frac{1}{\sqrt{7}} \quad \text{and} \quad \frac{DE_2}{DF_1} = \frac{DE}{DD_1} \implies \frac{DE_2}{1} = \frac{3}{\sqrt{7}}.$$

Noting that $FD_2 = DE_2$, we see that $D_2E_2 = \sqrt{7} - 1/\sqrt{7} - 3/\sqrt{7} = 3/\sqrt{7}$. This implies that $D_2E_2/DE = 1/\sqrt{7}$, and therefore

$$\frac{\text{Area}(\triangle D_2E_2F_2)}{\text{Area}(\triangle DEF)} = \left(\frac{D_2E_2}{DE}\right)^2 = \left(\frac{1}{\sqrt{7}}\right)^2 = \frac{1}{7}.$$

Since the ratio of areas is invariant under affine transformations, the result follows. $\square$

The reader may recall that Example 76 was previously presented as Problem 5.22. A comparison of the solutions should reveal that the above solution is more straightforward than the one presented previously. In Problem 12.11, we consider a generalization of this example.

Example 77. Is there a non-regular pentagon with the property that each diagonal is parallel to one of its sides?

Solution: First, we note that it is easy to show that a regular pentagon, $\mathcal{P}_5$, has this property. (In Figure 12.10, $ABCDE$ is such a pentagon.) We leave this task to the reader.

Theorem 12.7(4) establishes that any affine transformation will map a pentagon to a pentagon. We wish to find an affine transformation f such that the image of $\mathcal{P}_5$ under f is not a regular pentagon. There are many such affine transformations. Consider, for example, an affine transformation under which three consecutive vertices of $\mathcal{P}_5$ are mapped to the vertices of an equilateral triangle. Then, the image of $\mathcal{P}_5$ under f is not regular, since one of the angles of $f(\mathcal{P}_5)$ has measure 60°. (In Figure 12.10, the regular pentagon $ABCDE$ is mapped to the pentagon $A'B'C'D'E'$ in which $\angle A'B'C'$ has measure 60°.) However, by Theorem 12.7(3), the property of parallelism among line segments is preserved under an affine transformation, so the image of $\mathcal{P}_5$ will be a non-regular pentagon having the desired property. $\square$

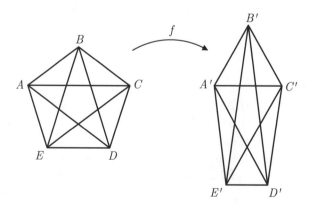

FIGURE 12.10.

12.5 Affine Transformations of Conic Sections

We have established that the image of an n-gon under an affine transformation is an n-gon, and the image of a parallelogram is a parallelogram. We next consider the effect of an affine transformation on the conic sections – ellipses, hyperbolas, and parabolas. Recall from Chapter 9 that any conic section can be represented by a second-degree equation having the general form

$$Ax^2 + Bxy + Cy^2 + Fx + Gy + H = 0,$$

where A, B, C, F, G, and H are real numbers. By Theorem 9.5, the equation represents an ellipse if $B^2 - 4AC < 0$, a parabola if $B^2 - 4AC = 0$, and a hyperbola if $B^2 - 4AC > 0$.

Theorem 12.12. *Let $f(\vec{x}) = \mathbf{A}\vec{x} + \vec{b}$, where $\mathbf{A}$ is an invertible 2×2 matrix, be an affine transformation. Then f maps an ellipse to an ellipse, a parabola to a parabola, and a hyperbola to a hyperbola.*

Proof. Suppose that the equation $Ax^2 + Bxy + Cy^2 + Fx + Gy + H = 0$ represents a non-degenerate conic, $\mathcal{F}$. If (x, y) is any point satisfying the equation, then the vector corresponding to this point, $\vec{x} = \begin{bmatrix} x \\ y \end{bmatrix}$, is mapped to $f(\vec{x}) = \vec{x}' = \begin{bmatrix} x' \\ y' \end{bmatrix} = \mathbf{A}\vec{x} + \vec{b}$. The inverse transformation, f^{-1}, is $\vec{x} \mapsto \mathbf{A}^{-1}\vec{x}' - \mathbf{A}^{-1}\vec{b}$. Therefore,

$$\vec{x} = \begin{bmatrix} x \\ y \end{bmatrix} = \begin{bmatrix} a & b \\ c & d \end{bmatrix} \begin{bmatrix} x' \\ y' \end{bmatrix} + \begin{bmatrix} t \\ u \end{bmatrix},$$

for real numbers a, b, c, d, t, and u. With these values, $x = ax' + by' + t$ and $y = cx' + dy' + u$. Substituting these expressions into the equation $Ax^2 + Bxy + Cy^2 + Fx + Gy + H = 0$ that represents $\mathcal{F}$ results in a second-degree equation in x' and y'. Thus, $\mathcal{F}$ is mapped to another conic, $\mathcal{F}'$.

Note that $\mathcal{F}'$ cannot be a degenerate conic. A degenerate conic can only be a pair of lines, a single line, a point, or the empty set. By Theorem 12.7, if $\mathcal{F}'$ were a degenerate conic, the image of $\mathcal{F}'$ under f^{-1} would again be a pair of lines, a single line, a point, or the empty set. This contradicts our assumption that $\mathcal{F}$ is non-degenerate.

Replacing x and y in the equation $Ax^2 + Bxy + Cy^2 + Fx + Gy + H = 0$ with $x = ax' + by' + t$ and $y = cx' + dy' + u$ yields a second-degree equation corresponding to $\mathcal{F}'$:

$$A(ax' + by' + t)^2 + B(ax' + by' + t)(cx' + dy' + u) + C(cx' + dy' + u)^2$$
$$+ F(ax' + by' + t) + G(cx' + dy' + u) + H = 0.$$

When reduced, the discriminant of this equation is found to be

$$(ad - bc)^2 (B^2 - 4AC),$$

where $B^2 - 4AC$ is the discriminant of the original conic, $\mathcal{F}$. As we've noted, the sign of the discriminant characterizes a non-degenerate conic. Since $(ad - bc)^2 > 0$, the sign of the discriminant is unchanged under affine transformation, and thus, the type of the conic is also unchanged. □

With Theorem 12.12, we have established that an affine transformation will send a conic to a conic of the same type. As with triangles and parallelograms, it turns out that we can actually do better than that: any ellipse can be mapped to <u>any</u> other ellipse under an affine transformation, and likewise for parabolas and hyperbolas.

Suppose that $\mathcal{E}$ is an ellipse with center at (h, k) and with major and minor axes of lengths $2a$ and $2b$. As discussed in the proof of Theorem 9.5, an ellipse is mapped to an ellipse under a translation or rotation. Under translation by $\vec{b} = \begin{bmatrix} -h \\ -k \end{bmatrix}$, the ellipse is mapped to a congruent ellipse with center at the origin. A rotation can be applied to the plane in order to align the major and minor axes of the ellipse with the x- and y-axes, respectively. The original ellipse, $\mathcal{E}$, has now been mapped to another ellipse, $\mathcal{E}'$, via the two specified affine transformations; $\mathcal{E}'$ can be represented by the equation

$$\frac{x^2}{a^2} + \frac{y^2}{b^2} = 1.$$

Now apply a third affine transformation, $f(\vec{x}) = \begin{bmatrix} 1/a & 0 \\ 0 & 1/b \end{bmatrix} \begin{bmatrix} x \\ y \end{bmatrix} = \begin{bmatrix} x/a \\ y/b \end{bmatrix}$.

Under this transformation, $\mathcal{E}'$ is mapped to an ellipse represented by the equation $x^2 + y^2 = 1$; that is, $\mathcal{E}'$ is mapped to the unit circle, $\mathcal{C}(O, 1)$. This proves the following theorem.

Theorem 12.13. *Given any ellipse, $\mathcal{E}$, there exists an affine transformation mapping $\mathcal{E}$ to the unit circle.*

From Theorem 12.13 follows Corollary 12.14, which establishes that all ellipses are affine equivalent.

Corollary 12.14. *Given any two ellipses, $\mathcal{E}_1$ and $\mathcal{E}_2$, there exists an affine transformation mapping $\mathcal{E}_1$ to $\mathcal{E}_2$.*

Proof. Consider ellipses $\mathcal{E}_1$ and $\mathcal{E}_2$. By Theorem 12.13, there are affine transformations f and g mapping $\mathcal{E}_1$ and $\mathcal{E}_2$, respectively, to $\mathcal{C}(O, 1)$. By the definition of inverse mappings, g^{-1} is an affine transformation mapping $\mathcal{C}(O, 1)$ to $\mathcal{E}_2$. By the definition of composition of mappings, $g^{-1} \circ f$ is an affine transformation mapping $\mathcal{E}_1$ to $\mathcal{E}_2$. □

Similar techniques can be applied to show that all hyperbolas are affine equivalent and that all parabolas are affine equivalent. See Problems S12.6 and 12.6.

Example 78. Given an ellipse, $\mathcal{E}$, consider a set of parallel chords of $\mathcal{E}$. Prove that the midpoints of these chords form a diameter of the ellipse and the tangent lines to $\mathcal{E}$ at the endpoints of the diameter are parallel to the chords.

Solution: Let $\mathcal{E}$ be an ellipse with a set of parallel chords, $c_1, c_2, \ldots, c_n$, as shown in Figure 12.11. By Theorem 12.13, there is an affine transformation mapping $\mathcal{E}$ to the unit circle, $\mathcal{C}$. Under this mapping, the chords of $\mathcal{E}$ are mapped to a set of parallel chords of $\mathcal{C}$. Furthermore, the midpoints of the chords of $\mathcal{E}$ are mapped to the midpoints of chords of $\mathcal{C}$.

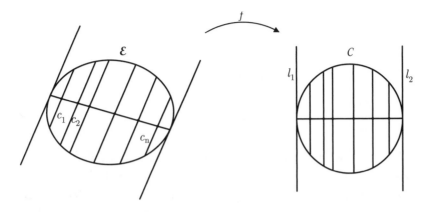

FIGURE **12.11.**

By Theorem 4.6(1), the midpoint of a chord of a circle lies on a diameter perpendicular to the chord. Corollary 4.8 implies that the tangent lines, l_1 and l_2, to C at the endpoints of the diameter are perpendicular to it, and hence parallel to the set of chords of C. This proves the theorem for C. By Theorem 12.7, the properties of a point bisecting a segment, segments being parallel, collinearity, and tangency are all invariant under an affine transformation, so the statement holds for $\mathcal{E}$ as well.   □

We invite the reader to compare this solution to that of Example 46 and Problem 9.12. In the case of the ellipse, which solution do you like more?

12.6 Problems

It's not that I'm so smart, it's just that I stay with problems longer.

— Albert Einstein (1879–1955)

12.1 Prove the uniqueness of the map in Theorem 12.8.

12.2 Given two trapezoids, is there always an affine transformation mapping one to the other?

12.3 Prove that the line joining the point of intersection of the extensions of the nonparallel sides of a trapezoid to the point of intersection of its diagonals bisects each base of the trapezoid.

12.4 Prove that all chords of an ellipse that cut off a region of constant area are tangent to a concentric similar (and similarly oriented) ellipse.

12.5 Complete the details of the proof of Theorem 12.11. (We note that the **centroid** of a polygon is the centroid of the set of its vertices.)

12.6 Given any parabola, $\mathcal{P}$, prove that there exists an affine transformation mapping $\mathcal{P}$ to the parabola given by the equation $y = x^2$.

12.7 Let A_1, B_1, and C_1 be points on the sides $\overline{BC}$, $\overline{CA}$, and $\overline{AB}$, respectively, of $\triangle ABC$, having the property that $BA_1/A_1C = CB_1/B_1A = AC_1/C_1B$. Prove that the centroids of $\triangle ABC$, $\triangle A_1B_1C_1$, and the triangle formed by lines AA_1, BB_1, and CC_1 coincide.

12.8 Let l be a line passing through the vertex M of parallelogram $MNPQ$ and intersecting lines NP, PQ, and NQ in points R, S, and T, respectively. Prove that $1/MR + 1/MS = 1/MT$.

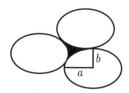

FIGURE **12.12.**

12.9 Prove that an ellipse with semi-axes of lengths a and b has area πab.

12.10 Suppose that an ellipse touches the sides AB, BC, CD, and DA of a parallelogram $ABCD$ at the points P, Q, R, and S, respectively. Prove that the lengths CQ, QB, BP, and CR satisfy $\frac{CQ}{QB} = \frac{CR}{BP}$. (This problem and its solution are from [7].)

12.11 Let A_1, B_1, and C_1 be points on the sides $\overline{BC}$, $\overline{CA}$, and $\overline{AB}$, respectively, of $\triangle ABC$, having the property that $BA_1/A_1C = \alpha$, $CB_1/B_1A = \beta$, and $AC_1/C_1B = \gamma$. Let $\triangle DEF$ be the triangle bounded by $\overline{AA_1}$, $\overline{BB_1}$, and $\overline{CC_1}$. Find $\dfrac{\text{Area}(\triangle DEF)}{\text{Area}(\triangle ABC)}$.

12.12 Consider three ellipses that are congruent, similarly oriented (that is, all major axes are parallel), and which touch externally in pairs. (See Figure 12.12.)

Prove that the area of the curvilinear triangle bounded by them (the shaded area in Figure 12.12) is independent of their position. Then, find the area of the curvilinear triangle if the length of each major axis is a and the length of each minor axis is b.

12.13 Prove that a necessary and sufficient condition for a triangle inscribed in an ellipse to have maximum area is that the centroid of the triangle coincides with the center of the ellipse. Generalize the problem for an inscribed n-gon with $n \geq 3$.

12.7 Supplemental Problems

$S12.1$ Suppose f is an affine transformation of $\mathbb{E}^2$ such that $f((2, 3)) = (3, -1)$, $f((2, 1)) = (1, 2)$ and $f((1, 0)) = (0, 1)$.

(a) Find $f((-2, 5))$.
(b) Let Φ be a figure having area 5 square units. What is the area of $f(\Phi)$?

$S12.2$ Each vertex of a triangle is joined to two points of the opposite side that divide the side into three congruent segments. Consider the hexagon formed by these six segments. Prove that the three diagonals joining opposite vertices of the hexagon are concurrent.

$S12.3$ Let $ABCD$ be a trapezoid with $\overline{BC} \parallel \overline{AD}$. Let the line through B parallel to the side CD intersect the diagonal AC at point P, and the line through C parallel to the side AB intersect the diagonal BD at point Q. Prove that $\overline{PQ}$ is parallel to the bases of the trapezoid.

$S12.4$ Is it always possible to use an affine transformation of a plane to map an altitude of a triangle to a bisector of the image of the triangle (not necessarily at the corresponding vertex)?

$S12.5$ Let $\mathcal{E}$ be an ellipse with center C. If f is any affine transformation, prove that $f(C)$ is the center of the ellipse $f(\mathcal{E})$.

$S12.6$ Given any hyperbola, $\mathcal{H}$, prove that there exists an affine transformation mapping $\mathcal{H}$ to the hyperbola given by $xy = 1$.

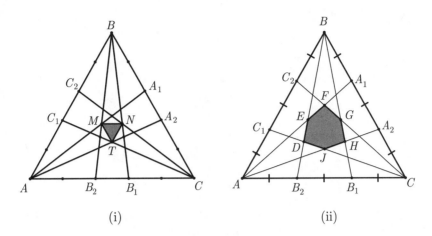

FIGURE 12.13.

*S*12.7 Let A_1, B_1, C_1, and D_1 be points on the sides CD, DA, AB, and BC, respectively, of a parallelogram $ABCD$ such that

$$\frac{CA_1}{CD} = \frac{DB_1}{DA} = \frac{AC_1}{AB} = \frac{BD_1}{BC} = \frac{1}{3}.$$

Show that the area of the quadrilateral formed by lines AA_1, BB_1, CC_1, and DD_1 is one thirteenth of the area of $ABCD$.

*S*12.8 Let n be a positive integer and consider an equilateral triangle ABC with unit side lengths. Let $\overline{A_1A_2}$, $\overline{B_1B_2}$, and $\overline{C_1C_2}$ be segments of length $1/(2n+1)$ lying on and centered at the midpoints of sides BC, AC, and AB, as shown in Figure 12.13(i).

(a) Let M be the intersection of segments BB_2 and AA_1, let N be the intersection of segments BB_1 and CC_2, and let T be the intersection of segments AA_2 and CC_1. Find MN and use it to find the area of $\triangle MNT$.

(b) Let P be the intersection of $\overline{AA_1}$ and $\overline{CC_2}$. Find the area of $\triangle MPN$.

(c) Suppose that each vertex of $\triangle ABC$ is joined to two points of the opposite side that divide the side into three congruent segments. Find the area of the hexagon formed by these six segments. (In Figure 12.13(ii), the hexagon is $DEFGHJ$.)

(d) Part (c) can be generalized.[12] For a positive odd integer m, divide each side of a triangle into m congruent segments and connect the endpoints of the middle segment on each side to the vertex opposite that side. These six segments bound a hexagonal region in the interior of the triangle. Determine the area of this hexagon as a fraction of the area of the original triangle. See Figure 12.14 for an illustration in the case where $m = 5$.

*S*12.9 How many ellipses can pass through four given points with no three of them being collinear? What if instead of the ellipses we consider parabolas?

*S*12.10 Given three non-collinear points in the plane. Find the locus of all points of all parabolas passing through them.

---

*S*12.11 Prove that any five points in a plane such that no three of them are collinear must lie on a unique conic that is either an ellipse, a hyperbola, or a parabola.

---

[12] The result given in part (c) is known as Marion's Theorem. The generalization given in part (d) was found by Ryan Morgan in 1994, when he was a tenth grader.

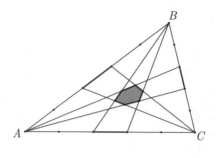

FIGURE 12.14.

$S12.12$ Suppose an affine transformation maps a circle to itself. Prove that the transformation is either a rotation or a symmetry with respect to a line.

$S12.13$ Prove that a necessary and sufficient condition for a triangle circumscribed around an ellipse to have minimum area is that the centroid of the triangle coincides with the center of the ellipse. Can you generalize the problem for an inscribed n-gon with $n \geq 3$?

13

Inversions

Mathematics seems to endow one with something like a new sense.

— Charles Darwin (1809–1892)[1]

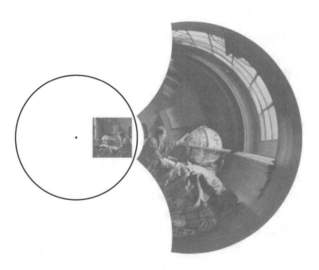

The Astronomer[2]

In this chapter we consider a rather strange transformation of the plane, called inversion with respect to a circle. Its main feature is that it "erases" the distinction between lines and circles, thus demonstrating that certain seemingly unrelated geometric properties of figures are actually equivalent. The role of inversion in modern mathematics is greater than this—for example, it is used in models of hyperbolic geometry and in complex analysis—but we will not discuss it here.

---

[1] As quoted in *Men of Mathematics* (1982) by Eric Temple Bell, p. 16

[2] The original painting of this image was created by Jan Vermeer. The inverted image was done by Daniel Piker. See `spacesymmetrystructure.wordpress.com`.

13.1   Introduction

The history of the inversion transformation is hard to trace. Ideas of inversion go back to Apollonius of Perga, who investigated one particular family of circles and straight lines; see Problem 6.13 and Example 37. Apollonius looked at the locus of points C such that $CA = kCB$, where A and B are fixed points in the plane, and k is a positive constant; varying k results in a family of loci. As we have seen, each curve in this family is a straight line if $k = 1$ and a circle otherwise. Apollonius proved that a circle with center O and radius r belongs to the family if and only if $AC \cdot BC = r^2$ and $B \in \overrightarrow{CA}$; the definition of inversion is based on this property. A systematic study of inversions started only in the 19th century. Jakob Steiner (1796–1863) was among the first to use extensively the technique of inversions in circles. For more on the history of inversion, see D. W. Henderson and D. Taimina [**28**] and references therein.

Let $\mathcal{C}(O, r)$ be a circle in a plane Π. Consider a mapping $I = I(O, r)$,

$$I : \Pi \setminus \{O\} \to \Pi \setminus \{O\},$$

which maps every point $A \neq O$ onto a point $A' = I(A)$, such that $A' \in \overrightarrow{OA}$ and $OA \cdot OA' = r^2$. (See Figure 13.1.) This map is called the **inversion of Π with respect to** $\mathcal{C}$. We call O, r, and $\mathcal{C}$ the **center of inversion**, the **radius of inversion**, and the **circle of inversion**, respectively. As always, the image Φ' of a figure Φ under I, denoted by $I(\Phi)$, is the set of images of all points of Φ.

Suppose points of Π are labeled by complex numbers in the usual way. If O corresponds to z_0, then

$$I = I(z_0, r) : \mathbb{C} \setminus \{z_0\} \to \mathbb{C}, \quad \text{where} \quad z \mapsto z' = \frac{r^2}{\overline{z - z_0}},$$

is the representation of the inversion via complex numbers.

It will be more convenient for us to consider not just the plane Π, but its extension Π_∞. The set of points of Π_∞ is the union of the set of points of Π with one additional element, called the **point at infinity** and denoted by the symbol ∞. The lines of Π_∞ are obtained from the lines of Π by joining the point ∞ to the set of points of each line of Π. Such an extension of a line l of Π may be denoted by l_∞, but often we will not use the subscript. The point at infinity is considered a point of intersection of all lines, so any two lines of Π_∞ intersect either at two points (if the corresponding lines of Π intersect), or at one point ∞ (if the corresponding lines of Π are parallel). In the latter case we will continue to refer to the lines as parallel. We will also assume that the angle between two lines in Π_∞ is the angle between the corresponding lines in Π. We assume that ∞ lies in the exterior of every circle in Π, and we take for granted that I restricted to $\Pi \setminus \{O\}$ is a continuous

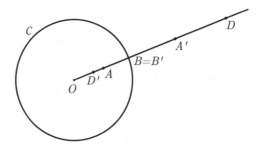

FIGURE 13.1.

map. We call Π_∞ the **inversion plane**. We also join the same symbol ∞ to $\mathbb{C}$, and denote this union by $\overline{\mathbb{C}}$.

We extend the definition of the inversion $I = I(O, r)$ from Π to Π_∞ by defining $I(O) = \infty$ and $I(\infty) = O$. Equivalently, $I = I(z_0, r)$ is extended from $\mathbb{C}$ to $\overline{\mathbb{C}}$ by defining $I(z_0) = \infty$ and $I(\infty) = z_0$. The main advantage of working in Π_∞ instead of Π is that many of the properties of inversion are simpler to state and correspond to our intuitive perception of a "point at infinity."[3]

13.2 Properties

Let us now state and prove some basic properties of inversions.

Theorem 13.1. *Let $I = I(O, r) : X \mapsto X'$ be an inversion of Π_∞ with respect to a circle $\mathcal{C} = \mathcal{C}(O, r)$ in Π. Then the following statements hold.*

(1) *I is a bijection on Π_∞, and $I^2(A) = I(I(A)) = I(A') = A$ for every point A. It maps the interior of $\mathcal{C}$ onto its exterior and vice versa. Points on $\mathcal{C}$ are the only ones of Π_∞ that are fixed by I.*

(2) *For every triangle OAB in Π, $\triangle OAB \sim \triangle OB'A'$.*

(3) *(Change of distance formula) For any pair of points A and B in Π, both distinct from O,*

$$A'B' = \frac{r^2}{OA \cdot OB} \cdot AB.$$

(4) *Let l be a line in Π_∞. If $O \in l$, then $I(l) = l$. If $O \notin l$, then $I(l)$ is a circle in Π passing through O whose diameter through O is on a line perpendicular to l.*

(5) *Let $\mathcal{S}$ be a circle in Π. If $O \in \mathcal{S}$, then $I(\mathcal{S})$ is a line in Π_∞ perpendicular to the line containing the diameter of $\mathcal{S}$ through O. If $O \notin \mathcal{S}$, then $I(\mathcal{S})$ is a circle.*

When we say that inversions "erase" distinctions between circles and lines, we mean that circles and lines can be mapped one to another by inversions, as specified by properties (4) and (5) of Theorem 13.1. In the inversion plane, lines can be intuitively thought of as circles with infinite radii. The notion of a generalized circle is introduced to stress this point: a **generalized circle in Π_∞** is just an ordinary circle in Π or any line in Π_∞. Hence, properties (4) and (5) can be combined in the following short statement:

Every inversion maps a generalized circle to a generalized circle.

Proof. (1) The proof of part (1) of Theorem 13.1 is left as an exercise.

(2) By the definition of inversion, $OA \cdot OA' = r^2$ and $OB \cdot OB' = r^2$. This yields $OA \cdot OA' = OB \cdot OB'$ or $OA/OB = OB'/OA'$. Since $\angle BOA \cong \angle B'OA'$, then $\triangle OAB \sim \triangle OB'A'$ by SAS. (See Figure 13.2.)

Remark. We do not claim that $I(\triangle OAB) = \triangle OA'B'$, which is always false: vertex O is mapped to ∞, and $I(\overline{AB})$ is a circular arc, not the segment $A'B'$.

---

[3] The same motivation leads to the notion of a projective plane when central projections are considered. Those who are familiar with projective planes also realize the difference: in the case of a projective plane, infinitely many "points at infinity" are joined to Π, one for each class of parallel lines, and all these points form the "line at infinity." The construction of the inversion plane is simpler.

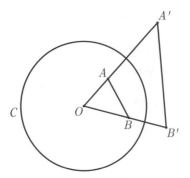

FIGURE 13.2.

(3) As $A' = I(A)$ and $B' = I(B)$, $OA' = r^2/OA$ and $OB' = r^2/OB$. If points O, A, and B are collinear, then so are O, A', and B'. If $AB = |OA \pm OB|$, then $A'B' = |OA' \pm OB'|$. In this case,

$$A'B' = |OA' \pm OB'| = \left| \frac{r^2}{OA} \pm \frac{r^2}{OB} \right| = \frac{r^2}{OA \cdot OB} |OA \pm OB| = \frac{r^2}{OA \cdot OB} AB.$$

Suppose points O, A, and B are not collinear. Using the similarity of $\triangle OAB$ and $\triangle OB'A'$ proved in part (2), we have $AB/A'B' = OB/OA'$. Since $OA' = r^2/OA$, we obtain

$$\frac{AB}{A'B'} = \frac{OB}{r^2/OA}, \qquad \text{which implies}$$

$$A'B' = \frac{r^2}{OA \cdot OB} \cdot AB.$$

Remark. We do not claim that $I(\overline{AB})$ is a segment, though this may happen sometimes.

(4) *Case 1:* Let $O \in l$. The result follows immediately from the definition of inversion. Note that the points O and ∞ of l_∞ are mapped to one another.

Case 2: Let a, b, c, and d be complex numbers corresponding to distinct points of l, and let a', b', c', and d' be their images with respect to I. We may assume that the number 0 corresponds to the center O of inversion and $r = 1$. As none of a, b, c, or d is 0, we have $a' = 1/\overline{a}, b' = 1/\overline{b}, c' = 1/\overline{c}$, and $d' = 1/\overline{d}$. By Corollary 10.5, a', b', c', and d' are cocyclic (lie on a circle or are collinear) if and only if

$$\frac{a' - c'}{a' - d'} \div \frac{b' - c'}{b' - d'}$$

is a real number. Substituting and simplifying we obtain

$$\frac{a' - c'}{a' - d'} \div \frac{b' - c'}{b' - d'} = \frac{1/\overline{a} - 1/\overline{c}}{1/\overline{a} - 1/\overline{d}} \div \frac{1/\overline{b} - 1/\overline{c}}{1/\overline{b} - 1/\overline{d}} = \overline{\left(\frac{c - a}{d - a} \right)} \div \overline{\left(\frac{c - b}{d - b} \right)}. \qquad (13.1)$$

As the fours points a, b, c, and d are collinear, the ratios $(c - a)/(d - a)$ and $(c - b)/(d - b)$ are both real and nonzero. Hence, the ratio of their conjugates is real, and the points corresponding to a', b', c', and d' are cocyclic.

As a circle is uniquely determined by any three of its points, fixing a, b, and c while letting d move along l gives (by continuity) that $I(l)$ is a circle passing through O. Note that $I(\infty) = O$.

Consider the line m through O which is perpendicular to l. It is obvious that $I(l)$ should be symmetric with respect to m. Hence, the center of the circle $I(l)$ is on m. Moreover, the following holds.

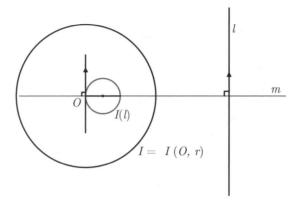

FIGURE 13.3.

Corollary 13.2. *Let $I = I(O, r)$, l be a line in Π_∞, $O \notin l$, and m be the line in Π_∞ containing O and perpendicular to l. Then the center of the circle $I(l)$ is on m, and the tangent to $I(l)$ at O is parallel to l. (See Figure 13.3.)*

---

(5) *Case 1:* Let $O \in S$. The result follows by reversing the argument of Case 2 in (4).

Case 2: Let $O \notin S$. Let a, b, c, and d be the numbers corresponding to four distinct points of S. We begin by repeating the argument of Case 2 in (4), but change the appearance of the last ratio in (13.1).

$$\frac{a' - c'}{a' - d'} \div \frac{b' - c'}{b' - d'} = \frac{1/\bar{a} - 1/\bar{c}}{1/\bar{a} - 1/\bar{d}} \div \frac{1/\bar{b} - 1/\bar{c}}{1/\bar{b} - 1/\bar{d}} = \overline{\left(\frac{a - c}{a - d}\right) \div \left(\frac{b - c}{b - d}\right)}. \qquad (13.2)$$

Now we use Corollary 10.5. As points corresponding to a, b, c, and d are cocyclic, the last ratio in (13.2) is real. Hence the first ratio is real, so the points corresponding to a', b', c', and d' are cocyclic. As a circle is uniquely determined by any three of its points, fixing a, b, and c, letting d move along S, and using continuity of I, we obtain that $I(S)$ is a circle. $\qquad \square$

Parts (4) and (5) of Theorem 13.1 can be proven without using complex numbers by using "conventional" methods of Euclidean geometry; see Problems 13.8, 13.9, S13.7, and S13.8.

The last theorem of the section regards another fundamental property of inversions, namely, that they preserve the measures of angles between generalized circles. We define the **tangent to a line** at any of its points as the line itself. For two intersecting generalized circles, the measure of the angle between them is the measure of the smaller (undirected) angle between their tangents at the intersection point.[4] If two lines share point ∞ only, the angle between them has measure zero. Two generalized circles are called **orthogonal** if the angle between them is a right angle.

---

[4] Recall that when we talk about the angle between two lines we do not specify the direction of the angle. Likewise, our angles, being generalized circles, have not been directed. If they were, it could be shown that inversions reverse the direction. Maps that preserve measures of angles between curves are called **conformal**, and inversions are such maps. Taking for granted that inversion preserves measures of angles between curves (see, e.g., [7]), in this text we restrict ourselves to generalized circles only.

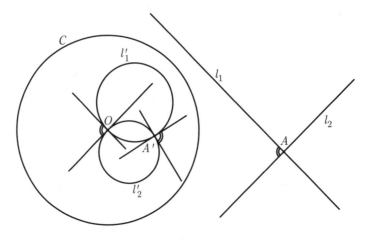

FIGURE 13.4.

Theorem 13.3. *The measure of the angle between any two intersecting generalized circles is invariant under inversion.*

Proof. Our proof follows closely the one in [7]. Let $I = I(\mathcal{C})$, where $\mathcal{C} = \mathcal{C}(O, r)$, $O \neq \infty$. In this proof we consider only the case when the two generalized circles are lines l_1 and l_2. Other cases are suggested as problems.

Case 1: Suppose ∞ is the only common point of these lines. Then their images are circles through O with tangent lines at O, parallel (in Π) to both l_1 and l_2 by Corollary 13.2. Thus, the measures of the angles between l_1 and l_2 and between $I(l_1)$ and $I(l_2)$ are zero.

Case 2: Suppose l_1 and l_2 share a point A, where $A \neq \infty$. If $O = A$, the lines coincide with their images under I, and the statement is obvious. If $O \neq A$, but O is on one of these lines, the statement follows from Corollary 13.2. So we assume that O is not on either of the lines. (See Figure 13.4.) Then the images of l_1 and l_2 are circles intersecting at points O and $I(A) = A'$. By Corollary 13.2, the angle between the circles at O is congruent to the one between the lines. Due to symmetry of these two circles with respect to the line passing through their centers, this angle is also congruent to the angle between the circles at A'. $\qquad\square$

Corollary 13.4. *If two generalized circles are tangent, then their images under inversion are tangent. If two generalized circles are orthogonal, their images under inversion are orthogonal.*

In the next section, we will see several examples of using inversions to solve geometry problems. Recall that when employing the coordinate method in Chapter 8, we invariably had several choices of how to choose a coordinate system to solve a problem, and each system had its relative merits. Similarly, when using the method of inversions to solve a geometry problem, one often has several choices for the circle of inversion, and it is not always immediately obvious which one will work best. Typically, one chooses the center of the inversion to be a point of tangency of two or more generalized circles.

We conclude this section with a problem for which an inversion solution seems appropriate. We will illustrate three different choices for the circle of inversion, each of which could be used to solve the problem.

Example 79. Let $A - K - B$ be points of line l. Let $\overline{AB}$ and $\overline{AK}$ be diameters of semicircles C_1 and C_2, respectively, tangent at A. (See Figure 13.5.) Let C_3 be a circle tangent to C_1, to C_2 at a point D, and to a line C_4 perpendicular to l at K.[5] Let m be a common tangent to C_2 and C_3 at D. Prove that m passes through B.

Solution: We consider three possible centers of inversion: A, D, and K. Notice that each of these points is a point of tangency of at least two of the generalized circles, C_1, C_2, C_3, and C_4. In Figures 13.6, 13.7, and 13.8 we present the image of the given figure under the inversions centered at A, D, and K, respectively. We invite the reader to check that the various lines and circles are mapped appropriately to lines or circles and to decide under which inversion it is easiest to prove that m' passes through B'. (See Problem S13.10.) □

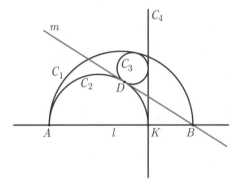

FIGURE 13.5. Original configuration for Example 79.

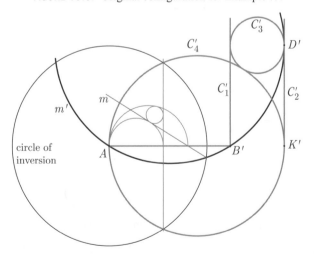

FIGURE 13.6. Inversion with center A.

---

[5] We use C_4 to denote the line perpendicular to l at K because it is a generalized circle.

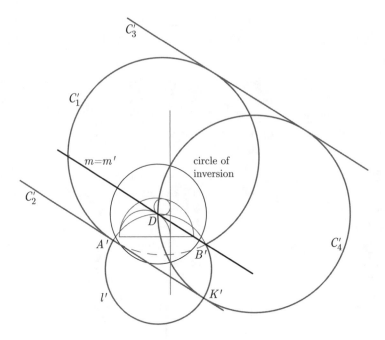

FIGURE 13.7. Inversion with center D.

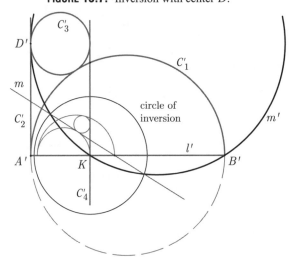

FIGURE 13.8. Inversion with center K.

13.3 Applications

We are ready to consider the first striking application of inversion. We begin with a theorem of Ptolemy, which has already appeared several times in this book.

Example 80. (Generalized Theorem of Ptolemy) Prove that for any four points A, B, C, and D in Π,

$$AB \cdot CD + BC \cdot DA \geq AC \cdot BD,$$

with equality if and only if the points are cocyclic in the order $A - B - C - D$.

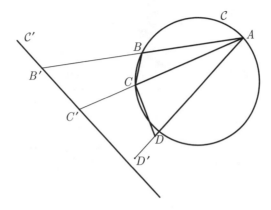

FIGURE **13.9.**

Proof. Consider an inversion $I(A, r)$ and let $\mathcal{C}$ be the circle through points A, B, and C. The image of circle $\mathcal{C}$ is a line $\mathcal{C}'$ containing the points B' and C', with $A' = \infty$. Figure 13.9 corresponds to the case where D is in the exterior of $\mathcal{C}$, but both $\mathcal{C}$ and D are in the interior of the circle of inversion. The argument we present will be independent of these assumptions.

Using the change of distance formula (Theorem 13.1 (3)), we obtain:

$$B'C' = \frac{r^2}{AB \cdot AC} \cdot BC,$$

$$C'D' = \frac{r^2}{AC \cdot AD} \cdot CD,$$

$$D'B' = \frac{r^2}{AD \cdot AB} \cdot DB.$$

By the triangle inequality,

$$D'C' + C'B' \geq D'B',$$

which is equivalent to

$$\frac{r^2}{AC \cdot AD} \cdot CD + \frac{r^2}{AB \cdot AC} \cdot BC \geq \frac{r^2}{AD \cdot AB} \cdot DB.$$

This yields

$$AB \cdot CD + BC \cdot AD \geq AC \cdot DB,$$

which proves the first statement.

Now consider points D and D'. Clearly,

$$B'C' + C'D' = D'B'$$

if and only if B', C', and D' are all on the line $\mathcal{C}'$. But B', C', and D' are collinear if and only if B, C, and D lie on a circle passing through the center of inversion A. Consequently,

$$AB \cdot CD + BC \cdot AD = AC \cdot DB$$

if and only if A, B, C, and D all lie on circle $\mathcal{C}$. $\qquad\square$

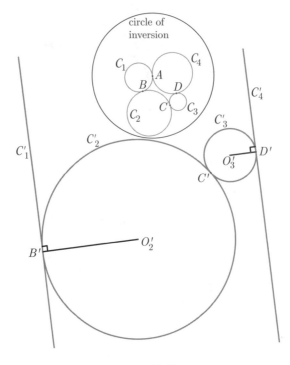

FIGURE 13.10.

We stress that applying inversion in the example above allowed us to reduce the very non-obvious relation between six distances defined by four points to the basic triangle inequality for three points, and, moreover, to investigate when the equality occurs. Don't you agree this is striking?

The invariance of tangency of generalized circles under inversions, as established in Corollary 13.4, is a very powerful property in solving problems. You will see this property utilized in the examples that follow.

Example 81. Consider four circles such that each is tangent to two others externally. Prove that the four points of tangency are cocyclic.

Solution: Consider four tangent circles, C_1, C_2, C_3, and C_4, and an inversion $I(A, r)$ where A is the point of tangency of C_1 and C_4. Figure 13.10 provides the actual correspondence between the original figure (shown in light) and its image (shown in dark) under inversion with respect to a circle centered at A and containing the four circles in its interior.

In practice, it really isn't necessary for scale and orientation to be preserved, and the presence of the circle of inversion in the figure isn't important. What is important is that the images of generalized circles are correctly drawn as either lines or circles, and that tangencies and relative positions are preserved. In solving problems, then, we may often simply draw a figure like the one in Figure 13.11, which shows the original figure on the left and representation of the image on the right.

Since both C_1 and C_4 pass through A, their images C_1' and C_4' are parallel lines. Since C_2 and C_3 do not pass through A, their images are circles. By Corollary 13.4, tangent circles are mapped to tangent generalized circles, as shown in the figures. To show that A, B, C, and D all

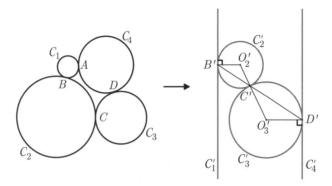

FIGURE 13.11.

lie on a circle, it is sufficient to show that B', C', and D' are collinear. Let O_2' and O_3' be the centers of C_2' and C_3', respectively. Then O_2', O_3', and C' are collinear.

As $\overline{O_2'B'} \perp C_1'$ and $\overline{O_3'D'} \perp C_4'$, we have $\overline{O_2'B'} \parallel \overline{O_3'D'}$. Then $\angle B'O_2'C' \cong \angle D'O_3'C'$ as alternate interior angles. Then, since $\triangle O_2'B'C'$ and $\triangle O_3'D'C'$ are isosceles and $\angle O_2' \cong \angle O_3'$, it follows that $\angle O_2'C'B' \cong \angle O_3'C'D'$. Hence, since O_2', C', and O_3' are collinear, $\angle O_2'C'B'$ and $\angle O_3'C'D'$ are vertical angles, and points B', C', and D' are collinear.

This implies that B, C, and D all lie on a circle that passes through the center of inversion A. $\qquad \square$

Here is another example which reduces the problem of proving that points are cocyclic to the problem of proving that their images are collinear.

Example 82. Consider a chain of tangent circles inscribed in a circular segment with a chord AB. (See Figure 13.12). Prove that all points of tangency of the circles from the chain are cocyclic.

Solution: Let C be the circle containing the circular segment, and let $I(A, r)$ be an inversion. Since line AB passes through the center of inversion, its image is the line through A and B'. Since C is a circle that passes through A, C' is a line that does not pass through A, but which intersects line AB' at B'.

Let C_1, C_2, C_3, C_4, ... be the circles from the chain, with A_1, A_2, A_3, A_4, ... being their respective points of tangency. Since none of the C_i pass through A, their images C_i' are all circles. By Corollary 13.4, the C_i' form a chain of tangent circles inscribed in an angle with the vertex at B' and sides on lines C' and AB'. Hence, A_1', A_2', A_3', A_4', ... are the points of tangency of the inverted circles. As the center of each C_i' lies on the bisector of the angle, all A_i' also lie on the angle bisector. Since the bisector does not pass through A, each preimage A_i of A_i' must lie on a circle that passes through the center of inversion, A. $\qquad \square$

Even though the centers of the inverted circles also lie on the angle bisector, we cannot draw any immediate conclusions about the centers of the original circles because the centers of the original circles are not the preimages of the centers of the inverted circles. Interestingly, though, the centers

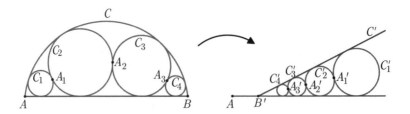

FIGURE 13.12.

of all circles of such a chain C_1, C_2, C_3, ..., lie on a parabola. This was previously shown in Example 14 in Chapter 6.

Often when a transformation is studied, the question of which figures are fixed by that transformation is of particular interest. The following theorem describes circles that are fixed by inversion.

Theorem 13.5. *Let C be a circle and $I(O, r)$ an inversion such that O is not on C. Let $\overrightarrow{OA}$ and $\overrightarrow{OB}$ be tangents to C at A and B, respectively. If $r = OA = OB$, then $I(C) = C$.*

Proof. Since $r = OA = OB$, then $I(A) = A' = A$ and $I(B) = B' = B$. We know that $I(C)$ must be a circle that passes through A' and B'. (See Figure 13.13.) In fact, since rays OA and OB are mapped to themselves by I, and C is tangent to them, C' is tangent to both of them as well. We have, then, two circles, C and C', each tangent to the same two rays OA and OB at the same two points. Therefore $C = C'$.   □

Remark. Though C is invariant under I, its center is not.

The solution of the following famous problem will make use of Theorem 13.5.

Example 83. (Pappus' Ancient Problem) Consider three semicircles S_1, S_2, and S_3, with segments AB, AF, and FB as diameters. (See Figure 13.14(i).) Let C_1 be a circle that is tangent to all three semicircles, and consider circles $C_i = C_i(O_i, r_i)$, $i \geq 2$, in order, such that C_{i+1} is tangent to S_1, S_2, and C_i. Let h_i be the distance from O_i to the diameter AB. Prove that $h_i = 2i \cdot r_i$, $i \geq 1$.

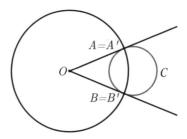

FIGURE 13.13.

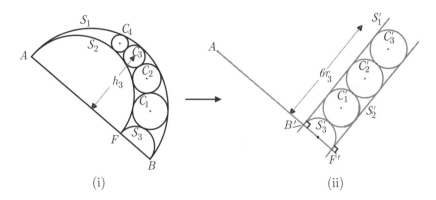

FIGURE 13.14.

Solution: For fixed i, consider the inversion $I(A, t_i)$, where t_i is the length of the tangent from A to the circle C_i. (The image under the inversion for $i = 1$ is shown in Figure 13.14(ii).) By Theorem 13.5, $I(C_i) = C_i$.

Since S_1 and S_2 are tangent at A, their images S_1' and S_2' must be parallel rays; also $I(\overrightarrow{AB}) = \overrightarrow{AB}$. S_3' must intersect $I(\overrightarrow{AB})$ at F' and B', and be tangent to lines S_1' and S_2'. Similarly, each circle C_i', $i \geq 2$, must be tangent to S_1', S_2', C_{i+1}', and C_{i-1}'.

Since all circles C_j' are tangent to the same pair of parallel lines, they must all have the same radius length, say r_i. As C_i is invariant under I, its radius also has length r_i. Since $\overleftrightarrow{AB} = \overleftrightarrow{B'F'}$ is invariant under I, the distance from the center of C_i' to $\overleftrightarrow{B'F'}$ is equal to the one from the center of C_i to $\overleftrightarrow{AB}$, which is h_i. As all C_j' have radius length r_i, we obtain $h_i = 2i \cdot r_i$. $\square$

The reader may recall Example 12 from Chapter 6, with its accompanying Figure 6.9. From that example, the centers of $C_1, C_2, \ldots$ will all lie on an ellipse whose foci are the centers of S_1 and S_2.

The last fundamental property of inversion we present is stated in the following theorem. An application will be demonstrated in Example 84.

Theorem 13.6. *Given two non-intersecting circles, there exists an inversion that maps them to concentric circles.*

Proof. Let C_1 and C_2 be the circles. We consider the case where each of them lies in the exterior of the other. Other cases can be dealt with similarly.

We first present an argument that makes the claim intuitively clear. Then we essentially repeat the argument more formally.

Consider the line that passes through the centers of the two circles, O_1 and O_2, intersecting C_1 at A and B and C_2 at C and D, as shown in Figure 13.15. Consider an inversion $I(O, r)$, where O lies on the line through the centers somewhere inside of circle C_1 and r is small enough to keep the circle of inversion, C, entirely inside C_1.

Suppose O is in the interior of $\overline{AO_1}$, close to A, and $r < \min\{OA, OO_1\}$. Applying the inversion $I(O, r)$, we get $I(C_2)$ in the interior of $I(C_1)$. In addition, the center P_1' of circle $I(C_1)$ will be in the interior of segment OA' while P_2', the center of circle $I(C_2)$, is in the interior of segment OB'.

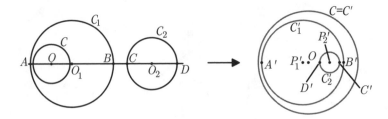

FIGURE 13.15.

Move O continuously along $\overline{AB}$ and into $\overline{O_1B}$, "close" to B, and consider a circle centered at O and of radius less than $\min\{OO_1, OB\}$. We now find that although P_2' is still in the interior of $\overline{OB'}$, P_1' is no longer in $\overline{OA'}$, but is, rather, in the interior of $\overline{P_2'B'}$. By continuity, there must exist a position of O along the diameter AB where P_1' and P_2' coincide. This completes the intuitive "proof."

Here is the promised formal argument. For the inversion $I = I(O, r)$, we have:

$$OA' \cdot OA = r^2 \text{ and } OB' \cdot OB = r^2.$$

Mark the coordinates of A, O, B, C, and D on a number line, as indicated in Figure 13.16.

Then $OA' \cdot (x - a) = r^2$ and $OB' \cdot (b - x) = r^2$. Consequently,

$$OA' = \frac{r^2}{x - a}, \qquad OB' = \frac{r^2}{b - x},$$

$$\text{and} \quad A' : \left(x - \frac{r^2}{x - a}\right), \qquad B' : \left(x + \frac{r^2}{b - x}\right).$$

Similarly, we find that:

$$C' : \left(x + \frac{r^2}{c - x}\right) \text{ and } D' : \left(x + \frac{r^2}{d - x}\right).$$

The center P_1' of circle C_1' is at the midpoint of $\overline{A'B'}$. Hence,

$$P_1' : \left(\frac{1}{2}\left(x - \frac{r^2}{x - a} + x + \frac{r^2}{b - x}\right)\right).$$

Similarly, the center P_2' of C_2' is:

$$P_2' : \left(\frac{1}{2}\left(x + \frac{r^2}{c - x} + x + \frac{r^2}{d - x}\right)\right).$$

When these two values are equal, the circles will be concentric. Hence, the problem is reduced to solving

$$\frac{r^2}{b - x} - \frac{r^2}{x - a} = \frac{r^2}{c - x} + \frac{r^2}{d - x},$$

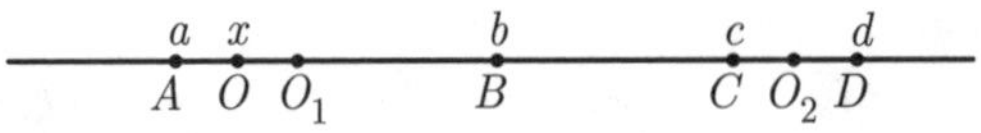

$$\begin{array}{ccccc} a\ x & & b & & c\quad d \\ \bullet\bullet & & \bullet & & \bullet\ \bullet \\ \hline A\ O\ O_1 & & B & & C\ O_2\ D \end{array}$$

FIGURE 13.16.

or equivalently, to showing that the following equation with respect to x has a solution in (a, b).

$$\frac{1}{b-x} - \frac{1}{x-a} = \frac{1}{c-x} + \frac{1}{d-x}.$$

As x changes from a to b, the value of the left side of the last equation changes from $-\infty$ to $+\infty$. The right side of the equation is always positive and finite. Hence, by the Intermediate Value theorem, there must exist a value of x for which the two sides are equal, and it is for this value of x that the two circles C_1' and C_2' will be concentric. □

Here is a great application of Theorem 13.6.

> **Example 84.** (Steiner's Porism) Consider circles C and C_0, with C_0 in the interior of C. Consider a chain of circles C_i, $i \geq 1$, between C_0 and C, such that C_i is tangent to C, to C_0, to C_{i-1}, and to C_{i+1}. We say that the chain *closes* if there exist i and j, $j > i + 1 > 1$ such that C_j is tangent to C_i. Suppose a chain of circles closes for the first time with C_{i+n} tangent to C_i. Prove that any other chain between C_0 and C also closes for the first time with the $(i + n)$-th circle of the chain tangent to the i-th circle of the chain.

Proof. Select an inversion $I(O, r)$ such that $C' = I(C)$ and $C_0' = I(C_0)$ are concentric circles. Such an inversion exists by Theorem 13.6. Then all $C_i' = I(C_i)$, $i \geq 1$, are congruent circles. (See Figure 13.17.)

Suppose the chain C_i', $i \geq 1$, closes for the first time with C_{i+n}' tangent to C_i'. It is obvious that any other chain between C_0' and C' will also close for the first time with the $(i + n)$-th circle tangent to the i-th circle, as that chain is obtained from the chain C_i' by a rotation around the common center of C_0' and C'. This implies that the same happens with any chain between C_0 and C. □

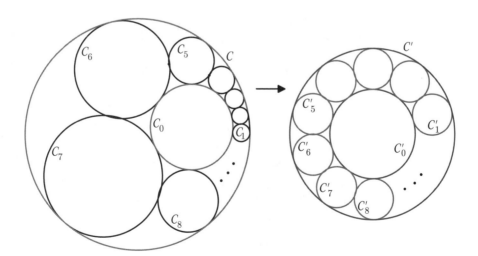

FIGURE 13.17.

13.4 Problems

Let us grant that the pursuit of mathematics is a divine madness of the human spirit.
— Alfred North Whitehead (1861–1947)

13.1 Consider a circle $\mathcal{C}$ with center O and radius r. Let $\overline{AB}$ be a chord of $\mathcal{C}$ with midpoint M. Let C be a point such that $\overline{CA}$ and $\overline{CB}$ are both tangent to $\mathcal{C}$. Prove that $I(M) = C$ under the inversion $I(O, r)$.

13.2 Suppose a circle $\mathcal{C}'$ is the image of a circle $\mathcal{C}$ with respect to an inversion $I = I(O, r)$. Let M be the center of $\mathcal{C}'$, $M \neq O$, and $M' = I(M)$. Can M' be the center of $\mathcal{C}'$?

13.3 Consider a Cartesian plane, and let $I = I((0, 0), 1)$ denote the inversion of the plane with respect to the unit circle centered at the origin. Suppose that for any point $P : (x, y)$ different from the origin, $I(P) = P'$, where $P' : (x', y')$. Express x' and y' in terms of x and y.

13.4 Prove Theorem 13.3 for the case where the generalized circles are a circle and a line that have at least one point of intersection.

13.5 Given triangle ABC, let points P, Q, and R be on segments $\overline{AB}$, $\overline{BC}$, and $\overline{AC}$, respectively. Construct circles $\mathcal{C}_1$ passing through points A, P, and R, $\mathcal{C}_2$ passing through P, B, and Q, and $\mathcal{C}_3$ passing through Q, C, and R. Prove that $\mathcal{C}_1$, $\mathcal{C}_2$, and $\mathcal{C}_3$ are concurrent.

13.6 Let $\mathcal{C}_1$, $\mathcal{C}_2$, $\mathcal{C}_3$, and $\mathcal{C}_4$ be four intersecting circles, with points of intersection A_1, A_2, A_3, A_4, B_1, B_2, B_3, and B_4, as shown in Figure 13.18. Prove that if A_1, A_2, A_3, and A_4 are cocyclic, then B_1, B_2, B_3, and B_4 are also cocyclic.

13.7 Consider an inversion with center O and radius k. Let $\mathcal{C}$ be a circle with center P that does not pass through O. Let r be the radius of $\mathcal{C}$ and r' the radius of $\mathcal{C}'$. Prove that $r' = \dfrac{rk^2}{|OP^2 - r^2|}$.

13.8 Prove part (4) of Theorem 13.1 without using complex numbers or the coordinate method.

13.9 Prove part (5) of Theorem 13.1 by using the coordinate method.

13.10 Consider triangle ABC where $AC \geq AB$ and $AC \geq BC$. Let M be any point in the plane. Prove $AM + CM \geq BM$, with equality if and only if $\triangle ABC$ is equilateral and M is a point on its circumcircle.

 The following statement is a partial generalization of the given problem. Let A_1, A_2, $\ldots$, A_n be consecutive vertices of a regular n-gon, where n is odd. Let M

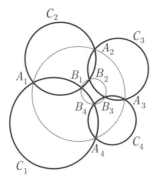

FIGURE 13.18.

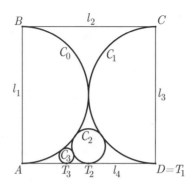

FIGURE **13.19.**

be a point on the circumcircle of the n-gon between A_n and A_1. Let $d_i = MA_i$. Then
$d_1 + d_3 + d_5 + \cdots + d_n = d_2 + d_4 + \cdots + d_{n-1}$.

13.11 (Ford-Rademacher Contour) Given a unit square $ABCD$, construct two tangent semicircles $\mathcal{C}_0$
and $\mathcal{C}_1$, one with $\overline{AB}$ as a diameter and the other with $\overline{CD}$ as a diameter. (See Figure 13.19.)
Inscribe a circle $\mathcal{C}_2$ touching $\mathcal{C}_0$, $\mathcal{C}_1$, and $\overline{AD}$. Inscribe the remaining circles $\mathcal{C}_i$, $i \geq 3$,
touching $\mathcal{C}_0$, $\mathcal{C}_{i-1}$, and $\overline{AD}$. Let T_i be the point of tangency of circle $\mathcal{C}_i$ with $\overline{AD}$.

(a) Prove that $AT_i = 1/i$.

(b) Use Problem 13.7 to determine the radius of $\mathcal{C}_i$.

13.12 Given an arc $\overset{\frown}{AB}$ of a circle and a connecting chord $\overline{AB}$, inscribe a chain of circles between
the chord and the arc. (See Figure 13.20.) Prove that the mutual tangents to adjacent inscribed
circles are concurrent.

13.13 Let A, B, C, and D be points on circle $\mathcal{C}$. Let tangents to $\mathcal{C}$ at A and C meet at a point that
lies on line BD. Then tangents to $\mathcal{C}$ at B and D meet at a point that lies on line AC.

13.14 Given triangle ABC, let $\mathcal{C}_1$ be the incircle with center O_1 and radius r. Let $\mathcal{C}_2$ be the
circumcircle with center O_2 and radius R. If $d = O_1 O_2$, prove that $d^2 = R^2 - 2Rr$.

13.15 Consider a pentagon $ABCDE$ such that the sum of any two adjacent interior angles is greater
than $180°$ (so that a star can be formed, as described next). Extend all five sides to form a
5-pointed star, with H, K, L, M, and N being the points of the star. Circumscribe circles

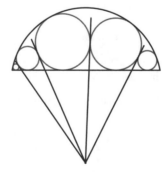

FIGURE **13.20.**

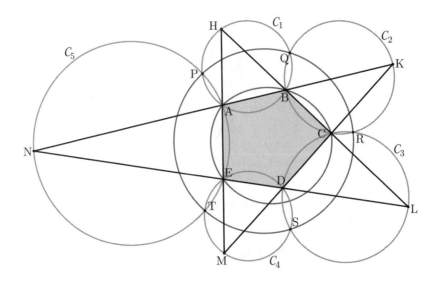

FIGURE 13.21.

C_1 on $\triangle HAB$, C_2 on $\triangle KBC$, C_3 on $\triangle LCD$, C_4 on $\triangle MED$, and C_5 on $\triangle NAE$. Label the points of intersections of the circles as P, Q, R, S, and T, where P is the intersection of C_1 and C_5, Q is the intersection of C_1 and C_2, and so on. Prove that if A, B, C, D, and E are cocyclic, then P, Q, R, S, and T are cocyclic. (See Figure 13.21. The pentagon is shaded for emphasis.)

13.5 Supplemental Problems

$S13.1$ Consider three circles, each externally tangent to the other two. Prove there are two circles that are tangent to all three of them.

$S13.2$ Prove part (1) of Theorem 13.1.

$S13.3$ State and prove necessary and sufficient conditions for an inversion $I = I(O, r)$ and a segment AB, $A \neq B$, for which

(a) $I(\overline{AB}) = \overline{AB}$;

(b) $A'B' = AB$.

$S13.4$ Let A, B, C, and D be four distinct points.

(a) Prove that $\dfrac{AC/BC}{AD/BD}$ is invariant under inversion.

(b) Is there an inversion that maps the vertices of a square, $ABCD$, to the vertices of a rhombus $A'B'C'D'$, where $m\angle A'B'C' = 60°$?

$S13.5$ Let $A_1 A_2 A_3 \ldots A_n$ be a regular n-gon. Let point M be anywhere on arc $\overset{\frown}{A_1 A_n}$ and let $d_i = MA_i$. Prove that

$$\frac{1}{d_1 d_2} + \frac{1}{d_2 d_3} + \cdots + \frac{1}{d_{n-1} d_n} = \frac{1}{d_1 d_n}.$$

$S13.6$ Prove Theorem 13.3 for the case where the generalized circles are two intersecting circles.

*S*13.7 Prove part (5) of Theorem 13.1 without using complex numbers or the coordinate method.

*S*13.8 Prove part (4) of Theorem 13.1 by using the coordinate method.

*S*13.9 Prove Theorem 13.6 for the case where one circle is in the interior of the other.

*S*13.10 Let $A - K - B$ be points of line l. Let $\overline{AB}$ and $\overline{AK}$ be diameters of semicircles $\mathcal{C}_1$ and $\mathcal{C}_2$, respectively, tangent at A. (See Figure 13.5.) Let $\mathcal{C}_3$ be a circle tangent to $\mathcal{C}_1$, to $\mathcal{C}_2$ at a point D, and to a line $\mathcal{C}_4$ perpendicular to l at K.[6] Let m be a common tangent to $\mathcal{C}_2$ and $\mathcal{C}_3$ at D. Prove that m passes through B.

*S*13.11 Prove that all circles passing through point A and orthogonal to a circle S, $A \notin S$, have another common point.

---

*S*13.12 Prove that if angles A and C in quadrilateral $ABCD$ are complementary, then

$$(AB)^2(CD)^2 + (AD)^2(BC)^2 = (AC)^2(BD)^2.$$

*S*13.13 Consider quadrilateral $ABCD$ with inscribed circle $\mathcal{C}_1$ and circumcribed circle $\mathcal{C}_2$. Let J, K, L, and M be points where $\mathcal{C}_1$ is tangent to quadrilateral $ABCD$. Prove that $\overline{JK} \perp \overline{KM}$.

*S*13.14 Consider a circular segment with chord AB and a circle $\mathcal{C}$ inscribed in it. Let P and Q be points of tangency of $\mathcal{C}$ with $\overline{AB}$ and $\widehat{AB}$, respectively. Prove there is a point W in the plane that will lie on $\overleftrightarrow{PQ}$, no matter how $\mathcal{C}$ is chosen.

---

[6] We use $\mathcal{C}_4$ to denote the line perpendicular to l at K because it is a generalized circle.

14

Coordinate Method with Software

In Chapter 8 we introduced the coordinate method and showed how analytic geometry could be used to solve many geometry problems. At times the algebra was a bit difficult, but mostly we considered problems that could be solved by hand. There are problems, however, for which a coordinate method solution seems very appropriate, but for which the necessary algebraic manipulations are too tedious to attempt by hand. In such cases, we may wish to use a computer algebra package to aid in the calculations. Below we list a variety of problems, most of which also appeared elsewhere in the book. In each case, the coordinate method will provide an alternative solution to the one given previously.

There is a Maple worksheet that accompanies this text, and it contains detailed instructions on how to use the program Maple. The reader who owns Maple will be able to use the worksheet interactively to solve geometry problems we pose. At the end of the worksheet we provide complete Maple solutions for all problems that appear below.

Problems

14.1 Prove that if in a triangle ABC, $m_c = c/2$, then $\angle C$ is a right angle.

14.2 Let $ABCD$ be a parallelogram, and let $F \in \overline{AD}$ be such that $AF = \frac{1}{5}AD$. Let E be the intersection of segments BF and AC. Prove that $AE = \frac{1}{6}AD$. Generalize the problem.

14.3 Prove that any two medians of a triangle intersect at a point that divides their length in ratio 2:1 (measuring from the vertex to the midpoint).

14.4 Let $\mathcal{P}$ be the parabola given by $y^2 = 4ax$. Prove that all chords of $\mathcal{P}$ that subtend a right angle at the vertex of $\mathcal{P}$ are concurrent.

14.5 Let A_1, B_1, and C_1 be points on the sides $\overline{BC}, \overline{CA}$, and $\overline{AB}$, respectively, of $\triangle ABC$, having the property that $BA_1/A_1C = CB_1/B_1A = AC_1/C_1B$. Prove that the centroids of $\triangle ABC$, $\triangle A_1 B_1 C_1$, and the triangle formed by lines AA_1, BB_1, and CC_1 coincide.

14.6 Let A, B, C, and D be four points such that $\overline{AB} \perp \overline{CD}$ and $\overline{BC} \perp \overline{AD}$.

  (a) Prove that $\overline{AC} \perp \overline{BD}$.
  (b) Prove also that in this case, $AB^2 + CD^2 = AC^2 + BD^2$. Does this equality imply that $\overline{AB} \perp \overline{CD}$?

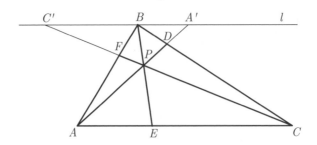

FIGURE 14.1.

14.7 Let l be a line passing through the vertex M of parallelogram $MNPQ$ and intersecting lines NP, PQ, and NQ in points R, S, and T, respectively. Prove that $1/MR + 1/MS = 1/MT$.

14.8 Prove that the three medians in any triangle are concurrent.

14.9 Prove that in a right triangle ABC where $\angle C$ is a right angle, $m_c = c/2$. (This is the converse to Problem 14.1.)

14.10 Consider three circles of distinct radii, each lying in the exterior of each of the others. For each pair of circles, take the intersection point of their common external tangents. Prove that the resulting three intersection points are collinear.

14.11 (Ceva's Theorem) Given $\triangle ABC$, let D, E, and F be interior points on its sides BC, CA, and AB, respectively. (See Figure 14.1.) Prove that $\overline{AD}$, $\overline{BE}$, and $\overline{CF}$ are concurrent if and only if $\dfrac{AE}{EC} \cdot \dfrac{CD}{DB} \cdot \dfrac{BF}{FA} = 1$.

14.12 Let $ABCD$ and $A_1B_1C_1D_1$ be parallelograms with $A_1B_1C_1D_1$ inscribed in $ABCD$, i.e., vertices A_1, B_1, C_1, and D_1 lie on $\overline{AB}, \overline{BC}, \overline{CD}$, and $\overline{DA}$, respectively. Prove that the centers of these parallelograms coincide.

14.13 Let $ABCD$ be a trapezoid with $\overline{BC} \parallel \overline{AD}$, $AD = a$, $BC = b$, and let O be the intersection of the diagonals AC and BD. A line through O that is parallel to the bases intersects lateral sides AB and CD at points E and F, respectively. Show that $EO = OF$.

14.14 Let A_1, B_1, C_1, and D_1 be the midpoints of sides $\overline{CD}, \overline{DA}, \overline{AB}$, and $\overline{BC}$, respectively, of a square $ABCD$ of area 1. Find the area of the quadrilateral bounded by segments AA_1, BB_1, CC_1, and DD_1. Can you generalize the problem to the case where $DA_1/DC = k$ $(k \neq 1/2)$ and likewise for the other three sides.

14.15 Let A, B, and C be complex numbers, and let $\triangle ABC$ denote the corresponding triangle in the complex plane with vertices ordered clockwise. Let ζ be the complex number of unit modulus and argument $60°$. Prove that $\triangle ABC$ is equilateral if and only if $A\zeta - B\zeta^2 - C = 0$.

14.16 On each side of a parallelogram, draw a square externally. Show that the centers of the squares form the vertices of a square.

14.17 Consider an equilateral triangle ABC. Let A_1, B_1, and C_1 be points on $\overline{BC}, \overline{CA}$, and $\overline{AB}$, respectively, such that $BA_1/A_1C = CB_1/B_1A = AC_1/C_1B = 1/2$. Let $A_2 = \overline{BB_1} \cap \overline{CC_1}$, $B_2 = \overline{CC_1} \cap \overline{AA_1}$, and $C_2 = \overline{BB_1} \cap \overline{AA_1}$. Find

$$\frac{\text{Area}(\triangle A_2 B_2 C_2)}{\text{Area}(\triangle ABC)}.$$

14.18 Let $ABCD$ be a trapezoid, and let P and Q be the midpoints of the bases BC and AD, respectively. Take a point M on ray AC such that M is outside of the trapezoid. Let lines MP and MQ intersect lateral sides AB and CD at points H and K, respectively. Prove that $\overline{HK}$ is parallel to the bases.

14.19 Let $ABCD$ be a trapezoid with $\overline{BC} \parallel \overline{AD}$, let $O = \overline{AC} \cap \overline{BD}$, and let $K = \overline{AB} \cap \overline{CD}$. Prove that line KO passes through the midpoints of the bases AD and BC.

14.20 Given an isosceles $\triangle ABC$ with $a = c$, prove that the sum of distances from every point of $\overline{AC}$ to lines AB and CB is the same.

14.21 Given an equilateral triangle ABC, prove that the sum of the three distances from every interior point or boundary point to its sides is the same.

14.22 (Generalization of Newton's Theorem) A quadrilateral is circumscribed around an ellipse. Prove that the center of the ellipse coincides with the midpoint of the segment joining the midpoints of the diagonals of the quadrilateral.

14.23 The midpoints of the sides AB and CD, and of sides BC and ED of a convex pentagon (5-gon) $ABCDE$ are joined by two segments. The midpoints H and K of these segments are joined. Prove that $\overline{HK} \parallel \overline{AE}$ and $HK = \frac{1}{4}AE$.

14.24 Two pirates decide to hide a stolen treasure on a desert island, which has a well (W), a birch tree (B), and a pine tree (P). To bury the treasure, one of the pirates starts at W and walks towards B and after reaching turns right $90°$ and walks the same distance as WB reaching the point Q. The second pirate starts at W and walks towards P, after which he turns left $90°$ degrees and walks the same distance as WP reaching the point R. Then they bury the treasure halfway between Q and R. Some months later the two pirates return to dig up the treasure only to discover that the well was gone. Can they find the treasure?

14.25 Suppose that four intersecting lines form four triangles. Prove that the circumcircles around these triangles share a common point.

14.26 Prove that in any triangle, a line passing through the bases of two of the altitudes is perpendicular to the line passing through the third vertex and the center of the circumcircle of the triangle.

14.27 In this problem we will discuss a reflection property of the hyperbola that is analogous to that of the ellipse. For additional information about the interior and exterior of a hyperbola, see the statement of Problem 6.9. Let $\mathcal{H}(F_1, F_2; d)$ be a hyperbola and let A be a point on $\mathcal{H}$. We define a tangent to $\mathcal{H}$ at A to be a line intersecting $\mathcal{H}$ at A but containing no interior points of $\mathcal{H}$. Suppose A is on the branch of $\mathcal{H}$ closest to F_2. Let m be perpendicular to the bisector of the angle made by $\overrightarrow{F_2A}$ and $\overrightarrow{AF_1}$ at A. (See Figure 14.2.) Show that m is the tangent to $\mathcal{H}$ at A.

14.28 Let a, b, c, and d be complex numbers corresponding to distinct points A, B, C, and D in the complex plane. Then $\overline{AD}$ and $\overline{BC}$ are perpendicular if and only if $(d - a)/(c - b) = 0$ is purely imaginary.

14.29 Prove that any five points in general position uniquely determine a non-degenerate conic.

14.30 Let A, B, C, and D be consecutive points on a parabola. Let C_1, C_2, and C_3 be the points of intersection of the pairs of lines A_1B_2 and A_2B_1, A_1B_3 and A_3B_1, and A_2B_3 and A_3B_2, respectively. Prove that C_1, C_2, and C_3 are collinear.

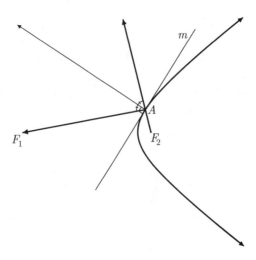

FIGURE **14.2.**

14.31 Suppose that three circles with equal radii intersect at a point. For each pair of these circles, consider the other point of their intersection. Prove that these three points lie on a circle of the same radius.

14.32 (Ptolemy's Theorem) Consider four points A, B, C, and D in the plane. Prove they lie on a circle if and only if $AB \cdot CD + BC \cdot AD = AC \cdot BD$.

14.33 Given three circles of radii r_1, r_2, and r_3, such that each is externally tangent to the other two, find the radius of a circle that passes through the points of tangency of these three circles.

Epilogue

If you enjoyed the topics of this book, you may wonder, "What's next?" There are other important and beautiful plane geometry topics that could be studied; however, we suggest you first explore Euclidean 3-dimensional geometry, often referred to as solid geometry. Your study of the coordinate method, vectors, conics, and affine transformations will help make the transition to analytic solid geometry easier. For example, the coordinate method can be used with vectors and matrices to solve problems involving lines and planes in both 2-space and 3-space. Affine transformations can likewise be used in the 3-dimensional setting. We also encourage you to pursue a study of projective transformations, a topic we omitted from this book, as a tool for solving problems in both plane and solid geometry.

While a first course in calculus deals with geometry of curves mainly in the plane, subsequent calculus courses incorporate geometry of surfaces in 3-dimensional space and the geometry of manifolds in the spaces of higher dimensions. Additionally, a study of linear algebra will facilitate a vocabulary for discussing geometrical concepts in higher dimensions.

Of course, there also exist non-Euclidean geometries, and they are often as important as the Euclidean ones. A good modern introduction is the book by Brannan, Esplen, and Gray ([7]). Finite and combinatorial geometries have been studied extensively in the last century, and resources abound. Numerous other topics, like geodesics, fractals, physics, and protein folding, have ties to geometry as well.

We end this book with the last lines from René Descartes' *La Géométrie*:

> *I hope that posterity will judge me kindly, not only as to the things which I have explained, but also as to those which I have intentionally omitted so as to leave to others the pleasure of discovery.*

Hints to Chapter Problems

A truly good book teaches me better than to read it. I must soon lay it down, and commence living on its hint.

— Henry David Thoreau (1817–1862)

Triangles

3.2.1 Use the Extended Pythagorean theorem. Answer: $\sqrt{89}$.

3.2.2 (a) Answers: $37/16, 8/3, 4$. If $CD = x$, then $BD = 8 - x$. To find x, apply the Pythagorean theorem to two triangles. To find AF, use the Triangle Bisector Property.
(b) Answers: $\sqrt{935}/16, \sqrt{34}/3, \sqrt{26}/2$. Use (a) and the Pythagorean theorem.

3.2.3 Notice that C is the largest angle. Let D be the base of the perpendicular segment from B on line AC. Consider the three cases where D is on $\overline{AC}$, $D = C$, and D is off of $\overline{AC}$.

3.2.4 Let $a_1/a_2 = b_1/b_2 = k$. Then $a_1 = ka_2$ and $b_1 = kb_2$.

3.2.5 Answer: $900°$.

3.2.6 Answers: (a) $360°$. (b) $180°$. (c) $180(n - 4)°$

3.2.7 Explain that if this happens that one of the interior angles has measure greater than $180°$. Alternatively, assume that the n-gon has a acute interior angles and $n - a$ nonacute interior angles; now obtain a bound on a.

3.2.8 Yes.

3.2.9 (a) Let A' be a point symmetric to A with respect to l. Consider $C = l \cap \overline{A'B}$.
(b) Use the triangle inequality.
(d) $\overline{AB} \cap l$.

3.2.10 Join the bases of the medians. Use the Triangle Midline theorem.

3.2.11 Use the result of Problem 3.2.10 or use Ceva's theorem.

3.2.12 Use congruency of triangles.

3.2.13 Use congruency of triangles.

3.2.14 Use the result of Problem 3.2.12.

3.2.15 Use the result of Problem 3.2.13 or Ceva's theorem.

3.2.16 Answer: $\sqrt{33}$.

3.2.17 One can use the result of Problem S3.2.5. Another solution can be obtained from the following construction. Let $\triangle ABC$, $m\angle C = 90°$, $m\angle A = 30°$, be our triangle. Let D be a point on ray BC such that $DC = CB$. Show that $\triangle DAB$ is equilateral.

3.2.18 Generalize computations for Problem 3.2.2. You can use any CAS (e.g., Maple) to assist you with algebraic transformations or to verify them.

3.2.19 (a) Let D be the base of the median at A, and let F be a point on ray AD such that $AD = DF$. Prove that $BF = AC$.
        (b) Use part (a).

3.2.20 (a) Consider intersection of lines l and AB. (b) No such C exists. (c) Consider what happens when A is reflected across l.

3.2.21 Prove and use the following inequality and two more like it: $c < AO + BO < a + b$.

3.2.22 Explain first that if $0 < x < 1$, then $x^3 < x^2$. Then use the result of Problem 3.2.3.

3.2.23 Construct $\overline{CD}$ parallel to $\overline{BB_1}$, D on $\overline{AB}$. The proof is similar to that of the Triangle Bisector theorem.

3.2.24 Use the expressions for the lengths of the bisectors found in the solution of Problem 3.2.18.

3.2.25 Generalize computations from Problem 3.2.2(a).

3.2.26 Show that $\triangle ABC \sim \triangle AB_1C_1 \sim \triangle A_1B_1C$.

3.2.27 Label the triangle so that B is the obtuse angle, and use a similar approach.

3.2.28 Show that this sum is equal to h_a $(= h_c)$. (At this point in the book, you should not use the notion of the area of a triangle.)

3.2.29 Use the result of Problem 3.2.28. (At this point you should not use the notion of the area of a triangle.)

3.2.30 Show that there is a diagonal of the polygon which lies in its interior. Then use induction on n.

Parallelograms and Trapezoids

3.3.1 We suggest showing $(2) \Rightarrow (1) \Rightarrow (4) \Rightarrow (3) \Rightarrow (2)$. Use standard techniques.

3.3.2 Assume $BC < AD$ and select E on $\overline{CD}$ so that $\overline{CE} \parallel \overline{AB}$ to create the parallelogram $ABCE$. To prove $(3) \Rightarrow (1)$, find two similar triangles in the trapezoid; use the constant ratios of their side lengths and the congruence of diagonals to show that the triangles are isosceles.

3.3.3 Use the Triangle Midline theorem.

3.3.4 Diagonals of $ABCD$ bisect each other. Then use Theorem 3.23.

3.3.5 A vertex at a nonacute angle has this property.

3.3.6 Use the fact that same side interior angles of a parallelogram are supplementary.

3.3.7 Consider the midline of the trapezoid. Explain that points M and N are on it.

3.3.8 Let ABC be our triangle. Through each vertex of the triangle draw a line parallel to the opposite side. Consider $\triangle A_1B_1C_1$ formed by these three lines.

3.3.9 Consider points on $\overline{BC}$ and $\overline{AD}$ which divide these segments in 5 congruent segments and consider lines through each of these points parallel to $\overline{BF}$.
        For a simple generalization, replace 5 by n and 6 by $n + 1$. For a broader generalization, let $AF = a$ and $AD = b$.

3.3.10 Consider $Z = \overrightarrow{D_1C_1} \cap \overrightarrow{BC}$. Show that $\angle DD_1C \cong \angle C_1ZB_1 \cong \angle BB_1A_1$.

3.3.11 Compute EO in terms of a and b. In order to do this, look for pairs of similar triangles.

3.312 Show $\triangle MNR$ is similar to $\triangle SQM$ and use this fact to prove that $(MR)(ST) = (MT)(MS)$.

3.3.13 Let E and F be the points of intersection of line MN with lateral sides AB and CD, respectively. Construct a line through M that is parallel to $\overline{AB}$; use the resulting parallelogram to show that $\overline{EF}$ is the midline of the trapezoid.

3.3.14 Use Problem 3.3.11(a).

3.3.15 Let M, N, P, R, and L be the midpoints of sides AB, BC, CD, DE, and diagonal BE, respectively. Prove that P, K, and L are collinear. Join M and L. Use the Triangle Midline theorem.

3.3.16 Let T be the midpoint of $\overline{QR}$. Drop perpendiculars from Q, R, W, and T on line BP. Consider congruent triangles. See that the possible position for T is defined only by B and P.

Circles

4.1 Use Theorem 4.17.

4.2 Let $A_1A_2 \cdots A_n$ denote the regular n-gon. Let O be the point of intersection of the bisectors of angles A_1 and A_2. Show that $\triangle A_1OA_2$ and $\triangle A_2OA_3$ are congruent isosceles triangles. Then show that all vertices of the polygon lie on a circle with center at O. Finally, show that O is also a center of a circle which touches all sides of the polygon.

4.3 Let M and N be the points of tangency on $\overrightarrow{BA}$ and $\overrightarrow{BC}$, respectively. Then $ED = EM$ and $FD = FN$.

4.4 Let A be a point of tangency of the circles. Join O_1 and O_2 with A and consider two isosceles triangles.

4.5 Refer to the proof of part (1) of the theorem, making sure to check all cases.

4.6 Let H' be the reflection of H with respect to line AC. Show that $m\angle AH'C + m\angle C'BC = 180°$.

4.7 (a) Show that no matter which line through A distinct from line AB is chosen, the measures of angles D_1 and D_2 do not change.

(b) Part (a) implies that all triangles D_1BD_2 (for various B) are similar by AA.

4.8 (a) Answer: $2\sqrt{r_1r_2}$. Apply the Pythagorean theorem to $\triangle O_1NO_2$, as shown in Figure 1.

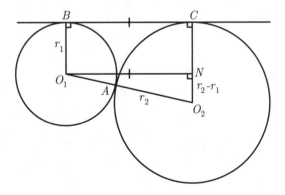

FIGURE 1.

(b) Answer: $r_1 r_2 / (\sqrt{r_1} + \sqrt{r_2})^2$. Use part (a). Call the unknown radius x. $MN = MO + ON$. Express MO and ON from right triangles $O_1 MO$ and $O_2 NO$.

4.9 Introduce the internal common tangent to these circles at A. Let it intersect $\overline{BC}$ at point D. Show that $DA = DB = DC$. (Compare to Problem S3.2.5.)

4.10 Use Ceva's theorem.

4.11 Show that there exists a point in the plane such that the distances from this point to points of the set are all distinct.

4.12 Use Theorem 4.21.

4.13 Use Theorem 4.15.

4.14 The sequence of radii values is geometric; C_n has a radius of r^n.

4.15 Consider two circles tangent to each other with the same centers as the given circles and use Example 4. You should obtain $R_1 \cdot O_1 O_2 / (R_2 - R_1)$.

4.16 Consider the line passing through the centers of the circles.

4.17 Consider the circumcircle for the triangle. Continue the bisector, the median and the altitude until they intersect the circle. Draw the perpendicular bisector through the base of the median and see where it intersects the circle.

4.18 Use Corollary 4.13(1).

4.19 Join O with C and F. Prove that $m\angle COF = 180°$.

4.20 Consider $\triangle ABC$. Let A_1 and C_1 be the bases of the altitudes from A and C, respectively. Let $\overline{BD}$ be a diameter, and let E be the intersection of lines BD and $A_1 C_1$. Consider $\triangle A_1 BC_1$. Show that $\angle BDC \cong \angle A$.

4.21 Suppose $MA \leq MB \leq MC$. Let D be a point on $\overline{MC}$ such that $MD = MA$ and show that $DC = MB$. Another solution follows immediately from Ptolemy's theorem. (See Problem S4.4.)

4.22 Note that the circumcircles of $\triangle CFE$ and $\triangle CAB$ both pass through C and P. Prove that P lies on the circumcircles of $\triangle BDF$ and $\triangle AED$.

4.23 Introduce the point of intersection of line AB with the common tangent of the circles at M.

4.24 Consider two triangles and prove that they are congruent. The vertices of the first triangle are the centers of the circles. The vertices of the second are the three points of intersection of the circles, distinct from their common point.

4.25 (a) Let O be the center of the incircle and let $\overline{AA_1}$ and $\overline{CO}$ be angle bisectors, respectively, as shown in Figure 2. Let $\overline{BD}$, $\overline{A_1 D_1}$, and $\overline{OK}$ be altitudes in triangles ABC, $AA_1 C$,

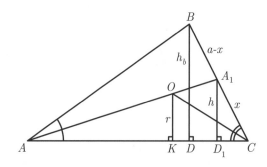

FIGURE 2.

and AOC, respectively. For convenience, let $BD = h_b$, $A_1D_1 = h$, $A_1C = x$, $OK = r$, $AA_1 = l_a$, and $AO = y$. Use a property of angle bisectors and similarity of two pairs of triangles to relate these parameters. Then use the formula for h_b in terms of a, b, and c.

(b) Use the fact that the measure of the inscribed angle is half of the measure of the corresponding central angle; introduce an altitude; use similarity of right triangles and the formula for the length of the altitude.

4.26 Answer: $\sqrt{\frac{r_1 r_2 r_3}{r_1 + r_2 + r_3}}$. Let O_1, O_2, and O_3 be the centers of the circles, and let A, B, and C be the points of tangency. Prove that the circumcircle of $\triangle ABC$ is the incircle of $\triangle O_1 O_2 O_3$. Then use the formula for the length of the radius of the incircle from Problem 4.25(a).

4.27 Let A_1, B_1, and C_1 be the midpoints of the sides BC, AC, and AB, respectively. Use the facts that $\triangle A_1 B_1 C_1 \sim \triangle ABC$ with coefficient of similarity 2 and that sides of $\triangle A_1 B_1 C_1$ are parallel to the corresponding sides of $\triangle ABC$.

4.28 Use Ptolemy's theorem.

Length and Area

5.1 Use formulae provided in the theorems.

5.2 In both (a) and (b) use additivity of area.

5.3 (a) Let A_1 be the midpoint of $\overline{BC}$ and let G be the centroid of $\triangle ABC$. Compare the areas of three triangles: ABC, AA_1C, and AGC.

(b) Use part (a).

5.4 Use the property of a midline in a triangle.

5.5 Use the Pythagorean theorem.

5.6 Introduce two altitudes: $\overline{AH}$ in $\triangle ABC$ and $\overline{C_1G}$ in $\triangle C_1BH$.

5.7 Answer: Yes.

5.8 Use the relation between area and semiperimeter (Theorem 5.3(2)).

5.9 If $ABCD$ is the quadrilateral and $\overline{AC}$ is the diagonal, consider two perpendiculars dropped from B and D to $\overline{AC}$.

5.10 (a) It is a square. Use algebra; specifically, complete the square in a quadratic trinomial. There is no rectangle of smallest area (dimensions must be positive numbers).

(b) Similar to (a).

5.11 Answer: $(\sqrt{3} - \frac{\pi}{2})r^2$. Centers of the circles are vertices of an equilateral triangle.

5.12 Answer: $EF = \sqrt{\frac{a^2 + b^2}{2}}$. Let h_1, h_2, and h be altitude lengths in the three trapezoids, where $h = h_1 + h_2$. Use algebra.

5.13 Use Heron's formula.

5.14 Rotate $ABCD$ around its center O by $90°$ clockwise. Prove that the segments partition $ABCD$ into four congruent triangles, four congruent trapezoids, and the square $EFGH$ formed by segments AA_1, BB_1, CC_1, and DD_1.

5.15 Use the fact that the semicircle built on the hypotenuse passes through the vertex C.

5.16 Answer: $(\sqrt{S_1} + \sqrt{S_2})^2$. Use similarity of triangles AOD and BOC.

5.17 (a) Use the fact that the ratio of areas of triangles with a common altitude is equal to the ratio of their base lengths.

(b) Use part (a).

5.18 Answer: $1 + \frac{\pi}{3} - \sqrt{3}$. The configuration contains many equilateral triangles.

5.19 There is no line for which Area($\triangle ABC$) is the greatest. The smallest area will be attained only in the case when M bisects $\overline{AC}$.

5.20 Prove that each triangle contains the center of the circle.

5.21 (a) Let $2x$, y, and y be the side lengths of an isosceles triangle of a given perimeter, $2p$. Then $x + y = p$ and the area of the triangle is

$$A(x) = x\sqrt{y^2 - x^2} = x\sqrt{(p-x)^2 - x^2} = x\sqrt{p^2 - 2px}.$$

Show that $A(x)$ attains its maximum on $(0, p)$ at $x = p/3$, which implies that the triangle is equilateral. There is no triangle of minimum area.

(b) Let $2x$, y, and y be the side lengths of an isosceles triangle of a given area A, and let h be the length of the altitude from a base vertex to a lateral side. Then $xh = x\sqrt{y^2 - x^2} = A$ and the perimeter of the triangle is $P(x) = 2x + 2y = 2x + 2\frac{\sqrt{A^2 + x^4}}{x}$. Show that $P(x)$ attains its minimum on $(0, +\infty)$ at $x = \sqrt{A/\sqrt{3}}$, which implies that the triangle is equilateral. There is no triangle of maximum perimeter.

Alternatively, part (b) can be reduced to part (a).

5.22 Answer: $1/7$. Both algebraic and dissection solutions are possible. For an algebraic solution, rotate $\triangle ABC$ around its center by $120°$ clockwise. Prove that the segments partition $\triangle ABC$ into three congruent triangles, three congruent quadrilaterals, and an equilateral triangle formed by segments $\overline{AA_1}$, $\overline{BB_1}$, and $\overline{CC_1}$. The problem can be generalized to an arbitrary triangle, and the answer will be the same.

5.23 There are several ways to carry out such a dissection. One way is to cut the larger square into four congruent quadrilaterals using two perpendicular lines, and then "surround" the smaller square by these quadrilaterals.

5.24 From each vertex A_i drop two perpendiculars to those sides of $A'_1 \ldots A'_n$ which are parallel to the sides $A_{i-1}A_i$ and A_iA_{i+1} (all indices are modulo n). Let α_i be the measure of the angles formed by these perpendiculars. Then $\alpha_1 + \cdots + \alpha_n = 2\pi$.

5.25 (a) Answer: an equilateral triangle. First show that if the sum of three positive numbers is fixed, then their product is the greatest if and only if they are all equal. Use Theorem 5.3(1).

(b) Answer: an equilateral triangle. Use part (a).

5.26 Let $ABCD$ be the quadrilateral. Suppose A, A_1, A_2, A_3, A_4, and B are consecutive vertices of congruent segments on $\overline{AB}$, and D, C_1, C_2, C_3, C_4 and C are the corresponding points on $\overline{CD}$. Join subsequently points D and A_1, C_1 and A_2, C_2 and A_3, C_3 and A_4, C_4 and B. Consider the areas of the resulting triangles.

5.27 Express the area A of the triangle and the inradius r in terms of the side lengths, a, b, and c.

5.28 The second inequality is easy. To establish the first one, explain that if the centers of the coin are preserved, but all radii are doubled, then the resulting larger coins will completely cover a concentric round table of radius $R - r$ (with overlaps).

5.29 Let ABC be an inscribed nonequilateral triangle. Explain how the points can be labeled to ensure there is a point B' such that $A - B - B'$ (order on the circle), $m\overset{\frown}{AB'} = 120°$, and Area($\triangle AB'C$) > Area($\triangle ABC$). Now repeat the argument on $\triangle AB'C$.

5.30 A circle of radius r inscribed in one of the triangles has the property that $r + R = S$, where R is the circumradius and S is the sum of the (signed) lengths for the triangle as determined by Carnot's theorem. Now sum over all triangles in the triangulation.

Loci

6.1 Recognize that a rotational ellipsoid has two "foci".

6.2 See Problem S3.2.5 or use the Pythagorean theorem.

6.3 See Example 11.

6.4 Fix a segment $\overline{AB}$ and for a given point C, $\triangle ABF$ is a right triangle. Compare the length of the median from F with AB.

6.5 (a) Let d be the distance between l_1 and l_2 and consider separately the cases $d < k$, $d = k$, and $d > k$.
   (b) Use the same approach as for part (a).
   (c) The locus is a pair of lines.
   (d) The locus is a pair of lines parallel to l_1 and l_2.

6.6 (a) The locus is a rectangle.
   (b) The locus consists of the extensions of the sides of the rectangle obtained in (a).
   (c) Suppose $k = 1$. What would the locus be in this case? Now generalize.
   (d) See what experimentation suggests.

6.7 (a) The locus is the union of two arcs.
   (b) You don't need a hint for this one!
   (c) The locus is a circle.
   (d) See Problem 4.6.

6.8 The locus is the whole plane, with l_1 and l_2 removed.

6.9 Let P be a point in the plane and consider the intersections of $\overrightarrow{PF_1}$ and $\overrightarrow{PF_2}$ with the hyperbola.

6.10 Let X be a point on m and employ the result in Problem 6.9. It may help to construct a line l through F_2 parallel to the bisector b of the angle formed by $\overrightarrow{F_2A}$ and $\overrightarrow{AF_1}$. Consider the points of intersection of l with $\overline{AF_1}$ and m and use the triangle inequality.

6.11 The property should be similar to those of Corollaries 6.2 and 6.4.

6.12 See Example 9.

6.13 If $0 < k \neq 1$, then the locus is a circle, sometimes called "the circle of Apollonius."

6.14 Drop a perpendicular ZX' from Z to line CX. Consider the locus of points X' and show the locus of points Z is a line segment.

6.15 Experiments suggest that the locus is a line perpendicular to line OA. One way to prove this is to drop perpendicular $\overline{XY}$ on line OA. If the given chord is perpendicular to this line, then OY can be easily computed, and $OY = r^2/OA$.

6.16 Reflect F_2 across the tangent lines AS and BS to obtain points M and N, respectively. Show that S is the circumcenter of the isosceles triangle MNF_2.

Trigonometry

7.1 Use the definitions of sine and cosine on an appropriate triangle.

7.2 Answer: $\sin 3\alpha = -9\sqrt{10}/50$ and $\cos 3\alpha = 13\sqrt{10}/50$.

7.3   Use the fact that $m\angle A + m\angle B/2 = 180° - m\angle C - m\angle B/2$.

7.4   Use the Cosine theorem.

7.5   Rewrite in terms of $\sin\theta$ and $\cos\theta$.

7.6   Write $\cos 3\alpha$ as $\cos(2\alpha + \alpha)$ (and $\tan 3\alpha$ as $\tan(2\alpha + \alpha)$), and then make use of angle sum formulas.

7.7 $a \approx 7.475$, $m\angle B \approx 40.25°$, and $m\angle C \approx 64.75°$.

7.8   Use the fact that $\sin\alpha = \cos(90° - \alpha)$.

7.9   Expand $\cos(\alpha + \beta)\cos(\alpha - \beta)$.

7.10 Use Corollary 7.5.

7.11 Find the cosine of each angle and use the identity $2\cos^2 A - 1 = \cos 2A$.

7.12 Let $\alpha = \arctan(1/5)$ and $\beta = \arctan(1/239)$ and find $\tan(4\alpha - \beta)$.

7.13 $\cos 60° = 1/2$ and $\sin 60° = \sqrt{3}/2$.

7.14 Write a product of trigonometric functions as a sum.

7.15 Use the results of Problem 7.14.

7.16 Use the Cosine theorem and the double angle formula for cosine to get $x = (5 + \sqrt{13})/2$.

7.17 Consider x^2 and x^3.

7.18 Multiply the sum by $\sin(\beta/2)$ and rewrite the product of sines as a sum.

7.19 Multiply the sum by $\sin(\beta/2)$ and rewrite the product of sines as a sum.

7.20 Use the formulas from Problems 7.18 and 7.19.

7.21 Use the Sine theorem on $\triangle ABP$ and $\triangle BPC$.

7.22 Note that one can rearrange the sides of the hexagon so that consecutive sides have lengths a and b around the hexagon.

7.23 Use the Sine theorem on six triangles, along with the fact that $\sin(\pi - x) = \sin x$.

7.24 Use the Sine theorem.

7.25 Use the Cosine theorem and the formula $2\sin^2 x = 1 - \cos 2x$.

7.26 Use the Sine theorem.

7.27 Show that $m\angle BPC = 120°$. Then use the Sine theorem.
      Answer: $28/\sqrt{37}$.

7.28 Let the inscribed circle have radius r and the sector have radius R. Find a trigonometric expression (in terms of α) for the ratio R/r.

7.29 To prove the formula for $\sin(\alpha - \beta)$, consider a quadrilateral $ABCD$ inscribed in a circle of diameter 1, with $\overline{AD}$ being a diameter, $m\angle BAD = \alpha$, and $m\angle CAD = \beta$.

7.30 Answer: $30°$. Use the Sine theorem on $\triangle EFC$.

7.31 Transform the equality into a quadratic equation with respect to $\cos\left(\frac{A+B}{2}\right)$.

7.32 Suppose the bisector at A and the bisector at B meet at K, while the bisector at C and the bisector at B meet at K'. Let the bisector at B intersect side AC at E, and use the Sine theorem to show that $KE = KE'$. Problem 7.3 may be helpful.

7.33 Let O be the center of the circumcircle for a $\triangle ABC$. Let $m\angle BOC = x$, $m\angle COA = y$, and $m\angle AOB = z$. Explain that for a triangle of the greatest area, each of the angles has to be at most $180°$. Reduce the problem to showing that
$$\sin x + \sin y + \sin z \le 3\sqrt{3}/4,$$
with equality if and only if $x = y = z = 120°$.

7.34 Reduce the problem to minimizing $\dfrac{1}{\tan\frac{A}{2}} + \dfrac{1}{\tan\frac{B}{2}} + \dfrac{1}{\tan\frac{C}{2}}$, where A, B, and C are the measures of the angles in a triangle.

Coordinatization

8.1 Choose a convenient coordinate system and calculate the coordinates of the midpoint of each diagonal.

8.2 Introduce a coordinate system as you would for a parallelogram.

8.3 Place C at the origin and let A and B lie on the coordinate axes.

8.4 If P, Q, R, and S are the midpoints of the original quadrilateral, prove that the midpoints of the diagonals of $PQRS$ coincide.

8.5 (a) Find the coordinates of the midpoint of $\overline{AB}$ and use the distance formula.
   (b) Use the formula for the distance from a point to a line.

8.6 Follow the method of Example 30.

8.7 Place A at the origin and give B coordinates $(2x_A, 2x_B)$. Calculate midpoints and use the distance formula.

8.8 Let $A : (-a, 0)$ and $C : (a, 0)$. Find the equations of the perpendicular bisectors of $\overline{AB}$ and $\overline{BC}$.

8.9 Choose $A : (0, 0)$ and $D : (5, 0)$. Use the equations for lines to find the coordinates of the relevant interior point.

8.10 Introduce a coordinate system, not necessarily Cartesian, such that A and D lie on the x-axis, and the y-axis passes through the midpoints of the bases.

8.11 Choose $A : (-a, 0)$, $B : (-b, 1)$, $C : (b, 1)$, and $D : (a, 0)$.

8.12 Position A at the origin and D on the x-axis.

8.13 Choose $A : (-1, 0)$, and $C : (1, 0)$. Let $P : (s, t)$ be a point inside or on the boundary of the triangle.

8.14 First determine the area of the rectangle that circumscribes the triangle and whose sides are parallel to the coordinate axes. Another approach is to use the fact that Area($\triangle ABC$) $= \frac{1}{2}ab\sin C = \frac{1}{2}ab\sqrt{1 - \cos^2 C}$, the Cosine theorem, and the distance formula.

8.15 A point (x_0, y_0) lies on or in the exterior of a circle $x^2 + y^2 = R^2$ if and only if $x_0^2 + y_0^2 \geq R^2$.

8.16 Introduce a (not necessarily Cartesian) coordinate system OXY such that $E : (0, 0)$, $A : (a, 0)$, $B : (0, 1)$, $C : (c, 0)$, $a < 0 < c$. Then follow the approach used in the second proof of Theorem 8.8.

Conics

9.1 Follow the method of Example 39 or of Example 41.

9.2 Consider different combinations of signs of A, C, and H. Consider the case when $H = 0$.

9.3 Complete the square.

9.4 Use the translation and rotation formulae.

9.5 Introduce a coordinate system and use the equation of a parabola.

9.6 The locus is a circle.

9.7 Use the ideas presented in Examples 49 and 50.

9.8 If $A : (x_A, y_A)$, then $B : (-x_A, -y_A)$.

9.9 Consider the intersection of a line and a curve. Require that the corresponding quadratic equation have two equal roots.

9.10 The locus is an arc of an ellipse.

9.11 Use the expressions for A', B', and C' derived in the proof of Theorem 9.5.

9.12 Similar to Example 46.

9.13 Consider the point at which a chord from our set of chords intersects the axis of the parabola.

9.14 Answers: (a) No. (b) Yes.

9.15 Let F_1 and F_2 represent equations for the parabolas in an appropriate coordinate system. Consider $a_2 F_1(x, y) + a_1 F_2(x, y) = 0$.

9.16 Collect what you know about the interior points, the definition of convexity, and the intersection of a line with an ellipse.

9.17 For an ellipse and a hyperbola, take the foci at $(\pm ae, 0)$ and the directrix at $x = \pm a/e$, respectively.

9.18 Choose a Cartesian coordinate system such that the equation of the parabola is $y = 4x^2$.

9.19 When an ellipse is tangent to two parallel lines, the points of tangency are symmetric with respect to the center of the ellipse.

9.20 Use the fact that a quadratic function $f(x) = Ax^2 + Bx + C$ takes the value zero at two distinct points if its discriminant $D = B^2 - 4AC > 0$; and $f(x) > 0$ for all x, if $A > 0$ and $D < 0$.

9.21 For an ellipse and hyperbola, the answer is 'YES.' For parabolas, the answer is 'NO.' Use the fact that every five points no three of which are collinear define a unique ellipse, or a unique hyperbola, or a unique parabola. For the cases of ellipses, or of hyperbolas, use the continuity of the discriminant of the curve.

Complex Numbers

10.1 (a) Answer: $-1 + i$.

    (b) Answer: $-64 - 64i$.

    (c) Answer: 0. Add the geometric series.

10.2 Let $z = \cos 120° + \sin 120° \, i = -1/2 + \sqrt{3}/2 \, i$.

10.3 It is helpful to interpret $|z_1 - z_2|$ as the distance between two points in the complex plane which correspond to complex numbers z_1 and z_2.

    (a) A circle.

    (b) The exterior of a circle.

    (c) An ellipse.

    (d) The empty set.

    (e) A hyperbola.

    (f) Right angle with its interior.

10.4 A quadrilateral is a parallelogram if and only if its diagonals bisect each other.

10.5 Represent the complex numbers z and $\bar{z}$ either in standard form or polar form.

10.6 Represent the complex numbers z, z_1, and z_2 either in standard form or polar form.

10.7 Use the property of the modulus of complex numbers: $|z_1||z_2| = |z_1 z_2|$.

10.8 Modify the argument used in Example 52.

10.9 Use De Moivre's formula and the expansions of $(a + b)^3$ and $(a + b)^6$ (the Binomial theorem can be used for the latter) to obtain $\sin 3\alpha = 3\sin\alpha - 4\sin^3\alpha$, $\cos 6\alpha = 32\cos^6\alpha - 48\cos^4\alpha + 18\cos^2\alpha - 1$.

10.10 Use the isosceles triangle with vertices at 0, 1, and z_k.

10.11 Use the product $(p + q + i)(p^2 + pq + 1 + qi)$.

10.12 Use the proof of the statement of Example 55.

10.13 Consider the real and imaginary part of the sum of the geometric series $1 + z + \cdots + z^n$, where $z = \cos\alpha + \sin\alpha\, i$.

10.14 Represent the vertices of the regular n-gon by complex numbers.

10.15 Answer: $|R^2 - a^2|$. Use the solution of Example 56.

10.16 Consider the arguments of $(5 + i)^4$, $-239 + i$, and $(5 + i)^4(-239 + i)$.

10.17 Use the property about the argument of a product of complex numbers and determine that $\triangle ABC$ is equilateral if and only $C - B = (A - B)\zeta$.

10.18 (a) Think of the complex number $a + bi$ as the vector from $(0, 0)$ to (a, b).
       (b) Use Proposition 10.3.
       (c) (i) Show that the slopes are distinct if and only if the lines intersect at one point.
            (ii) Consider the arguments of κ_1 and κ_2.

10.19 Use Problem 10.18.

10.20 Just check that the line through a and h is perpendicular to the line through b and c.

10.21 Use an approach similar to the one in the proof of Example 55.

10.22 Use Problem 10.21. In addition, prove that the segments joining the centers of opposite squares bisect each other.

Vectors

11.1 Assign coordinates to the points (for example, $A : (a_1, a_2)$), and use Theorem 11.6.

11.2 In triangle ABC, show that the vector corresponding to the segment joining the midpoints of sides AB and BC is given by $\frac{1}{2}AC$.

11.3 Use Problem 11.2 on each half of the quadrilateral.

11.4 Show that the parallelogram with sides determined by the vectors $|\vec{a}|\vec{b}$ and $|\vec{b}|\vec{a}$ is a rhombus.

11.5 Use the Polygon Rule.

11.6 Most of these will be straightforward calculations once the vectors are written in component form (i.e., $\vec{a} = \langle a_1, a_2 \rangle$).

11.7 Use the Triangle Law for vector addition.

11.8 Label the sides of the quadrilateral with vectors $\vec{a}, \vec{b}, \vec{c}$, and $-(\vec{a} + \vec{b} + \vec{c})$.

11.9 Show that $\overrightarrow{AC} + \frac{2}{3}\overrightarrow{CC'} = \frac{1}{3}\left(\overrightarrow{AB} + \overrightarrow{AC}\right)$, along with some similar results.

11.10 Use Problem 11.9 to find expressions for $\overrightarrow{MA}, \overrightarrow{MB}$, and $\overrightarrow{MC}$.

11.11 No. Consider a triangle where two of the angles are close to $90°$.

11.12 Use Problem 11.S3.

11.13 First show that $\overrightarrow{BC} \cdot \overrightarrow{AC} = -2c^2$.

11.14 Use that $\overrightarrow{BC}$ is perpendicular to $\overrightarrow{AB} + \overrightarrow{BC} + \overrightarrow{CD}$ and the fact that the dot product of perpendicular vectors is 0. The answer to the last question is "no."

11.15 Inscribe triangle ABC in a circle centered at O. Now form parallelograms $AOCX$ and $XHBO$. Show H is the desired point of concurrency.

11.16 Note that for any i, $\overrightarrow{XA_i} = \overrightarrow{XO} + \overrightarrow{OA_i}$. Now expand the sum and use the result from Example 67.

11.17 The point X will be the median of the triangle.

11.18 If a, b, c, and d are the side lengths and d_1 and d_2 are the diagonal lengths, the distance between the midpoints of the diagonals will be $\sqrt{(a^2 + b^2 + c^2 + d^2 - d_1{}^2 - d_2{}^2)}/2$. To show this, let the terminal points of vectors $\vec{a}$, $\vec{b}$, $\vec{c}$, and $\vec{d}$ correspond to the vertices A, B, C, and D of the quadrilateral, where these vectors all share some common initial point. Now express the desired length as a dot product and show it is equivalent to the value given above.

11.19 Let $\vec{x}$ be the vector on the left hand side of the equality. Show that $\vec{x} \cdot \overrightarrow{OA} = \vec{x} \cdot \overrightarrow{OB} = 0$.

Affine Transformations

12.1 Let h_1 and h_2 be affine transformations mapping the three given points to the three desired points, and prove $h_1 = h_2$. Compare the images of $\vec{i}$ and $\vec{j}$ under $h_1 \circ f$ and $h_2 \circ f$.

12.2 Answer: No.

12.3 Show there is an affine transformation f mapping trapezoid $ABCD$ to an isosceles trapezoid.

12.4 Use an affine transformation to map the ellipse to a circle.

12.5 Let $A_1, \ldots, A_n$ be vertices of the polygon, $\mathcal{P}$, and $A'_1, \ldots, A'_n$ be their images under an affine transformation. Find $\sum \overrightarrow{G'A'_i}$, where G' is the image of the centroid of $\mathcal{P}$.

12.6 The proof is similar to the proof of Theorem 12.13.

12.7 Consider an affine transformation mapping $\triangle ABC$ to an equilateral triangle. Show that the images of the other two triangles will also be equilateral. Finally, use the fact that centroids get mapped to centroids by affine transformations.

12.8 Map the parallelogram to a square and cite Theorem 12.7(6).

12.9 Consider the rectangle R that is tangent to the four vertices of the ellipse. Find the areas of $f(R)$ and $f(\mathcal{C})$, where f is an affine transformation sending the ellipse to $\mathcal{C}$, a circle of radius a.

12.10 Let f be an affine transformation mapping the ellipse to a circle. Note that $f(ABCD)$ will be a parallelogram circumscribing this circle.

12.11 Consider an affine transformation that maps $\triangle ABC$ to a right triangle (with right angle at the image of C) with side lengths $\alpha + 1$ and $\beta + 1$.

12.12 Consider an affine transformation that maps one of the ellipses to a circle.

12.13 Solve the problem by assuming the ellipse is a circle. For the generalization, assume that the n-gon is not regular and find a contradiction.

Inversions

13.1 Observe that $\overline{OC}$ is the perpendicular bisector of $\overline{AB}$. Use Theorem 3.19(2) and solve for OA^2.

13.2 No, it cannot. Use Theorem 13.1(3).

13.3 Answer: $x' = \frac{x}{x^2+y^2}$, $y' = \frac{y}{x^2+y^2}$.

13.4 Let $I = I(O, r)$ denote the inversion. Suppose a circle $\mathcal{C}$ is tangent to a line l at point A. Consider the following cases: $O = A$, $O \neq A$ and $O \in l$, $O \neq A$ and $O \in \mathcal{C}$, O is neither on l nor on $\mathcal{C}$. Then consider the case when the line is a secant to the circle.

13.5 Use the fact that a convex quadrilateral is inscribed in a circle if and only if its opposite angles are supplementary.

13.6 Reduce it to Problem 13.5.

13.7 Use Theorem 13.1(3) and similar triangles.

13.8 Use the fact that any circle with the endpoints of a diameter removed can be thought of as a locus of all points from which a segment is seen under a $90°$ angle.

13.9 Use Problem 13.3.

13.10 Use the idea of the solution in Example 80.

13.11 Consider the inversion $I = I(A, 1)$.

13.12 Prove that the common point lies on the circle containing the arc $\overset{\frown}{AB}$. Consider an inversion $I = I(A, r)$.

13.13 Let K and L be the midpoints of segments AC and BD, respectively. Consider an inversion with respect to $\mathcal{C}$. Use Problem 13.1.

13.14 Consider an inversion where $\mathcal{C}_1$ is the circle of inversion. Let K, L, and M be the points of tangency of $\mathcal{C}_1$ with segments AB, BC, and AC. Prove that $\triangle MLK \sim \triangle A'B'C'$ with coefficient $1/2$. Use Problem 13.7.

13.15 In order to show that five points are cocyclic, show that any four of the points are cocyclic. Consider the circle through N, D, and K and use Problem 13.6.

Solutions to Chapter Problems

We use 10% of our brains. Imagine how much we could accomplish if we used the other 60%.
— Ellen Lee DeGeneres (1958–)

Early History

1.1 The answers, clearly, can vary.

1.2 The answers, clearly, can vary.

1.3 If $a = c$ and $b = d$, then their formula simplifies to $\frac{1}{4}(2a)(2b) = ab$, which is our familiar formula for the area of a rectangle of length a and width b.

Brahmagupta's formula (the Indian Brahmagupta was born around 600 A.D.) asserts that a quadrilateral with side lengths a, b, c, and d will attain its maximum area if and only if it is inscribed in a circle, in which case the area is

$$A = \sqrt{(s - a)(s - b)(s - c)(s - d)},$$

where $s = a + b + c + d/2$ is the semiperimeter of the quadrilateral. This value will never exceed the value given by the Egyptian formula, because

$$A = \sqrt{\frac{(b + c + d - a)}{2} \frac{(a + c + d - b)}{2} \frac{(a + b + d - c)}{2} \frac{(a + b + c - d)}{2}}$$

$$= \frac{1}{4}\sqrt{(b + d + c - a)(b + d + a - c)(a + c + d - b)(a + c + b - d)}$$

$$= \frac{1}{4}\sqrt{\left((b + d)^2 - (c - a)^2\right)\left((a + c)^2 - (d - b)^2\right)}$$

$$\leq \frac{1}{4}(b + d)(a + c).$$

You will later be asked to use vectors to prove that this formula never underestimates the area of a quadrilateral. See Problem 11.12.

1.3 Suppose the vertices are labeled A, B, and C, where $AB = BC = 50$, and $AC = 60$. Let D be the midpoint of the base side AC. Observe that $\triangle BDC$ is a right triangle with one side length 30 and hypotenuse length 50. Using the Pythagorean Theorem, $BD = \sqrt{50^2 - 30^2} = 40$. We now use the Pythagorean Theorem on right triangle ODC, where O is the center of the circle. Clearly, $OC = r$, the radius of the circle. It is easy to argue that O is an interior point of $\overline{BD}$, which gives $OD = 40 - r$. In any case, $OD = |40 - r|$, and we obtain

$$30^2 + |40 - r|^2 = r^2 \quad \Leftrightarrow \quad 30^2 + (40 - R)^2 = r^2,$$

whence $r = \frac{125}{4}$.

1.4 Suppose that the circle circumscribed around a regular hexagon has radius r. Joining its center with two consecutive vertices of the hexagon, we obtain an isosceles triangle with the measure of the angle at the vertex $60°$. Hence, it is an equilateral triangle, and the length of each side of the hexagon is r. By definition of π, the length of the circumference is $2\pi r$, and by the Babylonian approximation, we have $6r/(2\pi r) \approx \frac{24}{25}$. The result follows immediately.

1.5 The second expression is correct. It simplifies to

$$h \cdot \left(\frac{a^2 + 2ab + b^2}{4} + \frac{a^2 - 2ab + b^2}{12} \right) = \frac{h}{12}(4a^2 + 4ab + 4b^2) = \frac{a^2 + ab + b^2}{3}.$$

1.6 It follows from results of Max Dehn in 1900 that no proof of this formula exists by "cutting and assembling" techniques. The only known proofs use some version of integration (calculus).

Axioms: From Euclid to Today

2.1 (a) Take any axiomatic system, pick an axiom or a theorem, construct its negation and add it to the system. Now the axiomatic system contains two opposite statements. For example, add to the Euclidean geometry system a statement: there exists a triangle whose angle sum is different than $180°$.

(b) Take any axiomatic system and a theorem in it which is not an axiom. Declare this theorem to be an axiom and add it to the system. The new set of axioms is dependent. For example, take Hilbert's axiomatic system for the Euclidean plane and add the following statement as a new axiom: three medians of any triangle are concurrent.

2.2 Yes, it is consistent. Take a line l with nine points on it, and a point P off the line. Consider nine 2-point lines, each joining P with a point of l. The resulting geometry satisfies Axioms A, B, and F.

Yes, it is independent. We can demonstrate independence by producing models for each case where one axiom is incorrect while the remaining axioms are correct.

- Take a geometry consisting of one line having five points. This geometry satisfies Axioms A and B, but not Axiom F.
- Take a geometry consisting of two intersecting lines, one having 5 points, and another 6 points. This geometry satisfies Axioms B and F, but not Axiom A.
- Take a geometry consisting of three lines. The lines, represented as sets of points, are the following: $a = \{1, 2, 3, 4, 5, 6, 7, 8, 9, 10\}$, $b = \{1\}$, $c = \{2\}$. This geometry satisfies Axioms A and F, but not Axiom B.

2.3 (a) We will construct several models for the Fano plane.

Model 1 is shown in Figure 3. It has seven lines, labeled as l_i, $i = 1, \ldots, 7$, and seven points, labeled as A, B, C, D, E, F, and G. A column labeled by l_i identifies the three points on l_i. Each of the five axioms can be easily checked. To check *Axiom 4* one has to consider all 21 sets of two distinct points. To check *Axiom 5* one has to consider all 21 sets of two distinct lines.

l_1	l_2	l_3	l_4	l_5	l_6	l_7
A	A	A	B	C	C	B
B	G	E	G	G	F	F
C	F	D	D	E	D	E

FIGURE 3.

Model 2 is given by the diagram in Figure 4. This diagram should be understood in the following way. There are only seven points. Each object that looks like a segment represents a line in the Fano plane, with three points each. We have six such lines. The seventh line does not really look like a segment and is represented by the circular arc; it contains the three points B, F, and E. All five axioms can be easily checked by examining the diagram.

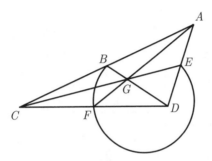

FIGURE 4.

Model 3. Let $\mathbb{Z}/2\mathbb{Z}$ denote the finite field of order 2. The reader who is not familiar with fields may simply view the elements of $\mathbb{Z}/2\mathbb{Z}$ as two "integers" 0 and 1, and then perform all computations by modulo 2. Consider the set P of seven ordered triples (x, y, z) of elements of $\mathbb{Z}/2\mathbb{Z}$ such that not all of x, y, and z are zeros. We call these ordered triples points, so P is a set of points. For a fixed triple $(a, b, c) \in P$, consider the set

$$L_{(a,b,c)} = \{(x, y, z) \in P : ax + by + cz = 0\},$$

where $0 \in \mathbb{Z}/2\mathbb{Z}$, and all computations take place in $\mathbb{Z}/2\mathbb{Z}$. We call each such set $L_{(a,b,c)}$ a line, and let L be the set of all seven lines. We say that a point (x, y, z) lies on a line $L_{(a,b,c)}$ if $ax + by + cz = 0$. It is straightforward to verify that the obtained geometry satisfies all five axioms of the Fano plane, and we leave the verification for the reader. For more models of the Fano plane, see [9].

(b) Yes, since it has models.

(c) Yes. We can demonstrate independence by producing models for each case where one axiom is incorrect while the remaining axioms are correct.

- A model in which only Axiom 1 is incorrect is the trivial geometry that contains no points or lines.
- A model in which only Axiom 2 is incorrect is one with a line l, containing exactly four points, and a point P, not on l, with a line between P and each of the four points on l.
- A model in which only Axiom 3 is incorrect is one with a line l containing exactly three points.

- A model in which only Axiom 4 is incorrect is one with a line l containing exactly three points, and a point P not on l with no lines connecting P to the points in l.
- A model in which only Axiom 5 is incorrect is much harder to construct.

  Let $\{1, 2, 3, 4, 5, 6, 7, 8, 9\}$ be the set of points. Here, we denote a line that is incident to points a, b, and c as abc. Consider the following 12 lines:

$$
\begin{array}{ccc}
123 & 456 & 789 \\
147 & 258 & 369 \\
159 & 267 & 348 \\
357 & 168 & 249
\end{array}
$$

One can check that every pair of distinct points is on exactly one line. This can be done directly, by considering (only!) $9 \cdot 8/2 = 36$ unordered pairs of points. Or it can be done by carefully examining the diagram in Figure 5.

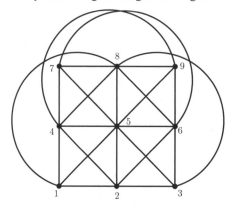

FIGURE 5.

The construction above is known as a Steiner triple system on 9 points or as an affine plane of order 3. More on affine planes or on Steiner triple systems can be found in [**11**] or [**27**].

(d) No, by Axiom 5 there are no two distinct parallel lines.

2.4 (a) By Axiom 5, any two distinct lines have at least one point in common. If two distinct lines had two distinct points in common, it would violate Axiom 4. Therefore any two distinct lines have exactly one point in common.

(b) There are many ways to arrive at this property. We choose one which will allow generalizations to finite geometries that are somewhat analogous to Fano's geometry.

First we prove that the Fano plane has even more uniformity: *There are exactly 3 lines on every point.*

Let P be a point. By Axiom 1, there exists a line l, which, by Axiom 2, contains exactly three points.

Case 1: P is not on l.
Then there are at least 3 lines on P, namely those defined by P and each of the points on l (Axiom 4). If there were a line on P distinct from these three, it would intersect l at a fourth point (Axiom 5), a contradiction with Axiom 2. Therefore, there are exactly 3 lines on P.

Case 2: P is on *l*.

Let *A* and *B* be two other points on *l*. As not all points are collinear, there is a point *C* not on *l*. Consider line *AC*, and let *D* be the third point on it. Clearly, *D* is not on *l*, and lines *PC*, *PD*, and *l* are three distinct lines on *P*. If there were another line on *P*, distinct from these three, it would intersect line *AC* at a point distinct from *A*, *D*, and *C*, so line *AC* would have at least four points. This proves that there are exactly three lines on *P*, and ends the proof that there are exactly three lines on any point.

Now we can easily count the total number of points and lines in the Fano plane. Pick a point *P* and consider all three lines on it. Every other point of the Fano plane is collinear with *P* by Axiom 4. Hence, all points lie on the three lines through *P*. As each of these lines contains exactly three points, we have $1 + 3 \cdot 2 = 7$ points in the Fano plane. Listing all lines through all points, we get $7 \cdot 3 = 21$ lines in the list. Clearly, each line appears exactly three times in the list. So there are exactly $21/3 = 7$ lines in the Fano plane.

2.5 (a) Let *S* denote the set of Axioms 1-5 and *S'* the set of Axioms A-D.

Let us show that *S* implies *S'*. Axiom A is the same as Axiom 4. Axiom B was shown to be implied by *S* in Problem 2.4 (a). Axioms 1 and 2 imply Axiom D.

Let us prove that *S* implies Axiom C. By Axioms 1, 2, and 3, there exists a line *l* with exactly 3 points, say *A*, *B*, and *C*, and there exists a point *D* not on *l*. Lines *DA*, *DB*, and *DC* exist by Axiom 4, they are distinct by Axiom 5, and each must contain one more point by Axiom 2. Let these be points *E*, *F*, and *G*, respectively. Note that *E*, *F*, and *G* are three distinct points: if two of them were equal, Axiom 4 would be violated. Then points *A*, *C*, *D*, and *F* obviously satisfy the property that no three of them are collinear, which proves that *S* implies Axiom C.

This ends the proof that *S* implies *S'*.

Does *S'* imply *S*? Clearly, *S'* implies Axioms 1, 3, 4, and 5. Does it imply Axiom 2? We begin with the following observation. The existence of four points, say *A*, *B*, *C*, and *D*, with no three on a line (Axiom C), implies that each of these four points is on at least three lines (defined by the remaining three points). Any line *m* of the geometry contains at most two of the points *A*, *B*, *C*, and *D*. Suppose *D* is not on *m*. Then three lines on *D* will intersect *m* in three distinct points. Hence, we have shown that

any line in the geometry is on at least three points.

Let us show that every line *m* is on exactly three points.

Consider a line *l* with exactly three points, say *K*, *M*, and *N*, and a point *P* not on it. (Such *l* and *P* exist by Axioms C and D.)

We begin with the case when *m* is not on *P*. If *m* were on more than three points, the lines through those points and *P* would intersect *l* in more than three distinct points, a contradiction. So, in this case, *m* is on exactly three points.

Suppose *m* is on *P*, and *K* is the intersection of *m* and *l*. Let *Q* be the point on *m* distinct from *P* and *K* (as *m* has at least three points, we know that such a point *Q* exists). Then lines *MQ* and *NP* intersect at a point *T*, which is neither on *m* nor on *l*. If *m* were on more than three points, the lines through those points and *T* would intersect *l* in more than three distinct points, a contradiction. So, *m* is on exactly three points in this case, too. Hence we have Axiom 2, and the proof that *S'* implies *S* is finished.

(b) Yes, it is consistent. A model can be built in a way that is analogous to Model 3 of the Fano plane.

Let $\mathbb{Z}/3\mathbb{Z}$ denote the finite field of order 3. (The elements of $\mathbb{Z}/3\mathbb{Z}$ can be thought of as three "integers," 0, 1, and 2, with all computations done by modulo 3.) Consider the set of all ordered triples (x, y, z) of elements of $\mathbb{Z}/3\mathbb{Z}$ such that not all of x, y, and z are zero. Then we identify a pair of triples if their components are proportional. More precisely, we identify triples (x, y, z) and (x', y', z') if there exists a nonzero element k in $\mathbb{Z}/3\mathbb{Z}$ such that $(x', y', z') = (kx, ky, kz)$. As every triple is identified with exactly two other triples, we get exactly $(3^3 - 1)/2 = 13$ distinct classes of identical triples. A class that contains a triple (x, y, z) will be denoted by $< x, y, z >$. We call these classes points, and we denote their set by P.

For a fixed class $< a, b, c > \in P$, consider the set

$$L_{<a,b,c>} = \{< x, y, z > \in P : ax + by + cz = 0\},$$

where $0 \in \mathbb{Z}/3\mathbb{Z}$, and all computations take place in $\mathbb{Z}/3\mathbb{Z}$. We call each such set $L_{<a,b,c>}$ a line, and let L be the set of all 13 lines. We say that a point $< x, y, z >$ lies on a line $L_{<a,b,c>}$, if $ax + by + cz = 0$. It is important to understand that with the way our classes of triples are defined, $< x, y, z > = < x', y', z' >$ if and only if the triples (x, y, z) and (x', y', z') are proportional, i.e. come from the same class. Similarly, $L_{<a,b,c>} = L_{<a',b',c'>}$ if and only if the triples (a, b, c) and (a', b', c') are proportional. It is easy to verify that the resulting geometry satisfies all four axioms A-D, and we leave the verification to the reader. For those who are familiar with vector spaces over finite fields, we just mention that this construction, as well as Model 3 for the Fano plane, has the set of all 1-dimensional subspaces as the set of its points and the set of all 2-dimensional subspaces as the set of its lines, with a point being on a line when the 1-dimensional subspace is contained in the 2-dimensional subspace.

Comment. Suppose Axiom D is replaced by Axiom E': *There is a line with exactly n points on it.*

Models for a finite plane defined by axioms Axiom A, B, C, and E' are known to exist only for n such that $n - 1$ is a prime power. Our examples correspond to $n = 3$ (Fano plane) and $n = 4$ (Problem 2.5 (b)). For more on this topic, see [**27**].

To see that Axioms A, B, C, and E are independent, see the four models in Figure 6.

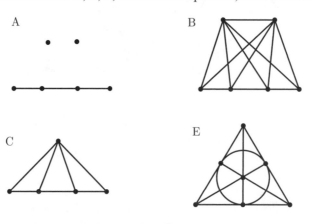

FIGURE 6.

In Model A, a model with only one line, all of the axioms except Axiom A are satisfied. In Model B, all of the axioms except Axiom B are satisfied, as the two "horizontal" lines do not meet. In Model C, all of the axioms except Axiom C are satisfied. In Model E (the Fano Plane), all of the axioms except Axiom E are satisfied.

Triangles

3.2.1 From the Extended Pythagorean theorem, using $s = 5$ and $r = 8$, we see that the angle bisector lengths of the three triangles satisfy $r^2 + s^2 = t^2$. Hence, $t = \sqrt{5^2 + 8^2} = \sqrt{89}$.

3.2.2 (a) Let $x = CD$; then $BD = 8 - x$. Since $\overline{AD} \perp \overline{BC}$, by the Pythagorean theorem, $AD^2 + BD^2 = AB^2$ and $AD^2 + CD^2 = AC^2$. It follows that $36 - (8 - x)^2 = AD^2 = 9 - x^2$, whence $16x = 37$. Thus, $CD = 37/16 \approx 2.31$.

By the Angle Bisector theorem, $BF/CF = AB/AC$. Therefore,

$$\frac{8 - CF}{CF} = \frac{AB}{AC} = 6/3 = 2 \implies CF = 8/3 \approx 2.67.$$

Since $\overline{AM}$ is a median, $CM = BM = BC/2 = 4$.

(b) By the Pythagorean theorem,

$$h_a = AD = \sqrt{AC^2 - CD^2} = \sqrt{3^2 - (37/16)^2} = \frac{\sqrt{935}}{16} \approx 1.91.$$

To find $l_a = AF$, first note that $DF = CF - CD = 8/3 - 37/16 = 17/48$. Then, using the Pythagorean theorem and part (a), we have

$$l_a = \sqrt{AD^2 + DF^2} = \frac{1}{16}\sqrt{935 + (17/3)^2} = \frac{\sqrt{34}}{3} \approx 1.94.$$

Similarly, $DM = CM - CD = 4 - 37/16 = 27/16$, so

$$m_a = AM = \sqrt{AD^2 + DM^2} = \frac{1}{16}\sqrt{935 + 17^2} = \frac{\sqrt{26}}{2} \approx 2.55.$$

3.2.3 Let D be the (unique) base of the perpendicular segment from B on the line AC. If C is on $\overline{AD}$, then $c^2 = (b + CD)^2 + BD^2 = (b + CD)^2 + a^2 - CD^2 = a^2 + b^2 + 2b(CD)$. On the other hand, if D is on $\overline{AC}$, then $c^2 = (b - CD)^2 + BD^2 = (b - CD)^2 + a^2 - CD^2 = a^2 + b^2 - 2b(CD)$. Thus, $a^2 + b^2 = c^2$ if and only if $CD = 0$. In this case, $C = D$, so $\triangle ABC$ is a right triangle by the Pythagorean theorem. The other two cases are mutually exclusive, so if $a^2 + b^2 < c^2$, then D is on $\overline{AC}$ and C is an acute angle; if $a^2 + b^2 > c^2$, then C is on $\overline{AD}$, so C is an obtuse angle.

3.2.4 Let $a_1/a_2 = b_1/b_2 = k$. Then $a_1 = ka_2$ and $b_1 = kb_2$. Therefore,

$$\frac{a_1 + ca_2}{a_1 + da_2} = \frac{(k + c)a_2}{(k + d)a_2} = \frac{(k + c)b_2}{(k + d)b_2} = \frac{b_1 + cb_2}{b_1 + db_2}.$$

3.2.5 Construct diagonals $\overline{AC}$, $\overline{CE}$, $\overline{CG}$, and $\overline{EG}$ to partition the polygon into five triangles. The sum of the interior angles of $ABCDEFG$ is the sum of the interior angles of the five triangles, which is $5(180) = 900°$. Note that although the polygon is not convex, the result from Theorem 3.4 holds. (See Problem 3.2.30.)

3.2.6 (a) The angles in question are the six angles of two distinct triangles. Therefore, the sum is $2(180) = 360°$.

(b) Note that this solution is not as obvious as the one in part (a). The answer is $180°$. See the general solution in part (c).

(c) The answer for the 7-star is $540°$ and the general answer is $180°(n - 4)$. Note that the n-star can be created from a convex n-gon $A_1 \ldots A_n$ by continuing its sides: its kth vertex is obtained by extending sides $\overline{A_{k-2}A_k}$ and $\overline{A_{k+1}A_{k+2}}$ of the given polygon.

For the interior angles of the n-gon, $\sum\limits_{i=1}^{n} a_i = 180(n - 2)$.

Label the angles of the stars as $s_1, s_2, \ldots, s_n, s_{n+1} = s_1$; then $m\angle s_i = 180°$ minus the measures of the two other angles of the triangle containing s_i. Therefore,

$$m\angle s_i = 180 - (180 - a_i) - (180 - a_{i+1}) = a_i + a_{i+1} - 180, \quad \text{so}$$

$$\sum_{i=1}^{n} m\angle s_i = 2\left(\sum_{i=1}^{n} a_i\right) - 180n$$

$$= 2(180n - 360) - 180n = 180n - 720 = 180°(n - 4).$$

3.2.7 Suppose that the n-gon has a acute interior angles and $n - a$ obtuse or right interior angles. Then the sum S of the measures of all n angles is

$$180(n - 2) < 90a + (n - a)(180), \quad \text{so } 180n - 360 < 180n - 90a.$$

Thus $90a < 360$, giving $a < 4$.

3.2.8 Yes. Consider a triangle with side lengths of 2 (miles), $1 + \epsilon$ (miles), and $1 + \epsilon$ (miles), where ϵ is sufficiently small. As ϵ tends to 0 (in fact, it is sufficient for ϵ to be less than an inch), the triangle gets closer to a degenerate triangle and all three altitudes approach 0 in length. For a more rigorous solution, see the construction of $\triangle ABC$ below.

Consider a right triangle ADC with $m\angle A = 90°$, $AD < 1$ inch, and $AC > 2$ miles. Let B' and B be midpoints of $\overline{AC}$ and $\overline{DC}$, respectively, and let $\overline{AA'}$ be an altitude of $\triangle ADC$. (See Figure 7, not drawn to scale.) Note that $\overline{BB'}$ is both the altitude h_b of $\triangle ABC$ and a midline of $\triangle ADC$. Thus, by the Triangle Midline theorem, $h_b = BB' = AD/2 < 1$ inch. Furthermore, since $\triangle ABC$ is isosceles, $h_c = h_a = AA' < AD < 1/2 < 1$ inch. Finally, since $AB = BC > AC/2 > 1$ mile, we see that $\triangle ABC$ has the desired dimensions.

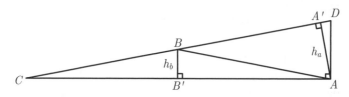

FIGURE 7.

3.2.9 (a) Let A' be a point symmetric to A with respect to l and consider C, the intersection of l and $\overline{A'B}$. (See Figure 8(i).) Let D be the intersection of l and $\overline{A'A}$, and note that $\triangle ADC \cong \triangle A'DC$ by SAS. Thus, $m\angle A'CD = m\angle ACD$, and since the vertical angles $\angle A'CD$ and $\angle BCX$ are congruent, so are the two angles formed by $\overline{AC}$ and $\overline{BC}$ with l.

(b) Let X be another point on l. Then by the triangle inequality, $A'B < A'X + BX$. (See Figure 8(ii).) It follows that $AC + BC = A'C + BC = A'B < A'X + BX = AX + BX$, as desired.

(c) This phenomenon occurs in nature when a light from a source at point A is reflected off of a mirror before arriving at point B. According to Fermat's Principle, the light will travel the shortest distance possible while making this trip. Snell's Law, which can be proven with

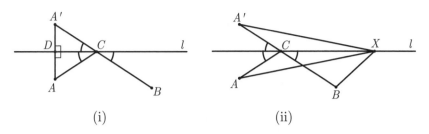

FIGURE 8.

Calculus, says the angle of incidence will equal the angle of refraction, which we showed in part (b).

(d) In this case, C will be the intersection of l and $\overline{AB}$. This follows easily from the triangle inequality: if D is any other point on l, then $AD + DB > AB = AC + BC$.

3.2.10 Consider the medians AA' and BB' in triangle ABC and label their intersection as M. (See Figure 9.)

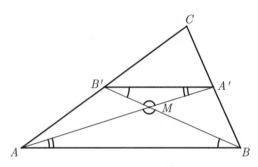

FIGURE 9.

By AAA, $\triangle B'A'M \sim \triangle ABM$. Then $\overline{A'B'} \parallel \overline{AB}$ and $A'B'/AB = 1/2$ by the Triangle Midline theorem. Thus, $A'M/AM = B'M/BM = 2$, so the median AA' meets the median BB' at the point M that separates each into segments in ratio $2:1$. A similar argument can be made with triangles AMC and $A'MC'$ to show that the median CC' is divided into the ratio $2:1$ at its intersection with AA'. It follows that $\triangle A'MB'$ is an image of $\triangle AMB$ under the homothety $H(M, -\frac{1}{2})$.

3.2.11 Using the setup in the solution of Problem 3.2.10, note that both $\overline{AA'}$ and $\overline{CC'}$ intersect $\overline{BB'}$ at a point M such that $BM : MB' = 2$. Hence, they cross $\overline{BB'}$ at the same point and all three medians are concurrent. How pretty!

Alternate proofs of Problems 3.2.10 and 3.2.11 are provided in Chapter 8 (see Theorem 8.8) and in Chapter 12 (see Theorem 12.10).

3.2.12 ($\Rightarrow$) First, consider a point P on the perpendicular bisector, l, of $\overline{AB}$, and let C be the point of intersection of l with $\overline{AB}$, as shown in Figure 10. By definition of bisector of a segment, $AC = CB$. If $P = C$, we are done; if not, then by the HL theorem for right triangles, $\triangle ACP \cong \triangle BCP$, so $AP = PB$.

($\Leftarrow$) Now, suppose that P is a point such that $AP = PB$. If P is on $\overline{AB}$, then the statement is obvious. If P is not on $\overline{AB}$, then let C be the midpoint of $\overline{AB}$. By SSS, $\triangle ACP \cong \triangle BCP$. Since $\angle ACP$ and $\angle BCP$ are both congruent and supplementary, they must be right angles. Hence, P is on the perpendicular bisector of $\overline{AB}$.

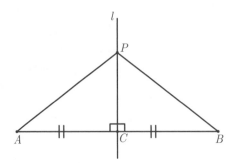

FIGURE 10.

3.2.13 ($\Rightarrow$) Assume that P is a point on the angle bisector of $\angle ABC$. Construct lines through P that are perpendicular to $\overrightarrow{BA}$ and $\overrightarrow{BC}$; let the points of intersection be D and E, respectively. By AAS, $\triangle PDB \cong \triangle PEB$, so we conclude that $PD = PE$.

($\Leftarrow$) Consider a point P that is equidistant from $\overrightarrow{BA}$ and $\overrightarrow{BC}$. If $P = B$, the statement is obvious. Otherwise, let D and E be the feet of perpendiculars from P to $\overrightarrow{BA}$ and $\overrightarrow{BC}$, respectively. (See Figure 11.) Then $\angle PDB$ and $\angle PEB$ are both right angles, and $PD = PE$, so by the HL theorem, $\triangle PDB \cong \triangle PBE$. Thus, $\angle PBD \cong \angle PEB$, and P is on the angle bisector of $\angle ABC$.

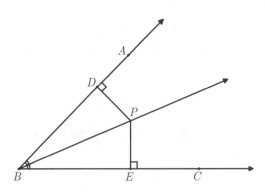

FIGURE 11.

3.2.14 For a triangle ABC, let l_1 and l_2 be the perpendicular bisectors of $\overline{AB}$ and $\overline{AC}$. (See Figure 12.) Because $\overleftrightarrow{AB}$ and $\overleftrightarrow{AC}$ are not parallel, neither are l_1 and l_2. Thus, l_1 and l_2 must intersect at a point O. By Problem 3.2.12, $AO = BO$ and $AO = CO$. Therefore, $BO = CO$, so O lies on the perpendicular bisector of $\overline{BC}$ as well.

3.2.15 *Solution 1:* Use absolutely the same pattern of explanation as used to prove that perpendicular bisectors are concurrent (Problem 3.2.14).

Solution 2: If $\overline{AA_1}$, $\overline{BB_1}$, and $\overline{CC_1}$ are the bisectors of $\triangle ABC$, then, by Theorem 3.16, $AC_1/C_1B = b/a$, $BA_1/A_1C = c/b$, and $CB_1/B_1A = a/c$. As $b/a \cdot c/b \cdot a/c = 1$, the bisectors are concurrent by Ceva's theorem.

Solution 3: Let $\overline{AA_1}$, $\overline{BB_1}$, and $\overline{CC_1}$ be the bisectors of $\triangle ABC$. Let $\overline{AA_1}$ intersect $\overline{BB_1}$ at point D. Denoting $AB_1 = x$, and using Theorem 3.16, we obtain $x/(b - x) = c/a$,

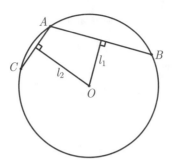

FIGURE 12.

hence, $x = bc/(a + c)$. Applying Theorem 3.16 again, this time for $\triangle B_1AB$, we obtain that $B_1D/DB = bc/(a + c)/c = b/(a + c)$.

Let $\overline{CC}_1$ intersect $\overline{BB}_1$ at point D'. Clearly, computing the ratio $B_1D'/D'B$ in a similar way, we obtain $b/(c + a)$. Since $b/(a + c) = b/(c + a)$, $D = D'$, and the proof is finished.

3.2.16   From the Extended Pythagorean theorem, we see that the median lengths of the three triangles satisfy $x^2 + y^2 = z^2$. Hence, $y = \sqrt{7^2 - 4^2} = \sqrt{33}$.

3.2.17   Let $\triangle ABC$, with m$\angle C = 90°$ and m$\angle A = 30°$, be our triangle. Let D be a point on $\overrightarrow{BC}$ such that $B - C - D$ and $DC = CB$. Then by the HL theorem, $\triangle ACB \cong \triangle ACD$; hence, $m\angle B = m\angle ADB = m\angle BAD = 60°$. Therefore, $\triangle DAB$ is equilateral, so $BC = BD/2 = AB/2$, as desired.

3.2.18   We begin by generalizing the solution used in Problem 3.2.2. Consider triangle ABC with altitude $\overline{AD}$, bisector $\overline{AF}$, and median $\overline{AM}$, as shown in Figure 13. Let $x = CD$; then by the Pythagorean theorem, $b^2 - x^2 = h_a{}^2 = c^2 - (a - x)^2$. Thus, $b^2 - c^2 + a^2 = 2ax$, so $x = \dfrac{b^2 - c^2 + a^2}{2a}$. Also, by the Angle Bisector theorem, $(a - CF)/CF = c/b$, whence

$$CF = \frac{ab}{b + c}.$$

We now use the above calculations to find the three desired values.

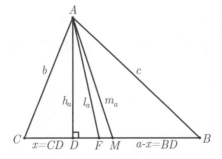

FIGURE 13.

(h_a:) We first show that p, $p - a$, $p - b$, and $p - c$ are factors of $h_a{}^2$.

$$h_a{}^2 = b^2 - \left(\frac{b^2 - c^2 + a^2}{2a}\right)^2 = \frac{4a^2b^2 - (b^2 - c^2 + a^2)^2}{4a^2}$$

$$= \frac{\left(2ab - b^2 + c^2 - a^2\right)\left(2ab + b^2 - c^2 + a^2\right)}{4a^2}$$

$$= \frac{\left(c^2 - (a - b)^2\right)\left((a + b)^2 - c^2\right)}{4a^2}$$

$$= \frac{(c - a + b)(c + a - b)(a + b - c)(a + b + c)}{4a^2}$$

$$= \frac{4}{a^2}\left(\frac{a + b + c}{2} - a\right)\left(\frac{a + b + c}{2} - b\right)\left(\frac{a + b + c}{2} - c\right)\left(\frac{a + b + c}{2}\right)$$

$$= \frac{4p(p - a)(p - b)(p - c)}{a^2}.$$

So, $h_a = \dfrac{2}{a}\sqrt{p(p - a)(p - b)(p - c)}$.

(l_a:) Using the Pythagorean theorem and the values above, we have

$$l_a{}^2 = h_a{}^2 + (CF - x)^2$$

$$= b^2 - x^2 + \left(x - \frac{ab}{b + c}\right)^2$$

$$= b^2 - \frac{2abx}{b + c} + \frac{(ab)^2}{(b + c)^2}$$

$$= \frac{b}{(b + c)^2}\left(b(b + c)^2 - (b^2 - c^2 + a^2)(b + c) + a^2b\right)$$

$$= \frac{b}{(b + c)^2}\left(b(b + c)^2 - (b - c)(b + c)^2 - a^2c\right)$$

$$= \frac{b}{(b + c)^2}\left(c(b + c)^2 - a^2c\right)$$

$$= \frac{bc}{(b + c)^2}(b + c - a)(b + c + a)$$

$$= \frac{4bc}{(b + c)^2}\left(\frac{b + c + a}{2} - a\right)\left(\frac{a + b + c}{2}\right)$$

$$= \frac{4}{(b + c)^2}bc(p - a)p.$$

So, $l_a = \dfrac{2}{b + c}\sqrt{bcp(p - a)}$.

(m_a:) Again, using the Pythagorean theorem and the fact that $BM = CM$, we obtain

$$l_a{}^2 = h_a{}^2 + (x - a/2)^2$$

$$= (b^2 - x^2) + (x^2 - ax + a^2/4)$$

$$= b^2 - a\left(\frac{b^2 - c^2 + a^2}{2a}\right) + a^2/4$$

$$= \frac{1}{4}\left(2(b^2 + c^2) - a^2\right).$$

So, $m_a = \dfrac{1}{2}\sqrt{2(b^2 + c^2) - a^2}$.

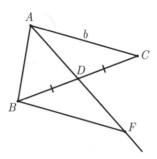

FIGURE **14.**

3.2.19 (a) Let D be the base of the median at A, and let F be the point on $\overrightarrow{AD}$ such that $AF = 2(AD)$. (See Figure 14.) Then by SAS, $\triangle FDB \cong \triangle ADC$, so $BF = AC = b$. By the triangle inequality, $m_a = \frac{1}{2}(AF) < \frac{1}{2}(BF + AB) = \frac{1}{2}(b + c)$, as desired.

(b) Again letting D be the base of the median at A, note that by the triangle inequality, $m_a + BD > c$, so $m_a > c - \frac{a}{2}$. Similarly, $m_b > a - \frac{b}{2}$ and $m_a > b - \frac{c}{2}$. Therefore,

$$m_a + m_b + m_c > a + b + c - \frac{a + b + c}{2} = p.$$

For the second inequality, use part (a) for each median to obtain

$$m_a + m_b + m_c < \frac{b + c}{2} + \frac{a + c}{2} + \frac{a + b}{2} = a + b + c = 2p.$$

3.2.20 (a) Let C be the intersection of lines l and AB and let D be any other point on l. Then by the triangle inequality, $|AD - BD| < AB$. Since A, B, and C are collinear, $AB = |AC - BC|$, which proves that $|AD - BD| < |AC - BC|$ when $D \neq C$. (See Figure 15.) So, the maximum value is obtained when C is chosen as above.

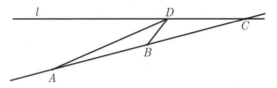

FIGURE **15.**

(b) No. If l and $\overline{AB}$ are parallel, then $|AC - BC|$ will approach AB as $BC \to \infty$, but the value will not be obtained for any point C. Below is a rigorous proof of this statement.

Consider points C and D on l such that $AC > BC$ and $AD > BD$. (See Figure 16.) Let O be the intersection of segments AD and BC. Then by the triangle inequality,

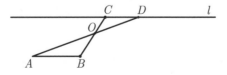

FIGURE **16.**

$OA + OC > AC$ and $OD + OB > BD$. So, $OA + OC + OD + OB > AC + BD$. This is equivalent to

$$AC - (OC + OB) < (OA + OD) - BD.$$

Hence, $AC - BC < AD - BD$, so the difference of distances grows when we move to the right (in our figure) on l.

Our claim can alternatively be proven by establishing $|AD - BD|$ as a function of the distance x shown in Figure 17. Calculus could then be used to verify that $|AD - BD|$ increases as D moves away from A, which shows that $|AD - BD|$ has no maximum for $x \in (0, \infty)$.

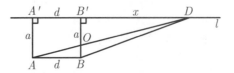

FIGURE 17.

(c) We consider two cases here. The first is where exactly one of the two points, say B, lies on l, and the second case is where A and B lie on opposite sides of l.

For the first case, note that by the triangle inequality, $|AC - BC| < AB$ for any C on l, so the maximum value is bounded by the length of segment AB. Since B lies on l, this value can actually be achieved by letting $C = B$.

Now suppose A and B lie on opposite sides of l. If l is a perpendicular bisector of $\overline{AB}$ then every point C of l is a solution, since $CA - CB = 0$ for all C on l; if not, simply consider the differences $|A'C - BC|$ as C varies over l, where A' is the reflection of A across l. In this case, $A'C = AC$, so from part (a), the maximum value of $|A'C - BC| = |AC - BC|$ will occur when C is the intersection of line $A'B$ and l.

3.2.21 The following inequalities (which we prove presently) provide an elegant solution:

$$a < BO + CO < b + c, \tag{17.1}$$
$$b < AO + CO < a + c,$$
$$c < AO + BO < a + b.$$

Adding these inequalities we obtain

$$a + b + c < 2AO + 2BO + 2CO < 2(a + b + c).$$

Dividing through by 2 gives the desired result.

We now prove the inequalities in (17.1). The lefthand sides come directly from the triangle inequality. To prove the righthand sides, by symmetry it suffices to show that $AO + BO < a + b$. Let P be the intersection of $\overline{AC}$ and $\overrightarrow{BO}$. (See Figure 18.) By the triangle inequality, $AO < OP + AP$ and $BP < BC + PC$. Hence,

$$AO + BO < BO + OP + AP = BP + AP$$
$$< BC + PC + AP = BC + AC = a + b.$$

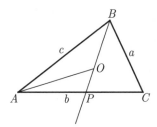

FIGURE **18.**

3.2.22 Clearly c is the longest side length. Thus $a/c < 1$, in which case $(a/c)^3 < (a/c)^2$. Similarly $(b/c)^3 < (b/c)^2$. Therefore,

$$1 = \left(\frac{a}{c}\right)^3 + \left(\frac{b}{c}\right)^3 < \left(\frac{a}{c}\right)^2 + \left(\frac{b}{c}\right)^2.$$

It follows that $c^2 < a^2 + b^2$, so by Problem 3.2.3, the triangle must be acute.

3.2.23 Assume first that $\overline{B_1B}$ bisects the exterior angle at B, and let D lie on $\overrightarrow{AB}$ so that $\overline{CD}$ is parallel to $\overline{BB_1}$. Then $\angle DCB \cong \angle CBB_1$ as alternate interior angles. Thus, since the sum of the angle measures in $\triangle BCD$ is $180°$ and $m\angle CBD = 180° - 2m\angle B_1BC$, we see that $\angle BDC \cong \angle BCD$. Therefore, triangle BCD is isosceles and $BC = BD$, as shown in Figure 19.

Now, since $\overline{BB_1}$ and $\overline{CD}$ are parallel, $AB_1/B_1C = AB/BD$. The latter ratio equals AB/BC, as we sought to prove.

To prove the result in the other direction, simply work backwards through the above proof: $BC = BD$ if and only if $\angle BDC \cong \angle BCD$ if and only if $\overline{BB_1}$ bisects the exterior angle at B. We leave the details to the reader.

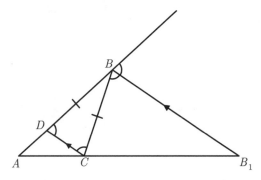

FIGURE **19.**

3.2.24 *Solution 1:* We prove this by contradiction. Suppose that $a < b$; we show that $l_a > l_b$. We use the results of Problem 3.2.18.

$$l_a > l_b \Leftrightarrow l_a{}^2 > l_b{}^2$$
$$\Leftrightarrow \frac{b}{(b+c)^2}\left[(b+c)^2 - a^2\right] > \frac{a}{(a+c)^2}\left[(a+c)^2 - b^2\right]$$
$$\Leftrightarrow b\left[1 - \left(\frac{a}{b+c}\right)^2\right] > a\left[1 - \left(\frac{b}{a+c}\right)^2\right] \tag{17.2}$$

So, we need to show that the last inequality holds. Since we have assumed $a < b$,

$$0 < \frac{a}{b+c} < \frac{b}{a+c} < 1.$$

Indeed, the numerator of the first fraction is less than that of the second while the denominator is greater. Hence,

$$1 - \left(\frac{a}{b+c}\right)^2 > 1 - \left(\frac{b}{a+c}\right)^2 > 0.$$

In the previous line, multiplying the first expression by b and the second expression by a, we obtain the desired inequality (17.2).

Solution 2: Suppose that $l_a = l_b$. From Problem 3.2.18,

$$\frac{2}{b+c}\sqrt{bcp(p-a)} = \frac{2}{a+c}\sqrt{acp(p-b)}.$$

Manipulating this equality by hand to show $a = b$ is tedious. Rather than doing so, we will employ the capabilities of the *Maple* software to demonstrate that $l_a = l_b$ if and only if $a = b$ by showing that $a - b$ is the only factor of $l_a{}^2 - l_b{}^2$. The *Maple* input used for this task is shown below along with the corresponding *Maple* output, all enclosed by horizontal dotted lines. An electronic supplement for this book is devoted to *Maple* and the ways in which it can be used for solving problems of Euclidean geometry.

$p := \dfrac{a+b+c}{2};$

$$\frac{1}{2}a + \frac{1}{2}b + \frac{1}{2}c$$

$l\_a := \dfrac{2(b\,c\,p\,(p-a))^{\left(\frac{1}{2}\right)}}{b+c};$

$$\frac{2\sqrt{b\,c\left(\frac{1}{2}a + \frac{1}{2}b + \frac{1}{2}c\right)\left(-\frac{1}{2}a + \frac{1}{2}b + \frac{1}{2}c\right)}}{b+c}$$

$l\_b := subs(\{a = b, b = a\}, l\_a);$

$$\frac{2\sqrt{a\,c\left(\frac{1}{2}a + \frac{1}{2}b + \frac{1}{2}c\right)\left(-\frac{1}{2}b + \frac{1}{2}a + \frac{1}{2}c\right)}}{a+c}$$

$factor(\ l\_a^2 - l\_b^2\ );$

$$\frac{c(a + b + c)(b - a)(a\,b^2 + 3b\,a\,c + b\,c^2 + b\,a^2 + c^3 + a\,c^2)}{(b+c)^2(a+c)^2}$$

Because each of a, b, and c, is positive, we see from the last line of output above that the only factor shown of $l_a{}^2 - l_b{}^2 = (l_a - l_b)(l_a + l_b)$ that will ever be zero is $(a - b)$. Then, since $l_a > 0$ and $l_b > 0$, $l_a = l_b$ if and only if $a = b$.

Solution 3: Let $\overline{AN}$ and $\overline{BP}$ be two congruent angle bisectors of $\triangle ABC$, i.e. $AN = BP$. Let $\overline{PQ} \parallel \overline{MN} \parallel \overline{AB}$ and assume that $P \neq M$ and $M - P - C$. (See Figure 20(i).)

Since $\angle 1 \cong \angle 2$ and $\angle 2 \cong \angle 3$ (alternate interior angles), $\angle 1 \cong \angle 3$; hence $AM = MN$. Similarly, $PQ = QB$.

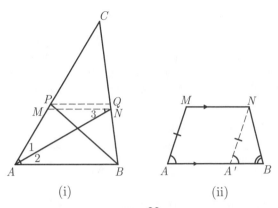

FIGURE 20.

Since $\overline{MN}$ is closer (by assumption) to the base $\overline{AB}$ than is $\overline{PQ}$, we know that $MN > PQ$. In the two isosceles triangles AMN and PQB, the bases are congruent, but $MN = AM > PQ = QB$. Hence, $m\angle 1 > m\angle QBP$, so $m\angle A > m\angle B$. Now choose A' on $\overline{AB}$ so that $\overline{A'N} \parallel \overline{AM}$, as shown in Figure 20(ii). By a property of parallelograms, $A'N = AM$. This implies that in $\triangle A'NB$, $m\angle A' > m\angle B$ (since $m\angle A' = m\angle A$). Therefore, $AM = A'N < BN$, so $AM < BN < BQ = PQ$ and $AM = MN < PQ$. This contradicts our assumptions that $P \neq M$ and $M - P - C$. Hence, $P = M$ and $\angle A \cong \angle B$. It follows that $AC = BC$, so $\triangle ABC$ is isosceles.[1]

3.2.25 Refer to the solution of Problem 3.2.2 (a). Let $\overline{AD}$, $\overline{AF}$, and $\overline{AM}$ be the altitude, angle bisector, and median, respectively, at A. As in that solution, $c^2 - (a - CD)^2 = (AC)^2 - (CD)^2$. Solving for CD, we obtain

$$CD = \frac{a^2 + b^2 - c^2}{2a}.$$

As before, $\dfrac{a - CF}{CF} = \dfrac{AB}{AC}$, so $CF = ab/(b + c)$. It follows that $CM = a/2$.

By obtaining the common denominator of $2a(b + c)$, we thus need to show that either $(a^2 + b^2 - c^2)(b + c) < ab(2a) < a^2(b + c)$ or

$$a^2(b + c) < ab(2a) < (a^2 + b^2 - c^2)(b + c).$$

Since $\triangle ABC$ is not isosceles, $b \neq c$; assume $b < c$. If $a < b < c$, then

$$\begin{aligned}
(a^2 + b^2 - c^2)(b + c) &= a^2b + a^2c + b^3 + b^2c - bc^2 - c^3 \\
&< a^2b + a^2c + b^3 + a^2b - bc^2 - c^3 \\
&= 2a^2b + (b^3 - c^3) + (a^2 - bc)c \\
&< 2a^2b.
\end{aligned}$$

If $b < a < c$, then

$$\begin{aligned}
(a^2 + b^2 - c^2)(b + c) &= a^2b + a^2c + b^3 + b^2c - bc^2 - c^3 \\
&< a^2b + a^2c + a^2b + b^2c - bc^2 - c^3 \\
&= 2a^2b + (a^2 - c^2)c + (b - c)bc \\
&< 2a^2b.
\end{aligned}$$

---

[1] This solution is based on the one from [**35**], where it is attributed to L. Kopeikina.

Finally, suppose that $b < c < a$. It suffices to show

$$(a^2 + b^2 - c^2)(b + c) - 2a^2 b < 0.$$

But the left-hand side of this inequality factors into $(a - b - c)(a + b + c)(c - b)$, and since $a - (b + c)$ is negative (by the triangle inequality) and the other two factors are positive, we are done.

3.2.26 First note that $\triangle AB_1 B \sim \triangle AC_1 C$, because each is a right triangle sharing $\angle A$. (See Figure 21.) Therefore, $\dfrac{AC_1}{AC} = \dfrac{AB_1}{AB}$. Then, by SAS, $\triangle ABC \sim \triangle AB_1 C_1$. Identical reasoning with the roles of A and C interchanged shows that $\triangle CBA \sim \triangle CB_1 A_1$. Transitivity of similarity for polygons proves that $\triangle AB_1 C_1 \sim \triangle A_1 B_1 C$. Hence, $\angle C_1 B_1 A \cong \angle A_1 B_1 C$, and since these two angles are the complements of $\angle C_1 B_1 B$ and $\angle A_1 B_1 B$, respectively, we have proven that $\angle C_1 B_1 B \cong \angle A_1 B_1 B$. Thus, $\overline{BB_1}$ bisects $\angle C_1 B_1 A_1$.

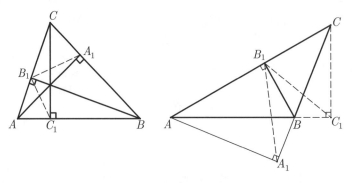

FIGURE 21. Figures for Problems 3.2.26 (left) and 3.2.27 (right).

3.2.27 The statement will still hold as long as the obtuse angle is at vertex B. To see this, we argue identically as in the proof of Problem 3.2.26 until we obtain that $\angle C_1 B_1 A \cong \angle A_1 B_1 C$. (See Figure 21.)

From that point, $m(\angle C_1 B_1 B) = m(C_1 B_1 A) - 90° = m(A_1 B_1 C) - 90° = m(A_1 B_1 B)$. Thus, $\angle C_1 B_1 B) \cong (A_1 B_1 B)$, so $\overline{BB_1}$ bisects $\angle C_1 B_1 A_1$, as desired.

3.2.28 Let M be a point on $\overline{AC}$. Clearly, if $M = A$ or $M = C$, the desired sum is simply $h_a = h_c$. Let M be an arbitrary point on the interior of segment AC. Let $\overline{CD}$ be the altitude of $\triangle ABC$ at C and let E be the point on $\overline{CD}$ such that $\overline{EM} \perp \overline{CD}$. (See Figure 22.) We wish to show that $EC = MM_a$, where $\overline{MM_a} \perp \overline{BC}$.

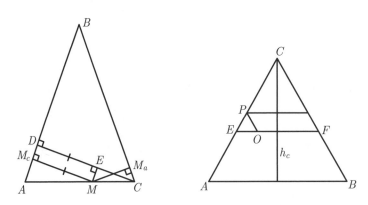

FIGURE 22. Figures for Problems 3.2.28 (left) and 29 (right).

Consider $\triangle CM_aM$ and $\triangle CEM$:

$$m\angle M_aMC = 90° - m\angle C = 90° - m\angle A = m\angle ECM.$$

Hence, $\triangle MEC \sim \triangle M_aCM$ by AA, so $MM_a = CE$. It follows that $CD = CE + ED = MM_c + MM_a$, where M_c is constructed similarly to M_a. So, for every point M of $\overline{AC}$, the sum of the distances is $h_c = CD$, the length of the altitude at C.

This problem will appear again as Problem 5.2.

3.2.29 Specifically, we wish to prove: *In an equilateral $\triangle ABC$, the sum of the distances from any interior point to the three sides is equal to the altitude of the triangle.*[2]

Solution 1: Throughout this solution we use the fact that corresponding altitudes in two congruent triangles are congruent.

Let O be a point in the interior of $\triangle ABC$. Construct points E and F on $\overline{AC}$ and $\overline{BC}$ such that $\overline{EF}$ contains O and $\overline{EF} \parallel \overline{AB}$. (See Figure 22.) Construct $P \in \overline{EC}$ such that $\overline{PO} \parallel \overline{BC}$ and note that triangles CEF and EOP are both equilateral. Then

$$d(O, \overline{AC}) = d(O, \overline{EP}) = d(E, \overline{OP}) = d(P, \overline{EO}) \text{ and}$$
$$d(O, \overline{BC}) = d(E, \overline{BC}) - d(E, \overline{OP}) = d(C, \overline{EF}) - d(P, \overline{EO}).$$

Therefore,

$$\begin{aligned}
d(O, \overline{AC}) &+ d(O, \overline{BC}) + d(O, \overline{AB}) \\
&= d(P, \overline{EO}) + d(C, \overline{EF}) + d(O, \overline{AB}) - d(P, \overline{EO}) \\
&= d(C, \overline{EF}) + d(O, \overline{AB}) = d(C, \overline{AB}) = h_c.
\end{aligned}$$

Essentially the same proof works if O is a boundary point of $\triangle ABC$. Say, for example, that $O \in \overline{AB}$. Simply note that in this case $A = E$, $B = F$, and $d(O, \overline{AB}) = 0$.

Solution 2: We use the result of Problem 3.2.28. If O is an interior point of $\triangle ABC$ and lies on $\overline{EF}$, with $\overline{EF} \parallel \overline{AB}$, then

$$\begin{aligned}
d(O, \overline{AC}) &+ d(O, \overline{BC}) + d(O, \overline{AB}) \\
&= d(O, \overline{EC}) + d(O, \overline{FC}) + d(O, \overline{AB}) \\
&= d(C, \overline{EF}) + d(O, \overline{AB}) = d(C, \overline{AB}) = h_c.
\end{aligned}$$

Alternative solutions to this problem can be found in Problem 5.2(b) and Problem 8.13.

3.2.30 We proceed by induction on n, the number of sides of the polygon. Clearly the result holds for triangles ($n = 3$). Suppose now that it holds for any polygon with k sides, $3 \le k \le n - 1$, and consider the n-sided polygon $P = A_1A_2 \ldots A_{n-1}A_n$.

We first show that there is a diagonal of P that lies completely in its interior. This is the subtle part of the problem. Assume that if we begin walking around $A_1, A_2, \ldots, A_n$ in order, then the interior will always be on our left. Let us begin to rotate ray A_1A_2 counterclockwise. The set of vertices of the polygon swept by the rotating ray is not empty, as it contains A_n. Let A_i, $i > 2$, be the first vertex of P swept by the rotating ray (if more than one such vertex lies on a given ray, then choose the one closest to A_1). If $i < n$, as it is with A_4 in Figure 23(i), then A_1A_i is the desired diagonal. If $i = n$, the desired diagonal is A_2A_n, as in Figure 23(ii). This diagonal (let's call it A_1A_i) divides the n-sided polygon P into an i-sided polygon and an $(n - i + 2)$-sided polygon, and we may proceed with strong induction.

---

[2] This is sometimes referred to as Viviani's theorem, after the seventeenth century Italian engineer Vincenzo Viviani. The proof provided in Solution 1 follows the method shown in "Proof Without Words: Viviani's theorem," Volume 73, Number 3, of *Mathematics Magazine*.

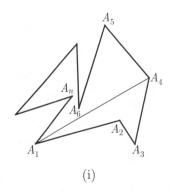

(i)

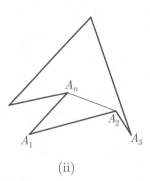

(ii)

FIGURE 23.

By the inductive hypothesis, the sum of the interior angle measures of polygon $\triangle A_1 A_2 \cdots A_i$ will be $180(i-2)°$, while the interior angle measure sum of the polygon $A_i A_{i+1} \ldots A_n A_1$ will equal $180(n-i+2-2)°$; their sum clearly gives the angle measure sum of the polygon $A_1 A_2 \cdots A_n$, and it is $180(n-2)$, as desired.

Parallelograms and Trapezoids

3.3.1 We will prove $(2) \Rightarrow (1) \Rightarrow (4) \Rightarrow (3) \Rightarrow (2)$.

$(2) \Rightarrow (1)$: Assume that $ABCD$ has a right angle; say $m\angle A = 90°$. Then since adjacent angles are supplementary, $\angle B$ and $\angle D$ are also right angles. Since opposite angles of parallelograms are congruent, $m\angle C = 90°$ as well. Thus, $ABCD$ is a rectangle.

$(1) \Rightarrow (4)$: Assume that $ABCD$ is a rectangle, so each angle is a right angle. Since opposite sides of this parallelogram are congruent, by the Pythagorean theorem,

$$AC^2 = AB^2 + BC^2 = AB^2 + AD^2 = BD^2.$$

This proves that the diagonals $\overline{AC}$ and $\overline{BD}$ are congruent.

$(4) \Rightarrow (3)$: Assume that the diagonals are congruent. Since opposite sides are also congruent, by SSS we immediately obtain that $\triangle ABD \cong \triangle DCA$.

$(3) \Rightarrow (2)$: Assume $\triangle ABD \cong \triangle DCA$. Then angles A and D are congruent, and since they are angles of the parallelogram, they are also supplementary. Therefore each is a right angle.

3.3.2 Without loss of generality, assume $BC < AD$. Select E on $\overline{CD}$ so that $\overline{CE} \parallel \overline{AB}$. This creates the parallelogram $ABCE$, in which case $AB = CE$. Furthermore, $m\angle A = m\angle BCE = m\angle CED = x$ and $m\angle B = m\angle CEA = 180 - x$. (See Figure 24(i).) We will prove $(1) \Rightarrow (2) \Rightarrow (3) \Rightarrow (1)$.

$(1) \Rightarrow (2)$: Assume the trapezoid is isosceles, so $AB = CD$. Then $CD = CE$, so triangle CED is isosceles; hence $\angle D \cong \angle CED$. Therefore, $\angle D$ and $\angle CEA$ are supplementary, as are angles A and CEA. This proves $\angle A \cong \angle D$. (It is also true that $\angle B \cong \angle C$, as we will see next.)

$(2) \Rightarrow (1)$: Assume that $m\angle A = m\angle D = x°$. We first note that angles B and C are congruent, as follows: from triangle CED, we see that $m\angle C = m\angle BCD = x + (180 - 2x) = 180 - x = m\angle B$. (If we assume that $\angle B$ and $\angle C$ are congruent, then we could similarly prove that $\angle A$ and $\angle D$ were congruent.)

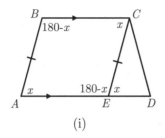

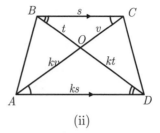

(i)   (ii)

FIGURE 24.

Then, since $AB = CE = CD$, by SAS it follows that $\triangle ACB \cong \triangle DBC$. This proves that $AC = BD$, so the diagonals are congruent.

(3) $\Rightarrow$ (1): Assume that $AC = BD$ and that the diagonals meet at O. We will prove $AB = CD$ by showing that triangles CBD and BCA are congruent. First note that since the diagonals serve as transversals for the parallel lines AD and BC, $\angle CBD \cong \angle ADB$ and $\angle BCA \cong DAC$. Hence, $\triangle BCO \sim \triangle DAO$. Suppose the corresponding sides of these triangles have ratio k; i.e., if the side lengths of triangle BCO are s, t, and v, then triangle DAO will have respective side lengths ks, kt, and kv. (See Figure 24(ii).) We then have

$$v(1 + k) = v + kv = AC = BD = t + kt = t(1 + k),$$

from which it follows that $v = t$. From here, the solution is clear: $\triangle BCO$ is isosceles $\Rightarrow \angle CBD \cong \angle BCA \Rightarrow \triangle CBD \cong \triangle BCA$ (by SAS).

3.3.3 Let W, X, Y, and Z be the midpoints of segments AB, BC, CD and DA, respectively. Then, by applying the Triangle Midline theorem to $\triangle ABC$ and $\triangle ADC$, we see that $\overline{WX} \parallel \overline{AC}$ and $\overline{YZ} \parallel \overline{AC}$, so $\overline{WX} \parallel \overline{YZ}$. Similarly, $\overline{XY} \parallel \overline{ZW}$, which proves that $WXYZ$ is a parallelogram.

3.3.4 Because the diagonals of quadrilateral $ABCD$ bisect each other, it is a parallelogram by Theorem 3.22. Therefore, $AB = CD$ and $BC = AD$. Furthermore, by Theorem 3.23, $AC^2 + BD^2 = AB^2 + BC^2 + CD^2 + DA^2$. As a result,

$$\begin{aligned} m_b &= \frac{BD}{2} = \frac{\sqrt{AB^2 + BC^2 + CD^2 + AD^2 - AC^2}}{2} \\ &= \frac{\sqrt{2AB^2 + 2BC^2 - AC^2}}{2} \\ &= \frac{1}{2}\sqrt{2a^2 + 2c^2 - b^2}. \end{aligned}$$

3.3.5 Let $ABCD$ be a parallelogram. If it is a rectangle, the statement is obvious. Let $\angle A$ and $\angle C$ be acute angles and $\angle B$ and $\angle D$ be obtuse angles of the parallelogram. Clearly, one of the angles BDA or BDC is acute (otherwise $m\angle D > 180°$). If the acute angle is $\angle BDA$, then the altitude at B in $\triangle ABD$ falls on $\overline{AD}$. If it is $\angle BDC$, then the altitude at B in $\triangle DBC$ falls on $\overline{CD}$.

3.3.6 In parallelogram $ABCD$, suppose that the angle bisectors from A and B meet at E and angles are labeled as in Figure 25(i).

Because adjacent interior angles of a parallelogram are supplementary, $m\angle A + m\angle B = 180°$, whence $m\angle 1 + m\angle 2 = 90°$. Since the sum of the angle measures in $\triangle ABE$ is $180°$, $m\angle AEB = 90°$. Thus, the rays $\overrightarrow{AE}$ and $\overrightarrow{BE}$ meet at a right angle. Similarly, the other pairs

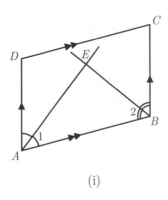

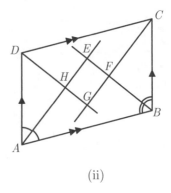

(i)   (ii)

FIGURE 25.

of angle-bisecting rays meet at right angles; thus, a rectangle is formed – in Figure 25(ii), the rectangle is $EFGH$.

3.3.7 Let $\overline{EF}$ be the midline of the trapezoid. (See Figure 26.) As it is parallel to $\overline{BC}$, by Theorem 3.13, it passes through M and N. Hence, $\overline{EM}$ is the midline in $\triangle ABC$, and $EM = b/2$. Similarly, $\overline{FN}$ is the midline in $\triangle DBC$, and $FN = b/2$. So $MN = EF - EM - FN = (a + b)/2 - b/2 - b/2 = (a - b)/2$.

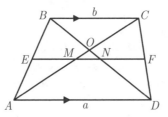

FIGURE 26.

3.3.8 Let ABC be our triangle. Through each vertex of the triangle draw a line parallel to the opposite side. Consider $\triangle A_1 B_1 C_1$ formed by these three lines. (See Figure 27.) As AC_1BC and ABA_1C are parallelograms, $C_1B = AC = BA_1$. Hence, B is the midpoint of $\overline{C_1A_1}$. Therefore, the altitude of $\triangle ABC$ at B lies on the perpendicular bisector of $\overline{C_1A_1}$. Similar statements hold for two other altitudes of $\triangle ABC$. As three perpendicular bisectors to the sides of a triangle are concurrent (see Problem 3.2.14), the lines defined by three altitudes of a triangle are concurrent.

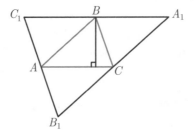

FIGURE 27.

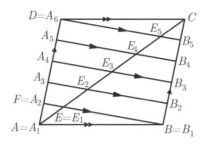

FIGURE 28.

3.3.9 *Solution 1 (sketch):* Construct points $A_1 = A$, $A_2 = F$, A_3, A_4, A_5, and $A_6 = D$ on $\overline{AD}$ that partition it into five congruent segments. Similarly, partition $\overline{BC}$ with points $B = B_1$, $B_2, \ldots, B_6 = C$. Now construct five segments $\overline{A_2B_1}$, $\overline{A_3B_2}, \ldots, \overline{A_6B_5}$. The latter four segments are each parallel to $\overline{BF}$, and it is easy to see that the five intersection points of the segments with $\overline{AC}$ (call them $E = E_1, E_2, \ldots E_5$) partition $\overline{AC}$ into six congruent segments (see Figure 28). It follows that $AE = AC/6$. One way to generalize this problem is to replace 5 by n and 6 by $n + 1$.

Solution 2 (generalization): Suppose $AF = a$ and $AD = b$, so $AF/AD = a/b$. We will show that $AE/AC = a/(a + b)$, which generalizes the given statement. To this end, let $D' \in \overline{BC}$ so that $\overline{DD'} \parallel \overline{FB}$, and let G be the intersection of segments AC and DD'. (See Figure 29.)

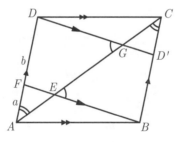

FIGURE 29.

Now, $\angle DAC \cong \angle BCA$ and $\angle DGA \cong \angle BCA$ (alternate interior angles in each case). Hence, since $DA = BC$, $\triangle DAG \cong \triangle BCE$ by SAA. From here it follows that $AG = EC$, so $AE = GC$. Next, because $\triangle AFE \sim \triangle ADG$, $\dfrac{a}{AE} = \dfrac{b}{AG}$. We conclude by noting that

$$\frac{AE}{AC} = \frac{AE}{AE + EC} = \frac{a}{a(1 + EC/AE)} = \frac{a}{a(1 + b/a)} = \frac{a}{a + b}.$$

The reader is invited to compare this solution with the one provided for Problem 8.9.

3.3.10 *Solution 1:* Let M be the center of $ABCD$; that is, $M = \overline{AC} \cap \overline{BD}$. Let $Z = \overrightarrow{D_1C_1} \cap \overrightarrow{BC}$. (See Figure 30.)

Since $\overleftrightarrow{BC} \parallel \overleftrightarrow{AD}$, angles DD_1C_1 and C_1ZB_1 are congruent alternate interior angles; since $\overleftrightarrow{A_1B_1} \parallel \overleftrightarrow{D_1C_1}$, angles C_1ZB_1 and A_1B_1B are congruent corresponding angles. Thus, $\angle DD_1C \cong \angle BB_1A_1$. Since $A_1B_1 = C_1D_1$ and $\angle A_1BB_1 \cong \angle C_1DD_1$ by Theorem 3.22, $\triangle DD_1C_1 \cong \triangle BB_1A_1$ by AAS.

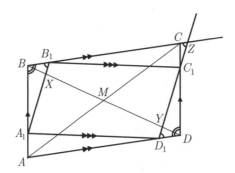

FIGURE **30.**

This implies that $DD_1 = BB_1$; furthermore, $MD = MB$ by Theorem 3.22. Let $X = \overline{A_1B_1} \cap \overline{BD}$ and $Y = \overline{C_1D_1} \cap \overline{BD}$. Then $\angle D_1DY \cong \angle B_1BX$ as alternate interior angles. Thus, $\triangle DD_1M \cong \triangle BB_1M$ by SAS.

Therefore, $B_1M = MD_1$ and $B_1 - M - D_1$. Therefore, M is the center of parallelogram $A_1B_1C_1D_1$.

Solution 2: Replicate Solution 1 up to the point where we conclude that $DD_1 = BB_1$. Consider now the homothety $H = H(M, -1)$. Since $H(\overline{BC}) = \overline{DA}$, $H(\overline{DA}) = \overline{BC}$, and $BB_1 = DD_1$, we see that $H(B_1) = D_1$ and $H(D_1) = B_1$. Hence, M is the center of $A_1B_1C_1D_1$.

3.3.11 (a) First note that $\triangle BOC \sim \triangle DOA$, $\triangle AOE \sim \triangle ACB$, and $\triangle COF \sim \triangle CAD$ by AA. (See Figure 31.)

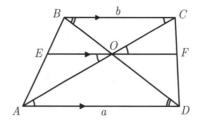

FIGURE **31.**

We thus obtain the following ratios:

$$\frac{AO}{CO} = \frac{a}{b}, \ \frac{AO}{EO} = \frac{AC}{b}, \ \text{and} \ \frac{CO}{FO} = \frac{AC}{a}.$$

Therefore, $EO = \dfrac{(AO)(b)}{AC} = \dfrac{(FO)(AO)(b)}{(CO)(a)} = FO$.

For another solution of this problem, see Solution 2 of Problem 3.3.14 and the comment following it.

(b) We now find EO using the ratios above.

$$EO = \frac{b(AO)}{AC} = \frac{b(AO)}{AO + CO} = \frac{b}{1 + b/a} = \frac{ab}{a + b}.$$

For an alternate solution, see Problem 8.10.

3.3.12 Since $\overline{MN}$ and $\overline{PQ}$ are parallel, $\angle MNT \cong \angle SQT$; also, $\angle MTN$ and $\angle STQ$ are congruent vertical angles. Thus, $\triangle MNT \sim \triangle SQT$. The congruent opposite angles of the parallelogram, $\angle MNR$ and $\angle SQM$, together with the congruent alternate interior angles NMR and QSM give us that $\triangle MNR \sim \triangle SQM$. Thus,

$$\frac{MT}{ST} = \frac{MN}{SQ}, \quad \text{and} \quad \frac{MR}{MS} = \frac{MN}{SQ}, \quad \text{which implies} \quad \frac{MR}{MS} = \frac{MT}{ST}.$$

Using the above equation and some algebra gives the result.

$$(MR)(ST) = (MT)(MS)$$
$$MR(MS - MT) = (MT)(MS)$$
$$(MT)(MR) + (MT)(MS) = (MR)(MS)$$
$$MT = \frac{(MR)(MS)}{MR + MS}$$
$$\frac{1}{MT} = \frac{MR + MS}{(MR)(MS)}$$
$$\frac{1}{MT} = \frac{1}{MR} + \frac{1}{MS}$$

This problem is solved using a different method in Chapter 12. See Problem 12.8.

3.3.13 Construct $\overline{HK}$ through M and parallel to $\overline{AB}$, yielding the parallelogram $ABHK$. (See Figure 32.) Then $\triangle AKM$ is isosceles, since $\angle MAK \cong \angle MAB \cong \angle AMK$. Similarly, $\triangle BHM$ is isosceles. Therefore, $KM = AK = BH = MH$. This proves that M is the midpoint of $\overline{HK}$. Similarly, N is the midpoint of the corresponding segment through N parallel to $\overline{CD}$.

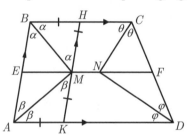

FIGURE 32.

It easily follows from Theorem 3.13 that the line containing $\overline{MN}$ is parallel to the bases of the trapezoid, and it intersects the lateral sides $\overline{AB}$ and $\overline{CD}$ at points E and F which bisect those sides. Thus, $\overline{EF}$ is the midline of $ABCD$. Therefore, $\overline{EF} = (AD + BC)/2$. It follows that

$$MN = EF - EM - NF = EF - AE - CF$$
$$= EF - \frac{1}{2}AB - \frac{1}{2}CD$$
$$= \frac{1}{2}(AD + BC - AB - CD).$$

3.3.14 Let $ABCD$ be a trapezoid with $\overline{AD} \parallel \overline{BC}$; let $Q = \overline{AC} \cap \overline{BD}$, and $P = \overline{AB} \cap \overline{CD}$.

Solution 1: Construct $\overline{EF}$ through point Q, parallel to the bases of the trapezoid. Suppose line PQ intersects the bases at points M and N as shown in Figure 33.

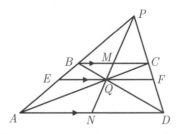

FIGURE **33.**

Then $\triangle PMB \sim \triangle PQE$ and $\triangle PMC \sim \triangle PQF$, by AA. Therefore, $EQ/BM = PQ/PM = FQ/CM$. From Problem 3.3.11 (a), $EQ = FQ$, which proves $BM = CM$. Hence, M is the midpoint of $\overline{BC}$. Replacing B, M, and C, with A, N, and D, respectively, proves that N is the midpoint of $\overline{AD}$.

Solution 2: We refer again to Figure 33. Let N and M be the midpoints of the bases $\overline{AD}$ and $\overline{BC}$, respectively. Consider the homothety $H_1 = H(Q, -\frac{BC}{AD})$. Then $H_1(\overline{AD}) = \overline{CB}$. Since any homothety maps the midpoint of a segment to the midpoint of its image, $H(N) = M$. As the center of homothety, a point, and the image of the point are always collinear, the points M, Q, and N are collinear.

Let P be the intersection of lines AB and CD. Then the homothety $H_2 = H(P, \frac{AD}{BC})$ maps $\overline{BC}$ to $\overline{AD}$. Arguing as above, we obtain that $H_2(M) = N$, and that points P, N, and M are collinear.

Hence, both P and Q belong to the line MN, so all four points are collinear.

Comment. Note that this solution does *not* use the result of Problem 3.3.11(a), but rather implies it. Indeed, consider the homothety $H = H(P, \frac{PQ}{PN})$. Then $H(\overline{BC}) = \overline{EF}$ and $H(M) = Q$; therefore Q is the midpoint of $\overline{EF}$.

3.3.15 Let M, N, P, R, and L be the midpoints of sides AB, BC, CD, DE, and diagonal BE, respectively. (See Figure 34.)

Note that since $LNPR$ is a parallelogram (by Problem S3.3.3), K is the midpoint of its diagonal $\overline{RN}$; hence K is also the midpoint of the parallelogram's diagonal $\overline{LP}$. Therefore P, K, and L are collinear. Since K is the midpoint of $\overline{LP}$ and H is the midpoint of MP, $\overline{HK}$ is the midline of triangle PLM. Thus, $HK = LM/2$. Similarly, since M and L are midpoints of $\overline{AB}$ and $\overline{BE}$, $\overline{ML}$ is the midline of $\triangle BAE$, and $ML = AE/2$. Since midlines of triangles are parallel to bases, $\overline{HK} \parallel \overline{ML}$ and $\overline{ML} \parallel \overline{AE}$. From here we see that $HK = AE/4$ and $\overline{KH} \parallel \overline{AE}$.

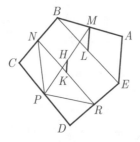

FIGURE **34.**

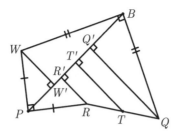

FIGURE 35.

3.3.16 Let T mark the treasure (the midpoint of $\overline{QR}$). Drop perpendiculars from Q, R, W, and T on line BP; label the intersections Q', R', W', and T' as in Figure 35. [Note: Our solution here assumes these vertices are ordered as in the figure; similar arguments can be made for other orders.]

Since $\overline{RR'}$ and $\overline{QQ'}$ are parallel and T is the midpoint of $\overline{RQ}$, $\overline{TT'}$ is the midline of trapezoid $R'RQQ'$. Therefore $R'T' = T'Q'$. Furthermore,

$$m\angle WPB + m\angle BPR = 90° = m\angle BPR + m\angle PRR',$$

so $\angle WPB \cong \angle PRR'$. It follows by AAS that $\triangle WW'P \cong \triangle PR'R$; thus, $WW' = PR'$ and $PW' = R'R$.

Similarly, $\angle PBQ \cong \angle BWW'$, so by AAS, $\triangle BWW' \cong \triangle QBQ'$; thus, $BQ' = WW' = PR'$ and $QQ' = BW'$. This allows us to show that T' is the midpoint of $\overline{BP}$:

$$PT' = PR' - R'T = BQ' - T'Q' = BT'.$$

Finally,

$$TT' = \frac{1}{2}(R'R + Q'Q) = \frac{1}{2}(PW' + BW') = \frac{1}{2}(PB).$$

Therefore, to find the treasure, one simply needs to walk halfway from P to B, turn right 90°, and then walk the same distance to the treasure.

The reader can also use coordinate geometry to find a solution to this problem. See Problem 14.24.

Circles

4.1 We use Theorem 4.17 to obtain the following.

(a) $PA \cdot PB = PC \cdot CD$, so $4(12 - 4) = PC^2$. Therefore, $PC = \sqrt{32}$, so $PD = 2PC = 2\sqrt{32}$.

(b) $CP \cdot PD = AP \cdot PB = 48$. Since $PD = PC$, $PD = \sqrt{48}$.

4.2 Let $A_1A_2 \cdots A_n$ denote the regular n-gon. (A portion of the n-gon is shown in Figure 36.) Consider bisectors of angles A_1 and A_2. They are not parallel since $m\angle A_1 + m\angle A_2 < 180°$. Let O be the point of their intersection. Join O with all other vertices. Then angles 1 and 2 are congruent as halves of congruent angles A_1 and A_2. Hence, $\triangle A_1OA_2$ is isosceles, so $OA_1 = OA_2$.

Now compare $\triangle A_1OA_2$ and $\triangle A_2OA_3$. They share the common side A_2O and $A_1A_2 = A_2A_3$ as two sides of a regular polygon. Since $\angle 2$ is a half of $\angle A_2$, $m\angle 2 = m\angle 3$. Hence,

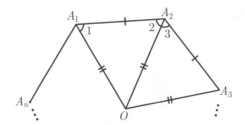

FIGURE 36.

$\triangle A_1 O A_2 \cong \triangle A_2 O A_3$ by SAS, so $O A_1 = O A_2 = O A_3$. Continuing comparing consecutive triangles, we obtain

$$O A_1 = O A_2 = \cdots = O A_n.$$

This implies that all vertices of the polygon lie on a circle with center O.

Since all n isosceles triangles with vertex O are congruent, their altitudes at O are congruent. Hence, all sides of the polygon touch a circle with center at O and radius congruent to these altitudes.

Thus, the centers of the incircle and circumcircle coincide. Clearly O must lie on the bisector of each angle of the polygon and on the perpendicular bisector to each side of the polygon. This implies uniqueness.

4.3 Let M and N be the points of tangency on $\overrightarrow{BA}$ and $\overrightarrow{BC}$, respectively. (See Figure 37.)

Since $ED = EM$ and $FD = FN$ as tangent segments to a circle from the same point, Perimeter $(\triangle EBF)$

$$\begin{aligned}
&= BE + EF + FB = BE + (ED + DF) + FE \\
&= BE + (EM + NF) + FB \\
&= (BE + EM) + (BF + FN) = BM + BN = 2BM.
\end{aligned}$$

Hence, the perimeter is independent of D and is always equal to twice the length of the tangent segment from B to the circle.

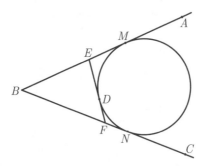

FIGURE 37.

4.4 Join the centers, O_1 and O_2, with A and consider two isosceles triangles. (See Figure 38.) Since A lies on $\overline{O_1 O_2}$, $\angle 1 \cong \angle 2$ as vertical angles. As the triangles $A O_1 B_1$ and $A O_2 B_2$ are isosceles, $\angle 1 \cong \angle 3$ and $\angle 2 \cong \angle 4$. Hence, $\angle 3 \cong \angle 4$. Since these are alternate interior angles for lines $O_1 B_1$, $O_2 B_2$, and secant $B_1 B_2$, we conclude that $\overline{O_1 B_1}$ is parallel to $\overline{O_2 B_2}$.

4.5 Recall that we assumed $r_1 \geq r_2$.

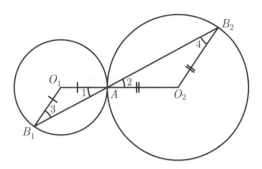

FIGURE **38.**

(2) ($\Rightarrow$) Suppose that $\mathcal{C}(O_1, r_1)$ and $\mathcal{C}(O_2, r_2)$ intersect (internally or externally) at one point, P. Then by Theorem 4.10, $\mathcal{C}(O_1, r_1)$ and $\mathcal{C}(O_2, r_2)$ are tangent to the same line, l, at P. Consequently, $\overline{O_1 P}$ and $\overline{O_2 P}$ are both perpendicular to l, and O_1, P, and O_2 are collinear. If the intersection is external (as in Figure 39(i)), then $O_1 - P - O_2$ so $O_1 O_2 = O_1 P + P O_2 = r_1 + r_2$.

If the intersection is internal (as in Figure 39 (ii)), then $O_1 - O_2 - P$ since we have assumed that $r_1 \geq r_2$. Thus, $r_1 = d + r_2$.

(i) $d = r_1 + r_2$ (ii) $r_1 = d + r_2$

FIGURE **39.**

($\Leftarrow$) Suppose first that $r_1 + r_2 = d$. Let P_1 and P_2 be points of intersection of $\overline{O_1 O_2}$ with $\mathcal{C}(O_1, r_1)$ and $\mathcal{C}(O_2, r_2)$, respectively. Then $O_1 P_1 = r_1$ and $O_2 P_2 = r_2$. Since O_1, P_1, P_2 and O_2 are all collinear, and by the supposition that $r_1 + r_2 = d$, it must be that $P_1 = P_2$. Recall that $\overleftrightarrow{O_1 O_2}$ is an axis of symmetry for both $\mathcal{C}(O_1, r_1)$ and $\mathcal{C}(O_2, r_2)$, so any additional point of intersection for $\mathcal{C}(O_1, r_1)$ and $\mathcal{C}(O_2, r_2)$ would result in three distinct points of intersection, an impossibility. Thus, $\mathcal{C}(O_1, r_1)$ and $\mathcal{C}(O_2, r_2)$ intersect at exactly one point.

Now, assume that $r_1 = d + r_2$. Reasoning similar to the preceding proves that the circles intersect at exactly one point.

(3) ($\Rightarrow$) Assume that $\mathcal{C}(O_1, r_1)$ and $\mathcal{C}(O_2, r_2)$ intersect at two points, P_1 and P_2. As before, $\overleftrightarrow{O_1 O_2}$ is an axis of symmetry for both circles. Thus, it must be that P_1 is mapped to P_2 under reflection about $\overleftrightarrow{O_1 O_2}$. In this case, O_1, O_2, and P_1 cannot be collinear, so must serve as vertices of a triangle. By the triangle inequality, the largest of the values $r_1 = O_1 P_1$, $r_2 = O_2 P_1$, and $d = O_1 O_2$ must be less than or equal to the sum of the other two.

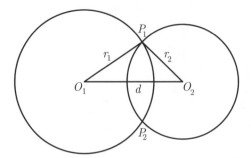

($\Longleftarrow$) We have already proven that $\mathcal{C}(O_1, r_1)$ and $\mathcal{C}(O_2, r_2)$ do not intersect if and only if the maximum of $\{r_1, r_2, d\}$ is greater than the sum of the other two values, and that $\mathcal{C}(O_1, r_1)$ and $\mathcal{C}(O_2, r_2)$ intersect in one point if and only if the maximum of $\{r_1, r_2, d\}$ is equal to the sum of the other two values. We can only conclude that if the maximum of $\{r_1, r_2, d\}$ is less than the sum of the other two values, then $\mathcal{C}(O_1, r_1)$ and $\mathcal{C}(O_2, r_2)$ must intersect in two points.

4.6 Consider $\triangle ABC$ with orthocenter H, inscribed in $\mathcal{C}(O, r)$. Let A', B' and C' be the feet of altitudes from A, B, and C, respectively, as shown in Figure 40. Let H' be the reflection of H across $\overline{AC}$.

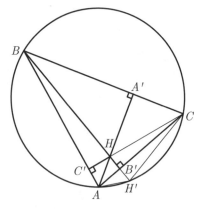

FIGURE 40.

By SAS, $\triangle AB'H$ is congruent to $\triangle AB'H'$. The right triangles $AB'H$ and $BA'H$ are similar. Thus, $m\angle AH'B' + m\angle HBA' = 90°$.

Likewise, $\triangle CB'H$ is congruent to $\triangle CB'H'$ and the right triangles $BC'H$ and $CB'H$ are similar. Thus, $m\angle CH'B' + m\angle HBC' = 90°$.

Therefore, $m\angle AH'C + m\angle C'BC = 180°$, which by Theorem 4.21 proves that quadrilateral $ABCH'$ is inscribed in $\mathcal{C}$. Therefore H' lies on $\mathcal{C}$.

4.7 (a) See Figure 41. For each line through A distinct from line AB, $\angle D_1$ is an inscribed angle of $\mathcal{C}$ whose sides subtend the same arc; hence its measure does not change. A similar statement holds for $\angle D_2$. Hence, $m\angle B = 180° - m\angle D_1 - m\angle D_2$, which does not depend on the line through A.

(b) Since points D_1, A, and D_2 are distinct, $BD_2 > BA$, and BD_2 decreases as AD_2 decreases. Therefore the smallest value of D_1D_2 is not achievable. We claim that the maximum value for D_1D_2 is achieved when $\overline{D_1D_2}$ is parallel to $\overline{O_1O_2}$, the segment connecting the circle

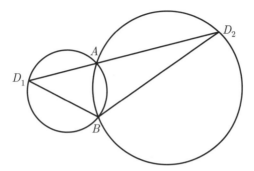

FIGURE 41.

centers. We present two proofs of this claim. The first proof will also demonstrate that the maximum occurs when chords BD_1 and BD_2 are diameters. The second proof will also demonstrate that the maximum value is $2 \cdot O_1O_2$, and that, in this case, $\overline{D_1D_2} \parallel \overline{O_1O_2}$.

Proof 1. Part (a) implies that all triangles of the form D_1BD_2 are similar by AA. Since sides of similar triangles are proportional, D_1D_2 is the greatest when BD_1 is the greatest, i.e., if $\overline{BD_1}$ is a diameter. This also implies that when $\overline{BD_1}$ is a diameter of $\mathcal{C}_1$, $\overline{BD_2}$ is a diameter of $\mathcal{C}_2$ and $\overline{BA} \perp \overline{D_1D_2}$. This is equivalent to the statement that $\overline{D_1D_2}$ is parallel to a line passing through the centers of the circles.

Proof 2. See Figure 42. Let M_1 and M_2 be the midpoints of the chords D_1A and D_2A, respectively. Then $\overline{O_1M_1} \perp \overline{D_1A}$ and $\overline{O_2M_2} \perp \overline{D_2A}$ by Theorem 4.6(1). Hence, $\overline{O_1M_1} \parallel \overline{O_2M_2}$. Let N be the base of the perpendicular from O_1 on $\overleftrightarrow{O_2M_2}$. Then $O_1M_1M_2N$ is a rectangle and $M_1M_2 = O_1N$. As $M_1M_2 = \frac{1}{2}D_1D_2$ and $O_1N \leq O_1O_2$, we see that $D_1D_2 \leq 2 \cdot O_1O_2$.

The length O_1O_2 does not depend on the line through A. Therefore D_1D_2 is maximum if and only if $N = O_2$. This is equivalent to the statement that $\overline{D_1D_2} \parallel \overline{O_1O_2}$.

An advantage of the first solution is that it does not require additional constructions and provides further information about the shape of $\triangle D_1BD_2$, in the extremal case. An advantage of the second solution is that it does not use similarity or inscribed angles. It also gives us the maximum of D_1D_2 in terms of the length of a fixed segment O_1O_2.

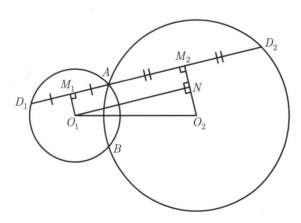

FIGURE 42.

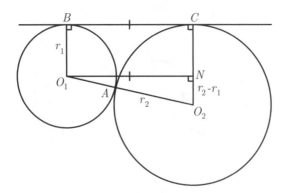

FIGURE 43.

4.8 Let $\overleftrightarrow{BC}$ be a line that is externally tangent to $\mathcal{C}_1 = \mathcal{C}(O_1, r_1)$ at B and $\mathcal{C}_2 = \mathcal{C}(O_2, r_2)$ at C.

(a) Construct $\overline{O_1 N}$ such that $\overleftrightarrow{O_1 N} \perp \overleftrightarrow{CN}$. (See Figure 43.) Applying the Pythagorean theorem to $\triangle O_1 N O_2$ we obtain

$$O_1 N = \sqrt{(O_1 O_2)^2 - (O_2 N)^2}$$
$$= \sqrt{(r_1 + r_2)^2 - (r_2 - r_1)^2} = \sqrt{4 r_1 r_2} = 2\sqrt{r_1 r_2}.$$

Clearly, $O_1 N = BC$.

(b) Let $\mathcal{C}(O, x)$ be the circle tangent to $\mathcal{C}_1$, to $\mathcal{C}_2$, and to their common tangent, $\overleftrightarrow{BC}$. Let $MB = NC = x$, the length of the unknown radius, as shown in Figure 44. Since MN is the length of the external tangent $\overline{BC}$, we have

$$MN = MO + ON = \sqrt{O_1 O^2 - O_1 M^2} + \sqrt{O_2 O^2 - O_2 N^2}$$
$$= \sqrt{(r_1 + x)^2 - (r_1 - x)^2} + \sqrt{(r_2 + x)^2 - (r_2 - x)^2}$$
$$= 2(\sqrt{r_1 x} + \sqrt{r_2 x}).$$

Using part (a) of this problem, $MN = 2\sqrt{r_1 r_2}$. Hence,

$$2\sqrt{r_1 r_2} = 2(\sqrt{r_1 x} + \sqrt{r_2 x}).$$

Solving for x we obtain $x = \dfrac{r_1 r_2}{(\sqrt{r_1} + \sqrt{r_2})^2}.$

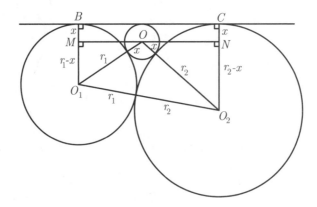

FIGURE 44.

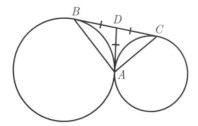

FIGURE **45.**

4.9 Introduce the common tangent to these circles at A. Let it intersect $\overline{BC}$ at point D, as shown in Figure 45.

Then $BD = AD$ and $CD = AD$, by Theorem 4.9. Therefore, in $\triangle ABC$, the median at A is half the length of $a = BC$. Thus, from Problem S3.2.5, $\angle A$ is a right angle. Another way to finish the solution is to observe that the circle with center D and radius DA circumscribes $\triangle ABC$, so segment BC is a diameter, in which case $\angle A$ is a right angle.

4.10 Let A' denote the intersection of the circles centered at B and C, B' the intersection of the circles centered at A and C, and C' the intersection of the circles centered at A and B. Clearly A', B', and C' lie on the segments joining the circle centers. (See Figure 46.)

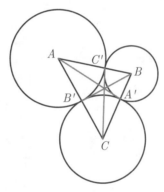

FIGURE **46.**

Therefore, since $AB' = AC'$, $BA' = BC'$, and $CA' = CB'$, we have

$$\frac{AC'}{C'B} \cdot \frac{BA'}{A'C} \cdot \frac{CB'}{B'A} = 1.$$

Thus, by Ceva's theorem, $\overline{AA'}$, $\overline{BB'}$, and $\overline{CC'}$ are concurrent.

4.11 Consider perpendicular bisectors to segments with endpoints at any two of these points. There are exactly $10^6(10^6 - 1)/2$ of such segments, but this is not important for us: what is important is that this number is finite. Hence, the number of distinct perpendicular bisectors is at most as large. Since the union of finitely many lines cannot be the whole plane, there is a point P of the plane which does not belong to any of these perpendicular bisectors. Thus, the distances from P to points of the set are all distinct. Let's arrange them in increasing order:

$$d_1 < d_2 < \cdots < d_{10^6}.$$

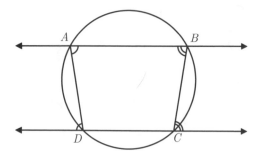

FIGURE **47.**

In order to find a circle that has exactly n points of the set in its interior, $1 \leq n \leq 10^6$, take a circle of radius d centered at P, where $d_n < d < d_{n+1}$.

A similar result can be obtained for any given finite set of points in the plane. For those who know the definition of a countable set: one can show that the result holds for any given countable set of points.

4.12 Let $ABCD$ be a quadrilateral with parallel sides $\overline{AB}$ and $\overline{CD}$.

($\Rightarrow$) Assume that a circumcircle exists for $ABCD$, as shown in Figure 47.

Because $\overleftrightarrow{AB} \parallel \overleftrightarrow{CD}$, $m\angle A + m\angle D = 180°$. Similarly, $m\angle B + m\angle C = 180°$, so $m\angle A + m\angle D = m\angle B + m\angle C$. By Theorem 4.21, we also know that $m\angle A + m\angle C = m\angle B + m\angle D$. The last two equalities imply $m\angle A = m\angle B$ and $m\angle C = m\angle D$, so $ABCD$ is an isosceles trapezoid or a rectangle (if $m\angle A = m\angle C$).

($\Leftarrow$) If we assume that $ABCD$ is an isosceles trapezoid or a rectangle, then $m\angle A = m\angle B$ and $m\angle C = m\angle D$. This immediately leads to the equation $m\angle A + m\angle C = m\angle B + m\angle D$ and the result follows from Theorem 4.21.

4.13 Let segments A_1C_1 and B_1D_1 intersect inside the circle at point F. Also, let $2t, 2x, 2y$, and $2z$ be the measures of the arcs AB, BC, CD and DA, respectively. Then $t + x + y + z = 180°$. Now, by Theorem 4.15, the measure of the vertical angles formed by two intersecting chords of a circle is the average of the measure of the two intercepted arcs. In this case, by considering

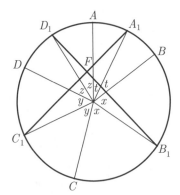

FIGURE **48.**

$\angle A_1 F B_1$ as one of the vertical angles, the intercepted arcs become $\overset{\frown}{A_1 B B_1}$ and $\overset{\frown}{C_1 D D_1}$. (See Figure 48.) Therefore,

$$m\angle A_1 F B_1 = \frac{1}{2}\big((t+x)+(y+z)\big)$$
$$= \frac{1}{2}(t+x+y+z) = \frac{1}{2}(180°) = 90°.$$

4.14 The center of the ith circle is equidistant (with distance r_i) from the two given lines. Therefore, the centers of the circles are collinear, as they all lie on the bisector of the angle formed by the two lines. Now for fixed i, $1 \leq i \leq n-1$, consider a right triangle with the following sides: its hypotenuse is the segment joining the centers of C_i and C_{i+1}, and has length $r_i + r_{i+1}$; its short leg is perpendicular to one of the lines and has length $r_{i+1} - r_i$; and its long leg is parallel to said line.

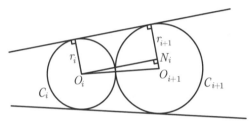

FIGURE 49.

The $n-1$ triangles formed in this way will all be similar, so for $2 \leq i \leq n-1$,

$$\frac{r_i + r_{i-1}}{r_i - r_{i-1}} = \frac{r_{i+1} + r_i}{r_{i+1} - r_i}.$$

Cross multiplying and simplifying yields the nice relationship $r_{i+1}/r_i = r_i/r_{i-1}$. Thus, we obtain a geometric sequence of radii for the n circles, $1, r, r^2, r^3, \ldots, r^n$, and C_n has radius r^n.

4.15 Let M be the point on $\overline{O_1 O_2}$ that is equidistant from the two circles; that is, if the distance between O_1 and O_2 is $R_1 + R_2 + 2d$, M is distance $R_1 + d$ from O_1 and distance $R_2 + d$ from O_2. (See Figure 50.) Construct circles C_3 and C_4, each going through M and centered at O_1 and O_2, respectively. Then C_3 and C_4 are mutually tangent circles with respective radii

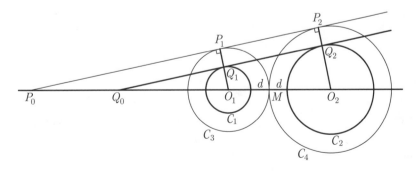

FIGURE 50.

$r_1 = R_1 + d$ and $r_2 = R_2 + d$. From Example 4, each external tangent of C_3 and C_4 meets line $O_1 O_2$ at a point P_0 that is distance $a = r_1(r_2 + r_1)/(r_2 - r_1)$ from O_1.

Now, let P_1 and P_2 be the respective intersection points of one of these tangent lines with circles C_3 and C_4. For $i = 1, 2$, let Q_i be the intersections of segment $O_i P_i$ with circle C_i. Then clearly line $Q_1 Q_2$ will be tangent to each of the original circles; let Q_0 be its intersection with line $O_1 O_2$. We need to determine the distance $Q_0 O_1$ to finish the problem.

Since $\triangle P_0 P_1 O_1 \sim \triangle Q_0 Q_1 O_1$, $P_0 O_1 / r_1 = Q_0 O_1 / R_1$. Therefore,

$$Q_0 O_1 = \frac{R_1}{r_1} \cdot a = R_1 \cdot \frac{r_2 + r_1}{r_2 - r_1} = R_1 \cdot \frac{R_2 + R_1 + 2d}{R_2 - R_1} = \frac{R_1 \cdot O_1 O_2}{R_2 - R_1}.$$

4.16 Let $C_i = C(O_i, r_i)$, $i = 1, 2$, be the circles. Consider line $O_1 O_2$. Let A_1 and B_1 and A_2 and B_2 be the points of intersection of $\overleftrightarrow{O_1 O_2}$ with C_1 and C_2, respectively, such that $B_1 - O_1 - A_1 - A_2 - O_2 - B_2$, as in Figure 51. Consider $\overline{D_1 D_2}$, where $D_i \in C_i$ for $i = 1, 2$.

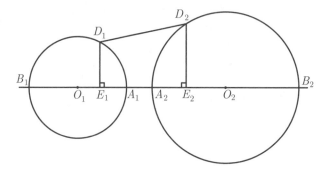

FIGURE 51.

We claim that the minimum value of $D_1 D_2$ is $A_1 A_2$. In order to prove this, drop perpendiculars from each D_i to line $O_1 O_2$, and let E_i be the bases of these perpendiculars. Then

$$D_1 D_2 \geq E_1 E_2 \geq A_1 A_2.$$

Clearly $D_1 D_2 = A_1 A_2$ if and only if $D_i = A_i$, $i = 1, 2$.

We next show that the maximum value of $D_1 D_2$ is $B_1 B_2$. Since $D_1 D_2$ is shorter than any other path between D_1 and D_2, we have

$$D_1 D_2 \leq D_1 O_1 + O_1 O_2 + O_2 D_2$$
$$= B_1 O_1 + O_1 O_2 + O_2 B_2 = B_1 B_2.$$

Obviously, equality will be obtained if and only if $D_1 = B_1$ and $D_2 = B_2$.

4.17 Label the triangle as $\triangle ABC$ and let C be its circumcircle. (See Figure 52.) Let $\overline{BM}$ be the median, $\overline{BK}$ the angle bisector, $\overline{BH}$ the altitude, l the perpendicular bisector to $\overline{AC}$, and K' the midpoint of the arc AC that does not contain B. Then by Theorem 4.6(2), l passes through K' since it contains the diameter perpendicular to the chord AC. Also line BK pass through K' since it divides the inscribed angle, $\angle ABC$, in half. Since $l \parallel \overline{BH}$, the statement follows. It is also easy to check that the result does not depend on whether the triangle is acute, right, or obtuse.

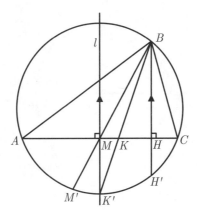

FIGURE 52.

4.18 Let u, v, w, x, y, and z be the measures of the arcs AB, BC, CD, DE, EF and FA, respectively. (See Figure 53.) Since $\overline{AB} \parallel \overline{DE}$, by Corollary 4.13(1), the measures of arcs BD and AE are equal. Hence, $v + w = y + z$. Since $\overline{BC} \parallel \overline{EF}$, the measures of arcs CE and FB are equal. Thus, $w + x = z + u$. Subtracting these two equalities we obtain $v - x = y - u$, which is equivalent to $u + v = x + y$. Hence, the arcs AC and DF have equal measure. This implies that $\overline{AF} \parallel \overline{CD}$.

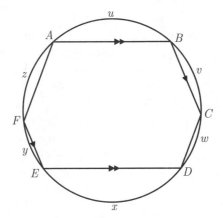

FIGURE 53.

4.19 Join O with C and F. Since O lies on $\overline{AD}$ and $\overline{BE}$, then by Corollary 4.7, $\overline{AD}$ bisects angles A and D, and $\overline{BE}$ bisects angles B and E. Similarly, segments FO and CO bisect angles F and C, respectively. Let $2a$, $2b$, $2c$, $2d$, $2e$, and $2f$ be the measures of the angles A, B, C, D, E, and F, respectively. (See Figure 54.)

Then $m\angle FOA + m\angle AOB + m\angle BOC$

$$= \left(180° - (f + a)\right) + \left(180° - (a + b)\right) + \left(180° - (b + c)\right)$$
$$= 540° - (2a + 2b + c + f).$$

Similarly we obtain that $m\angle FOE + m\angle EOD + m\angle DOC$

$$= \left(180° - (f + a)\right) + \left(180° - (a + b)\right) + \left(180° - (b + c)\right)$$
$$= 540° - (2d + 2e + f + c).$$

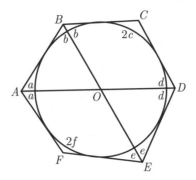

FIGURE 54.

We wish to show that $2a + 2b + c + f = 2d + 2e + f + c$. This will imply that the third diagonal CE passes through O. The equality is equivalent to $a + b = d + e$. Using Theorem 3.5 from Chapter 3, we obtain $m\angle AOE = a + b$ and $m\angle DOB = e + d$. Since these two angles are vertical, $a + b = d + e$, which finishes the proof.

4.20 Consider a triangle, $\triangle ABC$. Let A_1 and C_1 be the bases of the altitudes from A and C, respectively. Let $\overline{BD}$ be a diameter of the circumcircle, and let E be the intersection of lines BD and A_1C_1.

Case 1. Suppose $\triangle ABC$ is acute, as in Figure 55. Then $\angle BDC \cong \angle A$, as two inscribed angles subtending the same arc. Since the inscribed angle BCD subtends half of the circle, it is a right angle. Hence, the angles A and DBC are complementary.

Now we show that angles BA_1C_1 and A are congruent, which will imply that angles BA_1C_1 and DBC ($= EBA_1$) are complementary. The two right triangles AA_1B and CC_1B are similar as they share an angle A_1BC_1. Hence, $A_1B/C_1B = AB/BC$. Therefore $\triangle ABC \sim \triangle A_1BC_1$ by SAS. Hence, $\angle BA_1C_1 \cong \angle A$. (The reader may recall that a similar argument was used for Problem 3.2.26.) Hence, $\angle BA_1C_1$ and $\angle EBA_1$ are complementary, and $\overline{BE} \perp A_1C_1$.

Cases 2, 3. Suppose $\triangle ABC$ is right or obtuse. We must also consider the different cases depending on which of the angles A or B is right or obtuse. These cases clearly cover all possibilities, and we leave their solutions to the reader.

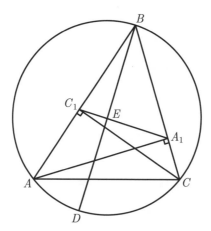

FIGURE 55.

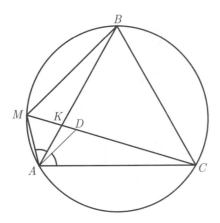

FIGURE 56.

4.21 *Solution 1:* Suppose $MA \leq MB \leq MC$. Let D be a point on $\overline{MC}$ such that $MD = MA$. (See Figure 56.) Inscribed angles CMA and ABC subtend the same arc. Hence, $m\angle CMA = m\angle ABC = 60°$, and $\triangle AMD$ is equilateral, so $AD = AM$. Now $m\angle DAC = m\angle BAM = 60° - m\angle BAD$. Therefore $\triangle ABM \cong \triangle ACD$ by SAS. Hence, $DC = MB$. This gives

$$MA + MB = MD + DC = MC.$$

Remark. In the statement $m\angle DAC = m\angle BAM = 60° - m\angle BAD$, we actually used the fact $K - D - C$, i.e., that D is on $\overline{KC}$, where K is the intersection of $\overline{AB}$ and $\overline{MC}$. This can be easily justified:

$$m\angle KAM < m\angle ACB = 60°,$$

since $\angle KAM$ subtends a smaller arc than $\angle ACB$, and $m\angle MKA = m\angle BKC = m\angle KBC + m\angle KCB > 60°$. This proves $AM > MK$ and justifies our argument.

Solution 2: Another solution follows immediately from Ptolemy's theorem. (See Theorem 4.23.) Let $AB = BC = AC = a$. Then

$$AM \cdot a + MB \cdot a = MC \cdot a \iff MA + MB = MC.$$

4.22 Consider the circumcircles of $\triangle CFE$ and $\triangle CAB$. They both pass through C, and we let P be the second point of their intersection. The particular arrangement of lines in Figure 57 implies that P and D are on opposite sides of line BC. We must prove that P lies on the circumcircles of $\triangle BDF$ and $\triangle ADE$.

In order to prove that P lies on the circumcircle for $\triangle BDF$, it is sufficient to show that $m\angle BPF + m\angle D = 180°$. Join P with C. Then $m\angle BPF = m\angle BPC + m\angle CPF$. Since A, B, P, and C lie on a circle, $m\angle BPC = 180° - m\angle BAC$. Hence,

$$m\angle BPF = (180° - m\angle BAC) + m\angle CPF.$$

Next we observe that $\angle CPF \cong \angle CEF$, since they are both inscribed angles and subtend the same arc. It follows that

$$m\angle BPF = (180° - m\angle BAC) + m\angle CEF.$$

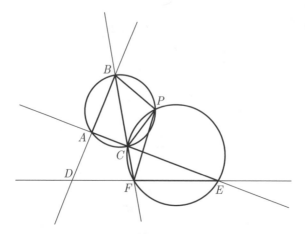

FIGURE 57.

Finally we observe that $\angle BAC$ is an exterior angle for $\triangle ADE$, which implies $m\angle BAC = m\angle D + m\angle CEF$. Therefore,

$$m\angle BPF = 180° - (m\angle D + m\angle CEF) + m\angle CEF = 180° - m\angle D.$$

This proves $m\angle BPF + m\angle D = 180°$, as desired.

A similar argument shows that P lies on the circumcircle for $\triangle ADE$.

4.23 Let C be the point of intersection of line AB with the common tangent to the circles at M. (See Figure 58.) Then $\angle BMC \cong \angle MAB$, since they subtend the same arc MB (by Corollary 4.14). Call their measure α. Also $CM = CT$ as two tangent segments to the smaller circle from C. Hence, $\angle CMT \cong \angle CTM$. Call their measure β. Using the fact that $\angle CTM$ is an exterior angle of $\triangle MAT$, we obtain

$$m\angle AMT = m\angle CTM - m\angle MAT = \beta - \alpha.$$

At the same time,

$$m\angle BMT = m\angle CMT - m\angle BMC = \beta - \alpha.$$

Hence, $\overrightarrow{MT}$ bisects $\angle AMB$.

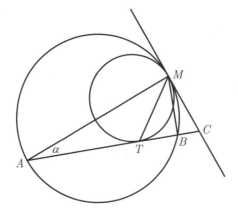

FIGURE 58.

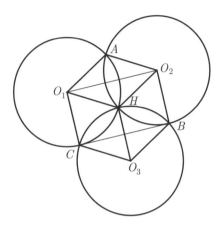

FIGURE 59.

4.24 See Figure 59 and consider two triangles: the vertices of the first are O_1, O_2, and O_3, the centers of the circles, while the vertices of the second are A, B, and C, the three points of intersection distinct from their common point H. We show that these triangles are congruent.

The quadrilaterals O_1CO_3H and HO_2BO_3 are rhombi since all sides are radii of the congruent circles. Hence, the segments O_1C, O_3H, and BO_2 are parallel and congruent, in which case O_1O_2BC is a parallelogram, since a pair of its opposite sides is both parallel and congruent. Therefore $\overline{O_1O_2} \cong \overline{BC}$. Similarly we can show that $\overline{O_1O_3} \cong \overline{AB}$ and $\overline{O_2O_3} \cong \overline{AC}$. Hence, $\triangle O_1O_2O_3 \cong \triangle ABC$ by SSS. Since the length of the radius of the circumcircle for $\triangle O_1O_2O_3$ equals the one for the circles, and since it also matches that of the circumcircle of $\triangle ABC$, the desired result has been proven.

4.25 (a) Let O be the center of the incircle. (See Figure 60.) Then $\overline{AA_1}$ and $\overline{CO}$ are bisectors of the angles at A and C, respectively. Let $\overline{BD}$, $\overline{A_1D_1}$, and $\overline{OK}$ be altitudes in triangles ABC, AA_1C, and AOC, respectively. For convenience, let $BD = h_b$, $A_1D_1 = h$, $A_1C = x$, $OK = r$, $AA_1 = l_a$, and $AO = y$. Using Theorem 3.16 and the similarity of two pairs of triangles ($\triangle CD_1A_1 \sim \triangle CDB$ and $\triangle AKO \sim \triangle AD_1A_1$), we obtain the following four equations:

$$\frac{x}{a-x} = \frac{b}{c}, \qquad \frac{h}{h_b} = \frac{x}{a}, \qquad \frac{r}{h} = \frac{y}{l_a}, \qquad \frac{y}{l_a - y} = \frac{b}{x}.$$

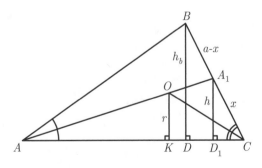

FIGURE 60.

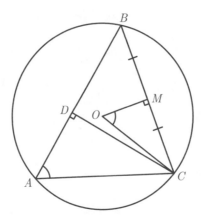

FIGURE 61.

From the first equation we obtain $x = \dfrac{ab}{b+c}$. From the last equation we get

$$\frac{y}{l_a} = \frac{b}{x+b} = \frac{b+c}{a+b+c} = \frac{b+c}{2p}.$$

From the third and the second equations and the obtained expressions for x and y/l_a above, we get

$$r = h\frac{y}{l_a} = \left(\frac{h_b x}{a}\right)\frac{y}{l_a} = \frac{h_b b}{b+c}\frac{b+c}{2p} = \frac{bh_b}{2p}.$$

Finally, using the formula for h_b obtained in Problem 4.18, namely

$$h_b = \frac{2}{b}\sqrt{p(p-a)(p-b)(p-c)},$$

we obtain the desired formula for r.

(b) Let O be the center of the circumcircle, M the midpoint of $\overline{BC}$, and $\overline{CD}$ an altitude of the triangle. (See Figure 61.)

Then $CD = h_c$, $OC = R$, and $MC = a/2$. Furthermore, $m\angle COM = \frac{1}{2}m\angle COB = m\angle CAB$, since the central and inscribed angles subtend the same arc. Hence, $\triangle OMC \sim \triangle ADC$ by AA. Therefore $b/h_c = R/(a/2)$. Hence, $R = ab/(2h_c)$. Using the formula

$$h_c = \frac{2}{c}\sqrt{p(p-a)(p-b)(p-c)},$$

we obtain the desired formula for R.

4.26 Let O_1, O_2, and O_3 be the centers of the circles, and let A, B, and C be the points of tangency. (See Figure 62.)

First we show that the circumcircle of $\triangle ABC$ is equal to the incircle of $\triangle O_1 O_2 O_3$. Let O be the center of the circumcircle of $\triangle ABC$. Then O lies on the perpendicular bisectors to the sides of $\triangle ABC$. Since the triangles $O_1 AB$, $O_2 BC$, and $O_3 AC$ are isosceles, these perpendicular bisectors are the angle bisectors in $\triangle O_1 O_2 O_3$. Hence, O is also the center of the incircle of $\triangle O_1 O_2 O_3$.

Let A', B', and C' be the points of tangency of the incircle of $\triangle O_1 O_2 O_3$; then $O_1 O_3 = r_1 + r_3 = O_1 A' + O_3 A'$, $O_1 O_2 = r_1 + r_2 = O_1 B' + O_2 B'$, and $O_2 O_3 = r_2 + r_3 = O_2 C' + O_3 C'$. Since $O_1 A' = O_1 B'$, $O_2 B' = O_2 C'$, and $O_3 C' = O_3 A'$, we obtain $O_1 A = O_1 A' = r_1$,

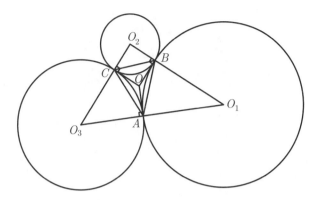

FIGURE 62.

$O_2B = O_2B' = r_2$ and $O_3C = O_3C' = r_3$. Therefore $A = A'$, $B = B'$, and $C = C'$. Hence, the circumcircle of $\triangle ABC$ and the incircle of $\triangle O_1O_2O_3$ coincide.

To find the length of the radius of the incircle of $\triangle O_1O_2O_3$, we use the formula from Problem 4.25(a). The semiperimeter of $\triangle ABC$ is $p = r_1 + r_2 + r_3$. Hence,

$$r = \frac{\sqrt{p(p - (r_1 + r_2))(p - (r_2 + r_3))(p - (r_1 + r_3))}}{p} = \sqrt{\frac{r_1 r_2 r_3}{r_1 + r_2 + r_3}}.$$

4.27 Let A_1, B_1, and C_1 be the midpoints of the sides BC, AC, and AB, respectively, as in Figure 63.

Since the sides of $\triangle A_1B_1C_1$ are midlines in $\triangle ABC$, they are parallel to the corresponding sides of $\triangle ABC$, and $\triangle A_1B_1C_1 \sim \triangle ABC$ with similarity coefficient 2. Therefore G is the centroid of both triangles. The perpendicular bisectors to the sides of $\triangle ABC$ are altitudes in $\triangle A_1B_1C_1$. Hence, O is also the orthocenter in the $\triangle A_1B_1C_1$. Thus, $\triangle OGB_1 \sim \triangle HGB$ with similarity coefficient 2. Therefore, $\angle OGB_1 \cong \angle BGH$. This makes points O, G, and H collinear, and $GH = 2GO$.

An interesting consequence of the above argument is a proof of the following striking fact.

Showing that the altitudes of a triangle are concurrent is equivalent to showing that the perpendicular bisectors of the sides of a triangle are concurrent.

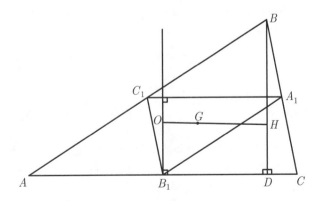

FIGURE 63.

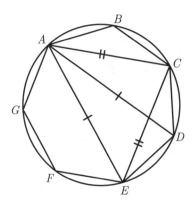

FIGURE 64.

4.28 Let E, F, and G be the other three vertices of the regular 7-gon. (See Figure 64.) Since $ACDE$ is an inscribed quadrilateral, applying Ptolemy's theorem we obtain

$$AC \cdot ED + AE \cdot CD = AD \cdot EC.$$

Then, since $ED = AB = CD$, $AE = AD$, and $EC = AC$, we note that

$$AC \cdot AB + AD \cdot AB = AD \cdot AC,$$

which is equivalent to the statement of the problem.

Length and Area

5.1 (a) The semiperimeter is $p = (8 + 3 + 6)/2 = 17/2$. By Theorem 5.3(1),

$$\text{Area}(\triangle ABC) = \sqrt{(17/2) \cdot (17/2 - 8) \cdot (17/2 - 3) \cdot (17/2 - 6)} = \sqrt{935}/4.$$

(b) By Theorem 5.1(2), $h_a = (2 \cdot \text{Area}(\triangle ABC))/a = \sqrt{935}/16$.

(c) Using Theorem 5.3, parts (2) and (3),

$$r = (\text{Area}(\triangle ABC))/p = (\sqrt{935}/4)/(17/2) = \sqrt{935}/34.$$

$$R = \frac{abc}{(4 \text{ Area}(\triangle ABC))} = \frac{144}{\sqrt{935}}.$$

5.2 (a) Let M be a point of $\overline{AC}$; let d_{AB} and d_{BC} be the distances from M to the sides $\overline{AB}$ and $\overline{BC}$, respectively. Then

$$\text{Area}(\triangle ABC) = \text{Area}(\triangle ABM) + \text{Area}(\triangle BCM)$$

$$= \frac{1}{2}AB \cdot d_{AB} + \frac{1}{2}BC \cdot d_{BC} = \frac{1}{2}AB(d_{AB} + d_{BC}).$$

On the other hand, $\text{Area}(\triangle ABC) = h_a BC/2$. Since $BC = AB$, $d_{AB} + d_{BC} = h_a$ for every point M on $\overline{AC}$.

(b) The solution is similar to the one in part (a). Generalization: The sum of the distances from any interior point to the lines containing the sides of a regular polygon will always be nr, where r is the length of the inradius.

　　These problems were previously solved by another method; see Problems 3.2.28 and 3.2.29 in Chapter 3. The solution provided in this section is clearly easier.

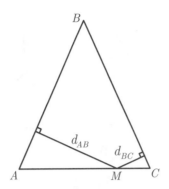

FIGURE **65.**

5.3 (a) Let A_1 be the midpoint of $\overline{BC}$, G be the centroid of $\triangle ABC$, and points E and F be the bases of the perpendiculars from A and G, respectively, on line BC. (See Figure 66.)

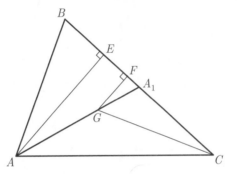

FIGURE **66.**

As G is the centroid of $\triangle ABC$, by Problem 3.2.10, $AG/GA_1 = 2$. Since $\triangle AEA_1 \sim \triangle GFA_1$, by AA, $AE/GF = 3$. Therefore,

$$\text{Area}(\triangle A_1 GC) = \frac{1}{3}\,\text{Area}(\triangle A_1 AC)$$
$$= \frac{1}{3}\left(\frac{1}{2}\,\text{Area}(\triangle BAC)\right) = \frac{1}{6}\,\text{Area}(\triangle BAC).$$

Similarly, we can show that the area of any other of the six triangles at G is one sixth the area of $\triangle ABC$.

(b) The statement we have to prove is equivalent to the statement that each of the areas is one third of the Area($\triangle ABC$). On the other hand, the statement of part (a) implies that if G is the centroid of $\triangle ABC$, then

$$\text{Area}(\triangle AGB) = \text{Area}(\triangle BGC) = \text{Area}(\triangle CGA) = \frac{1}{3}\,\text{Area}(\triangle ABC).$$

If $X \neq G$, then the distance from X to at least one of the sides of $\triangle ABC$ is smaller than the distance from G to this side. Suppose the side for which this is true is $\overline{AB}$ (for other sides the argument will be similar). Then Area($\triangle AXB$) < Area($\triangle AGB$) = $\frac{1}{3}$Area($\triangle ABC$), a contradiction. Therefore, $X = G$.

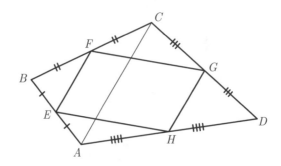

FIGURE 67.

5.4 Let E, F, G, and H be the midpoints of the sides $\overline{AB}$, $\overline{BC}$, $\overline{CD}$, and $\overline{DA}$, respectively. (See Figure 67.)

As $\overline{EF}$ is a midline in $\triangle ABC$, $EF = \frac{1}{2}AC$ and $\overline{EF} \parallel \overline{AC}$. Hence, $\triangle EBF \sim \triangle ABC$ with similarity coefficient $1/2$. Therefore, $\text{Area}(\triangle EBF) = \frac{1}{4}\text{Area}(\triangle ABC)$. Similarly we obtain $\text{Area}(\triangle FCG) = \frac{1}{4}\text{Area}(\triangle BCD)$, $\text{Area}(\triangle GDH) = \frac{1}{4}\text{Area}(\triangle CDA)$, and $\text{Area}(\triangle HAE) = \frac{1}{4}\text{Area}(\triangle DAB)$.

This implies that $\text{Area}(\triangle EBF) + \text{Area}(\triangle GDH)$

$$= \frac{1}{4}\text{Area}(\triangle ABC) + \frac{1}{4}\text{Area}(\triangle CDA)$$

$$= \frac{1}{4}\text{Area}(ABCD).$$

Similarly, $\text{Area}(\triangle FCG) + \text{Area}(\triangle HAE) = \frac{1}{4}\text{Area}(ABCD)$.

Therefore, $\text{Area}(EFGH)$

$$= \text{Area}(ABCD) - (\text{Area}(\triangle EBF) + \text{Area}(\triangle FCG)$$

$$+ \text{Area}(\triangle GDH) + \text{Area}(\triangle HAF)$$

$$= \text{Area}(ABCD) - \left(\frac{1}{4}\text{Area}(ABCD) + \frac{1}{4}\text{Area}(ABCD)\right)$$

$$= \frac{1}{2}\text{Area}(ABCD).$$

5.5 If a, b, and c are the side lengths of the right triangle, then the areas of the semicircles are $\pi(\frac{a}{2})^2/2$, $\pi(\frac{b}{2})^2/2$, and $\pi(\frac{c}{2})^2/2$, respectively. Since $c^2 = a^2 + b^2$, the result follows.

The statement holds for any three similar figures that contain the sides of the triangle as corresponding elements of the figures.

5.6 Let $\overline{AH}$ and $\overline{C_1 G}$ be altitudes in $\triangle ABC$ and $\triangle C_1 BA_1$, respectively. (See Figure 68.)

Then $\triangle C_1 GB \sim \triangle AHB$ by AA. Hence, $C_1 G/AH = C_1 B/AB = \lambda$, so

$$\frac{\text{Area}(\triangle C_1 BA_1)}{\text{Area}(\triangle ABC)} = \frac{\frac{1}{2}C_1 G \cdot BA_1}{\frac{1}{2}AH \cdot BC} = \lambda \cdot \mu.$$

5.7 If the height of the extended rope above the equator is h, and the radius of the Earth (thought of as a sphere) is R, then we have

$$2\pi(R + h) - 2\pi R = 3.$$

Solving for h, we obtain $h = 3/(2\pi) \approx .477$ feet.

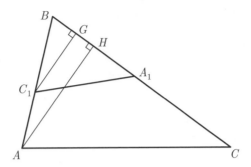

FIGURE 68.

So h is close to $\frac{1}{2}$ foot, and no house cat will have a problem getting under it. Note that h does not depend on the radius of the Earth, R. Many people find this result counterintuitive.

5.8 Let $S = \text{Area}(\triangle ABC)$. Then, by Theorems 5.1 and 5.3,

$$S = ah_a/2 = bh_b/2 = ch_c/2 = pr,$$

where $p = (a + b + c)/2$. Hence,

$$\frac{1}{r} = \frac{p}{S} = \frac{a+b+c}{2S} = \frac{\frac{2S}{h_a} + \frac{2S}{h_b} + \frac{2S}{h_c}}{2S} = \frac{1}{h_a} + \frac{1}{h_b} + \frac{1}{h_c}.$$

5.9 Let $ABCD$ be the quadrilateral and $\overline{AC}$ be a diagonal that divides the area of $ABCD$ in half. Let $\overline{BE} \perp \overline{AC}$, $E \in \overline{AC}$, and let $\overline{DF} \perp \overline{AC}$, $F \in \overline{AC}$. (See Figure 69.)

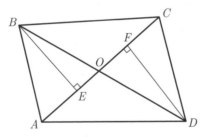

FIGURE 69.

Since triangles ABC and CDA have equal areas and a common side $\overline{AC}$, $BE = DF$. Let O be the intersection of $\overline{BD}$ and $\overline{AC}$. Then $\triangle BEO \cong \triangle DFO$ by AAS (corresponding legs are congruent and $\angle BOE \cong \angle DOF$), and therefore $BO = OD$.

The converse follows from the congruency of triangles $\triangle BEO$ and $\triangle DFO$ by AAS again (hypotenuses are congruent). Hence, $BE = DE$, so $\overline{AC}$ divides the quadrilateral into two triangles of equal areas.

5.10 (a) If the perimeter is $2p$, and one dimension of the rectangle is x, then the area of the rectangle is $x(p - x) = px - x^2$. Completing the square with respect to x, we get

$$x(p - x) = px - x^2 = p^2/4 - (p/2 - x)^2 \le p^2/4,$$

with equality if and only if $x = p/2$. Then $p - x = p/2$, which implies that the area is maximum if and only if the rectangle is a square.

There is no rectangle of the smallest area, since the function $x(p - x)$ attains no minimum on $(0, p)$. Of course, one could arrive at the same conclusions by using calculus.

(b) *Solution 1:* If the area is A, and one dimension of the rectangle is x, then its semiperimeter is $x + A/x$. Completing the square, we obtain

$$x + A/x = (\sqrt{x} - \sqrt{A/x})^2 + 2\sqrt{A} \geq 2\sqrt{A},$$

with equality if and only if $x = A/x$. Then $A = x^2$, which shows that the perimeter is minimum if and only if the rectangle is a square.

There is no rectangle of the largest area, since the function $x + A/x$ attains no maximum on $(0, \infty)$.

Solution 2: This solution demonstrates that the problem can be reduced to the one in part (a). Here we will take for granted that a rectangle of a given area A and minimum perimeter exists.[3] Let us denote the desired rectangle by R. If R is not a square, then its perimeter is less than $4\sqrt{A}$. Consider a (larger) rectangle R' which is similar to R and with the perimeter $4\sqrt{A}$. Clearly, Area(R') > Area(R). According to part (a), out of all rectangles with a fixed perimeter $4\sqrt{A}$, the only one with the largest area is a square. Since R' is not a square (being similar to R), we arrive at a contradiction.

5.11 Let O_1, O_2, and O_3 be the centers of the three circles. As the circles are externally tangent, $O_1O_2 = O_2O_3 = O_1O_3 = 2r$, and $\triangle O_1O_2O_3$ is equilateral. (See Figure 70.)

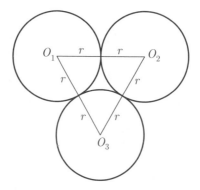

FIGURE 70.

The area of the figure we are interested in is clearly equal to the difference between Area($\triangle O_1O_2O_3$) and the total area of three congruent sectors, each with the central angle $60°$ and radius r. Hence, the area is

$$\frac{\sqrt{3}}{4}(2r)^2 - 3 \cdot \frac{\pi r^2}{6} = \left(\sqrt{3} - \frac{\pi}{2}\right)r^2.$$

5.12 Let h_1, h_2, and h be altitude lengths in the trapezoids $EBCF$, $AEFD$, and $ABCD$, respectively. (See Figure 71.)

---

[3] Though this claim may look obvious, it must be proved. In some other problems such an extremal object may not exist, and the assumption that it does may lead to errors. For example, consider the statement "*The number one is the largest natural number.*" This may seem like a ridiculous statement, but if we assume that such a number, n, exists, indeed we can prove that $n = 1$ as follows. By assumption, $n \geq n^2$, so $n - n^2 = n(1 - n) \geq 0$. Since n is a positive integer, $1 - n \geq 0$, which forces $n = 1$.

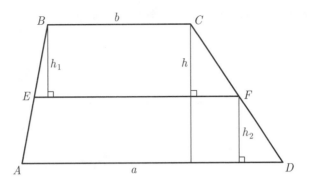

FIGURE 71.

Let $EF = x$, $AD = a$, and $BC = b$. Then we have the following system of equations:

$$h = h_1 + h_2, \quad (a+x)h_1/2 = (b+x)h_2/2 = (a+b)h/4.$$

Solving for x gives

$$x = \sqrt{\frac{a^2 + b^2}{2}},$$

i.e., EF is the quadratic mean of AD and BC.

5.13 The point of tangency of any two of the circles clearly lies on the segment joining their centers. Therefore $AB = a + b$, $AC = a + c$, and $BC = b + c$, so the semiperimeter of $\triangle ABC$ is $s = a + b + c$. From Heron's formula, we obtain an area of $\sqrt{(a+b+c)(abc)}$.

Now suppose that $\mathcal{C}(A, a)$ and $\mathcal{C}(B, b)$ are externally tangent to each other and both are internally tangent to $\mathcal{C}(C, c)$. Then $AB = a + b$, $AC = c - a$, and $BC = c - b$, so the semiperimeter of $\triangle ABC$ is $s = c$. From Heron's formula, we obtain an area of

$$\sqrt{(c)(c - a - b)(c - c + b)(c - c + a)} = \sqrt{abc(c - a - b)}.$$

Note that if the three centers are collinear, then $a + b = c$ and the triangle is degenerate with zero area.

5.14 Consider a unit square $ABCD$ with midpoints A_1, B_1, C_1, D_1, as specified. Let E be the intersection of $\overline{AA_1}$ and $\overline{BB_1}$, F the intersection of $\overline{BB_1}$ and $\overline{CC_1}$, G the intersection of $\overline{CC_1}$ and $\overline{DD_1}$, and H the intersection of $\overline{DD_1}$ and $\overline{AA_1}$, as shown in Figure 72. We wish to find the area of $EFGH$.

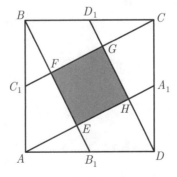

FIGURE 72.

Rotate $ABCD$ around its center by 90° clockwise. Then

$$A \to B \to C \to D \to A,$$

hence

$$\overline{AB} \to \overline{BC} \to \overline{CD} \to \overline{DA} \to \overline{AB}.$$

This implies

$$A_1 \to B_1 \to C_1 \to D_1 \to A_1,$$

so

$$\overline{AA_1} \to \overline{BB_1} \to \overline{CC_1} \to \overline{DD_1} \to \overline{AA_1}.$$

Hence,

$$E \to F \to G \to H \to E \text{ and } EFGH \text{ is a square.}$$

Consequently,

$$\triangle BFC_1 \to \triangle CGD_1 \to \triangle DHA_1 \to \triangle AEB_1 \to \triangle BFC_1,$$

and

$$GFBD_1 \to HGCA_1 \to EHDB_1 \to FEAC_1 \to GFBD_1.$$

Let Area($\triangle BFC_1$) $= x$ and Area($GFBD_1$) $= y$. As Area($\triangle C_1BC$) $= 1/4 = 2x + y$, and, Area($\triangle CGD_1$) $= \frac{1}{4}$Area($\triangle CFB$) ($\overline{GD_1}$ is a midline in $\triangle CFB$), we have $x = \frac{1}{4}(x + y)$, or, equivalently, $y = 3x$. Together with $1/4 = 2x + y$, this gives $x = 1/20$ and $y = 3/20$. Finally,

$$\text{Area}(EFGH) = 1 - (4x + 4y) = 1 - 4/5 = 1/5.$$

The proof in the general case is similar. The quadrilateral in question will still be a square and the following relationships are found:

$$\frac{x}{x + y} = k^2 \text{ and } 2x + y = \frac{k}{2}, \text{ so } x = \frac{k^3}{2(1 + k^2)}.$$

Then, the area is $1 - (4x + 4y) = (k - 1)^2/(k^2 + 1)$.

5.15 We refer to Figure 5.20, shown in Problem 5.5. The sum of the areas of the non-shaded regions is equal to

$$\frac{1}{2}\pi\left(\frac{b}{2}\right)^2 + \frac{1}{2}\pi\left(\frac{a}{2}\right)^2 - \left(\frac{1}{2}\pi\left(\frac{c}{2}\right)^2 - \text{Area}(\triangle ABC)\right)$$

$$= \frac{1}{8}\pi(a^2 + b^2 - c^2) + \text{Area}(\triangle ABC)$$

$$= \frac{1}{8}\pi \cdot 0 + \text{Area}(\triangle ABC) = \text{Area}(\triangle ABC).$$

5.16 Let $AD = a$, $BC = b$, and let h_1 and h_2 be the lengths of the altitudes of triangles AOD and BOC at O. (See Figure 73.)

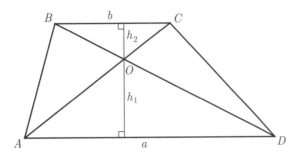

FIGURE 73.

Then $ah_1 = 2S_1$ and $bh_2 = 2S_2$. Since $\triangle AOD \sim \triangle BOC$, we have

$$\frac{a}{b} = \frac{h_1}{h_2} = \sqrt{\frac{S_1}{S_2}},$$

and this is probably the key point of this solution. It follows that

$$\begin{aligned}
\text{Area}(ABCD) &= \frac{1}{2}(a+b)(h_1+h_2) \\
&= \frac{1}{2}bh_2\left(\frac{a}{b}+1\right)\left(\frac{h_1}{h_2}+1\right) \\
&= \frac{1}{2}(2S_2)\left(\sqrt{S_1/S_2}+1\right)^2 = \left(\sqrt{S_1}+\sqrt{S_2}\right)^2.
\end{aligned}$$

5.17 (a) See Figure 74.

Since triangles ABB_1 and B_1BC share a common altitude at B,

$$\frac{\text{Area}(\triangle ABB_1)}{\text{Area}(\triangle B_1BC)} = \frac{AB_1}{B_1C}.$$

Since triangles AOB_1 and B_1OC share a common altitude at O,

$$\frac{\text{Area}(\triangle AOB_1)}{\text{Area}(\triangle B_1OC)} = \frac{AB_1}{B_1C}.$$

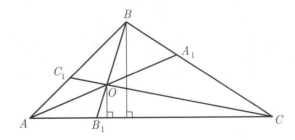

FIGURE 74.

Therefore,

$$\text{Area}(\triangle AOB) = \text{Area}(\triangle ABB_1) - \text{Area}(\triangle AOB_1)$$

$$= \frac{AB_1}{B_1C} \cdot \text{Area}(\triangle B_1BC) - \frac{AB_1}{B_1C} \cdot \text{Area}(\triangle B_1OC)$$

$$= \frac{AB_1}{B_1C} \cdot (\text{Area}(\triangle B_1BC) - \text{Area}(\triangle B_1OC))$$

$$= \frac{AB_1}{B_1C} \cdot (\text{Area}(\triangle BOC)).$$

Hence, the ratio of the areas of the triangles in question is AB_1/B_1C, as desired.

The argument can be shortened if we refer to the following property of proportions:

$$\text{if } \frac{a_1}{a_2} = \frac{b_1}{b_2} = k, \text{ then } \frac{a_1-b_1}{a_2-b_2} = k,$$

provided that no denominator is zero.

(b) The result follows from multiplication of three equalities based on the statement of part (a):

$$\frac{\text{Area}(\triangle AOC)}{\text{Area}(\triangle BOC)} = \frac{AC_1}{C_1B},$$

$$\frac{\text{Area}(\triangle AOB)}{\text{Area}(\triangle AOC)} = \frac{BA_1}{A_1C},$$

$$\frac{\text{Area}(\triangle BOC)}{\text{Area}(\triangle AOB)} = \frac{CB_1}{B_1A}.$$

5.18 Let $ABCD$ be the square having side length 1; the portions of the four circles of radius 1 are as shown in Figure 75.

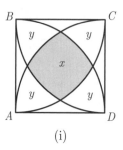
(i)

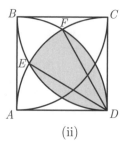
(ii)

FIGURE 75.

The letters x and y in Figure 75(i) denote the areas of the regions in which they lie. We wish to determine the value of x. (We leave it for the reader to justify the congruence of regions corresponding to area y. This can be easily proven by rotating the square 90° around its center.)

The shaded region shown in Figure 75(ii) can be partitioned into a sector EDF and two congruent circular segments.

Triangles AFD and DEC are equilateral triangles with sides of length 1. Therefore the angle of sector EDF is 30°. The area of each shaded circular segment in Figure 75(ii) is the difference of the areas of a sector FAD and Area($\triangle FAD$). This gives

$$x + y = \text{Area(sector } EDF) + 2(\text{area of a shaded circular segment})$$

$$= \frac{1}{2} \cdot \frac{\pi}{6} \cdot 1^2 + 2 \cdot \left(\frac{1}{2} \cdot \frac{\pi}{3} \cdot 1^2 - \frac{\sqrt{3}}{4} \right)$$

$$= \frac{\pi}{12} + \left(\frac{\pi}{3} - \frac{\sqrt{3}}{2} \right) = \frac{5}{12}\pi - \frac{\sqrt{3}}{2}.$$

The area of the figure bounded by the circles with centers at A and C is $x + 2y$. It can be found by subtracting the areas of two congruent shaded regions from the area of the square. The area of each such shaded region is the difference between the area of the square and the area of a sector at A or at C with $90°$ angle and unit radius. Thus, we have

$$x + 2y = 1^2 - 2 \cdot \left(1^2 - \frac{1}{2} \cdot \frac{\pi}{2} \cdot 1^2 \right) = \frac{\pi}{2} - 1.$$

Having found $x + y$ and $x + 2y$, we solve for x:

$$x = 2(x + y) - (x + 2y) = \left(5\pi/6 - \sqrt{3} \right) - \left(\pi/2 - 1 \right) = \frac{\pi}{3} - \sqrt{3} + 1.$$

5.19 Consider the line $\overleftrightarrow{AC}$ through M such that M bisects $\overline{AC}$. It is easy to show that such a line exists. One can construct it by first drawing a line through M parallel to one side of the angle until it intersects the other side at a point, say X, and then taking a point $A \in \overrightarrow{BX}$ such that $B - X - A$ and $BX = XA$. (See Figure 76(i).) The line $\overleftrightarrow{AM}$ is the one we are looking for: if it intersects the other side of $\angle B$ at C, then M bisects $\overline{AC}$.

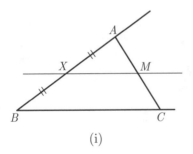   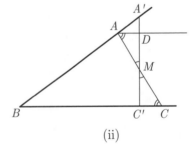

(i)   (ii)

FIGURE 76.

We claim that the triangle ABC just constructed has the minimum possible area among all triangles cut from $\angle B$ by a line through M. Consider any other line $A'C'$ through M. Suppose $B - A - A'$. If it happens that $B - A' - A$, the argument is similar. (See Figure 76(ii).)

Consider a line through A parallel to $\overline{BC}$. This line must intersect $\overline{AC}$ at some point D such that $A' - D - C'$. Then $\angle MCC' \cong \angle MAD$ as alternate interior angles, and $\angle CMC' \cong \angle AMD$ as vertical angles. Since $AM = MC$, $\triangle AMD \cong \triangle CMC'$ by ASA, and hence, their areas are equal. Therefore, we have

$$\begin{aligned}
\text{Area}(\triangle A'BC') &= \text{Area}(ABC'M) + \text{Area}(\triangle AMD) + \text{Area}(\triangle ADA') \\
&> \text{Area}(ABC'M) + \text{Area}(\triangle AMD) \\
&= \text{Area}(ABC'M) + \text{Area}(\triangle CMC') \\
&= \text{Area}(\triangle ABC).
\end{aligned}$$

Thus, Area($\triangle A'BC'$) > Area($\triangle ABC$).

There is no line for which Area($\triangle ABC$) is the greatest. Indeed, as the line through M approaches being parallel to a side of the angle, the area of $\triangle ABC$ grows without bound.

5.20 We prove that any triangle of area greater than 1 inside the circle of radius 1 contains the center of the circle in its interior. This will imply the desired result.

Suppose this is not the case, and let $\triangle ABC$ be a counterexample. First we assume that the triangle is inscribed in the circle, as shown in Figure 77(i). As the center of the circle is not in the interior of $\triangle ABC$, one of the angles of the triangle, say $\angle A$, has measure at least $90°$. Then $\overline{BC}$, a chord of the circle, has length less than or equal to 2, and $h_a \le 1$, so

$$\text{Area}(\triangle ABC) = (BC \cdot h_a)/2 \le (2 \cdot 1)/2 = 1,$$

a contradiction.

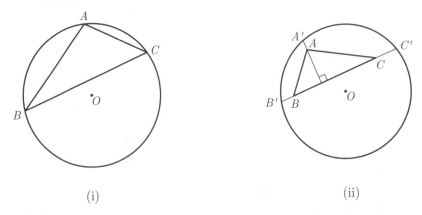

FIGURE 77.

If $\triangle ABC$ is not inscribed in the circle (as shown in Figure 77(ii)), then, continuing $\overline{BC}$ if needed, we can first replace it with a $\triangle AB'C'$ such that both B' and C' are on the circle. Additionally, if A is not on the circle, we can extend the altitude at A in $\triangle AB'C'$ beyond A until it hits the circle at A'. In this way we obtain

$$\text{Area}(\triangle ABC) = \frac{BC \cdot h_a}{2} \le \frac{B'C' \cdot h_a}{2} \le \frac{B'C' \cdot h_{a'}}{2} = \text{Area}(\triangle A'B'C').$$

But we have already shown that Area($\triangle A'B'C'$) ≤ 1. We conclude that Area($\triangle ABC$) ≤ 1, a contradiction again.

5.21 (a) Let $2x$, y, and y be the side lengths of an isosceles triangle of a given perimeter $2p$. Then $0 < x < p/2$, $x + y = p$, and the area of the triangle is

$$A(x) = x\sqrt{y^2 - x^2} = x\sqrt{(p-x)^2 - x^2} = x\sqrt{p^2 - 2px}.$$

We wish to show that the function $A(x)$ attains its maximum on $[0, p/2]$ at $x = p/3$, which will imply that $y = 2p/3$ and that the triangle is equilateral. We suggest two ways of doing this.

Method 1. The first method is for those who know one-variable calculus. Taking the derivative with respect to x, and simplifying the result, we get

$$A'(x) = \frac{p(p - 3x)}{\sqrt{p^2 - 2px}}.$$

Therefore, $x = p/3$ is the only critical point of $A(x)$ on $(0, p/2)$. As $A(p/3) > 0$ and $A(0) = A(p/2) = 0$, we have $\max\{A(x) : x \in [0, p/2]\} = A(p/3)$.

As $A(x)$ attains no minimum on $(0, p/2)$, there is no triangle of the minimum area.

Another solution could be obtained by using multivariable calculus.

Method 2. An alternate, calculus-free, way of establishing the fact uses the inequality between the arithmetic mean and the geometric mean of three nonnegative numbers:

$$\frac{a+b+c}{3} \geq \sqrt[3]{abc}, \tag{17.3}$$

with the equality occurring if and only if $a = b = c$.

This inequality and its generalization is very useful in mathematics. We will make use of it several times in this book, so we pause to discuss it in more detail.

Inequality (17.3) is a particular case ($n = 3$) of a more general inequality between the arithmetic and geometric means of n nonnegative numbers, the so-called AGM inequality: for all nonnegative numbers $a_1, \ldots, a_n$,

$$\frac{1}{n}(a_1 + a_2 + \cdots + a_n) \geq (a_1 \cdot a_2 \cdots a_n)^{\frac{1}{n}}, \tag{17.4}$$

with equality occurring if and only if $a_1 = a_2 = \cdots = a_n$.

Moreover, inequality (17.4) is a particular case of yet another, more general inequality, Jensen's inequality.[4]

For $n = 3$, the AGM inequality can be proven by the following ingenious and quite tricky argument, which uses only elementary algebra. The proof we provide below will make our solution in *Method 2* self-contained.

For every three nonnegative numbers x, y, and z,

$$x^3 + y^3 + z^3 - 3xyz$$
$$= (x + y + z)(x^2 + y^2 + z^2 - xy - yz - zx)$$
$$= (x + y + z)\left(\frac{1}{2} \cdot [(x - y)^2 + (y - z)^2 + (z - x)^2] \right) \geq 0.$$

Therefore $x^3 + y^3 + z^3 \geq 3xyz$, and as x, y, and z are nonnegative, the equality occurs if and only $x = y = z$. By taking x, y, and z to be the cube roots of a, b, and c, respectively, we get (17.3).

Let us now show how one uses (17.3) to prove $\max\{A(x) : x \in [0, p/2]\} = A(p/3)$. Taking $a = b = x$ and $c = p - 2x$, and applying (17.3), we obtain

$$\frac{x + x + (p - 2x)}{3} \geq \sqrt[3]{x \cdot x \cdot (p - 2x)} = \sqrt[3]{(A(x))^2/p}.$$

Therefore, $p \geq 3\sqrt[3]{(A(x))^2/p}$, with the equality occurring if and only if $x = p - 2x$, which is equivalent to $x = p/3$. Thus,

$$A(x) \leq \frac{p^2}{3\sqrt{3}} = A(p/3).$$

(b) *Solution 1:* We use one variable calculus. Let $2x$, y, and y be the side lengths of an isosceles triangle of a given area A, and let h be an altitude at the vertex. (See Figure 78.)

---

[4] For a proof of Jensen's inequality and its applications, see
http://www.math.udel.edu/~lazebnik/papers/jensen.pdf

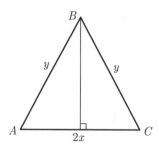

FIGURE 78.

Then $xh = x\sqrt{y^2 - x^2} = A$ and the perimeter of the triangle is $P(x) = 2x + 2y = 2x + 2\frac{\sqrt{A^2 + x^4}}{x}$. Now

$$P'(x) = 2 \cdot \frac{x^2\sqrt{A^2 + x^4} + x^4 - A^2}{x^2\sqrt{A^2 + x^4}}.$$

Hence, $P'(x) = 0$ on $(0, +\infty)$ only at $x = x_0 = \sqrt{A/\sqrt{3}}$. Since $P'(x) < 0$ on $(0, x_0)$ and $P'(x) > 0$ on (x_0, ∞),

$$\min\{P(x) : x \in (0, \infty)\} = P(x_0).$$

For $x = x_0$, $y = \sqrt{A^2 + x_0^4}/x_0 = 2x_0$. This implies that the triangle of the minimum perimeter is equilateral.

Clearly, there is no triangle with maximum perimeter.

Solution 2: One can also reduce part (b) to part (a) using a similar technique to that of our Solution 2 of Problem 5.10(b); or solve it using multivariable calculus.

5.22 *Solution 1:* See Figure 79(i). Rotate $\triangle ABC$ around its center by $120°$ clockwise. Then

$$A \rightarrow B \rightarrow C \rightarrow A,$$

hence

$$\overline{AB} \rightarrow \overline{BC} \rightarrow \overline{CA} \rightarrow \overline{AC},$$

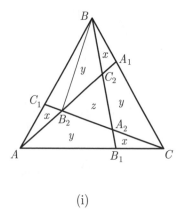

(i)

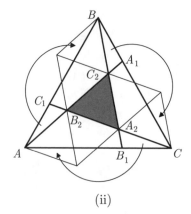

(ii)

FIGURE 79.

hence

$$A_1 \to B_1 \to C_1 \to A_1,$$

hence

$$\overline{A_1 B_1} \to \overline{B_1 C_1} \to \overline{C_1 A_1} \to \overline{A_1 C_1},$$

hence

$$A_2 \to B_2 \to C_2 \to A_2 \text{ and } \triangle A_2 B_2 C_2 \text{ is equilateral.}$$

In addition,

$$\triangle B_2 A C_1 \to \triangle C_2 B A_1 \to \triangle A_2 C B_1 \to \triangle A_2 A C_1,$$

and

$$B_2 C_1 B C_2 \to C_2 A_1 C A_2 \to A_2 B_1 A B_2 \to B_2 C_1 B C_2.$$

Let $\text{Area}(\triangle B_2 A C_1) = x$, $\text{Area}(B_2 C_1 B C_2) = y$, $\text{Area}(\triangle A_2 B_2 C_2) = z$, and $\text{Area}(\triangle ABC) = s$. Then

$$3x + 3y + z = s, \text{ and } 2x + y = \frac{s}{3}.$$

In order to obtain a third equation, we connect B to B_2. Then, as $\text{Area}(\triangle C B_2 A_1) = 2\,\text{Area}(\triangle B B_2 A_1)$, and $\text{Area}(\triangle C A A_1) = 2\,\text{Area}(\triangle B A A_1)$, we obtain

$$\text{Area}(\triangle C A B_2) = 2\,\text{Area}(\triangle B A B_2).$$

Since $\text{Area}(\triangle B B_2 C_1) = 2\,\text{Area}(\triangle A B_2 C_1)$ and $\text{Area}(\triangle A B_2 C_1) = x$, we have $x + y = 2(x + 2x) = 6x$. This yields $y = 5x$, the desired third equation. Solving this system of three equations, we get $x = s/21$, $y = 5s/21$, and $z = s/7$. Thus, $z/s = 1/7$.

Solution 2: The construction shown in Figure 79(ii) is self-explanatory, and provides a solution of the problem by a dissection method. Note that this solution holds for an *arbitrary* triangle. In Chapter 12, we will explain that solving this problem for an equilateral triangle is sufficient for solving it for an arbitrary triangle. In Example 76, we present an argument for the case of an equilateral triangle.

Solution 3: For another solution, which holds for an arbitrary triangle, and uses nothing more than similarity of triangles (though in a very clever way), see [**47**], Part I, Problem 4.11.

5.23 Two different solutions are exhibited in Figure 80. The drawings were taken from the web site of V. B. Balayoghan.[5]

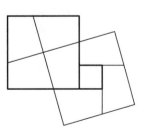 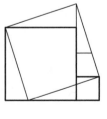

FIGURE 80.

[5] http://www.cs.utexas.edu/ vbb/misc/gd/index.html

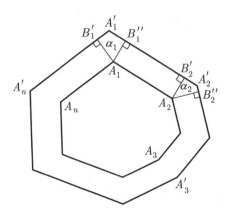

FIGURE 81.

5.24 From each vertex A_i drop two perpendiculars to those sides of $A'_1 \ldots A'_n$ which are parallel to the sides $A_{i-1}A_i$ and A_iA_{i+1} (all indices are modulo n). Let B'_i, B''_i be their bases. Let $\alpha_i = m\angle B'_iA_iB''_i$. (See Figure 81.)

Let $\angle A_i$ denote the interior angle of $A_1 \cdots A_n$ at A_i. Then

$$
\begin{aligned}
\alpha_1 &+ \ldots + \alpha_n \\
&= (2\pi - (\pi/2 + m\angle A_1 + \pi/2)) + \cdots + (2\pi - (\pi/2 + m\angle A_n + \pi/2)) \\
&= 2n\pi - [(m\angle A_1 + \cdots + m\angle A_n) + n\pi] = n\pi - (n-2)\pi = 2\pi,
\end{aligned}
$$

as the sum of interior angles of a convex n-gon is $(n-2)\pi$.

Consider a sector S_i with the center at A_i, radius 1, and angle α_i. It is clear that $\mathrm{Area}(S_i) < \mathrm{Area}(A_iB'_iA'_iB''_i)$ for all i, and the sum of all areas of all these sectors is $\pi \cdot 1^2 = \pi$. Therefore the sum of the areas of all quadrilaterals $A_iB'_iA'_iB''_i$ is greater than π. Note that the area of each rectangle $A_iB''_iB'_{i+1}A_{i+1}$ is $A_iA_{i+1} \cdot 1 = A_iA_{i+1}$. Therefore, adding the areas of all these rectangles and all 'corner' quadrilaterals $A_iB'_iA'_iB''_i$, we obtain

$$
\mathrm{Area}(A'_1 \ldots A'_n) > \mathrm{perimeter}(A_1 \ldots A_n) + \pi.
$$

5.25 (a) We use inequality (17.3), which compares the arithmetic and geometric means (AGM) of three positive numbers x, y, and z:

$$
\frac{x+y+z}{3} \geq \sqrt[3]{xyz}. \tag{17.5}
$$

This fact was established in Solution 2 of Problem 5.21(a). Let p and A denote the semiperimeter and the area of the triangle, respectively. Then the triangle inequality ensures that $p-a$, $p-b$ and $p-c$ are all positive. Applying the AGM to these numbers, and using $A = \sqrt{p(p-a)(p-b)(p-c)}$ (Theorem 5.3(a)), we obtain the following equivalent statements:

$$
\frac{(p-a) + (p-b) + (p-c)}{3} \geq \sqrt[3]{(p-a)(p-b)(p-c)},
$$

$$
\frac{3p - 2p}{3} \geq \sqrt[3]{(p-a)(p-b)(p-c)},
$$

$$
\frac{p^3}{27} \geq (p-a)(p-b)(p-c),
$$

$$
\frac{p^4}{27} \geq A^2, \text{ and}
$$

$$
\frac{p^2}{3\sqrt{3}} \geq A.
$$

Consequently, the area is never greater than $p^2/(3\sqrt{3})$. As we know, the equality sign in the AGM inequality is achieved if and only if the numbers are equal. Hence, $p - a = p - b = p - c$, which implies $a = b = c$. We conclude that the triangle with maximum area is equilateral.

One could also solve this problem using multi-variable calculus.

(b) *Solution 1:* Proceeding as in part (a), we obtain $\frac{p^2}{3\sqrt{3}} \geq A$. Thus, the minimum value of the semiperimeter p is $\sqrt[4]{27A^2}$, and this minimum value is attained if and only if the triangle is equilateral. This implies that the triangle of minimum area is equilateral.

Solution 2: One can also reduce part (b) to part (a) using a similar technique to that of our Solution 2 of Problem 5.10(b); or solve it using multivariable calculus.

5.26 Let $ABCD$ be the quadrilateral. Suppose A, A_1, A_2, A_3, A_4, and B are consecutive points on $\overline{AB}$, and D, C_1, C_2, C_3, C_4 and C are the corresponding points on $\overline{CD}$. Join subsequently points D and A_1, C_1 and A_2, C_2 and A_3, C_3 and A_4, C_4 and B. (See Figure 82.)

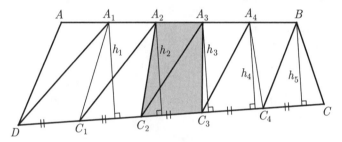

FIGURE 82.

Let h_1, h_2, h_3, h_4, and h_5 be the altitude lengths in $\triangle DA_1C_1$, $\triangle C_1A_2C_2$, $\triangle C_2A_3C_3$, $\triangle C_3A_4C_4$, and $\triangle C_4BC$, respectively. Then $h_2 = (h_1 + h_3)/2$, $h_3 = (h_2 + h_4)/2$, and $h_4 = (h_3 + h_5)/2$, as the lengths of midlines in the corresponding trapezoids. This implies that

$$h_3 = \frac{1}{5}(h_1 + h_2 + h_3 + h_4 + h_5).$$

Since the bases in the corresponding triangles are congruent, we have that Area($\triangle C_2A_3C_3$) is equal to one-fifth of the sum of areas of the five triangles. Similarly one can show that Area($\triangle A_2C_2A_3$) is equal to one-fifth of the sum of areas of the five other triangles, which have congruent bases on $\overline{AB}$. This implies that

$$\text{Area}(A_2A_3C_3C_2) = \frac{1}{5}\text{Area}(ABCD).$$

5.27 We use inequality (17.3) between the arithmetic and geometric means (AGM) of three positive numbers x, y, and z,

$$\frac{x + y + z}{3} \geq \sqrt[3]{xyz},$$

which was established in Solution 2 of Problem 5.21(a). Let p be the semiperimeter of the triangle, A be its area, and r be the inradius. The triangle inequality ensures that $p - a$, $p - b$, and $p - c$ are all positive. Applying the AGM to these numbers and using $A = \sqrt{p(p - a)(p - b)(p - c)} = pr$ (Theorem 5.3), we get

$$\frac{(p - a) + (p - b) + (p - c)}{3} \geq \sqrt[3]{(p - a)(p - b)(p - c)},$$

which is equivalent to

$$3p - 2p \geq 3\sqrt[3]{(p - a)(p - b)(p - c)} \quad \Leftrightarrow \quad p^3 \cdot p \geq 27\, p(p - a)(p - b)(p - c)$$

$$\Leftrightarrow p^4 \geq 27S^2 \quad \Leftrightarrow \quad A^4/r^4 \geq 27A^2 \quad \Leftrightarrow \quad A \geq 3\sqrt{3}\, r^2.$$

5.28   Since the coins do not overlap and are contained in a circle of radius R, we have

$$n \cdot \pi r^2 < \pi R^2,$$

which gives the second inequality.

To establish the first inequality, note that if the centers of the coins are preserved, but all radii are doubled, then the resulting larger coins will completely cover a concentric round table of radius $R - r$ (with overlaps). This is almost obvious: if they do not, then there would exist a point on the table of radius R such that the distance from this point to the edge of the table and to any other coin on the table is at least r. This implies that one more coin could be put on the table, a contradiction.

Therefore the total area of n coins of radius $2r$ should be at least as large as the area of the round table of radius $R - r$:

$$n \cdot \pi (2r)^2 > \pi (R - r)^2.$$

This gives the first inequality. Both inequalities are strict since it is easy to argue that finitely many coins cannot tile a round table.

5.29   Let ABC be an inscribed nonequilateral triangle. We want to show that the area of an inscribed equilateral triangle exceeds the area of $\triangle ABC$.

Since $\triangle ABC$ is not equilateral, one of the arcs, say $\widehat{AB}$, is less than $120°$ and another, say $\widehat{BC}$, is greater than $120°$, i.e., $m\angle ABC < 60°$ and $m\angle BAC > 60°$. Consider the point B' on the circle such that $A - B - B'$ and $m\widehat{AB'} = 120°$. Consider also the point B'' on the circle which is symmetric with B with respect to the perpendicular bisector of $\overline{AC}$. Then clearly $\text{Area}(\triangle ABC) = \text{Area}(\triangle AB''C)$ and $B - B' - B''$ since $m\widehat{AB''} > 120°$.

Now, since B' lies between B and B'', it is further from $\overline{AC}$ than B, so $\text{Area}(\triangle AB'C) > \text{Area}(\triangle ABC)$. If $\triangle AB'C$ is equilateral, we are done. If not, we consider $\overline{AB'}$ as a base, just as we did $\overline{AC}$ before. Repeating the argument for $\triangle AB'C$, we get to a triangle $AB'C'$ having area exceeding that of $\triangle AB'C$, with $m\angle CAB' = m\angle B'AC' = 60°$. But in this case, $\triangle AB'C'$ must be equilateral, and we are done.[6]

5.30   First of all, observe that any triangulation of an n-gon with its diagonals consists of $n - 2$ triangles. Number the triangles from 1 to $n - 2$, and denote their respective incircle radii as $r_1, r_2, \ldots, r_{n-2}$. Label the sum of the signed distances from the center of the ith circle to its edges as S_i. Then, observing that each triangle in the triangulation has the given circle of radius R as its circumcircle, Carnot's theorem states that $r_i + R = S_i$. Thus, the desired sum is given by

$$r_1 + r_2 + \cdots + r_{n-2} = S_1 + S_2 + \cdots + S_{n-2} - (n - 2)R. \tag{17.6}$$

In the sum $S_1 + S_2 + \cdots + S_{n-2}$, the lengths of the perpendicular segments from the polygon to the circle incenters are each counted once, while the lengths of the perpendiculars to the diagonals are each counted twice (each diagonal serves as a common side of adjacent triangles in the triangulation). However, with regard to the internal perpendicular segments, each is used once with a positive length and once with a negative length; indeed each triangle except the one actually containing the center of the given circle must be obtuse and will therefore have one side that contributes a negative distance to the sum (See the $+$ and $-$ signs

---

[6] This solution came from H. Rademacher and O. Toeplitz, [**49**], pages 19–22. They write that "it is likely that this problem was discussed, if not solved, at the time of Plato, a century before Euclid. However, neither Euclid nor more modern books give the following solution, which could easily have been understood and discovered by Greeks."

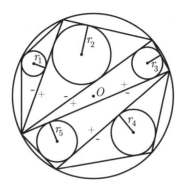

FIGURE 83.

on the internal diagonals in Figure 83.). Therefore, this sum is actually equal to the sum of the lengths of the perpendicular segments from the circumcenter O to the sides of the polygon, which proves that the right-hand side in (17.6) is constant. Thus, the left-hand side $r_1 + r_2 + \cdots + r_{n-2}$ is constant as well, which completes the proof.

Loci

6.1 The two whispering locations are positioned at the foci of the elliptical room. These two points are actually the foci of the ellipse formed by the intersection of the floor with the walls; since a rotational ellipsoid is formed by rotating an ellipse about its major axis, these two points will also be the foci of any cross-sectional piece (elliptical) of the room that contains them. Because of the reflective properties of an ellipse (see the comment after Corollary 6.2), all sound waves will bounce off the wall and pass through the other location. Furthermore, the waves will arrive simultaneously because the distance traveled will be constant. Thus, depending on how much sound is absorbed by the wall, the sound will be heard clearly rather than dissipate throughout the room. Presumably the echo will also bounce back to the whisperer, although it might arrive too quickly to be distinguished from the original sound.

6.2 For a fixed position of the ladder, label its upper and lower endpoints as X and Y, respectively, and let C be the point on the ground directly beneath X.

Solution 1: Let M be the midpoint of $\overline{XY}$. By Problem S3.2.5, since $\overline{CM}$ is the median of the right triangle XCY, $c = 2MC = XY$. Hence, the locus is the quarter circle of radius $\frac{c}{2}$ (half the length of the ladder), centered at C, and lying inside the angle formed by the ground and the side of the building.

Solution 2: Consider the right triangle XCY with side lengths $CX = y$, $CY = x$, and $XY = c = \sqrt{x^2 + y^2}$. Let M be the midpoint of XY and note that $d(M, \overline{CX}) = y/2$ and $d(M, \overline{CY}) = x/2$. By the Pythagorean theorem, it follows that $MC = \sqrt{(x/2 + y/2)^2} = c/2$. As in the first solution, the locus is the quarter circle of radius $\frac{XY}{2}$, centered at C, and lying inside the angle formed by the ground and the side of the building.

6.3 Notice that if A, B, C, and D are collinear (which happens when $D - A - C - B$ or $A - D - B - C$), then $\overleftrightarrow{AD}$ and $\overleftrightarrow{BC}$ are coincident, so their intersection is the line AB.

Suppose that the points are not all collinear. By SSS, $\triangle ACB \cong \triangle CAD$. Therefore, $\angle ACB \cong \angle CAD$, so $\triangle AFC$ is isosceles. (See Figure 84.) Thus, $FA = FC$. Therefore,

$$FA - FB = FC - FB = BC,$$

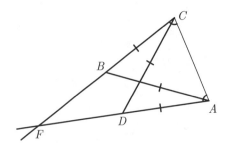

FIGURE 84.

where $BC = d$ is a fixed value. Therefore, F is on the hyperbola $\mathcal{H} = \mathcal{H}(A, B; d)$ (with axis of symmetry $\overleftrightarrow{AB}$).

We have shown that all points meeting the given criteria lie on the hyperbola. The ambitious student can show that the locus is all of $\mathcal{H}(A, B; d)$ by showing that all points on $\mathcal{H}$ satisfy the given criteria.

6.4 Let M be the midpoint of $\overline{AB}$. Then, $\triangle BFA$ is a right triangle with median $\overline{FM}$ from F. Therefore, by Problem S3.2.5, $FM = AB/2$. Thus, F lies on the circle $\mathcal{C} = \mathcal{C}(M, AB/2)$.

Now, let P be any point on the $\mathcal{C}$. If $P \neq A$ or B, then the parallelogram $ABCP$ (i.e., let $P = D$ and find the fourth vertex C) satisfies the criteria. If $P = A$ or $P = B$, then any rectangle $ABCD$ suffices. Therefore, the whole circle is part of the locus. Thus, the locus is the circle $\mathcal{C}(M, AB/2)$.

6.5 (a) Let $\mathcal{L}$ be the desired locus. If the distance, d, between the lines exceeds k, $\mathcal{L}$ is empty. If $d = k$, then for any point X between l_1 and l_2 or on either line, $d(X, l_1) + d(X, l_2) = k$. Thus, in this case, $\mathcal{L}$ consists of l_1, l_2, and all points between l_1 and l_2.

Suppose $d < k$. Any point on l_1, on l_2, or between l_1 and l_2 cannot be in $\mathcal{L}$. Consider a point X such that $d(X, l_2) = \frac{1}{2}(k - d)$. Then $d(X, l_1) + d(X, l_2) = \frac{1}{2}(k - d) + d + \frac{1}{2}(k - d) = k$. The same is true for a point Y such that $d(Y, l_1) = \frac{1}{2}(k - d)$. Thus, $\mathcal{L}$ contains two lines parallel to l_1 and l_2, one at distance $\frac{1}{2}(k - d)$ from l_1 and one at distance $\frac{1}{2}(k - d)$ from l_2. Clearly no other points are in $\mathcal{L}$.

(b) If $k < d$, the locus is the union of two lines between l_1 and l_2, one at distance $\frac{k+d}{2}$ from l_1 and the other at distance $d - \frac{k+d}{2} = \frac{d-k}{2}$ from l_1. Note that if $k = d/2$, these two lines will collapse into one, equidistant from l_1 and l_2.

If $k \geq d$, there will be no locus points between l_1 and l_2. In fact, for any point X not between these two lines, $k = |d(X, l_1) - d(X, l_2)| = d$. Thus, when $k = d$ the locus contains all points in the plane not between l_1 and l_2, and when $k > d$, the locus is empty.

(c) No solution is provided, as this problem also appears in the Supplemental Problems.

(d) Let $\mathcal{L}$ be the desired locus. We denote the following distances as specified: $d = d(l_1, l_2)$, $d_1 = d(X, l_1)$, and $d_2 = d(X, l_2)$. Note that $d_1 \cdot d_2 = k$ can be rewritten as $d_2 = k/d_1$.

Case 1: $X \in \mathcal{L}$ and X is between l_1 and l_2.

First, observe that if X is a point between l_1 and l_2, then the maximum value of $d_1 \cdot d_2$ is $(d/2)(d/2) = d^2/4$. Thus, if $k > d^2/4$, $\mathcal{L}$ contains no points between l_1 and l_2. If $k = d^2/4$, then $\mathcal{L}$ consists of a single line l that is parallel to l_1 and l_2 such that $d_1 = d_2 = d/2$.

Suppose $k < d^2/4$. In addition, since we are considering the case when X is between l_1 and l_2, then $d_1 + d_2 = d$; that is $d_1 + k/d_1 = d$. This gives that $d_1^2 - d \cdot d_1 + k = 0$, which in turn implies that

$$d_1 = \frac{d \pm \sqrt{d^2 - 4k}}{2}.$$

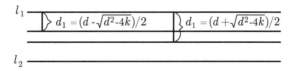

FIGURE 85.

Thus, when $k < d^2/4$, $\mathcal{L}$ contains two lines between l_1 and l_2 that are parallel to l_1 and l_2.

Case 2: $X \in \mathcal{L}$ and X is beyond l_1 or beyond l_2.

Suppose that X is a point in the locus that is beyond l_2, so that $d_1 - d_2 = d$ (as shown in Figure 86); that is, $d_1 - k/d_1 = d$. Solving the quadratic equation $d_1{}^2 - d \cdot d_1 - k = 0$, we see that

$$d_1 = \frac{d \pm \sqrt{d^2 + 4k}}{2}.$$

Recognizing that d_1 must be a positive value, we see that $\mathcal{L}$ contains a line parallel to l_1 and l_2 at distance $(d + \sqrt{d^2 + 4k})/2$ from l_1.

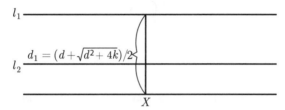

FIGURE 86.

Similarly, if X is beyond l_1, we solve the quadratic equation

$$-d_1{}^2 - d \cdot d_1 + k = 0$$

to obtain

$$d_1 = \frac{d \pm \sqrt{d^2 + 4k}}{-2} = \frac{-d \pm \sqrt{d^2 + 4k}}{2}.$$

Since d_1 must be positive (and k is positive), we conclude that $\mathcal{L}$ contains a line parallel to l_1 and l_2 at distance $(-d + \sqrt{d^2 + 4k})/2$ from l_1.

6.6 (a) Let $\mathcal{L}$ be the desired locus.

Solution 1: Let X be any point in $\mathcal{L}$. Find points A and B on l_1 and l_2, respectively, such that $A - X - B$ and $AO = BO$. Then $\triangle AOB$ is isosceles. By Problem 3.2.28, all of $\overline{AB}$ is in $\mathcal{L}$. Points A and B can be used to identify corresponding points D and C such that $A - O - C$, $B - O - D$, and $AO = BO = CO = DO$. We conclude that $\mathcal{L}$ consists of all points of the rectangle $ABCD$.

Solution 2: Find points A and C on l_1 and points B and D on l_2 such that $A - O - C$, $B - O - D, d(A, l_2) = d(C, l_2) = k$, and $d(B, l_1) = d(D, l_1) = k$. Then A, B, C, and D are in $\mathcal{L}$.

Take X on $\overline{AB}$. Let the line through X perpendicular to l_1 intersect l_1 at R; let the line through X perpendicular to l_2 intersect l_2 at Q. Drop perpendicular segments AP and BS from A and B to l_2 and l_1, respectively. (See Figure 87.)

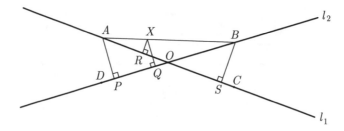

FIGURE **87.**

Then $\triangle XQB \sim \triangle APB$ and $\triangle XRA \sim \triangle BSA$ yield the proportions

$$\frac{XQ}{k} = \frac{XB}{AB} \text{ and } \frac{XR}{k} = \frac{XA}{XB} \, .$$

Adding both sides of these equations gives

$$\frac{XQ}{k} + \frac{XR}{k} = \frac{XB}{AB} + \frac{XA}{AB} \, ,$$

so that $XQ + XR = k$, and X is on $\mathcal{L}$. Likewise, the segments $\overline{BC}$, $\overline{CD}$, and $\overline{AD}$ are on $\mathcal{L}$, so the locus is a rectangle.

(b) No solution is given, as this problem appears in the Supplemental Problems.

(c) Let $\mathcal{L}$ be the desired locus. Take a line parallel to l_1 at distance k and a line parallel to l_2 at distance 1. Let X be the point of intersection of these two lines. Then $X \in \mathcal{L}$.[7]

Let P be any point on $\overleftrightarrow{OX}$, $P \neq O$. Drop perpendicular segments XA and PC to l_2 and perpendicular segments XB and PD to l_1. (See Figure 88.) By AA, $\triangle AXO \sim \triangle CPO$ and $\triangle BXO \sim \triangle DPO$. Therefore $XA/XB = PC/PD$. Thus, P is on $\mathcal{L}$ also, so $\mathcal{L}$ contains the line OX with O deleted.

Finally, notice that l_1 and l_2 divide the plane into four regions, and $\overleftrightarrow{XO}$ passes through two of them. By constructing X in a region adjacent to that of our original choice, we can find another punctured line in $\mathcal{L}$ passing through O. The complete locus contains these two punctured lines which "meet" at O.

(d) The authors have no solution that fits within the scope of this chapter. It is suggested as Problem S9.5, and its solution utilizes the Coordinate Method. We included the question here, however, because it "begged to be asked" in this series of questions!

6.7 (a) Assume first that B is on the major arc formed by A and C. For any choice of B on this arc, $m\angle B = 1/2m\widehat{AC}$; thus, the measure of the angle at B remains fixed as B traverses $\widehat{AC}$.

---

[7] We can also demonstrate the existence of a point $X \in \mathcal{L}$ in a nonconstructive fashion by using the completeness of the real numbers. Essentially we appeal to the fact that a nonnegative continuous unbounded function containing zero in its range must take on all real values. Let $X_1 \neq O$ and $X_2 \neq O$ be points on lines l_1 and l_2, respectively, and note that $\frac{d(X_1, l_1)}{d(X_1, l_2)} = 0$ while $\frac{d(X_2, l_2)}{d(X_2, l_1)} = 0$. Now, as X varies from X_1 to X_2 on $\overline{X_1 X_2}$, it follows that $d(X, l_1)/d(X, l_2)$ will take on all values in $[0, \infty)$, because this ratio is a continuous unbounded function of X. In particular, there is some point X for which $d(X_1, l_1)/d(X_1, l_2) = k$, as we sought to show.

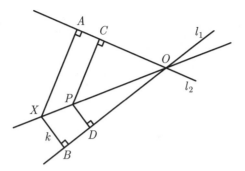

FIGURE 88.

Recall that in a triangle, the incenter is the point of concurrency of the angle bisectors of the triangle. Consequently,

$$m\angle AXC = 180^o - \left(\frac{m\angle A}{2} + \frac{m\angle A}{2}\right)$$

$$= 180° - (\frac{1}{2}(180° - m\angle B)) = 90° + \frac{m\angle B}{2},$$

a fixed value. By Corollary 4.4.16, the locus of the points X is thus an arc of a circle through A and C with A and C deleted.

When B is on the minor arc $\widehat{AC}$, we similarly obtain an arc through A and C with the endpoints deleted. See Figure 89 for the entire locus.

(b) If A, B, and C are all on the circle, then clearly $\mathcal{C}$ is the circumcircle for $\triangle ABC$. Thus X must be a single point, the center of this circle.

(c) No solution is given, as this problem appears in the Supplemental Problems.

(d) When $\triangle ABC$ is a right triangle with the right angle at C, the orthocenter, X, coincides with C. Likewise, when $\triangle ABC$ is a right triangle with the right angle at A, $X = A$.

When $\triangle ABC$ is acute, X lies in the interior of $\triangle ABC$. By Problem 4.6, the reflection of X across $\overline{AC}$, X', will lie on the minor arc $\widehat{AC}$; by symmetry, X must lie on an arc that is congruent to $\widehat{AC}$ and shares the chord $\overline{AC}$. This creates part of the locus.

When $\triangle ABC$ is obtuse, X lies in the exterior of $\triangle ABC$. In this case, X' lies on the major arc with endpoints at A and C. Thus, this portion of the locus forms an arc congruent to the major arc $\widehat{AC}$ and with endpoints at A and C.

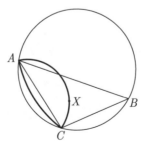

FIGURE 89.

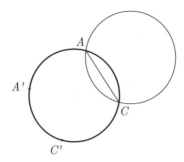

FIGURE 90.

The complete locus, shown in Figure 90, is the circle created by reflecting C across $\overleftrightarrow{AC}$.

6.8 First consider the case where the two lines are parallel. For any points X_1 and X_2 on l_1 and l_2, respectively, $\overline{X_1 X_2}$ will clearly have a midpoint that lies on the unique line m that is equidistant from l_1 and l_2 (i.e., m lies midway between l_1 and l_2, parallel to each of them). Thus, in this case, the locus is the line m.

Now suppose l_1 and l_2 intersect at O, and let X be a point in the plane not on either line. Construct a line m that is parallel to l_1 and contains X. Let $P = l_2 \cap m$. Now choose point $N \neq O$ on l_2 so that $OP = PN$ and let $M = l_1 \cap \overleftrightarrow{XN}$. (See Figure 91.) Then $\triangle ONM \sim \triangle PNX$ by AA. Since P is the midpoint of $\overline{ON}$, X is the midpoint of $\overline{MN}$. This proves that the locus of midpoints will be the entire plane, excluding the points on l_1 and l_2.

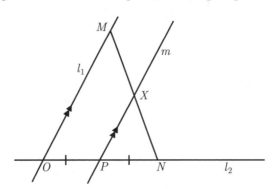

FIGURE 91.

6.9 No solution is provided, as this problem also appears in the Supplemental Problems.

6.10 Suppose first that A does not lie on segment $F_1 F_2$. Choose X on m, $X \neq A$. To prove that m is tangent to $\mathcal{H}$, it is sufficient to show that X lies in the exterior of $\mathcal{H}$, i.e., that $|XF_1 - XF_2| < AF_1 - AF_2$. We will assume $XF_2 < XF_1$, as similar methods may be used when $XF_2 > XF_1$.

Let b be the bisector of the angle formed by $\overrightarrow{F_2 A}$ and $\overrightarrow{A F_1}$, and construct line l through F_2 parallel to b; let D be the intersection of lines m and l and F the intersection of lines l and AF_1. (See Figure 92.)

Since m is perpendicular to b, $\angle DAF_2 \cong \angle XAF_1 = \angle DAF_1$. Thus, $\triangle FAD \cong \triangle F_2 AD$ by ASA. Therefore $DF = DF_2$ and $AF_2 = AF$. Furthermore, $\triangle XDF \cong \triangle XDF_2$ by SAS,

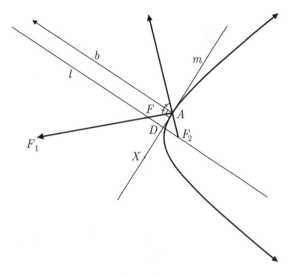

FIGURE 92.

so $XF = XF_2$. Using the triangle inequality in several key places, we then have

$$
\begin{aligned}
XF_1 - XF_2 &= XF_1 - XF \\
&< XF_1 - (XF_1 + FF_1) \\
&= -FF_1 < FF_1 \\
&= AF_1 - AF = AF_1 - AF_2.
\end{aligned}
$$

We leave the proof in the case where A lies on segment F_1F_2 to the reader.

6.11 Let $\mathcal{H}(F_1, F_2; d)$ be a hyperbola and suppose P is an exterior point of $\mathcal{H}$. Let Q be the intersection point of $\mathcal{H}$ with $\overleftrightarrow{PF_1}$. The tangent line to $\mathcal{H}$ at Q bisects $\angle F_1QF_2$.

6.12 This problem was partially solved in Example 9. In that example, C and X were chosen so that C lay on the major arc AB and $A - C - X$. In this problem, by only requiring that X lay on the line AC, we add five more possibilities.

(a) C lies on the minor arc AB and $A - C - X$
(b) C lies on the major arc AB and $A - X - C$
(c) C lies on the minor arc AB and $A - X - C$
(d) C lies on the major arc AB and $C - A - X$
(e) C lies on the minor arc AB and $C - A - X$

Although the figures will be slightly different, each case will be solved in a similar fashion to the solution in Example 9. We present several here and leave the rest to the reader. We note ahead of time that in each case the locus will be an open arc of a circle containing A and B.

The solution to (a) will start similarly to the one in Example 9, and again we will obtain that $m\angle AXB = (m\angle ACB)/2$. In this case, however, the measure of $\angle ACB$ will be half the measure of the **major** arc AB. As before, then, $m\angle AXB = m(\widehat{AB})/4$ (where $\widehat{AB}$ represents the major arc). Thus, $\angle AXB$ is independent of where C is chosen on the given arc, so by Corollary 4.16, X will again lie on an open arc of a circle containing A and B.

We now consider case (c). In this case, since $\triangle BCX$ is isosceles, $\angle CXB \cong \angle CBX$. (See Figure 93(i).) It follows that $m\angle AXB = m\angle BCA + (180 - m\angle BCA)/2$. Now, since C lies

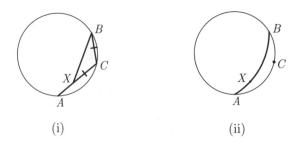

FIGURE 93.

on the minor arc AB, $m\angle BCA$ is half the measure of the major arc AB. It follows that $m\angle AXB = 90 + m(\widehat{AB})/4$. As before, since $\angle AXB$ takes on a single value, X must lie on an open arc of a circle through points A and B. This part of the arc is depicted in Figure 93(ii).

Finally, we consider case (d), where $C - A - X$ and C lies on the major arc AB. (See Figure 94(i).) Again, $\angle CXB \cong \angle CBX$. Therefore, $m\angle AXB = (180 - m\angle BCX)/2 = 90 - m\angle BCA/2 = 90 - m(\widehat{AB})/4$ (the minor arc, $\widehat{AB}$). As before, $m\angle AXB$ is independent of the location of C on the given arc, so X lies on the arc of a circle through A and B. The locus is shown in Figure 94(ii).

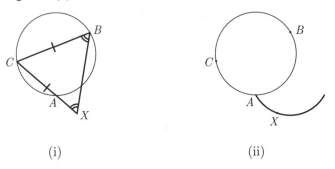

FIGURE 94.

We leave the other cases to the reader, since they are similar. We also leave it to the reader to show that the union of the six arcs obtained (one from Example 9 and five from the cases above) form two circles intersecting at A and B. This can be done by appealing to the congruency of the resulting arcs in Corollary 4.16 and using a continuity argument. The complete locus is shown in Figure 95.

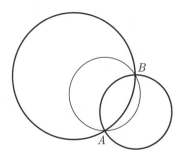

FIGURE 95.

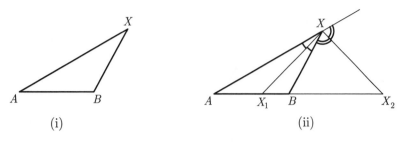

FIGURE 96.

6.13 For any point X in the locus not on $\overleftrightarrow{AB}$, consider triangle AXB, and assume that X is closer to B than to A, as in Figure 96(i). Let $\overline{XX_1}$ and $\overline{XX_2}$ be the respective bisectors of the interior and exterior angles at X, where X_1 and X_2 lie on $\overrightarrow{AB}$. (See Figure 96(ii).)

By the Triangle Bisector Property (Theorem 3.16) and the Triangle Exterior Angle Bisector Property (Problem 3.2.23),

$$\frac{AX_1}{X_1B} = \frac{AX}{XB} = \frac{AX_2}{X_2B} = k,$$

so the locations of X_1 and X_2 are independent of X. (These points are determined uniquely by points A and B and the number k.) Since $\angle X_1XX_2$ is clearly a right angle (it is formed by the angle bisectors at B), X belongs to the circle C with diameter X_1X_2.

Next we show that every point of C belongs to the locus. Let Q be a point of C distinct from X_1 and X_2. Then $\angle X_1QX_2$ is a right triangle.

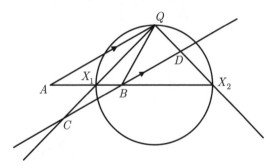

FIGURE 97.

Consider a line through B parallel to line AQ, and let C and D be the points of intersection of this line with lines QX_1 and QX_2, respectively. (See Figure 97.) Then $\triangle AX_1Q \sim \triangle BX_1C$ by AA, and so $AQ/BC = X_1A/X_1B = k$. As $\triangle QX_2A \sim \triangle DX_2B$ by AA again, $AQ/BD = X_2A/X_2B = k$. This implies that $BC = BD$, and hence B is the midpoint of $\overline{CD}$. Since $\triangle CQD$ is right, $QB = BC = BD$ by Problem S3.2.5. Therefore $QA/QB = k$, so Q is in the locus, and the locus is the entire circle C. This circle is sometimes called the Circle of Apollonius.

6.14 *Solution 1:* Let $\overline{ZX'}$ and $\overline{ZY'}$ be perpendiculars from Z to $\overrightarrow{CB}$ and $\overrightarrow{CA}$, respectively. (See Figure 98.)

Then $\angle Y'ZY \cong \angle X'ZX$, since the measure of each is the difference between $m\angle X'ZY$ (or $m\angle XZY'$, whichever is obtuse) and $90°$. It follows that right triangles $ZY'Y$ and $ZX'X$ are

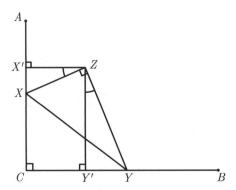

FIGURE 98.

similar by AA. Therefore $\dfrac{ZX'}{CX'} = \dfrac{ZX'}{ZY'} = \dfrac{ZX}{ZY}$ and is independent of the position of $\triangle XZY$. Thus Z lies on the line through C of slope $\dfrac{ZX'}{ZY'} = \dfrac{ZX}{ZY}$.

From our construction, the locus of points Z must be the segment joining the two limiting positions of Z. The limiting positions will correspond to the maximum and minimum distances of Z from C. This distance is

$$CZ = \sqrt{(X'Z)^2 + (Y'Z)^2} \le \sqrt{(XZ)^2 + (YZ)^2},$$

which is maximized when $X'Z = XZ$ and $Y'Z = YZ$; this occurs when $XZ \parallel CB$ and $YZ \parallel AC$, as shown in Figure 99(i). Assuming that $XZ \le YZ$ (as pictured in Figure 98), CZ is minimized when $X = C$ (as shown in Figure 99(ii)). Thus, the length of the segment that constitutes the locus is $\sqrt{(XZ)^2 + (YZ)^2} - YZ$.

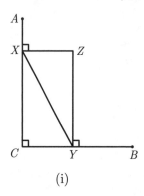

(i)

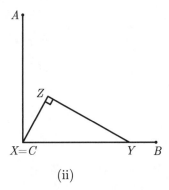

(ii)

FIGURE 99.

Solution 2: By Theorem 4.21, since

$$m\angle XCY + m\angle YZX = 90° + 90° = 180° = m\angle ZXC + m\angle CYZ,$$

there is a circumcircle for quadrilateral $XCYZ$. (See Figure 100.) In the circumcircle, $\angle ZXY$ and $\angle ZCY$ subtend the same angle; thus $\angle ZXY \cong \angle ZCY$. Therefore, Z is a point on a line through C having angle of inclination $\angle ZCY$. The limiting positions of the line segment are discussed in Solution 1.

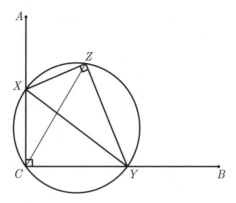

FIGURE **100.**

6.15 Label the intersection points of the chord and the circle as B and C and assume the tangents at B and C intersect at X. (We assume that line BC is not line OA, else the tangents would not meet.) Then segment OX intersects segment BC at M, the midpoint of segment BC. Now, drop the perpendicular segment XY to line OA, and note that $\triangle OMA$ and $\triangle OYX$ are similar right triangles, as are $\triangle OCX$ and $\triangle OMC$. (See Figure 101.) It follows that $\dfrac{OY}{OX} = \dfrac{OM}{OA}$ and $\dfrac{OX}{OC} = \dfrac{OC}{OM}$. Multiplying these equalities, we obtain $\dfrac{OY}{OC} = \dfrac{OC}{OA}$. Therefore

$$OY = \frac{(OC)^2}{OA} = \frac{r^2}{OA}$$

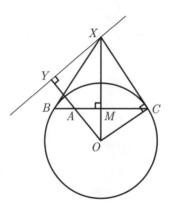

FIGURE **101.**

and thus does not depend on line BC. Hence, all points X lie on the line perpendicular to $\overleftrightarrow{OA}$ at the fixed point Y. This is the light line in the figure.

6.16 Reflect F_2 across the tangent lines AS and BS to obtain points M and N, respectively. Because of the reflective property of the ellipse (Corollary 6.2), $F_1 - A - M$ and $F_1 - B - N$. (See Figure 102.) Furthermore, $F_2B = BN$ and $F_2A = AM$.

From the definition of an ellipse, $F_1A + F_2A = F_1B + F_2A$. Therefore,

$$F_1M = F_1A + AM = F_1A + F_2A = F_1B + F_2B = F_1B + BN = F_1N,$$

and it follows that $\triangle MF_1N$ is isosceles.

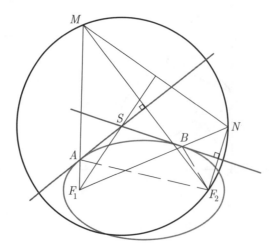

FIGURE 102.

Now, since $\overleftrightarrow{AS}$ and $\overleftrightarrow{BS}$ are the perpendicular bisectors of $\overline{F_2M}$ and $\overline{F_2N}$, S is the circumcenter of $\triangle MF_2N$. Therefore, S must also lie on the perpendicular bisector of $\overline{MN}$. But $\triangle MF_1N$ is isosceles, so this perpendicular bisector is also the angle bisector through S, namely $\overline{F_1S}$. Thus, $\angle AF_1S = \angle MF_1S \cong \angle NF_1S = \angle BF_1S$.

Trigonometry

7.1 Consider an equilateral triangle ABC with unit side lengths, and let D be the midpoint of $\overline{AC}$. Then $\overline{AD} \perp \overline{AC}$. Since $AD = 1/2$, by the Pythagorean theorem, $BD = \sqrt{1 - 1/4} = \sqrt{3}/2$. Then, since $\overline{BD}$ bisects the $60°$ angle at B, we have

$$\cos 60° = \sin 30° = \sin \angle ABD = AD/AB = 1/2.$$

Similarly,

$$\cos 30° = \sin 60° = \sin \angle BAD = BD/AB = \sqrt{3}/2.$$

Now consider an isosceles right triangle ABC with right angle at C and unit side lengths. The base angles at A and B will be $45°$ and by the Pythagorean theorem, $AB = \sqrt{2}$. Therefore,

$$\sin 45° = \cos 45° = \frac{AC}{AB} = \frac{1}{\sqrt{2}} = \frac{\sqrt{2}}{2}.$$

7.2 By considering a right triangle with leg lengths of 3 and 1, we obtain $\sin \alpha = 3/\sqrt{10}$. Then, using the triple angle formula for sine, we have

$$
\begin{aligned}
\sin 3\alpha &= 3\sin\alpha - 4\sin^3\alpha \\
&= 3\left(\frac{3}{\sqrt{10}}\right) - 4\left(\frac{3}{\sqrt{10}}\right)^3 \\
&= \frac{90 - 108}{10\sqrt{10}} = \frac{-9\sqrt{10}}{50}.
\end{aligned}
$$

Now, since $\sin 3\alpha < 0$ and $0° < \alpha < 90°$, we see that $180° < 3\alpha < 270°$. Therefore,

$$\cos 3\alpha = -\sqrt{1 - \sin^2 3\alpha} = -\sqrt{169/250} = -\frac{13\sqrt{10}}{50}.$$

7.3 Since $m\angle A + m\angle B/2 = 180° - m\angle C - m\angle B/2$, we have

$$\sin(A + B/2) = \sin(180° - (C + B/2)) = \sin(C + B/2).$$

7.4 By the Cosine theorem, $a^2 + b^2 - c^2 = 2ab \cdot \cos C$. The right-hand side of this equation will be negative when $\cos C < 0$, that is, when C is obtuse; it will be positive when $\cos C > 0$, that is, when C is acute. Thus, if $a^2 + b^2 > c^2$, the triangle is acute, and if $a^2 + b^2 < c^2$, then it is obtuse.

7.5 From the double angle formula for sine, $\sin 2\theta = 2\sin\theta\cos\theta$, so it suffices to reduce the right-hand side to this form. Rewriting in terms of sine and cosine, we obtain

$$\frac{2\tan\theta}{1 + \tan^2\theta} = \frac{(2\sin\theta)/(\cos\theta)}{(\cos^2\theta + \sin^2\theta)/(\cos^2\theta)} = 2\sin\theta\cos\theta.$$

Similarly,

$$\frac{1 - \tan^2\theta}{1 + \tan^2\theta} = \frac{\cos^2\theta - \sin^2\theta}{\cos^2\theta + \sin^2\theta} = \cos 2\theta.$$

7.6

$$\begin{aligned}
\cos 3\alpha &= \cos(2\alpha + \alpha) \\
&= \cos(2\alpha)\cos\alpha - \sin(2\alpha)\sin\alpha \\
&= (\cos^2\alpha - \sin^2\alpha)\cos\alpha - (2\sin\alpha\cos\alpha)\sin\alpha \\
&= \cos^3\alpha - 3\sin^2\alpha\cos\alpha \\
&= \cos^3\alpha - 3(1 - \cos^2\alpha)\cos\alpha \\
&= 4\cos^3\alpha - 3\cos\alpha.
\end{aligned}$$

In order to derive the formula for $\tan 3\alpha$, we use the angle sum formulas twice, as follows, letting $t = \tan\alpha$:

$$\begin{aligned}
\tan 3\alpha &= \tan(2\alpha + \alpha) \\
&= \frac{\tan 2\alpha + \tan\alpha}{1 - \tan 2\alpha \cdot \tan\alpha} \\
&= \frac{\frac{2t}{1-t^2} + t}{1 - \frac{2t}{1-t^2} \cdot t} \\
&= \frac{3t - t^3}{1 - 3t^2} \\
&= \frac{3\tan\alpha - \tan^3\alpha}{1 - 3\tan^2\alpha}.
\end{aligned}$$

7.7 From the Cosine theorem,

$$a^2 = b^2 + c^2 - 2bc\cos A = 49 + 25 - 70\cos 75° \approx 74 - 70(0.2588) \approx 55.88.$$

Therefore, $a \approx 7.475$.

We now use the Sine theorem to find angle B: $\sin B = b(\sin A)/a \approx 7(\sin 75°)/7.475 \approx$ 0.9045, so $m\angle B \approx 64.76°$ or $m\angle B \approx 115.24°$. The latter is not possible, as $m\angle A + m\angle B <$ 180°. From here, $m\angle C \approx 180° - 75° - 64.76° = 40.24°$.

7.8 Since $\beta = 90° - \alpha$ and $\sin \alpha = \cos(90° - \alpha)$, we have $\sin \alpha \cos \beta + \sin \beta \cos \alpha = \sin \alpha \sin \alpha + \cos \alpha \cos \alpha = 1$.

7.9 Using the formulas $\cos(\alpha \pm \beta) = \cos \alpha \cos \beta \mp \sin \alpha \sin \beta$, and $(x - y)(x + y) = x^2 - y^2$, we obtain:

$$\begin{aligned}\cos(\alpha + \beta) \cos(\alpha - \beta) &= \cos^2 \alpha \cos^2 \beta - \sin^2 \alpha \sin^2 \beta \\ &= \cos^2 \alpha \cos^2 \beta - (1 - \cos^2 \alpha)(1 - \cos^2 \beta) \\ &= 1 - (\cos^2 \alpha + \cos^2 \beta) = 1 - a.\end{aligned}$$

7.10 From Corollary 7.5, the area of $\triangle ABC$ may be expressed as $(1/2)bc \sin A$, as $(1/2)ac \sin B$, and as $(1/2)ab \sin C$. Setting the first two expressions equal yields $\dfrac{a}{\sin A} = \dfrac{b}{\sin B}$, and the equality of $c/\sin C$ follows similarly.

7.11 Label the triangle so $BC = 4$ and $AC = 6$. Then by the Cosine theorem, $16 + 25 - 40 \cdot \cos B = 36$, so $\cos B = 1/8$. Similarly, $\cos A = 3/4$. Note then that $\cos B = 2\cos^2 A - 1 = \cos 2A$. Since $2A$ and B are both in $(0°, 180°)$, $B = 2A$, as desired.

7.12 Let $\alpha = \arctan(1/5)$ and $\beta = \arctan(1/239)$, so that $\tan \alpha = 1/5$ and $\tan \beta = 1/239$. We will show that $\tan(4\alpha - \beta) = 1$.

Note that

$$\tan(4\alpha - \beta) = \frac{\tan 4\alpha - \tan \beta}{1 + \tan 4\alpha \tan \beta}.$$

To find $\tan 4\alpha$, we first find $\tan 2\alpha$:

$$\tan 2\alpha = \frac{2\tan \alpha}{1 - \tan^2\alpha} = \frac{2/5}{24/25} = \frac{5}{12}.$$

Therefore,

$$\tan 4\alpha = \frac{2(5/12)}{1 - (5/12)^2} = \frac{120}{119},$$

which proves that

$$\tan(4\alpha - \beta) = \frac{(120/119) - (1/239)}{1 + (120/119) \cdot (1/239)} = 1.$$

We're not quite done! Since $\tan(4\alpha - \beta) = 1$, we know that $4\alpha - \beta = \dfrac{\pi}{4} + \pi k$ for some integer k; we will show that $k = 0$ by bounding the value of $4\alpha - \beta$. Because $\arctan(1/5) > \arctan(1/239)$, $4\alpha - \beta > 0$. Also, $4\arctan(1/5) < 4\arctan 1 = \pi$. Therefore,

$$0 < 4\alpha - \beta < 4\alpha < \pi.$$

This proves that $4\alpha - \beta = \pi/4$.

7.13 The following statements are equivalent, which proves the result.

$$\frac{2\cos 40° - \cos 20°}{\sin 20°} = \sqrt{3},$$

$$\cos 40° = \frac{1}{2}\cos 20° + \frac{\sqrt{3}}{2}\sin 20°,$$

$$\cos 40° = \cos 60° \cos 20° + \sin 60° \sin 20°, \text{ and}$$

$$\cos 40° = \cos(60° - 20°).$$

7.14 Writing $\sin 3\alpha$ in terms of $\sin \alpha$, and using the product formula from Identity (7.10) for the product $\sin(60° - \alpha) \cdot \sin(60° + \alpha)$, we obtain:

$$4\sin\alpha \sin(60° - \alpha)\sin(60° + \alpha) - \sin 3\alpha$$
$$= \sin\alpha(\cos 2\alpha - \cos 120°) - (3\sin\alpha - 4\sin^3\alpha)$$
$$= \sin\alpha[2(1 - 2\sin^2\alpha + 1/2) - 3 + 4\sin^2\alpha]$$
$$= \sin\alpha[2(1 - 2\sin^2\alpha + 1/2) - 3 + 4\sin^2\alpha]$$
$$= (\sin\alpha) \cdot 0 = 0.$$

This proves the first identity. The second one can be proved in a similar way.

7.15 Using the identities of Problem 7.14, we obtain an analogous identity for tangents:

$$\tan\alpha \tan(60° - \alpha)\tan(60° + \alpha) = \tan 3\alpha.$$

Substituting $\alpha = 5°$ into it, we obtain

$$\tan 5° \tan 55° \tan 65° = \tan 15°.$$

This is equivalent to the original equality as $\tan 5° = (\tan 85°)^{-1}$ and $\tan 75° = (\tan 15°)^{-1}$.

7.16 Let α and β denote the angles shown in Figure 103. We will determine these angles and use the Cosine theorem to find x.

Using the Cosine theorem on the interior isosceles triangle, we obtain

$$7 + 7 - 14\cos\beta = 1.$$

Thus, $\cos\beta = 13/14$, in which case $\sin\beta = \sqrt{27}/14$.

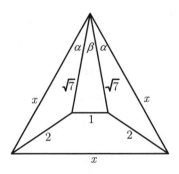

FIGURE **103.**

Now, in order to find $\cos \alpha$, we first find $\cos 2\alpha$ as follows:

$$\cos 2\alpha = \cos(60° - \beta)$$
$$= \cos 60° \cos \beta + \sin 60° \sin \beta$$
$$= \frac{1}{2} \cdot \frac{13}{14} + \frac{\sqrt{3}}{2} \cdot \frac{\sqrt{27}}{14} = \frac{11}{14}.$$

Hence, $\cos \alpha = \sqrt{(1 + \cos 2\alpha)/2} = 5/(2\sqrt{7})$. Using the Cosine theorem on the triangle with side lengths x, $\sqrt{7}$, and 2, we get:

$$7 + x^2 - 2\sqrt{7}x \cos \alpha = x^2 - 5x + 7 = 2^2.$$

Since $x > 1$, the root of the quadratic $x^2 - 5x + 3$ we seek is $x = (5 + \sqrt{13})/2 \approx 4.303$.

7.17 The problem is equivalent to showing that $x^3 = 3x - 2y$. Squaring both sides of the equality $\sin z + \cos z = x$, we get

$$\sin^2 z + 2 \sin z \cos z + \cos^2 z = x^2,$$

which implies

$$\sin z \cos z = \frac{x^2 - 1}{2}.$$

On the other hand, cubing both sides of $\sin z + \cos z = x$ gives

$$x^3 = \sin^3 z + 3 \sin^2 z \cos z + 3 \sin z \cos^2 z + \cos^3 z$$
$$= y + 3 \sin z \cos z (\sin z + \cos z)$$
$$= y + \frac{3}{2}(x^2 - 1)x.$$

Hence, $x^3 = y + \frac{3}{2}(x^3 - x)$, which is equivalent to $x^3 = 3x - 2y$.

7.18 Let $S := \sum_{k=0}^{n} \sin(\alpha + k\beta)$. Then, using the identity

$$\sin x \sin y = \frac{1}{2}[\cos(x - y) - \cos(x + y)],$$

we have:

$$S \cdot \sin \frac{\beta}{2} = \sum_{k=0}^{n} \sin(\alpha + k\beta) \cdot \sin \frac{\beta}{2}$$

$$= \frac{1}{2} \sum_{k=0}^{n} \left[\cos(\alpha + (k - 1/2)\beta) - \cos(\alpha + (k + 1/2)\beta) \right]$$

$$= \cos(\alpha - \beta/2) - \cos(\alpha + \beta/2) + \cos(\alpha + \beta/2) - \cos(\alpha + 3\beta/2) + \cos(\alpha - 3\beta/2)$$

$$+ \cos(\alpha + 5\beta/2) + \cdots + \cos(\alpha - (n - 1)\beta/2) + \cos(\alpha + (n + 1)\beta/2)$$

$$= \frac{1}{2} \left(\cos\left(\alpha - \frac{1}{2}\beta\right) + \cos\left(\alpha + \frac{n + 1}{2}\beta\right) \right)$$

$$= \sin\left(\alpha + \frac{n}{2}\beta\right) \sin\left(\frac{n + 1}{2}\beta\right).$$

Dividing both sides by $\sin \frac{\beta}{2}$, we obtain the required identity.

7.19 This solution is similar to the solution for Problem 7.18, and we leave it to the reader.

7.20 Substituting $n - 1$ instead of n, and $\beta = 2\pi/n$ in the right-hand side of the closed form for the sum in Problems 7.18 and 7.19, we obtain (in both cases) a factor of $\sin \pi$, which is equal to zero. Compare this solution to those for Example 53 in Chapter 10 and Example 67 in Chapter 11.

7.21 Suppose $AC = \sqrt{2}$, so $AP = CP = 1$. Let $m\angle ABP = x$, $m\angle CBP = y$, and $m\angle BAC = \alpha$. We wish to show $x = y$. (See Figure 104.) Using the Sine theorem on triangles ABP and BCP we have

$$\frac{1}{\sin x} = \frac{BP}{\sin(\alpha + 45°)} \quad \text{and} \quad \frac{1}{\sin y} = \frac{BP}{\sin(135° - \alpha)}.$$

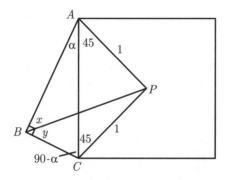

FIGURE **104.**

Therefore, $\sin x = \frac{\sin(\alpha+45°)}{BP}$ and $\sin y = \frac{\sin(135°-\alpha)}{BP}$. Since

$$\sin(\alpha + 45°) = \sin(180° - (\alpha + 45°)) = \sin(135° - \alpha),$$

we have $\sin x = \sin y$. Because both angles are acute, $x = y$.

7.22 Note that consecutive sides of the hexagon with lengths a and b can be switched to have lengths b and a by simply moving the vertex between the two sides an appropriate distance along the circle. One can repeat this process several times to yield a new hexagon inscribed in the same circle, where consecutive sides have lengths a and b around the hexagon. (See Figure 105.)

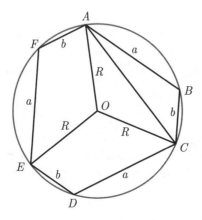

FIGURE **105.**

In this new hexagon $ABCDEF$, all six interior angles are congruent, and must therefore have a measure of $120°$. Furthermore, the central angles at the center, O, also have a measure of $120°$. Using the Cosine theorem on triangles ABC and ACO, we calculate AC^2 in two ways:

$$AC^2 = R^2 + R^2 - 2 \cdot R \cdot R \cos 120° = 3R^2, \text{ and}$$
$$AC^2 = a^2 + b^2 - 2 \cdot ab \cos 120° = a^2 + b^2 + ab.$$

Therefore, $R = \sqrt{(a^2 + b^2 + ab)/3}$.

When the hexagon is replaced by an $n - gon$, n even, a similar method will yield

$$R = \sqrt{\frac{a^2 + b^2 - 2ab \cos\left(\frac{n-2}{n} \cdot 180°\right)}{2\left(1 - \cos(720°/n)\right)}}.$$

7.23 Assume first that the three cevians intersect at P. Let $\alpha = m\angle PEC$, $\beta = m\angle BDP$, $\gamma = m\angle AFP$, $\phi = m\angle BPF = m\angle EPC$, and $\theta = m\angle BPD = m\angle APE$, as shown in Figure 106.

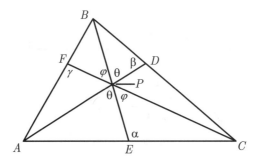

FIGURE 106.

We now use the Sine theorem on triangles BFP, BDP, and CDP, along with the fact that $\sin(180° - x) = \sin x$:

$$\frac{BF}{BP} = \frac{\sin\phi}{\sin\gamma}, \quad \frac{CD}{CP} = \frac{\sin(\phi + \theta)}{\sin\beta}, \text{ and } \frac{AE}{AP} = \frac{\sin\theta}{\sin\alpha}.$$

Therefore,

$$\frac{BF \cdot CD \cdot AE}{BP \cdot CP \cdot AP} = \frac{\sin\phi \sin(\phi + \theta) \sin\theta}{\sin\gamma \sin\beta \sin\alpha}.$$

Similarly, using the other three triangles present, we obtain

$$\frac{AF \cdot BD \cdot CE}{AP \cdot BP \cdot CP} = \frac{\sin(\phi + \theta) \sin\theta \sin\phi}{\sin\gamma \sin\beta \sin\alpha}.$$

These last two relationships prove that $BF \cdot CD \cdot AE = AF \cdot BD \cdot CE$.

The converse may be proved by contradiction, as was done in previous chapters. This problem also appears as Theorem 3.17, Problem 5.17, Problem 8.16, and Problem 12.11.

7.24 By the Sine theorem, we have

$$\frac{AB_1}{\sin(\angle ABB_1)} = \frac{c}{\sin(\angle BB_1A)} \text{ and } \frac{B_1C}{\sin(\angle CBB_1)} = \frac{a}{\sin(\angle BB_1C)}.$$

Dividing one equality by the other, we obtain

$$\frac{AB_1}{B_1C} \cdot \frac{\sin(\angle CBB_1)}{\sin(\angle ABB_1)} = \frac{c}{a} \cdot \frac{\sin(\angle BB_1C)}{\sin(\angle BB_1A)}.$$

As the angles BB_1A and BB_1C are supplementary, their sines are equal. Therefore,

$$AB_1/B_1C = c/a \quad \Leftrightarrow \quad \sin(\angle CBB_1) = \sin(\angle ABB_1).$$

Since the angles ABB_1 and CBB_1 are acute, their sines are equal if and only if they are congruent, i.e., if $\overline{BB_1}$ bisects $\angle B$.

The reader is invited to compare this solution with the proof of the same result in Chapter 3 (Theorem 3.16).

7.25 From the Cosine theorem,

$$\cos A = \frac{b^2 + c^2 - a^2}{2bc} = \frac{1}{2}\left(\frac{b}{c} + \frac{c}{b}\right) - \frac{a^2}{2bc}.$$

As $b/c + c/b \geq 2 \Leftrightarrow (b - c)^2 \geq 0$, then

$$\cos A \geq 1 - \frac{a^2}{2bc}.$$

Therefore

$$\sin^2 \frac{A}{2} = \frac{1 - \cos A}{2} \leq \frac{a^2}{4bc},$$

and the statement follows. The equality sign appears if and only if

$$b/c + c/b = 2 \Leftrightarrow (b - c)^2 = 0 \Leftrightarrow b = c.$$

7.26 Let $m\angle A = x$, $m\angle B = y$, and $m\angle C = z$. Then $x + y + z = 180°$. As bisectors of interior angles of a triangle are concurrent, $\overrightarrow{BO}$ bisects angle B. Using exterior angles of triangles AEC and CDA, respectively, we find $m\angle BEO = x + z/2$ and $m\angle BDO = x/2 + z$. Applying the Sine theorem to $\triangle BOE$ and $\triangle BOD$, we obtain:

$$\frac{BO}{\sin(x + z/2)} = \frac{OE}{\sin(y/2)} \quad \text{and} \quad \frac{BO}{\sin(x/2 + z)} = \frac{OD}{\sin(y/2)}.$$

As $OD = OE$, this implies

$$\frac{\sin(x + z/2)}{\sin(y/2)} = \frac{\sin(x/2 + z)}{\sin(y/2)}.$$

As $y/2 \in (0, 180°)$, $\sin(y/2) \neq 0$, so the last equation is equivalent to

$$\sin(x + z/2) = \sin(x/2 + z).$$

Both $x + z/2$ and $x/2 + z$ are in $(0, 180°)$, so the equality of their sines is equivalent to

$$x + z/2 = x/2 + z \quad \text{or} \quad (x + z/2) + (x/2 + z) = 180° \quad \Leftrightarrow$$

$$x = z \quad \text{or} \quad x + z = 120°.$$

The former is equivalent to $\triangle ABC$ being isosceles, and the latter to $y = m\angle B = 60°$.

7.27 Using the Cosine theorem on $\triangle AC'C$, we obtain

$$CC' = \sqrt{4^2 + 7^2 - 2 \cdot 4 \cdot 7 \cdot \cos 60°} = \sqrt{37}.$$

Next let $\alpha = m\angle C'CB$ and $\beta = m\angle B'BC$, as in Figure 107.

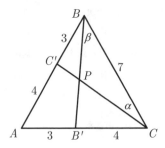

FIGURE 107.

Applying the Sine theorem to $\triangle BCC'$ and $\triangle B'BC$ yields

$$\frac{\sin 60°}{\sqrt{37}} = \frac{\sin \alpha}{3} \quad \text{and} \quad \frac{\sin 60°}{\sqrt{37}} = \frac{\sin \beta}{4},$$

hence $\sin \alpha = (3/2)\sqrt{3/37}$ and $\sin \beta = 2\sqrt{3/37}$.

Using the Sine theorem once again on $\triangle BCP$, we get

$$\frac{PC}{\sin \beta} = \frac{7}{\sin \angle BPC} = \frac{7}{\sin(180° - \alpha - \beta)} = \frac{7}{\sin(\alpha + \beta)}.$$

Therefore the problem is solved if we know $\sin(\alpha + \beta)$. Since we know sines of acute angles α and β, we can easily find their cosines: $\cos \alpha = 11/(2\sqrt{37})$, and $\cos \beta = 5/\sqrt{37}$. Then

$$\sin(\alpha + \beta) = \sin \alpha \cos \beta + \sin \beta \cos \alpha$$

$$= \frac{3}{2}\sqrt{\frac{3}{37}} \cdot \frac{5}{\sqrt{37}} + \frac{2\sqrt{3}}{\sqrt{37}} \cdot \frac{11}{2\sqrt{37}} = \frac{\sqrt{3}}{2}.$$

Having $\sin(\alpha + \beta) = \sqrt{3}/2$, we conclude that

$$PC = \frac{7 \sin \beta}{\sin(\alpha + \beta)} = \frac{7\left(2\sqrt{3}\right)}{\sqrt{37}\sqrt{3}/2} = \frac{28}{\sqrt{37}} \approx 4.6.$$

Comment. The value of $\sin(\alpha + \beta)$ could be obtained faster by noticing that $\triangle ABB' \cong \triangle BCC'$ by SAS. Hence, $m\angle ABB' = \alpha$, and so $\alpha + \beta = m\angle ABC = 60°$. This gives $\sin(\alpha + \beta) = \sqrt{3}/2$.

7.28 Let the inscribed circle C have radius r and the sector have radius R. Consider the right triangle whose vertices are the center of C, the center of the circular sector, and a point of tangency of C with a radius of the sector. (See Figure 108.)

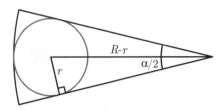

FIGURE 108.

The hypotenuse of this triangle has length $R - r$ and bisects the central angle, so $\sin(\alpha/2) = r/(R - r)$. Therefore

$$\frac{R}{r} = \frac{1}{\sin(\alpha/2)} + 1.$$

Now, the area of the sector is $\alpha R^2/2$ and the area of C is πr^2. Therefore, the desired ratio of their areas is

$$\frac{\alpha}{2}\left(\frac{R}{r}\right)^2 = \frac{\alpha}{2}\left(\frac{1}{\sin(\alpha/2)} + 1\right)^2.$$

7.29 If $\alpha = \beta$, the formula for $\sin(\alpha - \beta)$ is obviously correct. To prove it for $0° < \beta < \alpha < 90°$, consider a quadrilateral $ABCD$ inscribed in a circle of diameter 1, with $\overline{AD}$ being a diameter, $m\angle BAD = \alpha$ and $m\angle CAD = \beta$. (Clearly such a quadrilateral can be constructed– see Figure 109).

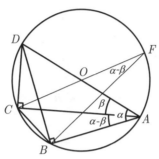

FIGURE 109.

Now, angles ACD and ABD are right angles, so $\cos\beta = AC$, $\sin\beta = DC$, $\cos\alpha = AB$, and $\sin\alpha = BD$. We claim that $\sin(\alpha - \beta) = BC$. To see this, construct diameter CF and note that triangle CBF is a right triangle with hypotenuse of length 1. Since $m\angle F = \alpha - \beta$ (the angle intercepts the same arc as angle ACB, which has measure $\alpha - \beta$), $\sin(\alpha - \beta) = BC$. We now may apply Ptolemy's theorem to quadrilateral $ABCD$ to obtain the desired identity.

7.30 Clearly $m\angle A = m\angle C = 80°$, so $m\angle EAF = 30°$ and $m\angle AEC = 40°$. Using the Sine theorem for $\triangle AEC$ and $\triangle AFC$, we get $EC/AC = \sin 80°/\sin 40°$ and $FC/AC = 1$. (See Figure 110.) Therefore, $EC/FC = \sin 80°/\sin 40°$. On the other hand, from $\triangle EFC$,

$$\frac{EC}{FC} = \frac{\sin(160° - x)}{\sin x} = \frac{\sin(20° + x)}{\sin x}.$$

Therefore, $\dfrac{\sin 80°}{\sin 40°} = \dfrac{\sin(20° + x)}{\sin x}$, which by the double angle formula for sine reduces to

$$2\cos 40° \sin x = \sin(20° + x). \tag{17.7}$$

Since $\cos 40° = \sin 50°$, (17.7) is equivalent to

$$\sin x = \frac{1}{2} \cdot \frac{\sin(20° + x)}{\sin 50°},$$

which clearly has a solution of $x = 30°$. To see that this is the only solution, we use the fact that $\sin(20° + x) = \sin 20° \cos x + \sin x \cos 20°$ and (17.7) to obtain

$$\frac{\cos x}{\sin x} = \frac{2\cos 40° - \cos 20°}{\sin 20°}.$$

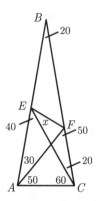

FIGURE 110.

Since $(\cos x)/(\sin x)$ is a decreasing function for $x \in (0°, 90°)$, the solution $x = 30°$ in (17.7) is indeed unique.

7.31 We have the following equivalent statements:

$$3/2 = \cos A + \cos B + \cos C,$$

$$3/2 = 2\cos\frac{A+B}{2}\cos\frac{A-B}{2} + \cos(180° - A - B),$$

$$3/2 = 2\cos\frac{A+B}{2}\cos\frac{A-B}{2} - \cos(A + B),$$

$$0 = 2\cos\frac{A+B}{2}\cos\frac{A-B}{2} - 2\cos^2\frac{A+B}{2} + 1 - \frac{3}{2},$$

$$0 = 2\cos^2\frac{A+B}{2} - 2\cos\frac{A+B}{2}\cos\frac{A-B}{2} + \frac{1}{2}.$$

Let $x = \cos\frac{A+B}{2}$ and $y = \cos\frac{A-B}{2}$. Then we can rewrite the last equality as

$$4x^2 - 4xy + 1 = 0 \quad \Leftrightarrow \quad (2x - y)^2 + (1 - y^2) = 0.$$

As $1 - y^2 \geq 0$, and the sum of nonnegative numbers is zero if and only if both numbers are zeros, the latter is equivalent to $y^2 = 1$ and $2x = y$. Hence, $(x, y) = (1/2, 1)$ or $(x, y) = (-1/2, -1)$. Since A, B, and C are angles in a triangle, $|A - B|/2 \leq \pi/2$. Therefore $y \geq 0$. This implies $y = 1$ and $x = 1/2$. Therefore $A = B = 60°$, giving $C = 60°$ as well.

7.32 Consider $\triangle ABC$. Suppose the bisector at A and the bisector at B meet at K, while the bisector at C and the bisector at B meet at K'. Let the bisector at B intersect side AC at E, as shown in Figure 111. Let $m\angle BAC = 2\alpha$, $m\angle ABC = 2\beta$, and $m\angle AEB = \theta$.

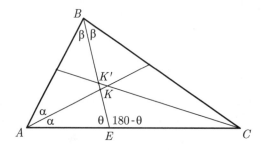

FIGURE 111.

From the Sine theorem,

$$\frac{BK}{\sin\alpha} = \frac{AK}{\sin\beta} \quad\text{and}\quad \frac{KE}{\sin\alpha} = \frac{AK}{\sin\theta}.$$

Therefore,

$$BE = KE + BK = KE + AK\left(\frac{\sin\alpha}{\sin\beta}\right)$$

$$= KE + KE\left(\frac{\sin\theta}{\sin\alpha}\frac{\sin\alpha}{\sin\beta}\right)$$

$$= KE\left(1 + \frac{\sin\theta}{\sin\beta}\right).$$

In a similar fashion, we find $BE = K'E\left(1 + \frac{\sin(180°-\theta)}{\sin\beta}\right)$. Since $\sin\theta = \sin(180° - \theta)$, we see that $KE = K'E$. Since K and K' both lie on $\overline{BE}$, this proves $K = K'$. Hence, the bisectors from A and C meet the bisector from B at a common point, as we sought to show.

7.33 Let O be the center of the circumcircle of a $\triangle ABC$. It is clear that we may assume that the circle has radius 1.

First we explain that it is sufficient to consider the problem for non-obtuse triangles only. If one of the triangle's angles, say $\angle A$, is obtuse, then O is in the exterior of $\triangle ABC$. Consider a chord $B'C'$ parallel to $\overline{BC}$, such that $BC = B'C'$. Then the area of $\triangle AB'C'$ is greater than the area of $\triangle ABC$, as they have congruent bases, but the altitude from A to $\overline{B'C'}$ is longer than the altitude from A to $\overline{BC}$. (See Figure 112(i).) Moreover, we claim that $\triangle AB'C'$ is acute. This follows from a simple computation (left for the reader), which demonstrates that $m\angle B'AC' = 180° - m\angle A, m\angle B' = m\angle B + m\angle A - 90°$, and $m\angle C' = m\angle C + m\angle A - 90°$. As $m\angle A > 90°$, and $m\angle A + m\angle B + m\angle C = 180°$, the angles of $\triangle AB'C'$ are all at most $90°$.

Therefore we may assume that O is in the interior of $\triangle ABC$ or on one of its sides. Let $m\angle BOC = x, m\angle COA = y$, and $m\angle AOB = z$. Then $x, y, z \in (0°, 180°]$, and $x + y + z = 360°$ as in Figure 112(ii). Considering the area of $\triangle ABC$, we obtain:

$$\text{Area}\triangle ABC = \text{Area}\triangle BOC + \text{Area}\triangle COA + \text{Area}\triangle AOB$$

$$= \frac{1}{2}1^2 \sin x + \frac{1}{2}1^2 \sin y + \frac{1}{2}1^2 \sin z$$

$$= \frac{1}{2}(\sin 2A + \sin 2B + \sin 2C),$$

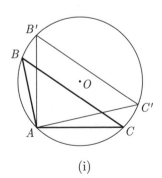

(i)

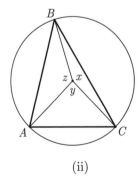

(ii)

FIGURE 112.

and the problem is reduced to maximizing the sum $\sin 2A + \sin 2B + \sin 2C$ over all A, B, and C satisfying the conditions $A, B, C \in (0°, 90°]$ and $A + B + C = 180°$. Each solution we know of this problem that uses elementary methods is rather challenging. Those who know the method of Lagrange multipliers (from multivariate calculus) can solve it easily by using that method.

But there exists a much better solution of this problem, one that uses Jensen's inequality applied to the function $f(x) = \sin x$. We remind the reader that a function $f : [a, b] \to \mathbb{R}$ is called **convex up** on $[a, b]$ if

$$f\left(\frac{x_1 + x_2}{2}\right) \le \frac{f(x_1) + f(x_2)}{2}$$

for all $x_1, x_2 \in [a, b]$, with equality if and only if $x_1 = x_2$. Jensen's inequality generalizes this result: if f is convex up on $[a, b]$, then a similar inequality holds for any $n \ge 2$ points from the interval. More precisely, we have:

Jensen's inequality: *If a function $f : [a, b] \to \mathbb{R}$ is convex up on $[a, b]$, then for all $x_1, x_2, \ldots, x_n \in [a, b]$,*

$$f\left(\frac{x_1 + x_2 + \cdots + x_n}{n}\right) \le \frac{f(x_1) + f(x_2) + \cdots + f(x_n)}{n},$$

and equality is attained if and only if $x_1 = x_2 = \cdots = x_n$.

Replacing the $\le$ sign in the definition of a convex up function with $\ge$, we obtain the definition of a **convex down** function. Making the same change in Jensen's inequality for convex up function, we obtain Jensen's inequality for a convex down function.

How does this all relate to our problem? First we note that the sine function is convex down on $[0°, 180°]$. Indeed, if $x_1, x_2 \in [0°, 180°]$,

$$\sin\left(\frac{x_1 + x_2}{2}\right) - \frac{\sin x_1 + \sin x_2}{2} = \sin\left(\frac{x_1 + x_2}{2}\right) - \sin\left(\frac{x_1 + x_2}{2}\right)\cos\left(\frac{x_1 - x_2}{2}\right)$$

$$= \sin\left(\frac{x_1 + x_2}{2}\right)\left(1 - \cos\left(\frac{x_1 - x_2}{2}\right)\right) \ge 0.$$

The last inequality is obvious: as $\frac{x_1+x_2}{2} \in [0°, 180°]$, $\sin(\frac{x_1+x_2}{2}) \ge 0$, and the cosine function does not take values greater than 1. It is also clear that the equality is attained only if $x_1 = x_2$.[8] As no angle of $\triangle ABC$ measures more than $90°$, we have each of $2A$, $2B$, and $2C$ lies in $[0°, 180°]$. Applying Jensen's inequality for $x_1 = 2A$, $x_2 = 2B$, and $x_3 = 2C$ gives

$$\sin\frac{2A + 2B + 2C}{3} \ge \frac{\sin 2A + \sin 2B + \sin 2C}{3}.$$

As $2A + 2B + 2C = 360°$, we obtain

$$\sin 2A + \sin 2B + \sin 2C \le 3\sin 120° = \frac{3\sqrt{3}}{2},$$

with the maximum value attained if and only if $2A = 2B = 2C = 120°$. Therefore the maximum is attained by the equilateral triangle.

Jensen's inequality is usually proved by induction on n, and the proof is calculus-free. This proof is readily available in the literature or on the web. The same approach can be used to

---

[8] Another way to prove that a function f is a convex up (down) on (a, b) is to show that $f''(x) > 0 \, (< 0)$ for all $x \in (a, b)$. This method can be used if the second derivative of f exists on (a, b).

prove a more general result: among all n-gons inscribed in a circle, $n \geq 3$, the regular one has the greatest area.

Please compare this solution with the one we give for Problem 5.29.

7.34 Let $\triangle ABC$ be inscribed in a circle of radius 1. The triangle may be divided into six right triangles, where one leg of each triangle is a radius of the incircle. The other leg of each of these six triangles will lie on the perimeter of $\triangle ABC$ and will have length $1/\tan \frac{A}{2}$, or $1/\tan \frac{B}{2}$, or $1/\tan \frac{C}{2}$ (each length will be obtained twice). Since the area of each triangle will be half the product of its leg lengths, we obtain that the area of $\triangle ABC$ is equal to

$$\frac{1}{\tan \frac{A}{2}} + \frac{1}{\tan \frac{B}{2}} + \frac{1}{\tan \frac{C}{2}},$$

and its perimeter is just twice this number.

Therefore the problem is reduced to minimizing the above sum, where A, B, and C are the measures of angles in a triangle. As each of $A/2$, $B/2$, and $C/2$ are measures of acute angles, we consider the function $f : (0, \pi/2) \to (0, \infty)$, where $f(x) = 1/\tan x = \cos x / \sin x$. The function is convex up on $(0, \pi/2)$. This can be seen by using the definition, as in the solution of the previous problem, but it is much faster to show this using calculus. As its second derivative is $f''(x) = 2 \cos x / \sin^3 x$, $f''(x) > 0$ on $(0, \pi/2)$, so f is convex up on the interval. By Jensen's inequality (which we discussed in the solution of the previous problem),

$$\frac{1}{\tan \frac{A/2 + B/2 + C/2}{3}} \leq \frac{1}{3}\left(\frac{1}{\tan \frac{A}{2}} + \frac{1}{\tan \frac{B}{2}} + \frac{1}{\tan \frac{C}{2}}\right),$$

or, since $A + B + C = \pi$,

$$\frac{3}{\tan(\pi/6)} \leq \frac{1}{\tan \frac{A}{2}} + \frac{1}{\tan \frac{B}{2}} + \frac{1}{\tan \frac{C}{2}}.$$

Equality in the line above is attained if and only if $A/2 = B/2 = C/2 = \pi/6$, so the statement follows.

The same approach can be used to prove a more general result: among all n-gons circumscribed around a circle, $n \geq 3$, the regular one has the smallest area and the smallest perimeter.

Coordinatization

8.1 Let $ABCD$ be a parallelogram. Assign coordinates to the vertices so that $A : (0, 0)$, $B : (x_B, y_B)$, and $D : (1, 0)$. Then $C : (x_B + 1, y_B)$. (See Figure 113.)

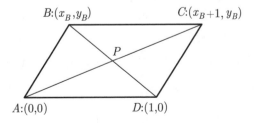

FIGURE 113.

Now we compute the coordinates of the midpoints of $\overline{AC}$ and $\overline{BD}$, and conclude that they are the same, namely

$$P : \left(\frac{x_B + 1}{2}, \frac{y_B}{2}\right).$$

Therefore the diagonals bisect each other.

8.2 Let $ABCD$ be a rhombus. We assign coordinates to the vertices as we would for a parallelogram, $A : (0, 0)$, $B : (x_B, y_B)$, $C : (x_B + 1, y_B)$, and $D : (1, 0)$, with the additional property that $y_B > 0$ and $\sqrt{x_B^2 + y_B^2} = AB = AD = 1$. This implies that $x_B \neq \pm 1$. Problem 8.1 establishes that $\overline{AC}$ and $\overline{BD}$ bisect each other. It remains to show that they are perpendicular.
The slopes of $\overleftrightarrow{AC}$ and $\overleftrightarrow{BD}$ are

$$m_{\overleftrightarrow{AC}} = \frac{y_B}{1 + x_B} \quad \text{and} \quad m_{\overleftrightarrow{BD}} = \frac{-y_B}{1 - x_B},$$

as $x_B \neq \pm 1$. Consequently,

$$m_{\overleftrightarrow{AC}} \cdot m_{\overleftrightarrow{BD}} = \left(\frac{y_B}{1 + x_B}\right)\left(\frac{-y_B}{1 - x_B}\right) = \frac{-y_B^2}{1 - x_B^2} = \frac{-y_B^2}{y_B^2} = -1,$$

proving that $\overline{AC}$ and $\overline{BD}$ are perpendicular.

8.3 Coordinatize the plane so that $A : (0, y_A)$, $B : (x_B, 0)$ and $C : (0, 0)$. Then $AB = \sqrt{x_B^2 + y_A^2}$. The median at C connects C with the midpoint of $\overline{AB}$, $(x_B/2, y_A/2)$. Therefore, the length of the median at C is

$$m_c = \sqrt{\left(\frac{x_B}{2} - 0\right)^2 + \left(\frac{y_A}{2} - 0\right)^2} = \frac{1}{2}\sqrt{x_B^2 + y_A^2} = \frac{1}{2}AB.$$

8.4 Let $ABCD$ be a convex quadrilateral with P, Q, R, S being the midpoints of sides AB, BC, CD, DA, respectively. Using the Midpoint Formula, we obtain

$$P : ((x_A + x_B)/2, ((y_A + y_B)/2),$$

$$Q : ((x_B + x_C)/2, ((y_B + y_C)/2),$$

$$R : ((x_C + x_D)/2, ((y_C + y_D)/2),$$

$$S : ((x_D + x_A)/2, ((y_D + y_A)/2).$$

Computing now the coordinates of the midpoint of $\overline{PR}$, we obtain

$$((x_A + x_B + x_C + x_D)/4, (y_A + y_B + y_C + y_D)/4).$$

Computing the coordinates of the midpoint of $\overline{QS}$ gives the same result, so the diagonals of $PQRS$ bisect each other. Hence, by Theorem 3.22, $PQSR$ is a parallelogram.

8.5 (a) Let $\overline{CM}$ be the median at C. Then $M : \left(\frac{6+2}{2}, \frac{0+4}{2}\right) = (4, 2)$, so

$$CM = \sqrt{(4 - 1)^2 + (2 - (-1))^2} = \sqrt{18} = 3\sqrt{2}.$$

(b) The slope of $\overleftrightarrow{AB}$ is $(4 - 0)/(2 - 6) = -1$. Hence, $\overleftrightarrow{AB} : y - 0 = (-1)(x - 6)$ or $x + y - 6 = 0$. The length of the altitude at C is the distance from C to $\overleftrightarrow{AB}$. From Theorem 8.7, this distance is

$$\frac{|1 + (-1) - 6|}{\sqrt{1^2 + 1^2}} = \frac{6}{\sqrt{2}} = 3\sqrt{2}.$$

Remark. The equality $m_c = h_c$ for the $\triangle ABC$ implies that $AC = BC$. (See Problem 3.7.) The latter can, of course, be checked easily by computing the length of each segment.

8.6 We choose a Cartesian coordinate system OXY in such a way that neither l_1 nor l_2 is vertical and both coordinate axes are axes of symmetry of $l_1 \bigcup l_2$. (See Figure 114.)

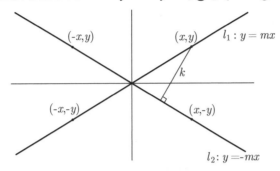

FIGURE 114.

Then $l_1 : y = mx$ for some $m > 0$, and $l_2 : y = -mx$. A point (x, y) belongs to the locus if and only if

$$d((x, y), l_1) - d((x, y), l_2) = \pm k.$$

Using Theorem 8.7, this is equivalent to

$$\frac{|mx - y|}{\sqrt{m^2 + 1}} - \frac{|-mx - y|}{\sqrt{(-m)^2 + 1}} = \pm k.$$

If $K = k\sqrt{m^2 + 1}$, the equation can be rewritten as

$$|mx - y| - |mx + y| = \pm K. \tag{17.8}$$

It is clear that if (x, y) satisfies this equation, then so do points $(-x, y)$, $(x, -y)$, $(-x, -y)$. Therefore, the graph of the equation is symmetric with respect to both coordinate axes, and it is sufficient to construct it in the 1st quadrant only, i.e., we may assume that $x, y \geq 0$.

If $0 \leq y \leq mx$, then

$$(17.8) \iff (mx - y) - (mx + y) = \pm K \iff y = \mp K/2.$$

The graph of this relation in the 1st quadrant is the horizontal ray

$$\{(x, K/2) : x \geq K/(2m)\} .$$

If $y > mx \geq 0$, then

$$(17.8) \iff -(mx - y) - (mx + y) = \pm K \iff x = \mp K/(2m).$$

The graph of this relation in the 1st quadrant is the vertical ray

$$\{(K/(2m), y) : y \geq K/2\}.$$

Therefore the locus is the union of eight rays as shown in Figure 115.

The reader is invited to compare this solution to the one given for Problem 6.6(b).

8.7 Let the coordinates of $ABCDE$ be $A : (0, 0)$, $B : (2x_B, 2y_B)$, $C : (2x_C, 2y_C)$, $D : (2x_D, 2y_D)$ and $E : (2, 0)$, as shown in Figure 116.

The midpoint of $\overline{AB}$ is $P : (x_B, y_B)$; the midpoint of $\overline{BC}$ is $Q : (x_B + x_C, y_B + y_C)$; the midpoint of $\overline{CD}$ is $R : (x_C + x_D, y_C + y_D)$; and the midpoint of $\overline{DE}$ is $S : (x_D + 1, y_D)$. It

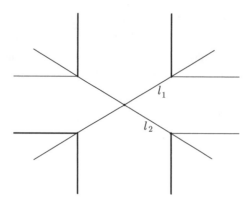

FIGURE 115.

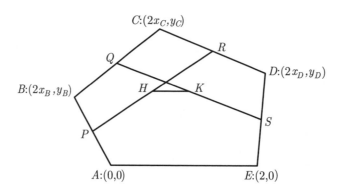

FIGURE 116.

follows that the midpoints of $\overline{PR}$ and $\overline{QS}$ are

$$H : \left(\frac{x_B + x_C + x_D}{2}, \frac{y_B + y_C + y_D}{2} \right) \text{ and}$$

$$K : \left(\frac{x_B + x_C + x_D + 1}{2}, \frac{y_B + y_C + y_D}{2} \right),$$

respectively. Since the y-coordinates of H and K agree, $\overline{HK}$ is parallel to $\overline{AE}$. In addition,

$$HK = |x_H - x_K| = \left| \frac{x_B + x_C + x_D + 1}{2} - \frac{x_B + x_C + x_D}{2} \right| = \frac{1}{2} = \frac{1}{4}AE.$$

8.8 Let $\triangle ABC$ be our triangle. Introduce a Cartesian coordinate system such that $A : (-a, 0)$, $B : (b, 1)$, and $C : (a, 0)$. Then the perpendicular bisector to $\overline{AC}$ coincides with the y-axis. If P is the midpoint of $\overline{BC}$, then $P : ((a + b)/2, 1/2)$. If $a = b$, then $\triangle ABC$ is isosceles and right, hence all perpendicular bisectors meet at the midpoint of the hypotenuse. Suppose $a \neq b$. As the slope of line BC is $1/(b - a)$, the slope of the perpendicular bisector at P is $a - b$, and its equation is

$$y - 1/2 = (a - b)(x - (a + b)/2).$$

Therefore its y-intercept is

$$\frac{1}{2}(b^2 - a^2) + \frac{1}{2}. \tag{17.9}$$

It is clear that the y-intercept of the perpendicular bisector to $\overline{AB}$ is obtained by replacing a by $-a$ in (17.9). As the expression (17.9) does not change under such a replacement, it proves that both perpendicular bisectors intersect the third one at the same point. Hence, they all are concurrent.

8.9 Let $ABCD$ be a parallelogram with vertices $A : (0, 0)$, $B : (x_B, y_B)$, $C : (x_B + 5, y_B)$, and $D : (5, 0)$, where $(x_B > 0, y_B \neq 0)$. Then $F \in \overline{AD}$ such that $AF = \frac{1}{5}AD$ means that $F : (1, 0)$. (See Figure 117.)

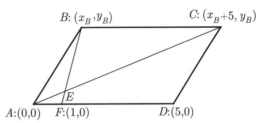

FIGURE 117.

The equation for $\overleftrightarrow{AC}$ is $y = \dfrac{y_B}{x_B + 5}x$, and for $x_B \neq 1$ the equation for $\overleftrightarrow{BF}$ is $y = \dfrac{y_B}{x_B - 1}(x - 1)$. In this case, equating these expressions for y to find the point of intersection,

$$\frac{y_B}{x_B + 5}x = \frac{y_B}{x_B - 1}(x - 1) \Rightarrow$$
$$(x_B - 1)x = (x_B + 5)(x - 1) \Rightarrow$$
$$x_B + 5 = 6x. \text{ Note this relation holds when } x_B=1 \text{ as well.}$$

Thus, the point of intersection of $\overline{AC}$ and $\overline{BF}$ is $E : \left(\frac{x_B+5}{6}, \frac{y_B}{6} \right)$.

$$AE = \sqrt{ \left(\frac{x_B + 5}{6} \right)^2 + \left(\frac{y_B}{6} \right)^2 } = \frac{1}{6}\sqrt{(x_B + 5)^2 + y_B^2} = \frac{1}{6}AC.$$

In order to generalize the problem, simply replace 5 with n.

8.10 Introduce a coordinate system (not necessarily Cartesian) such that points A and D are on the x-axis while the y-axis passes through the midpoints of each base. Choose the scale such that the y-coordinates of B and C are 1. Then $A : (-a/2, 0)$, $D : (a/2, 0)$, $B : (-b/2, 1)$, and $C : (b/2, 1)$. (See Figure 118.)

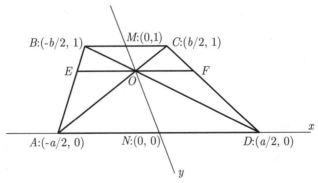

FIGURE 118.

Finding an equation for $\overleftrightarrow{AC}$ gives

$$\overleftrightarrow{AC}:\ y = \frac{1-0}{b/2-(-a/2)}(x-(-a/2)) \ \Leftrightarrow \ y = \frac{2}{a+b}x + \frac{a}{a+b}.$$

Hence, the y-intercept of $\overleftrightarrow{AC}$ is $a/(a+b)$. Clearly, in order to find the y-intercept of $\overleftrightarrow{BD}$, we can just substitute $-a$ for a and $-b$ for b in the y-intercept of $\overleftrightarrow{AC}$. Since $-a/(-a+(-b)) = a/(a+b)$, the two diagonals intersect the y-axis at the same point, namely $O : (0, a/(a+b))$. Finding equations for $\overleftrightarrow{AB}$ and $\overleftrightarrow{CD}$ in a similar way, we obtain

$$y = \frac{2}{a-b}x + \frac{a}{a-b} \ \text{ and } \ y = \frac{2}{b-a}x + \frac{a}{a-b},$$

respectively. Hence, these lines intersect on the y-axis, and the proof of part (c) is finished.

In order to find x_E, we solve the system of equations corresponding to $\overleftrightarrow{AB}$ and $\overleftrightarrow{EF}: y = a/(a+b)$. This gives $x_E = -ab/(a+b)$. Substituting $-a$ for a and $-b$ for b in the expression for x_E, we obtain $x_F = ab/(a+b)$. This proves (a) and also resolves (b):

$$EF = x_F - x_E = \frac{2ab}{a+b}.$$

8.11 Given trapezoid $ABCD$, introduce a (not necessarily Cartesian) coordinate system such that the x-axis passes through A and D, the y-axis passes through the midpoints of the bases, and $A : (-a, 0)$, $B : (-b, 1)$, $C : (b, 1)$ and $D(a, 0)$, $a > b > 0$. (See Figure 119.) Then $P : (0, 1)$ and $Q : (0, 0)$.

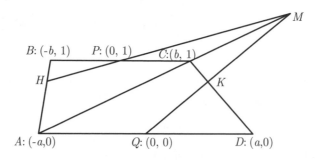

FIGURE 119.

The line $\overleftrightarrow{AC}$ has equation $y = \frac{1}{a+b}(x+a)$, so if $x_M = m$, then $M : (m, \frac{m+a}{a+b})$, with $m > b$ as M lies on $\overrightarrow{AC}$ but outside of the trapezoid. Finding equations of the other lines we obtain:

$$\overleftrightarrow{AB}:\ y = \frac{1}{a-b}(x+a), \quad \overleftrightarrow{CD}:\ y = \frac{1}{b-a}(x-a),$$

$$\overleftrightarrow{MP}:\ y-1 = \frac{m-b}{m(a+b)}x, \quad \text{and} \quad \overleftrightarrow{MQ}:\ y = \frac{m+a}{m(a+b)}x.$$

Simple algebra finds the y-coordinate of $\overleftrightarrow{AB} \cap \overleftrightarrow{MP}$ to be $y_H = \frac{m+a}{a-b+2m}$ and the y-coordinate of $\overleftrightarrow{CD} \cap \overrightarrow{MQ}$ to be $y_K = \frac{m+a}{a-b+2m}$. Since $y_H = y_K$, $\overline{HK}$ is parallel to the bases.

8.12 Let $ABCD$ be a trapezoid with $\overline{AD} \parallel \overline{BC}$; let $Q = \overline{AC} \cap \overline{BD}$, and $P = \overleftrightarrow{AB} \cap \overleftrightarrow{CD}$. Introduce a Cartesian coordinate system in such a way that $A : (0, 0)$, $B := (a, b)$, $C := (c, b)$, and $D := (1, 0)$. (See Figure 120.) Using these coordinates, we obtain the following equations for

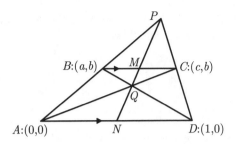

FIGURE 120.

the lines containing the sides and diagonals of $ABCD$.

$$\overleftrightarrow{AB}: \quad y = \frac{b}{a}x$$

$$\overleftrightarrow{BC}: \quad y = b$$

$$\overleftrightarrow{CD}: \quad y = \frac{b}{c-1}(x-1)$$

$$\overleftrightarrow{AD}: \quad y = 0$$

$$\overleftrightarrow{AC}: \quad y = \frac{b}{c}x$$

$$\overleftrightarrow{BD}: \quad y = \frac{b}{a-1}(x-1)$$

We solve $\dfrac{b}{a}x = \dfrac{b}{c-1}(x-1)$ to find P and $\dfrac{b}{c}x = \dfrac{b}{a-1}(x-1)$ to find Q, yielding

$$P: \left(\frac{a}{1+a-c}, \frac{b}{1+a-c}\right) \text{ and } Q: \left(\frac{c}{1+c-a}, \frac{b}{1+c-a}\right).$$

A straightforward calculation (preferably with a computer algebra system!) determines an equation for the line through P and Q:

$$\overleftrightarrow{PQ}: y = \frac{b}{a+c-1}(2x-1).$$

Letting M and N be the points of intersection of $\overleftrightarrow{PQ}$ with $\overline{BC}$ and $\overline{AD}$, respectively, we find that

$$M: \left(\frac{a+c}{2}, b\right) \text{ and } N: \left(\frac{1}{2}, 0\right).$$

Clearly, M and N are the midpoints of $\overline{BC}$ and $\overline{AD}$, as claimed.

This problem also appears as Problem 3.3.14 and as Problem 12.3.

8.13 We introduce a Cartesian coordinate system in such a way that $A: (-1, 0)$ and $C: (1, 0)$. Then $B: (0, \sqrt{3})$. Let $P: (s, t)$ be a point in the interior or on the boundary of $\triangle ABC$. Finding equations of $\overleftrightarrow{AB}$, $\overleftrightarrow{BC}$, and $\overleftrightarrow{AC}$, we obtain $\sqrt{3}x - y + \sqrt{3} = 0$, $-\sqrt{3}x - y + \sqrt{3} = 0$, and $y = 0$ respectively. Using Theorem 8.7 for the distance from a point to a line, we have

$$d(P, \overleftrightarrow{AB}) + d(P, \overleftrightarrow{BC}) + d(P, \overleftrightarrow{AC})$$
$$= \frac{|\sqrt{3}s - t + \sqrt{3}|}{2} + \frac{|-\sqrt{3}s - t + \sqrt{3}|}{2} + |t|.$$

As P is inside the triangle or on its boundary, then $t \geq 0$, $\sqrt{3}s + \sqrt{3} \geq t$ and $-\sqrt{3}s + \sqrt{3} \geq t$. Therefore the sum of the distances simplifies to

$$\frac{\sqrt{3}s - t + \sqrt{3}}{2} + \frac{-\sqrt{3}s - t + \sqrt{3}}{2} + t = \sqrt{3}.$$

Hence, the distance sum does not depend on P.

This problem has previously appeared as Problem 3.2.29 and as Problem 5.2b.

8.14 *Solution 1:* Let $A : (x_A, y_A)$, $B : (x_B, y_B)$, and $C : (x_C, y_C)$. Let us assume that the triangle is positioned in the plane such that $x_A \leq x_B \leq x_C$, and $y_A \leq y_C \leq y_B$. Consider the rectangle $APQR$ that circumscribes $\triangle ABC$ and whose sides are parallel to the coordinate axes. (See Figure 121.)

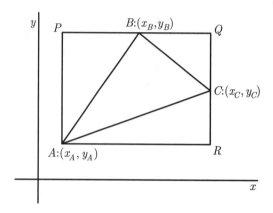

FIGURE 121.

Obviously, Area($\triangle ABC$)

$$= \text{Area}(APQR) - (\text{Area}(\triangle APB) + \text{Area}(\triangle BQC) + \text{Area}(\triangle CRA))$$
$$= (x_C - x_A)(y_B - y_A) - \frac{1}{2}\Big((y_B - y_A)(x_B - x_A)$$
$$+ (y_B - y_C)(x_C - x_B) + (y_C - y_A)(x_C - x_A)\Big)$$
$$= \frac{1}{2}(-x_A y_B - x_B y_C - x_C y_A + x_B y_A + x_C y_B + x_A y_C).$$

The last expression factors into
$\frac{1}{2}|(x_A - x_C)(y_B - y_C) - (x_B - x_C)(y_A - y_C)|.$[9] This proves the statement in this case. All other cases can be dealt with similarly, and we omit the details.

These formulas are useful, especially when software for symbolic computations is available. One can avoid going through many cases (which we skipped here anyway) by using the approach presented below, though it relies on its own tedious computation.

---

[9] This latter form for area may be familiar to those who know that the area of a parallelogram equals the absolute value of the determinant of a matrix whose columns are the spanning vectors of the parallelogram. This notion is discussed in Chapter 12.

Solution 2: Let $A; (x_A, y_A)$, $B : (x_B, y_B)$, and $C : (x_C, y_C)$. We have seen that the area of $\triangle ABC$ can be found as

$$\frac{1}{2}ab \sin C = \frac{1}{2}ab\sqrt{1 - \cos^2 C} = \frac{1}{2}ab\sqrt{1 - \left(\frac{a^2 + b^2 - c^2}{2ab}\right)^2}$$

$$= \frac{1}{4}\sqrt{4a^2b^2 - (a^2 + b^2 - c^2)^2} = \frac{1}{4}\sqrt{4a^2b^2 - (a^2 + b^2 - c^2)^2}.$$

Expressing the sides in terms of coordinates we rewrite the expression under the square root as:

$$4[(x_B - x_C)^2 + (y_B - y_C)^2][(x_A - x_C)^2 + (y_A - y_C)^2]$$

$$-\Big([(x_B - x_C)^2 + (y_B - y_C)^2] + [(x_A - x_C)^2 + (y_A - y_C)^2]$$

$$- [(x_A - x_B)^2 + (y_A - y_B)^2]\Big)^2.$$

What is left is to check that this expression is equal to

$$4\Big(x_A y_B + x_B y_C + x_C y_A - x_B y_A - x_C y_B - x_A y_C\Big)^2.$$

We leave this straightforward but tedious verification to the reader.

8.15 Choose coordinates so that $A : (-a, 0)$, $B : (a, 0)$, and $C : (c, d)$. The nine-point circle passes through the points $C' : (0, 0)$, which is the midpoint of $\overline{AB}$, $B' : ((-a + c)/2, d/2)$, which is the midpoint of $\overline{AC}$, and $D : (c, 0)$, the foot of the altitude from C. (See Figure 122.) Label the center of the nine-point circle as $O : (x_0, y_0)$. We need to show that $x_0^2 + y_0^2 \geq (a/2)^2$.

In order to find the coordinates of O, we use the fact that the perpendicular bisectors of chords of a circle pass through the center of the circle. Since the perpendicular bisector of chord $\overline{C'D}$ has equation $x = c/2$ and contains O, $x_0 = c/2$. We next find the equation of the perpendicular bisector m of chord $\overline{B'C'}$. The midpoint of $\overline{B'C'}$ will be $((-a + c)/4, d/4)$ and

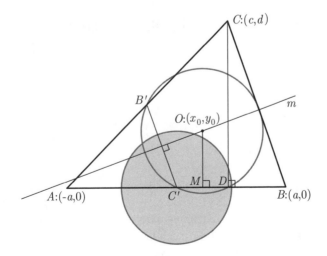

FIGURE 122.

the slope of $\overleftrightarrow{B'C'}$ is $d/(-a+c)$. Therefore, m will have the equation

$$y - \frac{d}{4} = \frac{a-c}{d}\left(x - \frac{c-a}{4}\right).$$

From here, we find

$$y_0 = \left(\frac{a-c}{d}\right)\left(\frac{c}{2}\right) + \frac{(a-c)^2 + d^2}{4d} = \frac{a^2 - c^2 + d^2}{4d}.$$

The following inequalities are then all equivalent, and, based on the last one, all true.

$$x_0^2 + y_0^2 \geq a^2/4$$

$$\left(\frac{a^2 - c^2 + d^2}{4d}\right)^2 \geq \frac{a^2}{4} - \frac{c^2}{4}$$

$$\left(a^2 - c^2 + d^2\right)^2 - (a^2 - c^2)(4d^2) \geq 0$$

$$(a^2 - c^2 - d^2)^2 \geq 0$$

To classify those triangles for which the nine-point circle will have radius $AB/4$, notice that since C' lies on that circle, we simply need to determine when $x_0^2 + y_0^2 = a^2/4$, i.e., when $a^2 = c^2 + d^2$. By the Pythagorean Theorem, $c^2 + d^2 = C'D^2 + CD^2 = C'C^2$, so we seek those triangles for which $a = C'C$. In other words, the desired triangles have the property that the median from C is half the length of AB. This will happen precisely when ABC is a right triangle with right angle at C (by Problem S3.2.5).

8.16 Introduce a coordinate system (not necessarily Cartesian) so that $E : (0,0)$, $A : (a, 0)$, $C : (c, 0)$, $a < 0 < c$, and $B : (0, 1)$.

Let $AF/FB = f$, $BD/DC = d$, and $CE/EA = c/(-a) = e$. By Theorem 8.2,

$$F : \left(\frac{a}{f+1}, \frac{f}{f+1}\right) \text{ and } D : \left(\frac{dc}{d+1}, \frac{1}{d+1}\right).$$

Then, as slopes of $\overleftrightarrow{AD}$ and $\overleftrightarrow{CF}$ exist, we have:

$$\overleftrightarrow{AD} : y = \frac{1}{cd - a - da}(x - a) \text{ and } \overleftrightarrow{CF} : y = \frac{f}{a - c - fc}(x - c).$$

The three segments are concurrent if and only if the y-intercepts of $\overleftrightarrow{AD}$ and $\overleftrightarrow{CF}$ are equal, or, equivalently,

$$\frac{-a}{dc - ad - a} = \frac{-fc}{a - fc - c} \quad \Leftrightarrow$$

$$\frac{1}{de + d + 1} = \frac{fe}{1 + fe + e} \quad \Leftrightarrow$$

$$fde(e+1) + fe = e + 1 + fe \quad \Leftrightarrow \quad fde = 1.$$

Conics

9.1 We prove part (a), and leave the remaining parts of the problem to the reader.

The parabola $\mathcal{P}$ given by $y = 4 - x^2$ can be rewritten in the form $(x - h)^2 = 4p(y - k)$, with vertex $(h, k) = (0, 4)$ and $p = -1/4$; the focus is $F : (0, 15/4)$ and the y-axis is the axis of symmetry. Let n be the line through the point $P : (3, -5)$ that is parallel to the axis of

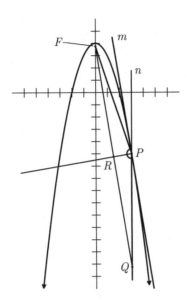

FIGURE 123.

symmetry; the line n is given by the equation $x = 3$. (See Figure 123.) By Theorem 6.3, the tangent line to $\mathcal{P}$ at P is perpendicular to the angle bisector of $\angle FPQ$.

We wish to find Q on n such that Q is interior to $\mathcal{P}$ and $FP = PQ$. Since $d(F, P) = \sqrt{(0-3)^2 + (3.75+5)^2} = 9.25$, $Q : (3, -14.25)$. Now $\triangle FPQ$ is an isosceles triangle. Let $R : (1.5, -5.25)$ be the midpoint of $\overline{FQ}$. Since the slope of $\overleftrightarrow{PR}$ is $1/6$, the slope of m, the line tangent to $\mathcal{P}$ at P, is -6.

9.2 Dividing both sides of the equation by AC, we obtain

$$x^2/C + y^2/A = -H/AC. \tag{17.10}$$

Suppose both $A > 0$ and $C > 0$.

If $H > 0$, the graph is, clearly, an empty set, as $x^2/C + y^2/A \geq 0$, and $-H/AC < 0$. If $H = 0$, then the graph is a single point, $(0, 0)$. If $H < 0$, then the graph is an ellipse, as the equation can be rewritten in the form

$$x^2/a_1{}^2 + y^2/b_1{}^2 = 1, \quad \text{where} \quad a_1{}^2 = -H/A, \ b_1{}^2 = -H/C.$$

The analysis of the case when $A < 0$ and $C < 0$ is similar, and we skip the details.

Suppose $A < 0$ and $C > 0$. If $H < 0$, then the graph of (17.10) is the hyperbola

$$y^2/b_1{}^2 - x^2/a_1{}^2 = 1, \quad \text{where} \quad a_1{}^2 = H/A \quad \text{and} \quad b_1{}^2 = -H/C.$$

If $H = 0$, then (17.10) is equivalent to

$$\left(\frac{x}{\sqrt{C}} - \frac{y}{\sqrt{-A}} \right) \left(\frac{x}{\sqrt{C}} + \frac{y}{\sqrt{-A}} \right) = 0,$$

hence the graph is the union of two intersecting lines.

If $H > 0$, then the graph of (17.10) is the hyperbola

$$x^2/a_1{}^2 - y^2/b_1{}^2 = 1, \quad \text{where} \quad a_1{}^2 = -H/A, \ b_1{}^2 = H/C.$$

The analysis of the case when $A > 0$ and $C < 0$ is similar, and we skip the details.

9.3 By completing the square with respect to x, the equation can be rewritten in the form $x'^2 = 4py'$, where $x' = x + F/(2A)$, $y' = y - (H - F^2/(4A))$, $p = 1/(4A)$. Hence, the graph is the parabola with vertex at $(-F/(2A), H - F^2/(4A))$, focus at $(-F/(2A), H - F^2/(4A) + 1/(4A))$, and the directrix $y = H - F^2/(4A) - 1/(4A)$ (in the OXY-system). This follows both from (9.1) and from the change of coordinate formulae.

9.4 We already know that each of the figures mentioned in the problem has an equation of the form (9.24) in some Cartesian coordinate system. Substituting $x = a_1x' + b_1y' + c_1$ and $y = a_2x + b_2y + c_2$ into (9.24), with all a_i, b_i and c_i being constants, always gives an equation of the same form. The change of coordinate formulae are of this form. If a Cartesian coordinate system $O'X'Y'$ is obtained from OXY by a translation and a rotation – and, as we know, each Cartesian system with the same scale is obtained this way – then the equation of the figures will again have form (9.24) in $O'X'Y'$. If the scale in $O'X'Y'$ is distinct from the one in OXY, replacing x' and y' by $x'' = kx'$ and $y'' = ky'$, for an appropriate $k > 0$, will make the scales in OXY and $O'X''Y''$ equal and keep the form of the equation of the figure as in (9.24).

9.5 Coordinatize the plane so that all points on the x-axis have y-coordinate zero and the y-axis is the axis of the parabola. Then the vertex of the parabola (the lowest point of the cable) has coordinates $(0, 20)$, and the tops of the two towers have coordinates $(-120, 60)$ and $(120, 60)$. Thus, using the equation $(x - 0)^2 = 4p(y - 20)$, the parabola can be described by the equation $x^2 = 360(y - 20)$.

A specified vertical distance from the roadway to the cable can be found using the obtained equation. As we are given the horizontal distance of 60 feet from a tower, we substitute $x = \pm 60$ into the equation; the corresponding y-value, $y = 30$, gives the vertical distance, in feet.

9.6 Choose a Cartesian coordinate system such that the endpoints of the hypotenuse are $(0, 0)$ and $(0, 1)$. Let (x, y) be the coordinates of the vertex of the right angle. From the distance formula and the Pythagorean theorem, we have

$$\left[(x - 0)^2 + (y - 0)^2\right] + \left[(x - 1)^2 + (0 - y)^2\right] = 1^2.$$

This simplifies to $(x - 1/2)^2 + y^2 = (1/2)^2$, which shows that the locus is a circle centered at $(1/2, 0)$ having radius $1/2$. In other words, the locus is the circle whose diameter is the given hypotenuse.

9.7 (a) The discriminant is $D = 6^2 - 4 \cdot 0 \cdot (-8) = 36 > 0$. Eliminating the degenerate cases, we conclude that the curve is either a hyperbola or a union of lines. Considering the intersection of the curve with the line $y = mx$ leads to the equation

$$(6m - 8m^2)x^2 - (12 - 26m)x + 11 = 0.$$

This equation is quadratic in x (for $m \neq 0, 3/4$) with its discriminant $(12 - 26m) - 44(6m - 8m^2)$ being negative for m in the interval (m_1, m_2), where m_1 and m_2 are $(111 - 3\sqrt{341})/257$ and $(111 + 3\sqrt{341})/257$, respectively. As there are infinitely many slopes m for which the line $y = mx$ does not intersect the curve, we conclude that the curve is a hyperbola.

(b) The discriminant is $D = 2^2 - 4 \cdot 1 \cdot 1 = 0$. Eliminating the degenerate cases, we conclude that the curve is either a parabola or a union of lines. Considering the intersection of the curve with the line $y = mx$, leads to the equation

$$(m + 1)^2x^2 - 8x + 4 = 0.$$

This equation is quadratic in x for $m \neq -1$. Hence, $m = -1$ is *the only* value of the slope when the line $y = mx$ intersects the curve at exactly one point. For any other values of m we have either no intersection ($|m + 1| > 2$), or two points of intersection ($|m + 1| \leq 2$). This allows us to conclude that the graph is a parabola.

(c) The discriminant is $D = (-4)^2 - 4 \cdot 9 \cdot 6 < 0$. The graph is not the whole plane as $(0, 0)$ is not on it. So the curve is an ellipse, or a point, or a union of lines. Considering the intersection of the curve with the line $y = mx$ leads to the equation

$$(9 - 4m + 6m^2)x^2 + (6 - 8m)x + 2 = 0.$$

This equation is quadratic in x for all m, since the leading coefficient, $9 - 4m + 6m^2$, is always positive. Solving for x, we get that for $m \in (-0.5, 4.5)$, there are no solutions, and that there are two solutions for all other values of m. As there are infinitely many slopes m such the line $y = mx$ does not intersect the curve, and infinitely many for which it does, we conclude that the curve is an ellipse.

9.8 Introduce a coordinate system so that the center of the ellipse is $(0,0)$ and the x- and y-axes are axes of symmetry for the ellipse. Write $A : (x_A, y_A)$, $B : (-x_A, -y_A)$, and $P : (x_P, y_P)$. The slopes of lines AP and BP, respectively, are

$$\frac{y_A - y_P}{x_A - x_P} \quad \text{and} \quad \frac{-y_A - y_P}{-x_A - x_P}.$$

Multiplying these slopes, we obtain

$$\frac{y_A^2 - y_P^2}{x_A^2 - x_P^2}. \quad (*)$$

Now, since A and P both lie on the ellipse, their coordinates satisfy the given standard equation. Therefore, we have

$$\frac{y_A^2}{b^2} - \frac{y_P^2}{b^2} = \left(1 - \frac{x_A^2}{a^2}\right) - \left(1 - \frac{x_P^2}{a^2}\right) = \frac{x_P^2 - x_A^2}{a^2}.$$

That is,

$$\frac{y_A^2 - y_P^2}{b^2} = \frac{x_P^2 - x_A^2}{a^2}.$$

Combining this last equation with (*), we see that the product of the slopes is $-b^2/a^2$, as desired.

9.9 (a) The vertical line $x = x_0$ through the point (x_0, y_0) is tangent to the ellipse if and only if the equation $x_0^2/a^2 + y^2/b^2 = 1$ (with respect to y) has two equal solutions. Since (x_0, y_0) is on the ellipse, $|x_0| \leq a$, and the equation has two equal solutions if and only if $x_0 = \pm a$ and $y_0 = 0$. In this case, $x = x_0 = \pm a$, which is equivalent to $(\pm a/a^2)x + (0/b^2)y = 1$. Therefore the statement is proven for $(x_0, y_0) = (\pm a, 0)$ and vertical tangent lines.

If the tangent is not vertical, its equation is $y - y_0 = m(x - x_0)$, for some m, with $|x_0| < a$, $y_0 \neq 0$. The x-coordinates of the intersection points of this line with the ellipse leads to a quadratic equation (in x):

$$\frac{x^2}{a^2} + \frac{(mx + (y_0 - mx_0))^2}{b^2} = 1 \quad \text{which is equivalent to}$$

$$(b^2 + m^2a^2)x^2 + 2m(y_0 - mx_0)a^2x + ((y_0 - mx_0)^2a^2 - a^2b^2) = 0.$$

This equation has equal roots if and only if its discriminant is zero. This is equivalent to

$$(m(y_0 - mx_0)a^2)^2 - (b^2 + m^2a^2)((y_0 - mx_0)^2a^2 - a^2b^2) = 0.$$

Using the fact that $x_0^2 b^2 + y_0^2 a^2 = a^2 b^2$, this gives

$$m = \frac{x_0 y_0}{x_0^2 - a^2}.$$

As $|x_0| < a$, the slope exists, and we get the equation of the tangent line in the form

$$y - y_0 = \frac{x_0 y_0}{x_0^2 - a^2}(x - x_0)$$

$$\Leftrightarrow \frac{x_0}{a^2}x + \frac{(-x_0^2 + a^2)}{a^2 y_0}y = 1.$$

Using $x_0^2 b^2 + y_0^2 a^2 = a^2 b^2$ again, we finally arrive at

$$\frac{x_0}{a^2}x + \frac{y_0}{b^2}y = 1.^{10}$$

(b) The argument we used to prove part (a) above can be replicated in this case almost verbatim.

9.10 For a given position of the ladder, let X represent the point at the top of the ladder and let Y represent the point at the bottom. For simplicity, let us also suppose that the ladder has length three units, so $XZ = 1$ and $YZ = 2$. Drop perpendicular segments from Z to $\overline{XC}$ and $\overline{YC}$ which meet those segments at points D and E, respectively. (See Figure 124.) Now, right triangles XDZ and XCY are similar, so $\dfrac{DZ}{XZ} = \dfrac{CY}{XY}$. Therefore, $CY = 3 \cdot DZ = 3 \cdot CE$; it follows that $EY = CY - CE = 2DZ$. By the Pythagorean theorem, $(EY)^2 + (EZ)^2 = (YZ)^2$, so by substitution, $4(DZ)^2 + (EZ)^2 = 4$. This relationship between the lengths DZ and EZ demonstrate that the locus is one fourth of an ellipse with major axis length 2 and minor axis length 1. See Chapter 9 for a discussion of the algebraic properties of ellipses.

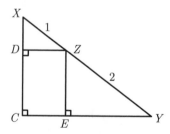

FIGURE 124.

---

[10] Though our solution is straightforward and does not use the notion of the derivative, one can appreciate the power of calculus, where it is easy to explain that an equation of the tangent line to a curve given by $F(x, y) = 0$ at its point (x_0, y_0) can be written as

$$F_x'(x_0, y_0)(x - x_0) + F_y'(x_0, y_0)(y - y_0) = 0.$$

This gives the required result much faster, and the method can be used for many other curves.

9.11 By substituting the expressions for A', B', and C' derived in the proof of Theorem 9.5 in $D' = B'^2 - 4A'C'$, we obtain

$$
\begin{aligned}
D' &= ((C - A)\sin 2\theta + B\cos 2\theta)^2 - \big[(A + C) + \big(B\sin 2\theta + (A - C)\cos 2\theta\big)\big] \\
&\quad \times \big[(A + C) - \big(B\sin 2\theta + (A - C)\cos 2\theta\big)\big] \\
&= ((C - A)\sin 2\theta + B\cos 2\theta)^2 - \big[(A + C)^2 - \big(B\sin 2\theta + (A - C)\cos 2\theta\big)^2\big] \\
&= (C - A)^2(\sin^2 2\theta + \cos^2 2\theta) + B^2(\cos^2 2\theta + \sin^2 2\theta) - (A + C)^2 \\
&= B^2 - 4AC = D.
\end{aligned}
$$

9.12 Denote the ellipse by $\mathcal{E}$ and choose a Cartesian coordinate system in such a way that

$$
\mathcal{E} : x^2/a^2 + y^2/b^2 = 1.
$$

If all chords are vertical (horizontal), then their midpoints lie on the x-axis (y-axis), and the proof is left to the reader.

Suppose each chord determines a line with same slope $m \neq 0$. Every such line l has an equation of the form $y = mx + t$. Finding the x-coordinates of the intersection of l with $\mathcal{E}$ amounts to solving the quadratic equation

$$
x^2/a^2 + (mx + t)^2/b^2 = 1 \iff
$$
$$
(b^2 + a^2m^2)x^2 + 2a^2mtx + (a^2t^2 - a^2b^2) = 0. \tag{17.11}
$$

If x_1 and x_2 are solutions of (17.11), then, by Vièta's theorem,

$$
x_M = (x_1 + x_2)/2 = -a^2mt/(b^2 + a^2m^2). \tag{17.12}
$$

Similarly, rewriting $y = mx + t$ as $x = y/m - t/m$ (permitted since $m \neq 0$), and substituting into $x^2/a^2 + y^2/b^2 = 1$, leads to the quadratic equation (in y)

$$
(y/m - t/m)^2/a^2 + y^2/b^2 = 1 \iff
$$
$$
(b^2 + a^2m^2)y^2 - 2tb^2y + (t^2 - a^2m^2)b^2 = 0. \tag{17.13}
$$

If y_1 and y_2 are solutions of (17.13), then, by Vièta's theorem,

$$
y_M = (y_1 + y_2)/2 = b^2t/(b^2 + a^2m^2). \tag{17.14}
$$

Comparing x_M and y_M, given by (17.12) and (17.14), respectively, we see that

$$
y_M = -\frac{b^2}{a^2m}\, x_M.
$$

Hence, the midpoints of the chords with slope $m \neq 0$ lie on the line

$$
y = -\frac{b^2}{a^2m} x.
$$

Remark. Analyzing the proof above, one can see that we actually proved a stronger statement:

> *The midpoints of a family of parallel chords of an ellipse lie on a diameter of the ellipse.*

For $a = b$, the statement above is a generalization of the corresponding property of a circle: the diameter that bisects a chord is perpendicular to the chord.

Given two parallel tangents, and considering their points of tangency as "degenerate" chords, we obtain a proof of another lovely property of ellipses:

Two parallel tangents to an ellipse touch it at the endpoints of a diameter of the ellipse.

A similar statement also holds for parabolas, though a parabola doesn't have a "diameter":

The midpoints of any family of parallel chords of a parabola lie on a line parallel to the axis of symmetry of the parabola.

Let $\mathcal{P}$ be a parabola. We choose a Cartesian coordinate system in such a way that $\mathcal{P} : y = x^2$. Then $\mathcal{P}$ has no vertical chords. Suppose each chord determines a line with the same slope m. Every such line l has an equation of the form $y = mx + t$. Finding the x-coordinates of the intersection of l with $\mathcal{P}$ amounts to solving the quadratic equation

$$x^2 - mx - t = 0.$$

If x_1 and x_2 are solutions of this equation, then by Vièta's theorem,

$$x_M = (x_1 + x_2)/2 = m/2.$$

This proves that the midpoints of all such chords are collinear, as they all lie on the vertical line given by $x = m/2$.

9.13 *Solution 1:* If all chords that subtend a right angle at the vertex, $(0, 0)$, of the parabola $\mathcal{P}$ are concurrent, then by the symmetry of $\mathcal{P}$ about the x-axis, the point of concurrency should lie on the x-axis. This can be verified by checking that the point of intersection of any such chord with the x-axis is independent of the endpoint of the chord.

Let the endpoints of an arbitrary chord subtending a right angle at $(0, 0)$ be (x_1, y_1) and (x_2, y_2), and let the x-intercept be $(x_0, 0)$. (See Figure 125.) Applying the Pythagorean theorem to the triangle with vertices at (x_1, y_1), (x_2, y_2), and $(0, 0)$ yields

$$(x_1{}^2 + y_1{}^2) + (x_2{}^2 + y_2{}^2) = (x_1 - x_2)^2 + (y_1 - y_2)^2.$$

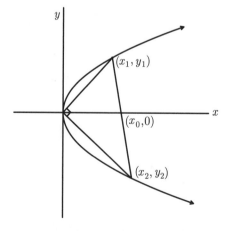

FIGURE 125.

Thus,

$$-2x_1x_2 - 2y_1y_2 = 0; \text{ equivalently, } x_1x_2 = -y_1y_2. \tag{17.15}$$

Because the endpoints of the chord lie on the parabola,

$$y_1{}^2 = 4ax_1 \text{ and } y_2{}^2 = 4ax_2. \tag{17.16}$$

Combining (17.15) and (17.16), we obtain

$$(y_1y_2)^2 = 16a^2x_1x_2 = -16a^2y_1y_2, \text{ so } y_1y_2 = -16a^2. \tag{17.17}$$

Notice that (17.16) also implies that

$$4ax_1y_2 - 4ax_2y_1 = y_1{}^2y_2 - y_2{}^2y_1 = y_1y_2(y_1 - y_2). \tag{17.18}$$

The chord with endpoints (x_1, y_1) and (x_2, y_2) passing through $(x_0, 0)$ has slope

$$\frac{y_1 - 0}{x_1 - x_0} = \frac{y_1 - y_2}{x_1 - x_2}.$$

Solving for x_0 and then substituting from equations (17.17) and (17.18), we obtain

$$x_0 = \frac{x_1y_2 - x_2y_1}{y_2 - y_1} = \frac{y_1y_2(y_1 - y_2)}{4a(y_2 - y_1)} = \frac{-y_1y_2}{4a} = \frac{16a^2}{4a} = 4a.$$

That is, all chords subtending a right angle at the vertex of the parabola contain the point $(4a, 0)$.

Solution 2: As above, let (x_1, y_1) and (x_2, y_2) be the endpoints of a chord subtending a right angle of the vertex of $\mathcal{P}$. Since the slopes of the lines from these endpoints to the origin are negative reciprocals,

$$\frac{y_1}{x_1} = -\frac{x_2}{y_2}, \text{ yielding } x_1x_2 + y_1y_2 = 0. \tag{17.19}$$

If the equation for the chord is given by $y = mx + b$, then the points (x_1, y_1) and (x_2, y_2) satisfy the equation $(mx + b)^2 = 4ax$; that is, x_1 and x_2 are roots of the quadratic equation $m^2x^2 + (2mb - 4a)x + b^2 = 0$. It follows that

$$x_1x_2 = \frac{b^2}{m^2}.$$

Since $y^2 = 4ax$, $y_1{}^2 = 4ax_1 \geq 0$ and $y_2{}^2 = 4ax_2 \geq 0$ and

$$y_1y_2 = (2\sqrt{ax_1})(2\sqrt{ax_2}) = 4a\sqrt{x_1x_2} = \frac{4ab}{m}.$$

Substituting the above expressions into the equation $x_1x_2 + y_1y_2 = 0$ from (17.19),

$$\frac{b^2}{m^2} + \frac{4ab}{m} = 0, \quad \text{so } b = -4am.$$

Therefore, every chord that subtends a right angle at the vertex of $\mathcal{P}$ has an equation of the form $y = mx - 4am$ and thus passes through the point $(4a, 0)$.

Remark. Investigations with *Geometer's Sketchpad*® indicate that the statement of this problem can be generalized: The chords subtending right angles at any point on a conic (parabola, hyperbola, or ellipse) are all concurrent. The reader is invited to explore this generalized statement.

9.14 (a) Consider a finite number of parabolas on the plane. Let l be a line *not* parallel to the axis of any of these parabolas. The intersection of l with the union of a parabola and its interior is either empty, or a point (if l is tangent to the parabola), or a segment. As finitely many points and segments cannot cover a line, the parabolas do not cover l. So they cannot cover the whole plane.

(b) Clearly one can do it. For example, consider the six hyperbolas given by

$$xy = 1, \ xy = -1, \ (x \pm 10)^2 - y^2 = 1, \ \text{and} \ x^2 - (y \pm 10)^2 = 1.$$

9.15 In a Cartesian coordinate system with axes parallel to those of the parabolas, $\mathcal{P}_1$ and $\mathcal{P}_2$ have equations in the form:

$$F_1(x, y) = a_1 x^2 + b_1 x + c_1 - y = 0 \quad \text{and}$$
$$F_2(x, y) = a_2 y^2 + b_2 y + c_2 - x = 0, \quad \text{respectively.}$$

Since $a_1 \neq 0$ and $a_2 \neq 0$, the equation $a_2 F_1(x, y) + a_1 F_2(x, y) = 0$ represents the equation of the circle passing through the points of intersection of the parabolas.

9.16 Let $\overline{\mathcal{E}}$ denote the union of an ellipse with its interior, and let $A, B \in \overline{\mathcal{E}}$. In Example 43 we showed that line AB intersects $\mathcal{E}$ in two points if both A and B are in the interior of $\overline{\mathcal{E}}$. Obviously, the same proof applies to the case when one or both of these points are on the ellipse. Examining the proof more closely, we realize that the intersection points are always *outside* of $\overline{AB}$, or coincide with A or B.

In order to prove the convexity of $\overline{\mathcal{E}}$, we have to show that every interior point of $\overline{AB}$ is in the interior of $\overline{\mathcal{E}}$. Suppose that it is not the case, and let C be an interior point of $\overline{AB}$ which is in the exterior of $\overline{\mathcal{E}}$. Then, using the same argument as in our proof in Example 43, each of the segments AC and CB intersects $\overline{\mathcal{E}}$. If P_1 and P_2 are the intersection points, then $A - P_1 - C$ and $C - P_2 - B$. This implies that $\overline{AB}$ intersects the ellipse at four distinct points, a contradiction. Therefore $\overline{\mathcal{E}}$ is convex.

9.17 The argument is given in many books. See [7].

9.18 Choose a Cartesian coordinate system such that the equation of the parabola is $y = 4x^2$. Then its focus is at $(0, 1)$. Any non-vertical line through the focus has an equation of the form $y = mx + 1$. It intersects the parabola at points with x-coordinates x_1 and x_2, which are the solutions of the equation

$$x^2 - (m/4)x - 1/4 = 0.$$

Let (x_m, y_m) be the midpoint of the chord resulted by intersecting the parabola and the line. Hence,

$$x_m = (x_1 + x_2)/2 = m/8.$$

Similarly we can find y_m. If $m = 0$, the midpoint of the resulting chord coincides with the focus. For $m \neq 0$, the equation of the line can be written as $x = y/m - 1/m$. It intersects the parabola at points with y-coordinates y_1 and y_2, which are the solutions of the equation

$$y^2 - \frac{8 + m^2}{4} y + 1 = 0.$$

Hence,

$$y_m = (y_1 + y_2)/2 = (8 + m^2)/8.$$

As $y_m = 1 + 8x_m^2$, points (x_m, y_m) lie on the parabola $y = 1 + 8x^2$. Note that the focus $(0, 1)$ also lies on this parabola, so the locus is the whole parabola.

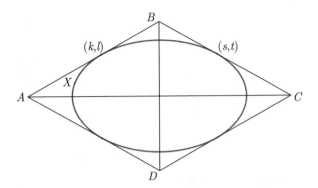

FIGURE 126.

9.19 Let $\mathcal{E}$ be an ellipse. Consider a Cartesian coordinate system OXY for which the equation of $\mathcal{E}$ is $x^2/a^2 + y^2/b^2 = 1$. Let $(k, l) \in \overline{AB}$ and $(s, t) \in \overline{BC}$ be the points of tangency of $\mathcal{E}$ with two consecutive sides of the rhombus $ABCD$. This implies that none of $k, l, s,$ or t is zero. (See Figure 126.)

First we note that when an ellipse is tangent to two parallel lines, the points of tangency are symmetric with respect to the center of the ellipse. There are many ways to prove this fact. It follows easily from an algebraic analysis of the intersection of an ellipse with a line, or from the equation of a tangent line to an ellipse at a given point (Problem 9.9), or from the remarks at the end of our solution of Problem 9.12 (a). It also becomes clear from the symmetry of the ellipse, by considering the homothety $H(O, -1)$, where O is the center of the ellipse. This implies that $(-k, -l) \in \overline{CD}$ and $(-s, -t) \in \overline{DA}$ are the other two points of tangency.

The tangent lines at (k, l) and at (s, t) have the equations $(k/a^2)x + (l/b^2)y = 1$ and $(s/a^2)x + (t/b^2)y = 1$, respectively. (See Problem 9.9.) They intersect at B, and it is easy to find that

$$B : \left(\frac{(t-l)a^2}{kt - sl}, \frac{(k-s)b^2}{kt - sl} \right).$$

(The condition that points (k, l) and (s, t) are on consecutive sides of the rhombus implies that $kt - sl \neq 0$). Changing signs of letters in this expression gives the coordinates of three other vertices of the rhombus:

$$D : \left(\frac{(-t+l)a^2}{kt - sl}, \frac{(-k+s)b^2}{kt - sl} \right), \quad C : \left(\frac{(t+l)a^2}{-kt + sl}, \frac{(-k-s)b^2}{-kt + sl} \right),$$

$$\text{and } A : \left(\frac{(-t-l)a^2}{-kt + sl}, \frac{(k+s)b^2}{-kt + sl} \right).$$

We have not fully used the condition of $ABCD$ being a rhombus so far. We have only used that it was a parallelogram. As we know, in a rhombus the diagonals are perpendicular. Hence, $\overline{BD} \perp \overline{AC}$. This condition is equivalent to the slopes of the corresponding lines, $m_{\overleftrightarrow{BD}}$ and $m_{\overleftrightarrow{AC}}$, being negative reciprocals of one another, provided that both slopes *exist* (so neither is zero).

Case 1: If $k = \pm s$, then $l = \mp t$ and vice versa, since $k^2/a^2 + l^2/b^2 = s^2/a^2 + t^2/b^2 = 1$. In this case, the opposite vertices of the rhombus lie on the coordinate axes.

Case 2: Suppose $k^2 \neq s^2$ and $l^2 \neq t^2$. Then slopes $m_{\overleftrightarrow{BD}}$ and $m_{\overleftrightarrow{AC}}$ both exist and are not zero:

$$m_{\overleftrightarrow{BD}} = \frac{(k-s)b^2}{(t-l)a^2}, \quad \text{and} \quad m_{\overleftrightarrow{AC}} = -\frac{(k+s)b^2}{(t+l)a^2}.$$

Therefore,

$$m_{\overleftrightarrow{BD}} \cdot m_{\overleftrightarrow{AC}} = -1 \quad \Leftrightarrow \quad \frac{(k-s)b^2}{(t-l)a^2} \cdot \frac{(k+s)b^2}{(t+l)a^2} = 1 \quad \Leftrightarrow$$

$$(k^2 - s^2)b^4 = (t^2 - l^2)a^4.$$

Why does this relation hold? Well, as $k^2/a^2 + l^2/b^2 = s^2/a^2 + t^2/b^2 = 1$, subtracting the equations we obtain $(k^2 - s^2)b^2 = (t^2 - l^2)a^2$. Together with $(k^2 - s^2)b^4 = (t^2 - l^2)a^4$, $k^2 \neq s^2$, and $l^2 \neq t^2$, it implies $a^2 = b^2$, i.e., the ellipse is a circle. Then, as the diagonals of the rhombus bisect its interior angles, they must contain the center of the circle. As the diagonals are orthogonal, they define two perpendicular axes of symmetry of the circle. Therefore the statement of the problem also holds in this case.

9.20 Let $\mathcal{E}$ denote the ellipse, and let P be the interior point. Choose a coordinate system so that $\mathcal{E}$ is given by $x^2/a^2 + y^2/b^2 = 1$. In this system, label $P : (s, t)$. Since P is an interior point,

$$s^2/a^2 + t^2/b^2 < 1. \tag{17.20}$$

Let l be a line through P. If l is vertical, the statement is trivial. Let l be a non-vertical line having slope m. Then $l : y = mx + (t - ms)$, and the x-coordinates of the intersection points of l and $\mathcal{E}$ are the solutions of the equation

$$x^2/a^2 + (mx + (t - ms))^2/b^2 = 1 \quad \Leftrightarrow$$

$$(b^2 + a^2m^2)x^2 + (2a^2mt - 2a^2m^2s)x + a^2t^2 - 2a^2tms + a^2m^2s^2 - a^2b^2 = 0.$$

The discriminant D_1 of this quadratic equation in x simplifies to

$$D_1 = (2tms - m^2s^2 + b^2 - t^2 + a^2m^2)b^2.$$

If we can prove that, for each fixed pair of s and t satisfying (17.20), $D_1 > 0$ for all m, we are done. As $b^2 > 0$, we get

$$D_1 > 0 \quad \Leftrightarrow \quad (a^2 - s^2)m^2 + 2stm + b^2 - t^2 > 0. \tag{17.21}$$

Inequality (17.21) is quadratic in m. The coefficient at m^2 is $a^2 - s^2$, and it is positive due to (17.20). Therefore (17.21) holds for all m, if the discriminant

$$D_2 = 4(s^2b^2 - a^2b^2 + a^2t^2)$$

is negative. But $s^2b^2 - a^2b^2 + a^2t^2 > 0 \Leftrightarrow$ (17.20), which completes the proof.

9.21 We restrict ourselves to describing only the idea of a proof for the affirmative answer in the case of ellipses. The affirmative answer for the case of hyperbolas can be proven similarly. We do not know any proof for the negative answer in the case of parabolas that is suitable for this book. A discussion of all cases, and a proof for the negative answer in the case of parabolas, can be found in [**45**].

Take six points on an ellipse $\mathcal{E}$, and move one of them slightly, in such a way that it is not on $\mathcal{E}$, and such that no three of the six points formed by the original five points (denoted from here on as (x_i, y_i), $i = 1, \ldots, 5$) and the new point are collinear. (The reader should verify why this can be done.)

Next, we may use the following fundamental fact.

Theorem. *Any five points in a plane such that no three of them are collinear must lie on a unique conic, which is either an ellipse, or a hyperbola, or a parabola.*

Here we just comment on an obvious approach to proving this theorem. One can introduce a Cartesian coordinate system XYO such that none of (x_i, y_i), $i = 1, \ldots, 5$, is at the origin, and then search for an equation of such a conic in the form $Ax^2 + Bxy + Cy^2 + Fx + Gy + H = 0$. Since the origin is not on the curve, we may assume that $H = 1$ (just divide both sides of the equation by $H \neq 0$ to obtain an equivalent equation). So we are searching for an equation of the form $Ax^2 + Bxy + Cy^2 + Fx + Gy + 1 = 0$. Substituting the coordinates of (x_i, y_i), $i = 1, \ldots, 5$, we obtain a system with five linear equations and five unknowns: A, B, C, F, and G. Then, eliminating variables, one by one, and using the fact that no three of the points (x_i, y_i) are collinear, one can solve the system uniquely.

Unfortunately, this approach requires a lot of patience and care with algebraic manipulations, as the expressions for solutions become very large. If you have a computer program that can do this, use it! It would also be helpful to have knowledge of determinants of matrices. The expressions for the coefficients A, B, C, F, and G that we obtain will be rational functions of x_i and y_i, $i = 1, \ldots, 5$. The solution can be greatly simplified with the knowledge of affine transformations; see Chapter 12 and Problem S12.11. The fact that the equation represents an ellipse, a hyperbola, or a parabola follows from Theorem 9.4(1) and the fact that the conic contains at least five points and no three of them are collinear.

Using the above theorem, we can conclude the following.

(a) Our six points do not lie on an ellipse. Indeed, the five points which we did not move lie on only one (!) conic, namely, our original ellipse $\mathcal{E}$. As the sixth (new) point is not on it, the six points are not on an ellipse.

(b) Every five of these six points lie on a unique conic. We claim that each of these conics is an ellipse.

The discriminant $B^2 - 4AC$ of an equation $Ax^2 + Bxy + Cy^2 + Fx + Gy + 1 = 0$ is a continuous function of A, B, and C. For the ellipse $\mathcal{E}$ this discriminant is negative. As A, B, and C are rational functions of x_i and y_i, the discriminant is a continuous function of these variables. Therefore, a small move of the sixth point could be performed in such a way that the discriminant of each of the six conics, defined by each 5-subset of the set of our six points, is also negative. As each of these six conics is either an ellipse, or a hyperbola, or a parabola, all of them are ellipses.

Complex Numbers

10.1 (a)

$$\frac{(1 - 3i)(2 + i)}{i - 3} + \frac{1}{i^{24}} = \frac{5 - 5i}{-3 + i} + \frac{1}{1}$$
$$= \frac{(5 - 5i)(-3 - i)}{(-3 + i)(-3 - i)} + 1$$
$$= \frac{-20 + 10i}{10} + 1 = -1 + i.$$

(b)

$$(1+i)^{13} = \left[\sqrt{2} \left(\cos \frac{\pi}{4} + \sin \frac{\pi}{4} i \right) \right]^{13}$$

$$= 64\sqrt{2} \left(\cos \frac{13\pi}{4} + \sin \frac{13\pi}{4} i \right)$$

$$= 64\sqrt{2} \left(\cos \frac{5\pi}{4} + \sin \frac{5\pi}{4} i \right) = -64 - 64i.$$

(c)

$$1 + i + i^2 + \ldots + i^{2007} = \frac{i^{2008} - 1}{i - 1} = \frac{(i^4)^{502} - 1}{i - 1}$$

$$= \frac{1^{502} - 1}{i - 1} = \frac{1 - 1}{i - 1} = 0.$$

10.2 Let M be the midpoint of $\overline{AB}$. Then $M : (3 + i)$, and $M - A = 1 - 2i$. Let $z = \cos 120° + \sin 120° \, i = -1/2 + \sqrt{3}/2 \, i$. The number which corresponds to the image of M under the rotation is

$$A + (M - A)z = (2 + 3i) + (1 - 2i) \left(\frac{-1}{2} + \frac{\sqrt{3}}{2} i \right)$$

$$= \left(\frac{3}{2} + \sqrt{3} \right) + \left(4 + \frac{\sqrt{3}}{2} \right) i.$$

10.3 In many of the solutions below, we interpret $|z_1 - z_2|$ as the distance between two points in the complex plane which correspond to complex numbers z_1 and z_2. Another approach is to write z in its standard form, $z = x + yi$, and rewrite the condition determining the locus in terms of real variables x and y. This second approach is sometimes harder.

(a) Think of $|z| = 3$ as $|z_1 - z_2|$ with $z_1 = z$ and $z_2 = 0$. Hence, we have a circle of radius 3 centered at the origin. We reach the same conclusion if we substitute $z = x + yi$, and rewrite $|z| = 3$ as $\sqrt{x^2 + y^2} = 3$, or, equivalently, as $x^2 + y^2 = 9$.

(b) Think of $|z + 2 - 3i| > 2$ as $|z_1 - z_2|$ with $z_1 = z$ and $z_2 = -2 + 3i$, yielding the exterior of a circle of radius 2 centered at $-2 + 3i$ (or at $(-2, 3)$).

(c) In both of the terms $|z - 2|$ and $|z - 1 + 3i|$, we let $z_1 = z$. In the former, $z_2 = 2$, while in the latter, $z_2 = 1 - 3i$. Hence, we are looking for the locus of all z such that the sum of distances from z to points 2 and $1 - 3i$ is 5. The distance between 2 and $1 - 3i$ is $\sqrt{(2 - 1)^2 + (0 - (-3))^2} = \sqrt{10} < 5$. So the locus is an ellipse with foci at 2 (or $(2, 0)$) and $1 - 3i$ (or $(1, -3)$) such that the sum of distances from each of its points to the foci is 5.

(d) The set is empty, since by the triangle inequality, the sum of distances from z to points 2 and $1 - 3i$ is at least $\sqrt{10}$, but $3 < \sqrt{10}$.

(e) As 1 is less than the distance between points $2i$ and $2 - 3i$, the locus is a branch of the hyperbola with foci at $2i$ (or $(0, 2)$) and $2 - 3i$ (or $(2, -3)$) such that the absolute value of the difference of distances from each of its points to the foci is 1. As $|0 - 2i| - |0 - 2 + 3i| = 2 - \sqrt{13} < 1$, this is the branch containing the origin in its interior.

(f) If $z = x + yi$ is the standard form of z, then the condition is equivalent to the system of inequalities: $x \geq -1$ and $y \leq 3$. The locus is, of course, the right angle with its interior. Its vertex is at $(-1, 3)$, and it contains the origin.

10.4 We use the fact that a convex quadrilateral is a parallelogram if and only if its diagonals bisect each other. The midpoint of the segment with the endpoints $1 + 2i$ and $3 - i$ is $2 + (1/2)i$. If $z = x + yi$ is the standard form of the fourth vertex, the midpoint of the diagonal defined by $x + yi$ and $2 + i$ is $(x + 2)/2 + ((y + 1)/2)i$. Solving the system $(x + 2)/2 = 2$ and $(y + 1)/2 = 1/2$, we obtain $z = 2 + 0i = 2$ as one solution.

Starting with $3 - i$ and $2 + i$ as a pair of diagonal points leads to $4 - 2i$ as the fourth vertex.

Starting with $1 + 2i$ and $2 + i$ as a pair of diagonal points leads to $4i$ as the fourth vertex.

Note that the triangle with vertices at the three solutions is similar to the original triangle, with similarity coefficient $k = 2$.

10.5 Let $x + yi$ be the standard form of z.

(a) $\bar{z} = z \Leftrightarrow x - yi = x + yi \Leftrightarrow -y = y \Leftrightarrow y = 0 \Leftrightarrow \operatorname{Im} z = 0$.
(b) $\bar{z} = -z \Leftrightarrow x - yi = -x - yi \Leftrightarrow x = -x \Leftrightarrow x = 0 \Leftrightarrow \operatorname{Re} z = 0$.
(c) Note that $z \neq 0 \Leftrightarrow \bar{z} \neq 0$. Then

$$\bar{z} = \frac{1}{z} \Leftrightarrow \bar{z} = \frac{\bar{z}}{z\bar{z}} \Leftrightarrow \bar{z} = \frac{\bar{z}}{|z|^2} \Leftrightarrow |z|^2 = 1 \Leftrightarrow |z| = 1.$$

10.6 We will present proofs of properties (1), (2), (7) and (9) only. Let $z_1 = a + bi$ and $z_2 = c + di$ be the standard forms of z_1 and z_2.

Proof of (1): We have

$$\overline{z_1 + z_2} = \overline{(a + bi) + (c + di)}$$

$$= \overline{(a + c) + (b + d)i} = (a + c) - (b + d)i$$

$$= (a - bi) + (c - di) = \overline{z_1} + \overline{z_2}.$$

Proof of (2): We have

$$\overline{z_1 z_2} = \overline{(a + bi)(c + di)}$$

$$= \overline{(ac - bd) + (bc + ad)i} = (ac - bd) - (bc + ad)i \quad \text{and}$$

$$\overline{z_1}\,\overline{z_2} = (a - bi)(c - di) = (ac - bd) - (bc + ad)i.$$

The statement follows.

Proof of (7):
    Using (2), it is sufficient to check that

$$\sqrt{(ac - bd)^2 + (bc + ad)^2} = \sqrt{a^2 + b^2}\sqrt{c^2 + d^2} \Leftrightarrow$$

$$(ac - bd)^2 + (bc + ad)^2 = (a^2 + b^2)(c^2 + d^2),$$

which is easily verified by just expanding the expressions in both sides.

Proof of (9):

$$|z_1 + z_2| \le |z_1| + |z_2| \Leftrightarrow |z_1 + z_2|^2 \le (|z_1| + |z_2|)^2 \quad \Leftrightarrow$$

$$(a + c)^2 + (b + d)^2 \le \left(\sqrt{a^2 + b^2} + \sqrt{c^2 + d^2} \right)^2 \quad \Leftrightarrow$$

$$ac + bd \le \sqrt{a^2 + b^2} \sqrt{c^2 + d^2} \quad \Rightarrow$$

$$a^2 c^2 + b^2 d^2 + 2acbd \le a^2 c^2 + a^2 d^2 + b^2 c^2 + b^2 d^2 \quad \Leftrightarrow$$

$$0 \le (ad - bc)^2.$$

The last inequality is obvious. It becomes an equality if and only if $ad = bc$. Let us analyze this case.

Suppose that $ad = bc$, but neither c nor d is zero. Then denoting the ratio $a/c = b/d$ by λ, we get $z_1 = \lambda z_2$. As $\lambda \ne 0$ is real, its argument is 0 (if $\lambda > 0$), or π (if $\lambda < 0$). In the former case, $|z_1 + z_2| = |z_1| + |z_2|$, but in the latter, $|z_1 + z_2| < |z_1| + |z_2|$.

Suppose $c = 0$. If $d = 0$, then $z_2 = 0$ and $|z_1 + z_2| = |z_1| + |z_2|$ holds. If $d \ne 0$, then $ad = bc$ implies that $a = 0$. Then $z_1 = bi$ and $z_2 = di$, and $|z_1 + z_2| = |z_1| + |z_2|$ holds if and only if b and d are both nonnegative or are both nonpositive. Therefore either z_1 or z_2 is zero, or $z_1 = \lambda z_2$ for some positive λ. In this case the arguments of z_1 and z_2 are equal.

Thus the equality is attained if and only if one of the numbers is zero or if the numbers have equal arguments.

Remark. The fact that $ad = bc$ does not imply $|z_1 + z_2| = |z_1| + |z_2|$ is caused by the fact that the transition between the third and the fourth lines in the proof above (squaring of both sides) is not an equivalent transformation. The triangle inequality and the condition for it becoming an equality can be trivially generalized to any $n \ge 2$ complex numbers. The inequality $(ac + bd)^2 \le (a^2 + b^2)(c^2 + d^2)$ is also known as a simple instance of the Cauchy-Schwartz inequality. (See Theorem 11.9.)

10.7 The property of the modulus of complex numbers, $|z_1||z_2| = |z_1 z_2|$, implies that for any real numbers a, b, c, and d,

$$(a^2 + b^2)(c^2 + d^2) = (ac - bd)^2 + (bc + ad)^2.$$

If a, b, c, and d are integers, then so are $ac - bd$ and $bc + ad$.

10.8 Find a proof in textbooks or on the web.

10.9 By De Moivre's formula,

$$(\cos\alpha + (\sin\alpha)i)^3 = \cos 3\alpha + i \sin 3\alpha.$$

On the other hand, expanding the cube of the sum, we obtain

$$\begin{aligned}
(\cos\alpha + i \sin\alpha)^3 &= \cos^3\alpha + (3\cos^2\alpha \sin\alpha)i - 3\cos\alpha \sin^2\alpha - (\sin^3\alpha)i \\
&= \cos^3\alpha + (3(1 - \sin^2\alpha)\sin\alpha)i - 3\cos\alpha \sin^2\alpha - (\sin^3\alpha)i \\
&= (\cos^3\alpha - 3\cos\alpha \sin^2\alpha) + (3\sin\alpha - 4\sin^3\alpha)i.
\end{aligned}$$

Equating the imaginary parts in the right hand sides of the previous equalities, we obtain

$$\sin 3\alpha = 3\sin\alpha - 4\sin^3\alpha.$$

The same method can be used to show that $\sin(2n + 1)\alpha$ can be written as a polynomial of $\sin\alpha$ for any integer n. Can $\sin(2n)\alpha$ be written as a polynomial of $\sin\alpha$ for <u>some</u> integer $n \ne 0$?

In a similar way, one can show that

$$\cos 6\alpha = 32\cos^6\alpha - 48\cos^4\alpha + 18\cos^2\alpha - 1.$$

Can $\cos n\alpha$ be written as a polynomial of $\cos\alpha$ for <u>each</u> integer n?

10.10 The triangle with vertices at 0, 1, and z_k is isosceles with two sides of length one and the angle at 0 measuring $2\pi/n$. The altitude from 0 to the base bisects both the angle at 0 and the base. Hence the length of the base is

$$|z_k - 1| = 2 \cdot \sin\left(\frac{1}{2} \cdot \frac{2\pi k}{n}\right) = 2\sin\frac{\pi}{n}k.$$

10.11 Note that $\arctan\frac{1}{2}$ and $\arctan\frac{1}{3}$ are arguments of $2+i$ and $3+i$, respectively. As $(2+i)(3+i) = 5+5i$, an argument of the product is $\arctan\frac{1}{2} + \arctan\frac{1}{3}$ on one hand and is $\arctan\frac{5}{5} = \frac{\pi}{4}$ on the other. Since

$$0 < \arctan\frac{1}{2} + \arctan\frac{1}{3} < 2\arctan\frac{1}{2} < 2\arctan 1 = \frac{\pi}{2},$$

both angles are in $(0, \pi/2)$, and hence differ by less than any nonzero multiple of 2π. Therefore they are equal.

One can compare this solution to the one in Example 25 from Chapter 7.

Now we consider the generalization. Computing the standard form of the product, we obtain

$$(p + q + i)(p^2 + pq + 1 + qi) = ((p+q)^2 + 1)p + ((p+q)^2 + 1))i.$$

Similarly to the particular case above ($p = q = 1$), this translates to the required equality for arguments of the numbers.

10.12 Using the notation and results of the solution of Example 55, we obtain that the centroid of $\triangle ABC$ is at $\frac{1}{3}(A + B + C)$, and the one of $\triangle O_A O_B O_C$ is at

$$\frac{1}{3}(O_A + O_B + O_C) =$$

$$\frac{1}{3}\left(\frac{2B + C + \omega(C - B)}{3} + \frac{2C + A + \omega(A - C)}{3} + \frac{2A + B + \omega(B - A)}{3}\right)$$

$$= \frac{1}{3}\left(A + B + C + \omega \cdot \frac{0}{3}\right) = \frac{1}{3}(A + B + C).$$

The statement follows.

10.13 Let $z = \cos\alpha + \sin\alpha\, i$. Using De Moivre's formula, we obtain

$$1 + z + z^2 + \cdots + z^{n-1} = \sum_{k=0}^{n-1}(\cos k\alpha + \sin k\alpha\, i)$$

$$= \left(\sum_{k=0}^{n-1}\cos k\alpha\right) + \left(\sum_{k=0}^{n-1}\sin k\alpha\right)i.$$

Hence, the sums in question are the real and the imaginary parts of the sum of the geometric series $\sum_{k=0}^{n-1} z^k$. Summing the geometric series, and applying the standard trigonometric identities, we obtain:

$$
\begin{aligned}
\sum_{k=0}^{n-1} z^k &= \frac{1-z^n}{1-z} = \frac{1 - \cos n\alpha - \sin n\alpha\, i}{1 - \cos \alpha - \sin \alpha\, i} \\
&= \frac{(1 - \cos n\alpha - \sin n\alpha\, i)(1 - \cos \alpha + \sin \alpha\, i)}{(1 - \cos \alpha)^2 + \sin^2 \alpha} \\
&= \frac{(1 - \cos n\alpha - \sin n\alpha\, i)(2 \sin^2 \frac{\alpha}{2} + 2 \sin \frac{\alpha}{2} \cos \frac{\alpha}{2} i)}{2\,(1 - \cos \alpha)} \\
&= \frac{(1 - \cos n\alpha - \sin n\alpha\, i)(\sin \frac{\alpha}{2} + \cos \frac{\alpha}{2} i)}{2 \sin \frac{\alpha}{2}}.
\end{aligned}
$$

The real part of this fraction, which corresponds to $\sum_{k=0}^{n} \cos k\alpha$, is

$$
\frac{(1 - \cos n\alpha) \sin \frac{\alpha}{2} + \sin n\alpha \cos \frac{\alpha}{2}}{2 \sin \frac{\alpha}{2}} = \frac{\sin \frac{\alpha}{2} + \sin(n - \frac{1}{2})\alpha}{2 \sin \frac{\alpha}{2}}
$$

$$
= \frac{\cos \frac{(n-1)\alpha}{2} \cdot \sin \frac{n\alpha}{2}}{\sin \frac{\alpha}{2}}.
$$

The last equality utilizes (7.12). This proves the first formula. Similarly, simplifying the imaginary part of the fraction, we obtain the second formula.

Comparing this solution with the solutions to Problems 7.18 and 7.19, we see that the multiplier $2 \sin \frac{\alpha}{2}$ appears more naturally in this case. Another proof of the formulae uses mathematical induction. The result of Example 53 is just a simple corollary to the formulae in this problem, using $\alpha = 2\pi/n$.

These formulae can be easily generalized in order to find closed form expressions for the sums $\sum_{k=0}^{n-1} \cos(\beta + k\alpha)$ and $\sum_{k=0}^{n-1} \sin(\beta + k\alpha)$, by considering

$$
\sum_{k=0}^{n-1} z_0 z^k = z_0 \sum_{k=0}^{n-1} z^k
$$

for $z_0 = \cos \beta + \sin \beta\, i$ and $z = \cos \alpha + \sin \alpha\, i$.

10.14 Let us solve the problem for $R = 1$ and then just multiply the obtained result by R^2. Consider a complex plane where the vertices of the regular n-gon correspond to the complex numbers $z_k = \cos \frac{2\pi}{n} k + \sin \frac{2\pi}{n} k\, i$, where $k = 0, \ldots, n - 1$.

We will use the following fact, which also appeared in the solution for Example 53:

$$
\sum_{k=0}^{n-1} z_k = \sum_{k=0}^{n-1} z_1^k = \frac{1 - z_1^n}{1 - z_1} = \frac{1 - 1}{1 - z_1} = 0.
$$

Next, considering the sum S of the squares of all segments sharing vertex z_0, we get

$$S = \sum_{k=0}^{n-1} |z_k - z_0|^2 = \sum_{k=0}^{n-1} (z_k - z_0)(\overline{z_k - z_0})$$

$$= \sum_{k=0}^{n-1} (z_k - z_0)(\overline{z_k} - \overline{z_0})$$

$$= \sum_{k=0}^{n-1} (z_k\overline{z_k} - z_0\overline{z_k} - z_k\overline{z_0} + z_0\overline{z_0})$$

$$= 2n - z_0 \left(\sum_{k=0}^{n-1} \overline{z_k} \right) - \overline{z_0} \left(\sum_{k=0}^{n-1} z_k \right)$$

$$= 2n - z_0 \cdot \overline{0} - \overline{z_0} \cdot 0 = 2n.$$

Clearly the same sum S is obtained if we change the shared vertex z_0 to any other z_k. Therefore, multiplying by n and dividing by 2 in order not to double count, we find the sum of the squares of the side lengths and diagonals is

$$\frac{1}{2} \cdot n \cdot 2n \cdot R^2 = n^2 R^2.$$

10.15 We use the idea of our solution of Example 56. We can assume that z_k is the complex number corresponding to A_k and that $z_1 = R$. Let $z = a$ be a real number which corresponds to X. Then we are interested in finding

$$|a - z_1| \cdot |a - z_2| \cdot \ldots \cdot |a - z_n|.$$

Note that each z_k is a root of the polynomial $z^n - R^n$. Hence,

$$z^n - R^n = (z - z_1)(z - z_2) \ldots (z - z_n).$$

This implies that

$$|z^n - R^n| = |(z - z_1)||(z - z_2)| \ldots |(z - z_n)|,$$

which, after substituting $z = a$, gives

$$|a^n - R^n| = |a - z_1| \cdot |a - z_2| \cdot \ldots \cdot |a - z_n|.$$

As the number $a^n - R^n$ is real and $0 \le a < R$, its absolute value is $R^n - a^n$.

It seems that the solution of Example 56 can be obtained from the solution above if we manage, somehow, to remove the extra factor $|a - z_1| = |a - R|$ from the product and set $a = R$. Restating the idea more precisely, we wish to compute the limit

$$\lim_{a \to R} \frac{|a^n - R^n|}{|a - R|},$$

since for every a, $0 \le a < R$, the value of the fraction $|a^n - R^n|/|a - R|$ is the product of $n - 1$ distances excluding $|a - R|$. Let's do it.

$$\lim_{a \to R} \frac{|a^n - R^n|}{|a - R|} = \lim_{a \to R} \left| \frac{a^n - R^n}{a - R} \right|$$

$$= \lim_{a \to R} \left| a^{n-1} + a^{n-2}R + \cdots + aR^{n-2} + R^{n-1} \right|$$

$$= nR^{n-1}.$$

For $R = 1$, the product is indeed n.[11] Notice also that Problem 10.15 is easier than Example 56 due to the greater "symmetry" in its statement: point A_1 is not that special anymore.

10.16 It is easy to verify that

$$(5 + i)^4(-239 + i) = -114244 - 114244i.$$

An argument of $5 + i$ is $\arctan \frac{1}{5}$, and of $-239 + i$ is $-\arctan \frac{1}{239}$. Therefore, $4 \arctan \frac{1}{5} - \arctan \frac{1}{239}$ is an argument of the product $(5 + i)^4(-239 + i)$. Since the product is $-114244 - 114244i$, and $\arctan 1 = \frac{\pi}{4}$ is its argument, the numbers $4 \arctan \frac{1}{5} - \arctan \frac{1}{239}$ and $\frac{\pi}{4}$ can differ only by a multiple of 2π. As $0 < \arctan \frac{1}{239} < \arctan \frac{1}{5} < \pi/6$, we have

$$0 < 4 \arctan \frac{1}{5} - \arctan \frac{1}{239} < 4 \arctan \frac{1}{5} < 4 \frac{\pi}{6} < \pi.$$

Hence, $4 \arctan \frac{1}{5} - \arctan \frac{1}{239} \in (0, \pi)$. Since $\pi/4$ is in the same range, the two numbers are equal.

Compare this solution to the one of Problem 7.12. The method used there may be more straightforward and proves the fact at least as quickly, but perhaps the approach via complex numbers offers something new. In particular it explains how such a relation might appear and how many other similar relations may be found. For example, the reader is invited to explore the relation between arctangents that follows from

$$(-1 + i)^3(3 - 2i)^2(5 - i) = 156 - 104i.$$

10.17 As $\zeta^3 = -1$, ζ is a root of the polynomial $x^3 + 1 = (x + 1)(x^2 - x + 1)$. As $\zeta \neq 1$, it is a root of $x^2 - x + 1$. Hence, $\zeta^2 = \zeta - 1$. Multiplication by ζ rotates every complex number counterclockwise around the origin by $180°/3 = 60°$. (Refer to Figure 127.) It is clear that $\triangle ABC$ is equilateral if and only if side BA rotated by $60°$ around B gives side BC. Equivalently,

$$C - B = (A - B)\zeta \quad \Leftrightarrow \quad A\zeta + B(1 - \zeta) - C = 0$$

$$\Leftrightarrow \quad A\zeta - B\zeta^2 - C = 0.$$

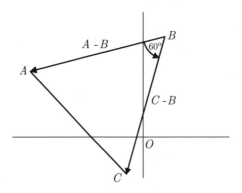

FIGURE 127. Problem 10.17

10.18 (a) The point corresponding to a complex number z belongs to the line passing through z_1 and z_2 if and only if $z - z_1 = \lambda(z_2 - z_1)$ for some *real* number λ. Taking conjugates of

---

[11] One could also compute the limit by noticing that it is the absolute value of the derivative of $f(a) = a^n$ at $a = R$, or by using L'Hôpital's rule.

both sides and using $\lambda = \overline{\lambda}$ gives $\overline{z - z_1} = \lambda(\overline{z_2 - z_1})$, or $\overline{z} - \overline{z_1} = \lambda(\overline{z_2} - \overline{z_1})$. The two equations imply (10.2).

Conversely, suppose z satisfies (10.2). If $z = z_1$, then z is on the line defined by z_1 and z_2. So, suppose $z \neq z_1$. Then

$$(z - z_1)/(\overline{z} - \overline{z_1}) = (z_2 - z_1)/(\overline{z_2} - \overline{z_1}) \quad \Leftrightarrow$$
$$(z - z_1)/\overline{(z - z_1)} = (z_2 - z_1)/\overline{(z_2 - z_1)}.$$

As $\arg \overline{u} = -\arg u$, and $\arg u/v = \arg u - \arg v$ for any complex numbers u and v, $v \neq 0$, we have

$$2 \arg(z - z_1) = 2 \arg(z_2 - z_1) \quad \Leftrightarrow \quad \arg(z - z_1) = \arg(z_2 - z_1).$$

The last implies that z is on the line defined by z_1 and z_2.

(b) We have:

$$\overline{\kappa} = \overline{\left(\frac{z_2 - z_1}{\overline{z_2} - \overline{z_1}} \right)} = \frac{\overline{z_2 - z_1}}{\overline{\overline{z_2} - \overline{z_1}}} = \frac{\overline{z_2} - \overline{z_1}}{z_2 - z_1} = \kappa^{-1}.$$

This implies $|\kappa|^2 = \kappa \overline{\kappa} = \kappa \kappa^{-1} = 1$. Hence, $|\kappa| = 1$.

(c) (i) Let l_1 and l_2 be the lines. Then there exist two distinct complex numbers u_1 and u_2 such that $l_1 : z - u_1 = \kappa_1(\overline{z} - \overline{u_1})$ and $l_2 : z - u_2 = \kappa_2(\overline{z} - \overline{u_2})$. If $\kappa_1 \neq \kappa_2$, solving the system of two equations for z, we obtain the unique common solution

$$z = \overline{\left(\frac{u_2 - u_1 + \kappa_2 \overline{u_2} - \kappa_1 \overline{u_1}}{\kappa_1 - \kappa_2} \right)}.$$

Hence, in this case, the lines intersect.

Conversely, suppose the lines intersect, but are distinct. Let z be the complex number corresponding to the common point. Then, as was shown in (a), $2 \arg(z - u_1) = \arg \kappa_1$ and $2 \arg(z - u_2) = \arg \kappa_2$. Since z, u_1, and u_2 are not collinear, $\arg(z - u_1) \neq \arg(z - u_2)$. Hence, $\kappa_1 \neq \kappa_2$, which ends the proof.

(ii) Let $\alpha \in (0, \pi/2]$ be the measure of the angle between two intersecting lines with equations $l_1 : z - u_1 = \kappa_1(\overline{z} - \overline{u_1})$ and $l_2 : z - u_2 = \kappa_2(\overline{z} - \overline{u_2}), u_1 \neq u_2$. Then, the arguments of $(z - u_1)$ and $(z - u_2)$ differ by $\pm \alpha$ and a multiple of 2π. Hence, the arguments of κ_1 and κ_2 differ by $\pm 2\alpha$ and a multiple of 2π. Therefore, $\alpha = \pi/2$ if and only if $\kappa_1 = -\kappa_2$.

10.19 Using the results of Problem 10.18, parts (a) and (b), we find the slopes of lines AD and BC to be

$$\kappa_{AD} = \frac{d - a}{\overline{d} - \overline{a}}, \quad \text{and} \quad \kappa_{BC} = \frac{c - b}{\overline{c} - \overline{b}}.$$

Using the result of Problem 10.18(c), we obtain

$$\overline{AD} \perp \overline{BC} \quad \Leftrightarrow \quad \kappa_{AD} = -\kappa_{BC} \quad \Leftrightarrow \quad \frac{d - a}{b - c} = -\frac{\overline{d} - \overline{a}}{\overline{c} - \overline{b}}.$$

Let $z := (d - a)/(b - c)$. The last equality can be rewritten as $\overline{z} = -z$, which is equivalent to z being purely imaginary. (See Problem 10.5.)

10.20 Let R denote the length of the circumradius. Then the slope of the line through a and h is equal to

$$\frac{h - a}{\overline{h} - \overline{a}} = \frac{b + c}{\overline{b} + \overline{c}} = \frac{b + c}{(R^2/b + R^2/c)} = \frac{bc}{R^2},$$

and the slope of the line through b and c is

$$\frac{c-b}{\bar{c}-\bar{b}} = \frac{c-b}{\bar{c}-\bar{b}} = \frac{c-b}{(R^2/c - R^2/b)} = -\frac{bc}{R^2}.$$

As the slopes are opposite, the lines are orthogonal (as was shown in Problem 10.18(c)ii). The arguments for the other two altitudes are similar. Hence, they all pass through h.

If we had to find h in terms of a, b, and c, how could we do it? The expression for h makes one think about the centroid g of the triangle, which, as we have proven many times by now, is given by $1/3 \cdot (a + b + c)$. This gives us an immediate proof of the second statement, which is often stated as follows:

In any triangle, the centroid lies on the segment connecting the orthocenter with the center of the circumcircle and it divides the segment's length in proportion $2 : 1$.

Compare this solution with the one of Problem 4.27.

10.21   Suppose the quadrilateral is in a complex plane. In what follows, the notation of a point and a number corresponding to it will be the same. Let A, B, C, and D be the complex numbers corresponding to the vertices of the quadrilateral taken in the clockwise direction, and let O_{AB}, O_{BC}, O_{CD}, and O_{DA} denote the complex numbers corresponding to the centers of the squares built on the corresponding sides. (See Figure 128.)

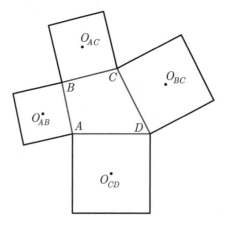

FIGURE 128.

As point O_{AB} can be obtained from point A by rotating it $-90°$ around the midpoint of $\overline{AB}$, we have

$$O_{AB} = \frac{1}{2}(A + B) + \frac{1}{2}(A - B)(-i) = \frac{1}{2}[(A + B) + (B - A)i].$$

Similarly,

$$O_{BC} = \frac{1}{2}[(B + C) + (C - B)i],$$

$$O_{CD} = \frac{1}{2}[(C + D) + (D - C)i], \text{ and}$$

$$O_{DA} = \frac{1}{2}[(D + A) + (A - D)i].$$

Then

$$z_1 := O_{AB} - O_{CD} = \frac{1}{2}[(A + B - C - D) + (B - A - D + C)i], \quad \text{and}$$

$$z_2 := O_{BC} - O_{DA} = \frac{1}{2}[(B + C - D - A) + (C - B - A + D)i].$$

As $z_1 = z_2 \cdot (-i)$, the segments $O_{AB}O_{CD}$ and $O_{BC}O_{DA}$ are congruent and perpendicular.[12]

10.22 We use the results of Problem 10.21 and the notation introduced in its solution. Complex number corresponding to the midpoints of the segments $O_{AB}O_{CD}$ and $O_{BC}O_{DA}$ are:

$$m_1 := \frac{1}{2}(O_{AB} + O_{CD}) = \frac{1}{4}[(A + B + C + D) + (B - A + D - C)i],$$

$$m_2 := \frac{1}{2}(O_{BC} + O_{DA}) = \frac{1}{4}[(B + C + D + A) + (C - B + A - D)i].$$

Since $ABCD$ is a parallelogram, $A - B = D - C$. This implies $m_1 = m_2$, so the segments bisect each other. Hence, using that they are congruent and orthogonal, we conclude that $O_{AB}O_{BC}O_{CD}O_{DA}$ is a square.

Vectors

11.1 Let $A : (a_1, a_2)$, $B : (b_1, b_2)$, $C : (c_1, c_2)$, and $D : (d_1, d_2)$ be four arbitrary points. Each pair of points forms a vector; for example, $\overrightarrow{AB} = \langle b_1 - a_1, b_2 - a_2 \rangle$. Thus, $\overrightarrow{AB} \cdot \overrightarrow{CD} + \overrightarrow{BC} \cdot \overrightarrow{AD} + \overrightarrow{CA} \cdot \overrightarrow{BD}$

$$= (b_1 - a_1)(d_1 - c_1) + (b_2 - a_2)(d_2 - c_2) + (c_1 - b_1)(d_1 - a_1)$$
$$+ (c_2 - b_2)(d_2 - a_2) + (a_1 - c_1)(d_1 - b_1) + (a_2 - c_2)(d_2 - b_2)$$
$$= b_1d_1 - a_1d_1 - b_1c_1 + a_1c_1 + b_2d_2 - a_2d_2 - b_2c_2 + a_2c_2$$
$$+ c_1d_1 - b_1d_1 - a_1c_1 + a_1b_1 + c_2d_2 - b_2d_2 - a_2c_2 + a_2b_2$$
$$+ a_1d_1 - c_1d_1 - a_1b_1 + b_1c_1 + a_2d_2 - c_2d_2 - a_2b_2 + b_2c_2$$
$$= 0.$$

11.2 Given triangle ABC, let M and N denote the midpoints of $\overline{AB}$ and $\overline{BC}$, respectively. Then $\overrightarrow{MN} = \frac{1}{2}\overrightarrow{AB} + \frac{1}{2}\overrightarrow{BC} = \frac{1}{2}\overrightarrow{AC}$. This proves that $\overrightarrow{MN}$ is parallel to $\overrightarrow{AC}$ and has half its length.

11.3 Given a quadrilateral $ABCD$, let M, N, P, and Q denote the midpoints of $\overline{AB}$, $\overline{BC}$, $\overline{CD}$, and $\overline{DA}$, respectively. Then $\overrightarrow{MN} = \frac{1}{2}\overrightarrow{AB} + \frac{1}{2}\overrightarrow{BC} = \frac{1}{2}\overrightarrow{AC}$. Similarly, $\overrightarrow{RP} = \frac{1}{2}\overrightarrow{AC}$, so $\overrightarrow{MN} = \overrightarrow{RP}$. In the same way, we can show $\overrightarrow{MR} = \overrightarrow{NP}$. Thus, opposite sides of quadrilateral $MNPR$ are parallel and have the same length, so it is a parallelogram.

11.4 *Solution 1:* Since $||\vec{a}|\vec{b}| = |\vec{a}||\vec{b}| = |\vec{b}||\vec{a}| = ||\vec{b}|\vec{a}|$, the parallelogram with sides determined by the vectors $|\vec{a}|\vec{b}$ and $|\vec{b}|\vec{a}$ is a rhombus, and $\vec{c}$ corresponds to a diagonal of the rhombus. (See Figure 129.) But a diagonal of a rhombus bisects its angle (because the two resulting triangles are congruent by SSS).

Clearly, the same argument proves a more general statement: the sum of two vectors of equal length bisects the angle between them.

---

[12] Have we used the convexity of $ABCD$ in this solution? What if the squares were built internally?

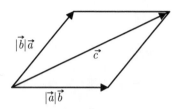

FIGURE **129.**

Solution 2: Here, we prove the more general statement: the sum $\vec{z} = \vec{x} + \vec{y}$ of two vectors, $\vec{x}$ and $\vec{y}$, of equal length bisects the angle between them. Let α denote the measure of the angle between $\vec{x}$ and $\vec{z}$ and let β denote the angle between $\vec{y}$ and $\vec{z}$. Since $|\vec{x}| = |\vec{y}|$,

$$\cos\alpha = \frac{\vec{x} \cdot \vec{z}}{|\vec{x}||\vec{z}|} = \frac{\vec{x} \cdot \vec{x} + \vec{x} \cdot \vec{y}}{|\vec{x}||\vec{z}|} = \frac{|\vec{x}|^2 + \vec{x} \cdot \vec{y}}{|\vec{x}||\vec{z}|}$$

$$= \frac{|\vec{y}|^2 + \vec{x} \cdot \vec{y}}{|\vec{y}||\vec{z}|} = \frac{\vec{y} \cdot \vec{x} + \vec{y} \cdot \vec{y}}{|\vec{y}||\vec{z}|} = \frac{\vec{y} \cdot \vec{z}}{|\vec{y}||\vec{z}|} = \cos\beta.$$

By definition of the angle between two vectors, both α and β are in $[0, \pi]$. Therefore, the equality of their cosines implies equality of the angles, and the proof is finished.

11.5   In the convex 6-gon $A_1 A_2 \ldots A_6$, let $\vec{v}_1 = \overrightarrow{A_1 A_2}, \vec{v}_2 = \overrightarrow{A_2 A_3}, \ldots, \vec{v}_6 = \overrightarrow{A_6 A_1}$. By the Polygon Rule, $\vec{v}_1 + \vec{v}_2 + \cdots + \vec{v}_5 + \vec{v}_6 = \vec{0}$.
Now, $\overrightarrow{M_1 M_2} = \frac{1}{2}\vec{v}_1 + \frac{1}{2}\vec{v}_2$, $\overrightarrow{M_3 M_4} = \frac{1}{2}\vec{v}_3 + \frac{1}{2}\vec{v}_4$, and $\overrightarrow{M_5 M_6} = \frac{1}{2}\vec{v}_5 + \frac{1}{2}\vec{v}_6$. Thus,

$$\overrightarrow{M_1 M_2} + \overrightarrow{M_3 M_4} + \overrightarrow{M_5 M_6} = \frac{1}{2} \sum_{i=1}^{6} \vec{v}_i = \vec{0}.$$

By the Triangle Rule, $\overrightarrow{M_1 M_2}$, $\overrightarrow{M_3 M_4}$, and $\overrightarrow{M_5 M_6}$ form the sides of a triangle. (In Figure 130, $\triangle QRS$ is such a triangle.)

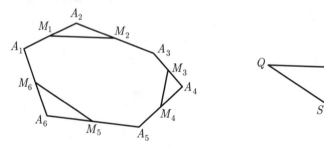

FIGURE **130.**

11.6 (1) $\vec{a} \cdot \vec{b} = \langle a_1, a_2 \rangle + \langle b_1, b_2 \rangle = \langle a_1 + b_1, a_2 + b_2 \rangle = \langle b_1 + a_1, b_2 + a_2 \rangle = \langle b_1, b_2 \rangle + \langle a_1, a_2 \rangle = \vec{b} \cdot \vec{a}$.

(2) $\vec{a} \cdot (\vec{b} + \vec{c}) = \langle a_1, a_2 \rangle \cdot \langle b_1 + c_1, b_2 + c_2 \rangle = \langle a_1(b_1 + c_1), a_2(b_2 + c_2) \rangle = \langle a_1 b_1, a_2 b_2 \rangle + \langle a_1 c_1, a_2 c_2 \rangle = \vec{a} \cdot \vec{b} + \vec{a} \cdot \vec{c}$.

(3) $k(\vec{a} \cdot \vec{b}) = k(a_1 b_1 + a_2 b_2) = (k a_1 b_1 + k a_2 b_2) = \langle k a_1, k a_2 \rangle \cdot \langle b_1, b_2 \rangle = k\vec{a} \cdot \vec{b}$. Similarly, $k(\vec{a} \cdot \vec{b}) = \vec{a} \cdot k\vec{b}$.

(4) $\vec{0} \cdot \vec{a} = \langle 0, 0 \rangle \cdot \langle a_1, a_2 \rangle = 0a_1 + 0a_2 = 0$.

(5) $\vec{a} \cdot \vec{a} = \langle a_1, a_2 \rangle \cdot \langle a_1, a_2 \rangle = a_1{}^2 + a_2{}^2 = |\vec{a}|^2$.

(6) Let θ represent the angle between $\vec{a}$ and $\vec{b}$. Then $\vec{a}$ and $\vec{b}$ are perpendicular if and only if $\cos\theta = 0$ if and only if $|\vec{a}||\vec{b}|\cos\theta = 0$ if and only if $\vec{a} \cdot \vec{b} = 0$.

11.7 (a) Let $\overrightarrow{OX} \in L$, so X is a point on l. Since A, B, and X are collinear, $\overrightarrow{AX} = k\overrightarrow{AB}$ for some scalar k. Then, from the Triangle Law for vector addition, $\overrightarrow{OX} = \overrightarrow{OA} + \overrightarrow{AX} = \overrightarrow{OA} + k\overrightarrow{AB}$. This proves $L \subseteq \{\overrightarrow{OA} + k\overrightarrow{AB} : k \in \mathbb{R}\}$. By reversing the argument, we easily obtain equality in the other direction.

   (b) Repeat the solution in part (a), noting that $k = 0$ corresponds to the position vector $\overrightarrow{OA}$ and $k = 1$ corresponds to the position vector $\overrightarrow{OB}$.

11.8 Suppose the four consecutive sides of the quadrilateral are represented by vectors $\vec{a}$, $\vec{b}$, $\vec{c}$, and $\vec{d} = -(\vec{a} + \vec{b} + \vec{c})$. Then the sum of the squares of the side lengths of the quadrilateral will be

$$
\begin{aligned}
S &= |\vec{a}|^2 + |\vec{b}|^2 + |\vec{c}|^2 + |-\vec{a} - \vec{b} - \vec{c}|^2 \\
&= |\vec{a}|^2 + 2\vec{a}\cdot\vec{b} + |\vec{b}|^2 + |\vec{a}|^2 + 2\vec{a}\cdot\vec{c} + |\vec{c}|^2 + |\vec{b}|^2 + 2\vec{b}\cdot\vec{c} + |\vec{c}|^2 \\
&= (\vec{a}+\vec{b})\cdot(\vec{a}+\vec{b}) + (\vec{a}+\vec{c})\cdot(\vec{a}+\vec{c}) + (\vec{b}+\vec{c})\cdot(\vec{b}+\vec{c}) \\
&= |\vec{a}+\vec{b}|^2 + |\vec{a}+\vec{c}|^2 + |\vec{b}+\vec{c}|^2.
\end{aligned}
$$

Now, the sum of the squares of the diagonal lengths of the quadrilateral will be $|\vec{a} + \vec{b}|^2 + |\vec{b} + \vec{c}|^2$, which will equal S if and only if $\vec{a} = -\vec{c}$. Recall that a quadrilateral is a parallelogram if and only if it has a pair of opposite sides that are parallel and congruent. But, this is exactly what is implied by $\vec{a} = -\vec{c}$! So, the quadrilateral is a parallelogram if and only if

$$S = |\vec{a}+\vec{b}|^2 + |\vec{b}+\vec{c}|^2.$$

11.9 Let $\vec{v} = \overrightarrow{AB}$ and $\vec{w} = \overrightarrow{AC}$; then $\overrightarrow{BC} = \vec{w} - \vec{v}$. Let $\overrightarrow{AA'}$, $\overrightarrow{BB'}$, and $\overrightarrow{CC'}$ be the medians of $\triangle ABC$. (See Figure 131.)

Consider three vectors having initial point at A:

$$
\begin{aligned}
\frac{2}{3}\overrightarrow{AA'} &= \frac{2}{3}\left(\overrightarrow{AB} + \overrightarrow{BA'}\right) = \frac{2}{3}\left(\overrightarrow{AB} + \frac{1}{2}\overrightarrow{BC}\right) \\
&= \frac{2}{3}\overrightarrow{AB} + \frac{1}{3}\overrightarrow{BC} \\
&= \frac{2}{3}\vec{v} + \frac{1}{3}(\vec{w} - \vec{v}) = \frac{1}{3}(\vec{v} + \vec{w}),
\end{aligned}
$$

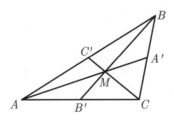

FIGURE 131.

$$\overrightarrow{AB} + \frac{2}{3}\overrightarrow{BB'} = \overrightarrow{AB} + \frac{2}{3}\left(\overrightarrow{BA} + \overrightarrow{AB'}\right)$$

$$= \overrightarrow{AB} + \frac{2}{3}\left(\overrightarrow{BA} + \frac{1}{2}\overrightarrow{AC}\right)$$

$$= \vec{v} + \frac{2}{3}(-\vec{v} + \frac{1}{2}\vec{w}) = \frac{1}{3}(\vec{v} + \vec{w}),$$

$$\overrightarrow{AC} + \frac{2}{3}\overrightarrow{CC'} = \overrightarrow{AC} + \frac{2}{3}\left(\overrightarrow{CA} + \overrightarrow{AC'}\right)$$

$$= \overrightarrow{AC} + \frac{2}{3}\left(\overrightarrow{CA} + \frac{1}{2}\overrightarrow{AB}\right)$$

$$= \vec{w} + \frac{2}{3}(-\vec{w} + \frac{1}{2}\vec{v}) = \frac{1}{3}(\vec{v} + \vec{w}).$$

Since all three vectors are equal and begin at A, they also have the same terminal point. Therefore, the point $2/3$ of the way from A to A', the point $2/3$ of the way from B to B', and the point $2/3$ of the way from C to C' must all be the same point. We conclude that the medians are concurrent at a point M that divides each median in ratio $2{:}1$.

This is a repeat of Problem 3.2.10.

11.10 Let $\overrightarrow{CC'}$ be a median of $\triangle ABC$, in which case $\overrightarrow{C'B} = -\overrightarrow{C'A}$. (Refer again to Figure 131.)
($\Rightarrow$) Suppose M is situated such that $\overrightarrow{MA} + \overrightarrow{MB} + \overrightarrow{MC} = \vec{0}$. Then we have

$$\vec{0} = \left(\overrightarrow{MC} + \overrightarrow{CA}\right) + \left(\overrightarrow{MC'} + \overrightarrow{C'B}\right) + \left(\overrightarrow{MC'} + \overrightarrow{C'A} + \overrightarrow{AC}\right)$$

$$= \overrightarrow{MC} + 2\overrightarrow{MC'}.$$

Therefore, $\overrightarrow{MC} = 2\overrightarrow{C'M}$, so $\overrightarrow{MC}$ and $\overrightarrow{C'M}$ are parallel. This means that M lies on the median $\overline{C'C}$. Similarly, M lies on the other two medians, so the medians must all meet at M.

($\Leftarrow$) Let M be the point of intersection of the medians. By Problem 11.9, $\overrightarrow{MC} = \frac{2}{3}\overrightarrow{C'C}$. Therefore, using vector addition,

$$\overrightarrow{MC} = \frac{2}{3}(\overrightarrow{C'B} + \overrightarrow{BC}) = \frac{2}{3}(\overrightarrow{BC} + \frac{1}{2}\overrightarrow{AB}).$$

Similarly,

$$\overrightarrow{MA} = \frac{2}{3}(\overrightarrow{CA} + \frac{1}{2}\overrightarrow{BC}) \text{ and } \overrightarrow{MB} = \frac{2}{3}(\overrightarrow{AB} + \frac{1}{2}\overrightarrow{CA}).$$

Then, since $\overrightarrow{AB} + \overrightarrow{BC} + \overrightarrow{CA} = \vec{0}$, we have

$$\overrightarrow{MA} + \overrightarrow{MB} + \overrightarrow{MC} = \frac{2}{3}\left(\frac{3}{2}\overrightarrow{CA} + \frac{3}{2}\overrightarrow{BC} + \frac{3}{2}\overrightarrow{AB}\right) = \vec{0}.$$

11.11 Consider an isosceles triangle ABC where $AB = BC$ and AC is very small so that the angles at A and C are close to being right. Then each of the bisectors at A and C will have length less than $\sqrt{2}AC$. However, the length of the bisector at B increases beyond any bound as $m\angle ABC$ approaches 0. Thus it is possible to construct triangle ABC so that its angle bisectors do not satisfy the triangle inequality, which means the angle bisectors (as vectors) do not necessarily form a triangle.

11.12 Let d_1 and d_2 be the midlines of the quadrilateral (join midpoints of opposite sides). See the solution to Problem 11.S3. Since $2\overrightarrow{MN} = \overrightarrow{BC} + \overrightarrow{AD}$, we have that $d_1 \leq (a + c)/2, d_2 \leq (b + d)/2$.

Therefore $d_1 d_2 \leq (1/4)(a + c)(b + d)$. Now, d_1 and d_2 are diagonals of the parallelogram with area equalling Area($ABCD$)/2. If α is an angle between them, then $d_1 d_2 \sin(alpha) =$ Area($ABCD$). Hence, Area($ABCD$) $\leq d_1 d_2 \sin(90) \leq (1/4)(a + c)(b + d)$.

We obtain equality if and only if $2MN = BC + AD$, i.e., if $\overline{BC}$ and $\overline{AD}$ are parallel, and $\alpha = 90°$. This is shown similarly for the other pair of sides. Therefore, equality holds if and only if $ABCD$ is a rectangle.

11.13 Suppose first that vectors $\vec{a} = \overrightarrow{BC}$, $\vec{b} = \overrightarrow{CA}$, and $\vec{c} = \overrightarrow{BA}$. Then $c^2 = \vec{c} \cdot \vec{c} = (\vec{a} + \vec{b}) \cdot (\vec{a} + \vec{b}) = a^2 + 2\vec{a} \cdot \vec{b} + b^2 = 5c^2 + 2\vec{a} \cdot \vec{b}$. This implies that $\vec{a} \cdot \vec{b} = -2c^2$.

Next, let $\overrightarrow{BB'}$ and $\overrightarrow{AA'}$ be the medians to the sides AC and BC, respectively. We must show $\overrightarrow{BB'} \cdot \overrightarrow{AA'} = 0$.

$$\overrightarrow{BB'} \cdot \overrightarrow{AA'} = (\vec{a} + \vec{b}/2) \cdot (-\vec{b} - \vec{a}/2)$$
$$= -\vec{a} \cdot \vec{b} - (a^2/2 + b^2/2) - (1/4)\vec{b} \cdot \vec{a}$$
$$= (-5/4)\vec{a} \cdot \vec{b} - (1/2)(a^2 + b^2)$$
$$= (-5/4)(-2c^2) - (1/2)(5c^2) = 0.$$

11.14 Let $\overrightarrow{AB} = \vec{u}$, $\overrightarrow{BC} = \vec{v}$, and $\overrightarrow{CD} = \vec{w}$. Then $\overrightarrow{AD} = \vec{u} + \vec{v} + \vec{w}$, $\overrightarrow{AC} = \vec{u} + \vec{v}$, and $\overrightarrow{BD} = \vec{v} + \vec{w}$. (See Figure 132.)

(a) If $\overline{AB} \perp \overline{CD}$ and $\overline{BC} \perp \overline{AD}$, then $\vec{u} \cdot \vec{w} = 0$ and $\vec{v} \cdot (\vec{u} + \vec{v} + \vec{w}) = 0$. Consequently,

$$0 = \vec{u} \cdot \vec{w} + \vec{v} \cdot (\vec{u} + \vec{v} + \vec{w})$$
$$= \vec{u} \cdot \vec{w} + \vec{v} \cdot \vec{u} + \vec{v} \cdot \vec{v} + \vec{v} \cdot \vec{w}$$
$$= (\vec{u} + \vec{v}) \cdot (\vec{v} + \vec{w}).$$

Thus, $\overline{AC} \perp \overline{BD}$.

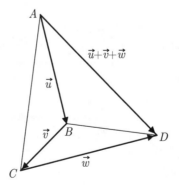

FIGURE 132.

(b) In the following string of equations, we make use of the fact that
$\vec{v} \cdot (\vec{u} + \vec{v} + \vec{w}) = 0$.

$$
\begin{aligned}
(AB)^2 + (CD)^2 &= \vec{u} \cdot \vec{u} + \vec{w} \cdot \vec{w} \\
&= \vec{u} \cdot \vec{u} + \vec{w} \cdot \vec{w} + 2\vec{v} \cdot (\vec{u} + \vec{v} + \vec{w}) \\
&= \vec{u} \cdot \vec{u} + 2\vec{u} \cdot \vec{v} + \vec{v} \cdot \vec{v} + \vec{v} \cdot \vec{v} + 2\vec{v} \cdot \vec{w} + \vec{w} \cdot \vec{w} \\
&= (\vec{u} + \vec{v}) \cdot (\vec{u} + \vec{v}) + (\vec{v} + \vec{w}) \cdot (\vec{v} + \vec{w}) \\
&= (AC)^2 + (BD)^2
\end{aligned}
$$

The partial converse is not true. Consider the familiar right triangle ABC where $AB = 3$, $AC = 4$, and $BC = 5$. Let D lie on the circle of radius $\sqrt{18}$ centered at B. Then $AB^2 + BC^2 = 34 = AC^2 + BD^2$, but our freedom in choosing D certainly keeps us from concluding that $\overline{AB} \perp \overline{CD}$.

11.15 Let O be the circumcenter of $\triangle ABC$, and let $\overrightarrow{OA} = \vec{u}$, $\overrightarrow{OB} = \vec{v}$, and $\overrightarrow{OC} = \vec{w}$. Since $\vec{u}$, $\vec{v}$, and $\vec{w}$ are all radii of the circumcircle of $\triangle ABC$, they all have the same length.

Let X be the fourth vertex of a parallelogram $AOCX$. (See Figure 133(i).) Then $AOCX$ is a rhombus whose diagonals, $\overline{AC}$ and $\overline{OX}$, are perpendicular and bisect each other. Note also that

$$
\overrightarrow{OX} = \overrightarrow{OA} + \overrightarrow{AX} = \overrightarrow{OA} + \overrightarrow{OC} = \vec{u} + \vec{w}.
$$

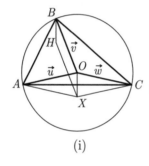
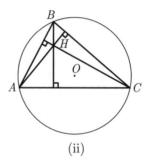

(i)   (ii)

FIGURE 133.

Let H be the fourth vertex of a parallelogram $BOXH$. Then

$$
\overrightarrow{OH} = \overrightarrow{OB} + \overrightarrow{BH} = \overrightarrow{OB} + \overrightarrow{OX} = \vec{v} + \vec{u} + \vec{w}.
$$

Since $\overline{OX}$ is perpendicular to $\overline{AC}$, $\overrightarrow{BH}$ is also perpendicular to $\overrightarrow{AC}$. Furthermore, the symmetry of the representation of $\overrightarrow{OH}$ implies that $\overrightarrow{AH}$ is perpendicular to $\overrightarrow{BC}$ and $\overrightarrow{CH}$ is perpendicular to $\overrightarrow{AB}$. Thus, H is a point of concurrency of the altitudes of $\triangle ABC$, as illustrated in Figure 133(ii).

11.16 Assume that the vertices $A_1, A_2, \ldots, A_n$ are ordered so that the interiors of $\angle XOA_1$ and $\angle A_1OA_2$ do not overlap. If α_1 is the angle between $\overrightarrow{XO}$ and $\overrightarrow{OA_1}$, it follows that $\alpha_{i+1} = \alpha_1 + \frac{2\pi}{n}i$ is the angle between $\overrightarrow{XO}$ and $\overrightarrow{OA_{i+1}}$.

Recall from Example 67 that

$$
\sum_{i=1}^{n} \cos\left(\alpha_1 + \frac{2\pi}{n}i\right) = 0.
$$

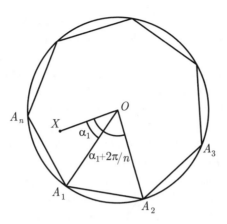

FIGURE 134.

Therefore, $(A_1X)^2 + (A_2X)^2 + \cdots + (A_nX)^2$

$$= |\overrightarrow{XA_1}|^2 + |\overrightarrow{XA_2}|^2 + \cdots + |\overrightarrow{XA_n}|^2$$

$$= |\overrightarrow{XO} + \overrightarrow{OA_1}|^2 + |\overrightarrow{XO} + \overrightarrow{OA_2}|^2 + \cdots + |\overrightarrow{XO} + \overrightarrow{OA_n}|^2$$

$$= \sum_{i=1}^{n} \left(|\overrightarrow{XO}|^2 + 2|\overrightarrow{XO}||\overrightarrow{OA_i}| \cos \alpha_i + |\overrightarrow{OA_i}|^2 \right)$$

$$= n\,d^2 + 2(nR)(0) + n\,R^2 = n(R^2 + d^2).$$

11.17 Let $\vec{u} = \overrightarrow{AB}$ and $\vec{w} = \overrightarrow{AC}$. For an arbitrary point X in the plane, define $\vec{a} = \overrightarrow{XA}$, $\vec{b} = \overrightarrow{XB}$, and $\vec{c} = \overrightarrow{XC}$, as in Figure 135(i). Let D correspond to the terminal point of $\vec{u} + \vec{w}$, creating the parallelogram $ABDC$, and let $\theta = m\angle DAX$.

We need to minimize $f(\theta) = \vec{a} \cdot \vec{a} + \vec{b} \cdot \vec{b} + \vec{c} \cdot \vec{c}$, which we calculate next.

$$f(\theta) = \vec{a} \cdot \vec{a} + (\vec{a} + \vec{u}) \cdot (\vec{a} + \vec{u}) + (\vec{a} + \vec{w}) \cdot (\vec{a} + \vec{w})$$

$$= 3\vec{a} \cdot \vec{a} + \vec{u} \cdot \vec{u} + \vec{w} \cdot \vec{w} + 2\vec{a} \cdot \vec{u} + 2\vec{a} \cdot \vec{w}$$

$$= 3(XA)^2 + (AB)^2 + (AC)^2 + 2\vec{a} \cdot (\vec{u} + \vec{w})$$

$$= (AB)^2 + (AC)^2 + 3(AX)^2 + 2(AX)(AD) \cos(180° - \theta)$$

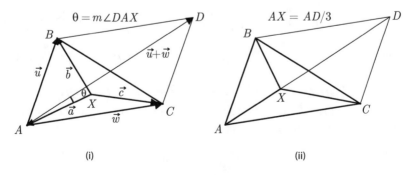

FIGURE 135.

From here, we note $f(\theta)$ will be minimized when $\cos(180° - \theta)$ takes on its minimum value of -1, i.e., when $\theta = 0°$. In this case, $\vec{a}$ and $\overrightarrow{AD}$ will be parallel, so $\vec{a} = k\overrightarrow{AD}$ for some scalar k. We then need to find k such that $(AB)^2 + (AC)^2 + 3(kAD)^2 - 2(kAD)(AD)$ is minimized. Since the quantity $3k^2 - 2k$ attains its minimum value at $k = 1/3$, we see that the original sum of squares will achieve its minimum value when $A - X - D$, where $AX = AD/3$. This choice of X is shown in Figure 135(ii); it is the centroid of $\triangle ABC$, since we have previously observed that the centroid lies $1/3$ of the way from a vertex to the midpoint of the opposite side and the diagonals of a parallelogram bisect each other.

11.18 Given quadrilateral $ABCD$, let $\vec{a} = \overrightarrow{OA}$, $\vec{b} = \overrightarrow{OB}$, $\vec{c} = \overrightarrow{OC}$, and $\vec{d} = \overrightarrow{OD}$, where O is an arbitrary point in the plane (You may think of O as the origin if you wish, but that is not necessary.). The midpoints of diagonals AC and BD can then be represented by the vectors $\frac{1}{2}(\vec{a} + \vec{c})$ and $\frac{1}{2}(\vec{b} + \vec{d})$, respectively. The length x of the segment joining these midpoints will then equal $|\frac{1}{2}(\vec{b} + \vec{d}) - \frac{1}{2}(\vec{a} + \vec{c})|$, and we see that it will be sufficient to find

$$\left(\vec{b} + \vec{d} - \vec{a} - \vec{c}\right) \cdot \left(\vec{b} + \vec{d} - \vec{a} - \vec{c}\right). \tag{17.22}$$

Our task is to express this dot product in terms of the lengths of the sides and diagonals of the quadrilateral. Notice that the squares of the side lengths will appear once we expand (17.22), which is promising. For example, $(AB)^2 = (\vec{b} - \vec{a}) \cdot (\vec{b} - \vec{a})$. Now, if we already knew the answer, we could simply manipulate both sides of the alleged identity to show they were, in fact, equal. Here we will show, though, how to actually find the proper simplification of (17.22). The vectors representing the sides of the quadrilateral will be $\vec{b} - \vec{a}, \vec{b} - \vec{c}, \vec{c} - \vec{d}$, and $\vec{d} - \vec{a}$, so the dot product of each with itself will give the square of the appropriate side. Algebraically, we have

$$(AB)^2 + (BC)^2 + (CD)^2 + (AD)^2 = (\vec{b} - \vec{a}) \cdot (\vec{b} - \vec{a})$$
$$+(\vec{b} - \vec{c}) \cdot (\vec{b} - \vec{c}) + (\vec{c} - \vec{d}) \cdot (\vec{c} - \vec{d}) + (\vec{d} - \vec{a}) \cdot (\vec{d} - \vec{a}). \tag{17.23}$$

Now, by comparing (17.23) with the terms one gets when expanding (17.22), we see that we have duplicated some terms and we have left out some terms. Specifically, in (17.23), $\vec{a} \cdot \vec{a}, \vec{b} \cdot \vec{b}, \vec{c} \cdot \vec{c}$, and $\vec{d} \cdot \vec{d}$ are additional terms, while $2\vec{b} \cdot \vec{d}$ and $2\vec{a} \cdot \vec{c}$ are missing. Thus, if we subtract

$$\vec{a} \cdot \vec{a} + \vec{b} \cdot \vec{b} + \vec{c} \cdot \vec{c} + \vec{d} \cdot \vec{d} - 2\vec{b} \cdot \vec{d} - 2\vec{a} \cdot \vec{c}$$

from (17.23), the resulting expression will equal (17.22). But this subtracted expression simplifies to

$$(\vec{a} - \vec{c}) \cdot (\vec{a} - \vec{c}) + \left(\vec{b} - \vec{d}\right) \cdot \left(\vec{b} - \vec{d}\right) = (AC)^2 + (BD)^2 \ !$$

Therefore, $x^2 = \frac{1}{4}\left((AB)^2 + (BC)^2 + (CD)^2 + (AD)^2 - (AC)^2 - (BD)^2\right)$, and x is readily found.

11.19 Let $\vec{x} = (\text{Area}(\triangle BOC))\overrightarrow{OA} + (\text{Area}(\triangle AOC))\overrightarrow{OB} + (\text{Area}(\triangle AOB))\overrightarrow{OC}$. In order to prove that $\vec{x} = \vec{0}$, it is sufficient (by Theorem 11.7) that $\vec{x} \cdot \overrightarrow{OA} = \vec{x} \cdot \overrightarrow{OB} = 0$.

We introduce the following notation:

$$\overrightarrow{OA} := \vec{a} \text{ with } |\vec{a}| := a, \ \overrightarrow{OB} := \vec{b} \text{ with } |\vec{b}| := b, \text{ and } \overrightarrow{OC} := \vec{c} \text{ with } |\vec{c}| := c;$$

$$\alpha := m\angle BOC, \ \beta := m\angle AOC, \text{ and } \gamma := m\angle AOB.$$

Then, using $\alpha + \beta + \gamma = 360°$, we obtain:

$$\vec{x} \cdot \vec{a} = (1/2) \left(bc \sin\alpha\, \vec{a} + ac \sin\beta\, \vec{b} + ab \sin\gamma\, \vec{c} \right) \cdot \vec{a}$$

$$= (1/2) \left(bc \sin\alpha\, ||\vec{a}||^2 + (ac \sin\beta)(ab\cos\gamma) + (ab\sin\gamma)(ac\cos\beta) \right)$$

$$= (1/2)a^2bc\, (\sin\alpha + \sin\beta\cos\gamma + \sin\gamma\cos\beta)$$

$$= (1/2)a^2bc\, (\sin\alpha + \sin(\beta+\gamma)) = (1/2)a^2bc\, (\sin\alpha + \sin(360° - \alpha))$$

$$= (1/2)a^2bc\, (\sin\alpha - \sin(\alpha)) = 0.$$

Similarly we can show that $\vec{x} \cdot \vec{b} = 0$. Since $\vec{x} \cdot \vec{a} = 0$ and $\vec{x} \cdot \vec{b} = 0$, but $\vec{a}$ and $\vec{b}$ are not parallel, $\vec{x} = \vec{0}$ by Theorem 11.7.

Affine Transformations

12.1 Suppose $h_1 = g \circ f^{-1}$ and h_2 are both affine transformations sending $\vec{p}, \vec{q},$ and $\vec{r}$ to $\vec{p}', \vec{q}',$ and $\vec{r}'$, respectively. We need to show $h_1 = h_2$. Now, $h_1 \circ f$ and $h_2 \circ f$ are both affine transformations mapping $\{\vec{0}, \vec{i}, \vec{j}\}$ to $\{\vec{p}', \vec{q}', \vec{r}'\}$. But since an affine transformation is uniquely determined by where it maps $\vec{0}, \vec{i},$ and $\vec{j}$, $h_1 \circ f = h_2 \circ f$. Composing each of these compositions on the right with f^{-1}, we obtain $h_1 = h_2$, which proves the uniqueness of the composition h_1.

12.2 No. Every affine transformation will preserve the ratio of lengths of the parallel sides. So if these ratios are different in two trapezoids, then they are not affine equivalent (i.e., there is no affine transformation mapping one to the other).

12.3 Let $ABCD$ be a trapezoid with bases AD and BC. Let P be the intersection of $\overleftrightarrow{AB}$ and $\overleftrightarrow{CD}$, and let Q be the intersection of the diagonals, $\overline{AC}$ and $\overline{BD}$. By Corollary 12.9, there is an affine transformation f mapping $\triangle APD$ to an isosceles triangle $A'P'D'$. (See Figure 136.)

Because f maps line segments to line segments, preserves the property of parallelism among line segments, and preserves ratios of lengths of collinear segments (all by Theorem 12.7), $ABCD$ is mapped to an isosceles trapezoid, $A'B'C'D'$, with $A'B' = C'D'$. It is easy to see that $\overleftrightarrow{P'Q'}$ bisects the bases of $A'B'C'D'$; we leave the details to the reader. We conclude that $\overleftrightarrow{PQ}$ bisects the bases of $ABCD$ also.

This problem also appeared as Problem 3.3.14 and as Problem 8.12.

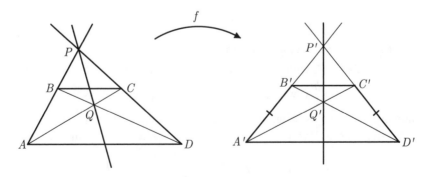

FIGURE 136.

12.4 Assume that $\overline{AB}$ and $\overline{CD}$ are chords of an ellipse, $\mathcal{E}$, having the property that the area determined by $\overline{AB}$ and the boundary of $\mathcal{E}$ is the same as the area determined by $\overline{CD}$ and the boundary of $\mathcal{E}$. Let f be an affine transformation mapping $\mathcal{E}$ to a circle, $\mathcal{C} = \mathcal{C}(O', r)$. (Recall that $O' = f(O)$, where O is the center of $\mathcal{E}$.) The chords of $\mathcal{E}$ are mapped by f to chords of $\mathcal{C}$; let A' designate the point on $\mathcal{C}$ that corresponds to $f(A)$ and likewise for B', C', and D'.

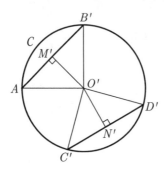

FIGURE 137.

By Theorem 12.7(7), the area of the region bounded by the chord $A'B'$ and the boundary of $\mathcal{C}$ will be equal to the area of the region bounded by the chord $C'D'$ and the boundary of $\mathcal{C}$. This implies that $\text{Area}(\triangle A'B'O') = \text{Area}(\triangle C'D'O')$ and that $A'B' = C'D'$. Let M' and N' be the midpoints of $\overline{A'B'}$ and $\overline{C'D'}$, respectively. By Theorem 4.6 of Chapter 4, $\overline{O'M'} \perp \overline{A'B'}$ and $\overline{O'N'} \perp \overline{C'D'}$. Since $\text{Area}(\triangle A'B'O') = \text{Area}(\triangle C'D'O')$ and $A'B' = C'D'$, we conclude that $O'M' = O'N'$. Hence, $\overline{A'B'}$ and $\overline{C'D'}$ are both tangent to the circle $\mathcal{C}(O', O'M')$. The pre-image of $\mathcal{C}(O', O'M')$ is an ellipse having center at O, which is tangent to $\overline{AB}$ and $\overline{CD}$.

12.5 Let $A_1, \ldots, A_n$ be the vertices of an n-gon, and $A_1', \ldots, A_n'$ be their images under an affine transformation f. Let G be the centroid of $\{A_1, \ldots, A_n\}$. Then by Example 65,

$$\sum \overrightarrow{GA_i} = \vec{0}.$$

Let $G' = f(G)$.

Suppose f is defined by $x \mapsto x' = \mathbf{A}(\vec{x}) + (\vec{b})$. Then $f(\overrightarrow{OA_i}) = \mathbf{A}(\overrightarrow{OA_i}) + (\vec{b})$, and $f(\overrightarrow{OG}) = \mathbf{A}\overrightarrow{OG} + \vec{b}$, so

$$\overrightarrow{G'A_i'} = f(\overrightarrow{GA_i}) = \mathbf{A}\overrightarrow{GA_i}.$$

This implies (by a generalization of Theorem 12.2(5)) that

$$\sum \overrightarrow{G'A_i'} = \vec{0},$$

which is equivalent to G' being the centroid of $\{A_1', \ldots, A_n'\}$.

12.6 Consider a parabola, $\mathcal{P}$, with vertex at (h, k). As was the case with ellipse and hyperbola, the translation $\vec{x} + \begin{bmatrix} -h \\ -k \end{bmatrix}$ will map $\mathcal{P}$ to a parabola whose vertex is the origin. Then apply a rotation so that the axis of the parabola aligns with the positive y-axis. Under these two affine transformations, $\mathcal{P}$ is mapped to a parabola $\mathcal{P}'$, represented by an equation of the form

$$y = ax^2,$$

where a is a positive real number.

Now apply a third affine transformation, $f(\vec{x}) = \begin{bmatrix} 1/a & 0 \\ 0 & 1/a \end{bmatrix} \begin{bmatrix} x \\ y \end{bmatrix} = \begin{bmatrix} x/a \\ y/a \end{bmatrix} = \begin{bmatrix} x' \\ y' \end{bmatrix}$.

The equation $y = ax^2$ can now be written $(y'/a) = a(x'/a)^2$, or $y' = (x')^2$. This proves the theorem.

Note that this establishes as a corollary that any two parabolas, $\mathcal{P}_1$ and $\mathcal{P}_2$, are affine equivalent. The proof is similar to that of Corollary 12.14.

12.7 Let A_2, B_2, and C_2 be the points of intersection of $\overline{BB}_1$ and $\overline{CC}_1$, $\overline{AA}_1$ and $\overline{CC}_1$, and $\overline{AA}_1$ and $\overline{BB}_1$, respectively. Let f be an affine transformation mapping $\triangle ABC$ to an equilateral triangle DEF, as illustrated in Figure 138. Because affine transformations preserve ratios of parallel line segments, the points $D_1 = f(A_1)$, $E_1 = f(B_1)$, and $F_1 = f(C_1)$ will divide the sides of $\triangle DEF$ in the same fixed ratio as A_1, B_1, and C_1 divide the sides of $\triangle ABC$. Consequently, $\triangle F_1ED_1 \cong \triangle D_1FE_1 \cong \triangle E_1DF_1$ by SAS, so $\triangle D_1E_1F_1$ is equilateral.

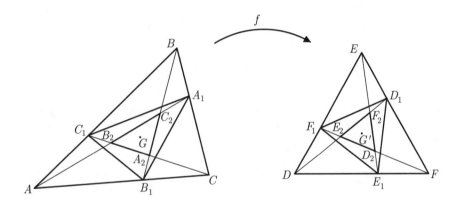

FIGURE 138.

Let D_2, E_2, and F_2 be the points of intersections of the segments EE_1 and FF_1, FF_1 and DD_1, and DD_1 and EE_1, respectively. Rotating $\triangle DEF$ clockwise by $120°$ around the centroid, G', of $\triangle DEF$ (as we did in Example 76) we see that $D_1 \mapsto E_1 \mapsto F_1 \mapsto D_1$, where "$\mapsto$" means "is mapped to." This implies that $\overline{DD}_1 \mapsto \overline{EE}_1 \mapsto \overline{FF}_1 \mapsto \overline{DD}_1$, and therefore $D_2 \mapsto E_2 \mapsto F_2 \mapsto D_2$. This proves that $\triangle D_2E_2F_2$ is equilateral.

Furthermore, under the $120°$ rotation about O', the equilateral triangle $D_1E_1F_1$ is mapped to itself, as is the equilateral triangle $D_2E_2F_2$. This will only happen if G' is the centroid of both $\triangle D_1E_1F_1$ and $\triangle D_2E_2F_2$. Since any affine transformation maps the centroid of one triangle to the centroid of another, the pre-image of G', $f^{-1}(G') = G$, is the centroid of $\triangle ABC$, $\triangle A_1B_1C_1$, and $\triangle A_1B_2C_2$.

12.8 By Theorem 12.7, collinear segments change their lengths under an affine transformation by the same ratio $k > 0$, so any affine transformation preserves the equality we seek to prove. Let f be an affine transformation mapping the parallelogram $MNPQ$ to a square $M'N'P'Q'$ having side length $a = 1$. (See Figure 139.) We now wish to prove that $1/M'R' + 1/M'S' = 1/M'T'$.

Let $m\angle Q'M'T' = \alpha$. Then, from the right triangles $M'N'R'$ and $M'Q'S'$, respectively, we see that

$$\frac{1}{M'R'} = \cos \angle N'M'R' = \cos(90° - \alpha) = \sin \alpha \quad \text{and} \quad \frac{1}{M'S'} = \cos \alpha.$$

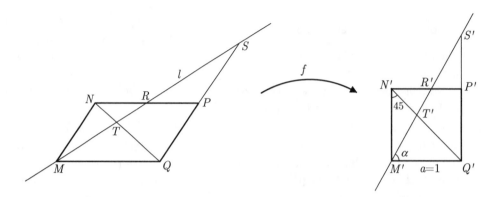

FIGURE 139.

In addition, note that $m\angle M'N'Q' = 45°$ since $\overline{N'Q'}$ is the diagonal of a square. Applying the Sine theorem to $\triangle M'T'N'$,

$$\frac{1}{\sin(45° + \alpha)} = \frac{M'T'}{\sin 45°}, \quad \text{so} \quad \frac{1}{M'T'} = \frac{\sin(45° + \alpha)}{\sin 45°}.$$

Thus, verifying that $1/M'R' + 1/M'S' = 1/M'T'$ reduces to checking that $\sin(45° + \alpha)/\sin 45° = \cos\alpha + \sin\alpha$. The validity of the latter equation follows from the sine sum formula:

$$\frac{\sin(45° + \alpha)}{\sin 45°} = \frac{(\sin 45°)(\cos\alpha) + (\cos 45°)(\sin\alpha)}{\sin 45°} = \cos\alpha + \sin\alpha.$$

An alternate solution is provided in Problem 3.3.12.

12.9 Suppose $\mathcal{E}$ is an ellipse given by the equation $x^2/a^2 + y^2/b^2 = 1$; then $\mathcal{E}$ has semi-axes of lengths a and b, as shown in Figure 140(i). Construct a rectangle, R, with center at the center of $\mathcal{E}$ and with sides parallel to the axes of $\mathcal{E}$. Hence, the side lengths of R are $2a$ and $2b$.

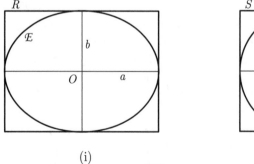
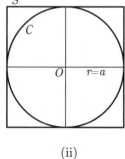

(i)   (ii)

FIGURE 140.

Consider the affine transformation $f(\vec{p}) = \mathbf{A}\vec{p}$, where $\mathbf{A} = \begin{bmatrix} 1 & 0 \\ 0 & a/b \end{bmatrix}$. Then f maps a vector $\langle x, y \rangle$ to $\langle x', y' \rangle = \langle x, ay/b \rangle$. Observe that the origin is mapped to the origin, and the coordinate axes are mapped to themselves. Now $(x')^2 + (y')^2 = x^2 + a^2 y^2/b^2 = a^2$, so $\mathcal{E}$ is mapped to a circle with radius a, as shown in Figure 140(ii). Obviously, $f(R)$ is a parallelogram that circumscribes the circle and has sides parallel to the coordinate axes. Therefore $f(R)$ is a square with side length $2a$.

Since an affine transformation preserves the ratio of areas of two figures,

$$\frac{\text{Area } \mathcal{E}}{\text{Area } R} = \frac{\text{Area } \mathcal{C}}{\text{Area } S} \implies \frac{\text{Area } \mathcal{E}}{4ab} = \frac{\pi a^2}{4a^2}.$$

From here, it follows that the area of $\mathcal{E}$ is πab.

12.10 Let $ABCD$ be a parallelogram with inscribed ellipse $\mathcal{E}$ tangent to the sides $AB, BC, CD,$ and DA at the points $P, Q, R,$ and S, respectively. Let f be an affine transformation mapping $\mathcal{E}$ to a circle $\mathcal{C} = \mathcal{C}(O', r)$. Under f, $ABCD$ is mapped to a parallelogram $A'B'C'D'$, which is tangent to the sides $A'B', B'C', C'D',$ and $D'A'$ of $\mathcal{C}$ at points $P', Q', R',$ and S', respectively. (See Figure 141.)

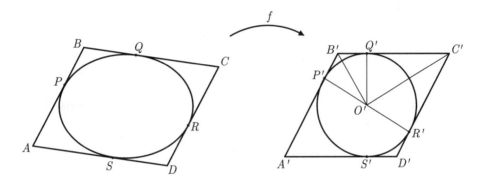

FIGURE 141.

Since the ratio of lengths of parallel segments is preserved, it is sufficient to show that $C'Q'/Q'B' = C'R'/B'P'$. As tangent segments to a circle from a point are congruent, $B'P' = B'Q'$ and $C'R' = C'Q'$. This implies the required equality of the ratios.

Remark. As $A'B' + C'D' = B'C' + D'A'$, the parallelogram $A'B'C'D'$ is a rhombus, but that fact is not used in the solution above.

12.11 Since affine transformations preserve the ratio of areas, we may assume that $\triangle ABC$ is a right triangle. Furthermore, we will choose a coordinatization such that $C : (0, 0)$, $A_1 : (1, 0)$, $B : (\alpha + 1, 0)$, $B_1 : (0, \beta)$, and $A : (0, \beta + 1)$, with intersections as shown in Figure 142.

We first find the coordinates, (x_{C_1}, y_{C_1}), of C_1. By similar triangles,

$$\frac{\gamma}{x_{C_1}} = \frac{\gamma + 1}{\alpha + 1} \quad \text{and} \quad \frac{y_{C_1}}{1} = \frac{\beta + 1}{\gamma + 1}.$$

Therefore,

$$x_{C_1} = \gamma \frac{\alpha + 1}{\gamma + 1} \quad \text{and} \quad y_{C_1} = \frac{\beta + 1}{\gamma + 1}.$$

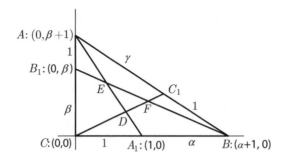

FIGURE 142.

We then have the following equations of lines (details omitted):

$$\overleftrightarrow{AA_1} : x + \frac{y}{\beta + 1} = 1,$$

$$\overleftrightarrow{BB_1} : \frac{x}{\alpha + 1} + \frac{y}{\beta} = 1,$$

$$\overleftrightarrow{CC_1} : y = \frac{(\beta + 1)x}{\gamma(\alpha + 1)}.$$

Solving the three systems of equations corresponding to the pairwise intersections of the lines above (details omitted), we obtain :

$$D : \left(\frac{\gamma(\alpha + 1)}{\gamma + \alpha\gamma + 1}, \frac{\beta + 1}{\gamma + \alpha\gamma + 1} \right),$$

$$E : \left(\frac{\alpha + 1}{\alpha + \alpha\beta + 1}, \frac{\alpha\beta(\beta + 1)}{\alpha + \alpha\beta + 1} \right),$$

$$F : \left(\frac{\beta\gamma(\alpha + 1)}{\beta + \beta\gamma + 1}, \frac{\beta(\beta + 1)}{\beta + \beta\gamma + 1} \right).$$

Now, to find the area of $\triangle DEF$, we use the results of Problem 8.14, where the area of a triangle is expressed in terms of the coordinates of its vertices. Again, we omit the calculations (if you wish, use a computer!):

$$\text{Area}(\triangle DEF) = \frac{1}{2} | - x_E y_D - x_D y_F - x_F y_E + x_D y_E + x_F y_D + x_E y_F |$$

$$= \frac{(\alpha\beta\gamma - 1)^2(\alpha + 1)(\beta + 1)}{2(\alpha + \alpha\beta + 1)(\alpha + \alpha\gamma + 1)(\beta + \beta\gamma + 1)}.$$

Since the area of the (right) triangle ABC is $(\alpha + 1)(\beta + 1)/2$, we have

$$\frac{\text{Area}(\triangle DEF)}{\text{Area}(\triangle ABC)} = \frac{(\alpha\beta\gamma - 1)^2}{(\alpha + \alpha\beta + 1)(\alpha + \alpha\gamma + 1)(\beta + \beta\gamma + 1)}.$$

Note that when $\alpha = \beta = \gamma = 2$, this ratio simplifies to $1/7$, which is the value found in Example 76.

Also, note that our calculation of the area of $\triangle DEF$ gives us yet another proof of Ceva's theorem (see Theorem 3.17, Problem 5.17, Problem 7.23, and Problem 8.16), since segments

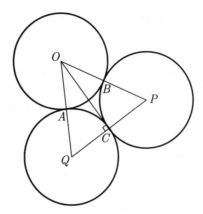

FIGURE 143.

BB_1 and CC_1 will be concurrent if and only if the area of $\triangle DEF$ is 0. Obviously, this is the case if and only if $\alpha\beta\gamma = 1$.

12.12 Let $\mathcal{E}$ be one of the three congruent and similarly oriented ellipses. Let f be an affine transformation that maps $\mathcal{E}$ to a circle, $\mathcal{C}$. By Theorem 12.7, each of the other three ellipses will be mapped to a circle that is congruent to $\mathcal{C}$ and the three circles will be externally tangent in pairs, as the three ellipses were. Assume that the three circles have centers O, P, and Q, and radius r; let the points of tangency be A, B, and C. (See Figure 143.)

The area of the curvilinear triangle ABC can be found by subtracting the area of the three congruent circular sectors from the area of the triangle OPQ:

$$\frac{1}{2}(OC)(QP) - 3\left(\frac{\pi}{6}r^2\right) = \frac{1}{2}(2r)(\sqrt{3}r) - \frac{\pi}{2}r^2 = r^2\left(\sqrt{3} - \frac{\pi}{2}\right).$$

This area is constant, determined only by the radii of the three congruent, externally tangent, circles. Since an affine transformation preserves ratios of areas, the area of the original curvilinear triangle bounded by the three congruent ellipses must also be independent of the position of the three ellipses. Furthermore, the ratio of the area of the elliptical curvilinear triangle to the area of the circular curvilinear triangle must be the same as the ratio of the area of one of the ellipses to one of the circles – which is $ab : r^2$, by Problem 12.9. In order to attain this ratio, the area of the elliptical curvilinear triangle must be $ab(\sqrt{3} - \pi/2)$.

12.13 Consider an affine transformation f that maps the given ellipse, $\mathcal{E}$, to the unit circle, which we denote $\mathcal{C}$. Then f will map the inscribed triangle T to a triangle T' inscribed in $\mathcal{C}$. Furthermore, if T' is a triangle with maximum area in $\mathcal{C}$, by Theorem 12.7(7), T will be a triangle with maximum area in the original ellipse. Now, since f will map the centroid of T to the centroid of T', it remains for us to show that the centroid of a T' with maximal area coincides with the center of the unit circle. This is true since T' must be equilateral. See the solution of Problem 5.28, the related discussion in [49] mentioned in the footnote, and the solution of Problem 7.33.

We now turn to the general case of an n-gon inscribed in an ellipse. What we present below is *not* a rigorous argument, but a very natural one which is often suggested by those who are unaware of the hidden problem with this kind of reasoning. We show that a non-regular n-gon $\mathcal{P}'$ inscribed in a circle does not have the maximum area. Moreover, we describe a way in which $\mathcal{P}'$ can be modified to create another inscribed n-gon with a greater area. This solution becomes rigorous if one can explain that an inscribed n-gon of the maximum area exists, but we do not know how to do this in a relatively short manner and without applying some facts from Real Analysis that many students may not know. See the solution of Problem 5.29, the related discussion in [49] mentioned in the footnote, and the solution of Problem 7.33. The solution we gave for Problem 5.29 and the discussion in [49] circumvent the necessity for such a proof by a direct comparison of the areas of $\mathcal{P}$ with the area of the regular inscribed n-gon. Here is our promised nonrigorous argument.

Consider a convex n-gon, $\mathcal{P}' = A_1 A_2 A_3 \ldots A_n$, inscribed in the unit circle, and assume that the n-gon is not regular. Then there must be a set of three consecutive vertices such that the pair of side lengths with endpoints at these vertices are not congruent; assume that $A_1 A_2 \neq A_2 A_3$, so that the three consecutive vertices are A_1, A_2, and A_3. The area of $A_1 A_2 A_3 \ldots A_n$ is the area of the $(n-1)$-gon $A_1 A_3 \ldots A_n$ together with the area of $\triangle A_1 A_2 A_3$. The area of $\triangle A_1 A_2 A_3$ is $(A_1 A_3)(A_2 X)/2$, where X is the foot of the altitude from A_2. For a fixed value of $A_1 A_3$, the area of the triangle is maximized when $A_2 X$ is maximized, which occurs when $\overline{A_2 X}$ bisects $\overline{A_1 A_3}$. Since this is not the case for $\triangle A_1 A_2 A_3$, we conclude that $\mathcal{P}'$ does not have maximum area among convex n-gons that are inscribed in the unit circle. That is, if $\mathcal{P}'$ is not regular, $\mathcal{P}'$ does not have maximum area.

Inversions

13.1 Consider the circle $\mathcal{C}(O, r)$ with chord $\overline{AB}$ having midpoint M, as shown in Figure 144. Clearly segment OC bisects chord AB. Hence, we are assured that points O, M, and C are collinear.

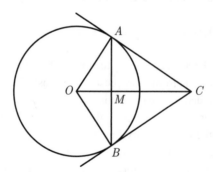

FIGURE 144.

Since $\triangle OAC$ is a right triangle and $\overline{AM}$ is an altitude to its hypotenuse, by Theorem 3.19(2), $AM^2 = (OM)(MC)$. Therefore,

$$OA^2 = AM^2 + OM^2 = (OM)(MC) + OM^2 = (OM)(OC).$$

But $OA = r$, which is the radius of inversion. Hence, $r^2 = (OM)(OC)$. This assures us that $I(M) = C$.

13.2 Let $I = I(O, r)$, and let line OM intersect $\mathcal{C}$ along the diameter $\overline{AB}$. Then line OM intersects $\mathcal{C}'$ along the diameter $\overline{A'B'}$. Using Theorem 13.1(3) we obtain

$$A'M' = \frac{r^2}{OA \cdot OM} AM, \quad \text{and} \quad B'M' = \frac{r^2}{OB \cdot OM} BM.$$

If M' is the center of C', then $A'M' = B'M'$. Since $AM = BM$, this implies that $OA = OB$. As points A and B are distinct, this would imply that $M = O$, a contradiction.

13.3 Let $P : (x, y)$ be distinct from the origin, and let $P' = I(P) : (x', y')$. Then $OP' = \lambda OP$ for some $\lambda > 0$, and $(x', y') = (\lambda x, \lambda y)$. As $OP \cdot OP' = 1^2$, we obtain $\lambda OP^2 = 1$. Hence, $\lambda = \frac{1}{OP^2} = \frac{1}{x^2+y^2}$, so

$$x' = \frac{x}{x^2 + y^2} \quad \text{and} \quad y' = \frac{y}{x^2 + y^2}.$$

13.4 Let $I = I(O, r)$ denote the inversion.

Case 1: Suppose a circle C is tangent to a line l at point A. Consider the following four sub-cases: $O = A$; $O \neq A$ and $O \in l$; $O \neq A$ and $O \in C$; and O is neither on l nor on C. Since the measure of the angle between a circle and a line tangent to it is zero, we will show that the measure of the angle between the images of the circle and its tangent line is also zero.

If $O = A$, then $I(l) = l$ and $I(C)$ is a line parallel to l. As the measure of the angle between l and C is zero, and the measure of the angle between parallel lines is zero, the statement is proven.

In the remaining three sub-cases, $I(l)$ and $I(C)$ are either a line and a circle sharing only one point $A' = I(A)$ or two circles sharing only one point $A' = I(A)$. Thus, in each case, the images will share exactly one point, which is necessarily a point of tangency (because at least one of the images is a circle). Therefore, the measure of the angle between the images is zero.

Case 2: Suppose now that l intersects C at two points, A and B. Let t be the line tangent to C at A. We have already established in Case 1 that the measure of the angle between t and C is preserved. Therefore, we simply need to verify that the measure of the angle between lines t and l is preserved under any inversion. This was done in the proof following the statement of Theorem 13.3.

13.5 Consider an inversion I with center A. We then obtain an image like the one shown in Figure 145.

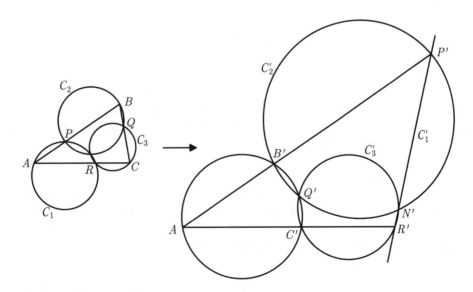

FIGURE 145.

Let N' be the point of intersection of circles C_2' and C_3'. We want to show that $N' \in \overleftrightarrow{P'R'} = C_1'$, which is equivalent to showing that $m\angle P'N'Q' + m\angle Q'N'R' = 180°$.

Since quadrilateral $B'P'N'Q'$ is inscribed in circle C_2', we know that $m\angle P'N'Q' + m\angle P'B'Q' = 180°$. Let $\alpha = m\angle P'N'Q'$. Then $m\angle P'B'Q' = 180° - \alpha$. Since $\angle P'B'A$ is a straight angle, $m\angle Q'B'A = \alpha$.

However, quadrilateral $AB'Q'C'$ is also an inscribed quadrilateral in circle $I(\overleftrightarrow{BC})$. Hence, $m\angle AB'Q' + m\angle Q'C'A = 180°$, $m\angle Q'C'A = 180 - \alpha$, and $m\angle Q'C'R' = \alpha$.

Finally, quadrilateral $Q'C'R'N'$ is inscribed in circle C_3, giving $m\angle Q'C'R' + m\angle R'N'Q' = 180°$ and $m\angle R'N'Q' = 180° - \alpha$.

We have now established that $m\angle P'N'Q' + m\angle Q'N'R' = 180°$, making $\angle P'N'R'$ a straight angle. Hence, N' lies on segment $P'R'$ and we see that C_1', C_2', and C_3' all intersect at the single point N'. Since concurrency is preserved under an inversion of the plane, C_1, C_2, and C_3 are concurrent at the preimage of N'.

13.6 Consider an inversion of the plane with center A_1. Since A_2, A_3, and A_4 lie on a circle passing through the center of inversion, it follows that A_2', A_3', and A_4' lie on a straight line. (See Figure 146.)

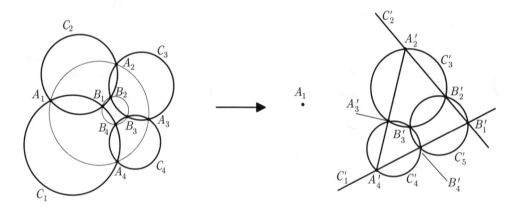

FIGURE 146.

Consider a circle C_5' passing through B_1', B_2', and B_3'. We know that B_4' is the intersection of C_1' and C_4'. Suppose C_5' intersects C_4' at Q. Then by the same argument used in Problem 13.5, A_4', Q, and B_1' are collinear. Hence, $Q = B_4'$, making B_1', B_2', B_3', and B_4' all lie on C_5'. By Theorem 13.1, we know that B_1, B_2, B_3, and B_4 are cocyclic.

Remarks.

(a) If A_1, A_2, A_3, and A_4 are collinear rather than cyclic, the above argument is still valid.
(b) In the case where C_5' happens to pass through the center of inversion, A_1, the points B_1, B_2, B_3, and B_4 will be collinear rather than cyclic.
(c) The converse of the problem statement is also true. Simply choose B_1 as the center of inversion and use an argument similar to the one above.

13.7 Let $\overrightarrow{OK}$ be tangent to C at K. Since tangency is preserved under inversion, $\overrightarrow{OK}$ is tangent to C' at K'. Let Q' be the center of C'. Then O, Q', and P are collinear, while $\overline{Q'K'}$ and $\overline{PK}$

are both perpendicular to $\overrightarrow{OK}$, so $\triangle OK'Q' \sim \triangle OKP$. Therefore,

$$\frac{r'}{r} = \frac{OK'}{OK} = \frac{k^2/OK}{OK} = \frac{k^2}{|OP^2 - r^2|}.$$

The result follows.

13.8 *Case 1:* Let $O \in l$. Then, it follows immediately from the definition of an inversion that $I(l) = l$.

Case 2: Let $O \notin l$. Construct $\overleftrightarrow{OK} \perp l$ where $K \in l$. Then for any point D on l we have previously proved that $\triangle OKD \sim \triangle OD'K'$. See Figure 147.

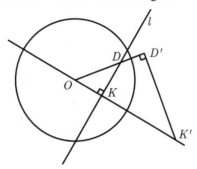

FIGURE 147.

Hence, $m\angle OD'K' = 90°$. Consequently, all points D' lie on the circle whose diameter is $\overline{OK'}$. By continuity, the complete circle will be obtained.

13.9 Choose a Cartesian coordinate system such that the circle of inversion becomes the unit circle centered at the origin. Let $\mathcal{C} : (x - a)^2 + (y - b)^2 = r^2$. Using the result of Problem 13.3, we find an equation of $I(\mathcal{C})$ as

$$\left(\frac{x'}{x'^2 + y'^2}\right)^2 + \left(\frac{y'}{x'^2 + y'^2}\right)^2 - \frac{2ax'}{x'^2 + y'^2} - \frac{2by'}{x'^2 + y'^2} + (a^2 + b^2 - r^2) = 0. \quad (17.24)$$

If $\mathcal{C}$ passes through the origin, $a^2 + b^2 - r^2 = 0$, and (17.24) is equivalent to

$$1 - 2ax' - 2by' = 0.$$

As at least one of a and b is not zero, this is an equation of a line. So we obtain that $I(\mathcal{C})$ is a subset of this line. By continuity, it must be the whole line.

Suppose $\mathcal{C}$ does not pass through the origin. Let $\delta = a^2 + b^2 - r^2$. Then $\delta \neq 0$, and (17.24) is equivalent to

$$x'^2 + y'^2 - \frac{2a}{\delta}x' - \frac{2b}{\delta}y' + \frac{1}{\delta} = 0.$$

Completing the square we obtain

$$\left(x' - \frac{a}{\delta}\right)^2 + \left(y' - \frac{b}{\delta}\right)^2 = \left(\frac{r}{\delta}\right)^2.$$

Hence, $I(\mathcal{C})$ is a subset of a circle. By continuity it must be the whole circle.

13.10 Consider an inversion of the plane with center M and radius r. Then by the Triangle inequality, $A'C' \leq A'B' + B'C'$. Using the change of distance formula, we obtain:

$$\frac{r^2}{(MA)(MC)} AC \leq \frac{r^2}{(MA)(MB)} AB + \frac{r^2}{(MB)(MC)} BC.$$

Therefore,

$$(MB)(AC) \le (MC)(AB) + (MA)(BC)$$
$$\le (MC)(AC) + (MA)(AC) = (MC + MA)(AC).$$

Hence, $MB \le MC + MA$.

Equality will occur if and only if A', B', and C' are collinear and $AB/AC = BC/AC = 1$, which implies that A, B, C, and M are cyclic. Hence, $\triangle ABC$ would have to be an equilateral triangle and M a point on its circumscribed circle.

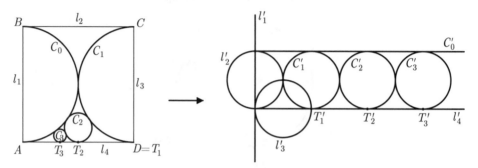

FIGURE 148.

13.11 Consider the inversion $I = I(A, 1)$.

(a) If $l_1 = \overleftrightarrow{AB}$, $l_2 = \overleftrightarrow{BC}$, $l_3 = \overleftrightarrow{CD}$, and $l_4 = \overleftrightarrow{AD}$, then l_1 and l_4 are their own images under I. We see that l_2' is a circle passing through A and orthogonal to l_1' at B'. Similarly, l_3' is a circle passing through A and orthogonal to l_4' at D'. (See Figure 148.)

C_0' is a line orthogonal to l_1' and tangent to l_2' at B'. C_1' is a circle tangent to C_0', l_2', and l_4'. C_2' is a circle tangent to C_0', C_1', and l_4'. Similarly, for $i \ge 3$, C_i' is a circle tangent to C_0', C_{i-1}', and l_4'. Since C_0' and l_4' are parallel lines, circles l_2', C_1', C_2', C_3' ... are all congruent.

T_i' will be the point of tangency of circle C_i' with l_4'. Thus $T_1' = D'$. Since the square has sides of length 1 and the radius of inversion was chosen to be 1, then $D = D'$. Hence $AT_1' = 1$. By the congruence of the circles we see that $AT_i' = i$.

Using the definition of an inversion we have $(AT_i)(AT_i') = r^2$. Therefore, $(AT_i)i = 1$, so $AT_i = 1/i$.

(b) Since $AT_1 = AD = 1$ and $I(T_1) = T_1$, then $AT_1' = 1$. The circles C_i' are congruent to each other, so each has a radius of $r_i' = 1/2$.

Let r_i be the radius of C_i and let Q_i be the center of C_i. Then, applying Problem 13.7 to our circle of inversion,

$$\frac{1}{2} = \frac{(r_i)(1^2)}{|AQ_i^2 - r_i^2|} = \frac{r_i}{AT_i^2} = \frac{r_i}{(1/i)^2} = i^2 r_i.$$

Therefore, $r_i = \dfrac{1}{2i^2}$.

13.12 Let C be the circle containing arc AB and l be the line containing chord AB. Let $I(A, r)$ be an inversion. Since l passes through the center of inversion, l' is a line through A. Since C is a circle that passes through the center of inversion, C' is a line that does not pass through A, but which intersects l' at B'.

Let C_1, C_2, C_3, and C_4 be inscribed, tangent, circles, with A_1, A_2, and A_3, their respective points of tangency. (See Figure 149.)

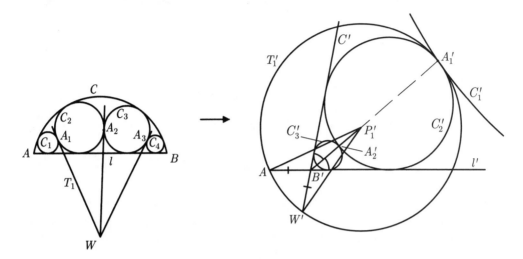

FIGURE 149.

Since none of the inscribed circles pass through the center of inversion, their images are all circles. Also, since any inversion preserves tangency of generalized circles, A'_1, A'_2, and A'_3 are points of tangency of the inverted circles.

Consider the line T_1 which is tangent to C_1 and C_2 at A_1. Clearly T_1 does not pass through A. Hence, its inverse T'_1 is a circle which is tangent to C'_1 and C'_2 at A'_1 and passes through A. This forces the bisector of the angle formed by l' and C' to pass through the center of T'_1.

Let P'_1 be the center of circle T'_1. Select point W' on C' so that $B'A = B'W'$. Then $\triangle AB'P'_1 = \triangle W'B'P'_1$ by SAS, making $AP'_1 = W'P'_1$. Now since T'_1 passes through A, it must also pass through W'.

We can use exactly the same argument for each tangent and show that the image of each tangent passes through W'. Since concurrency is preserved under inversion, we have all of the tangents concurrent at W, the preimage of W'.

Remark. In addition to showing that the tangents are concurrent, we can establish the exact placement of the point of concurrency. We know already that W is some point on C. Furthermore, the points of tangency of the circles lie on a circle S for which S' is the angle bisector pictured in the inversion. Since $\overline{AW'} \perp S'$, the line AW is orthogonal to S at A. If we duplicate the whole inversion procedure using B as the center instead of A, we would get $\overleftrightarrow{BW}$ orthogonal to S' at B. Clearly, W must be the center of circle S. Since the centers of S and C must both fall on the perpendicular bisector of $\overline{AB}$, we conclude that W falls on the intersection of the perpendicular bisector of $\overline{AB}$ and C.

13.13 Let K be the midpoint of $\overline{AC}$ and L be the midpoint of $\overline{BD}$. Consider an inversion I with $C = C(O, r)$ as the circle of inversion. Then by Problem 13.1, the tangents to C at A and C meet at $I(K) = K'$. Likewise, the tangents to C at B and D meet at $I(L) = L'$. Let L_1 be the intersection of lines OL and AC. We will show that $L_1 = I(L) = L'$. (See Figure 150.)

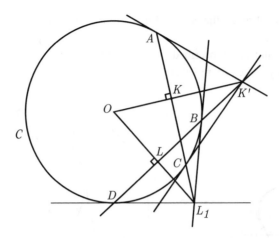

FIGURE 150.

Consider $\triangle OLK'$ and $\triangle OKL_1$. As K and L are the midpoints of chords AC and BD, respectively, $m\angle OLK' = 90° = m\angle OKL_1$ and $m\angle L_1 OK = m\angle LOK'$. Thus, $\triangle OLK' \sim \triangle OKL_1$ by AA.

Using corresponding parts of these triangles we have $\dfrac{OL}{OK'} = \dfrac{OK}{OL_1}$.

Therefore, $(OL)(OL_1) = (OK')(OK) = r^2$, by the definition of an inversion. Hence, $L_1 = I(L) = L'$. Since L_1 lies on $\overleftrightarrow{AC}$ by construction, we have L' on $\overleftrightarrow{AC}$.

13.14 Consider an inversion where $\mathcal{C}_1$ is the circle of inversion. Let K, L, and M be the points of tangency of $\mathcal{C}_1$ with segments AB, BC, and AC, respectively. (See Figure 151(i).)

Since $\overline{BA}$ and $\overline{BC}$ are tangents to the circle of inversion, we know (from Problem 13.1) that B' is the midpoint of $\overline{KL}$. Similarly C' is the midpoint of $\overline{LM}$ and A' is the midpoint of $\overline{MK}$. (See Figure 151(ii).)

As $A'B' = ML/2$, $B'C' = KM/2$ and $A'C' = KL/2$, triangles $A'B'C'$ and MLK are similar by SSS, with coefficient $1/2$. Now $\mathcal{C}_1$ and $\mathcal{C}_2'$ are circumcircles for $\triangle MLK$ and $\triangle A'B'C'$, respectively. If R' is the length of the radius of $\mathcal{C}_2'$, then $R'/r = 1/2$.

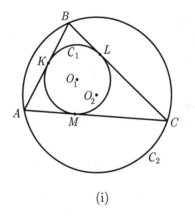

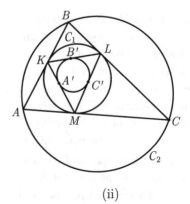

(i)   (ii)

FIGURE 151.

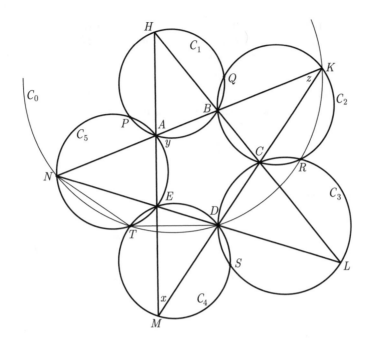

FIGURE **152.**

Applying Problem 13.7 to C_2, with circle of inversion C_1, we obtain

$$\frac{r}{2} = R' = \frac{Rr^2}{|d^2 - R^2|}.$$

Clearly $R > d$ when C_1 is in the interior of C_2, hence

$$\frac{r}{2} = \frac{Rr^2}{d^2 - R^2} \qquad \text{so}$$

$$R^2 - 2Rr = d^2.$$

13.15 Construct circle C_0 circumscribed about $\triangle NKD$. We want to show that quadrilateral $TNKD$ is cyclic.

Let $m\angle KMA = x$, $m\angle MAK = y$, and $m\angle AKM = z$. Clearly $x + y + z = 180°$. We note that $\angle NAE$ and $\angle MAK$ are supplementary. Since quadrilateral $NAET$ is inscribed in C_5, $\angle NTE$ and $\angle NAE$ are also supplementary. Therefore, $m\angle NTE = m\angle EAK = m\angle MAK = y$. (See Figure 152.)

Since $\angle ETD$ and $\angle EMD$ are both inscribed angles in C_4 that intercept the same arc, $m\angle ETD = m\angle EMD = m\angle AMK = x$. Hence,

$$m\angle NTD = m\angle NTE + m\angle ETD = y + x.$$

Since $m\angle NKD = m\angle AKM = z$, $\angle NTD$ and $\angle NKD$ are supplementary. This makes quadrilateral $TNKD$ cyclic and forces point T to lie on C_0, which was constructed to pass through points D, N, and K.

With a similar argument we can show that C_0 must also pass through point R. Thus, C_0 intersects C_5 at N and T and intersects C_2 at R and K.

Therefore, we have a set of four circles, C_0, C_1, C_2, and C_5, each intersecting two others. By a special case of Problem 13.6 (see remark (b) at the end of our solution to that problem), since points N, A, B, and K are collinear, points P, Q, R, and T are cyclic.

To complete the problem, one needs to repeat the entire argument with still another circle, call it C_6, circumscribed on $\triangle HLE$. The result will be that points P, Q, R, and S are cyclic. These two results taken together assure us that points P, Q, R, S, and T all lie on one circle.

Bibliography

[1]   T. Andreescu and D. Andrica, *Complex Numbers from A to . . . Z*, Birkhauser, 2006.

[2]   B.I. Argunov and M.B. Balk, *Elementary Geometry*, Moscow, 1966. In Russian.

[3]   Astronomical Consultants & Equipment, Inc., "Telescope Optics," 2004, http://www.astronomical
      .com/TelescopeOptics.htm.

[4]   E.J. Barbeau, M.S. Klamkin, W.O.J. Moser, *Five Hundred Mathematical Challenges*, Mathematical
      Association of America, 1995.

[5]   V.G. Boltianskii, *Hilbert's third problem*, Halstead Press, 1978.

[6]   C. Boyer, revised by U. Merzbach, *A History of Mathematics*, Wiley, 1991.

[7]   D.A. Brannan, M.F. Esplen, J.J. Gray, *Geometry*, Cambridge University Press, 2002.

[8]   J. Britton, "Occurrence of the Conics," 2008, http://britton.disted.camosun.bc.ca/jbconics
      .htm

[9]   E. Brown, The Many Names of (7, 3, 1), *Mathematics Magazine*, April 2002, pp. 83–94.

[10] D. Burton, *The History of Mathematics: An Introduction*, 6th ed., McGraw Hill, 2006.

[11] P.J. Cameron, *Combinatorics: Topics, Techniques, Algorithms*, Cambridge University Press, 1994.

[12] J.L. Coolidge, *A History of Geometrical Methods*, Oxford, 1940.

[13] J.L. Coolidge, *A Treatise on the Circle and the Sphere*, Oxford, 1916. (Chelsea Publishing Company,
      1971).

[14] C. Dodge, *Euclidean Geometry and Transformations*, Dover, 1972.

[15] H. Dörrie, *100 Great Problems of Elementary Mathematics. Their History and Solutions*, Dover Publica-
      tions, Inc., 1965.

[16] M.D. Edwards, A Proof of Heron's Formula, *The American Mathematical Monthly*, Vol. 114, No. 10,
      December 2007, p. 937.

[17] J.H. Eves, *Great Moments of Mathematics, before 1650*, Dolciani Mathematical Expositions, No. 5,
      Mathematical Association of America, 1983.

[18] J.H. Eves, *An Introduction to the History of Mathematics*, 6th ed., Saunders College Publishing,
      1992.

[19] J.H. Eves, *A Survey on Geometry. Revised Edition*, Allyn and Bacon, Inc., Boston, 1972.

[20] R.J. Gardner and S. Wagon, At long last, the circle has been squared. *Notices of the American Mathematical
      Society* 36 (December), 1989, pp. 1338–1343.

[21] I.S. Gerasimova, V.A. Gusev, G.G. Maslova, Z.A. Skopetz, M.I. Yagodovsky, *Collection of Problems on
      Geometry, 9–10*, Prosveschenie, Moscow, 1977. In Russian.

[22] I. Grattan-Guinness, *The Rainbow of Mathematics: A History of the Mathematical Sciences*, W.W. Norton
      and Company, Inc., New York, 1997.

[23] M.J. Greenberg, Euclidean and Non-Euclidean Geometries: Development and History, 3rd ed., W.H. Freeman and Company, New York, 1994.

[24] V. Gutenmacher, N.B. Vasilyev, and A. Kundu, *Lines and Curves: A Practical Geometry Handbook*, Birhauser, 2005.

[25] J. Hadamard, *Elementary Geometry*, Moscow, 1948. In Russian.

[26] L. Hahn, *Complex Numbers & Geometry*, Spectrum Series, Mathematical Association of America, 1994.

[27] M. Hall, Jr., *Combinatorial Theory*, John Wiley & Sons; 2nd ed., 1986.

[28] D.W. Henderson, D. Taimina, How to use history to clarify common confusions in geometry, Ch. 6 in *From Calculus to Computers: Using Recent History in the Teaching of Mathematics* (A. Shell and D. Jardine, eds.), Mathematical Association of America Notes 68, 2005, pp. 57–73.

[29] D. Hilbert, *The Foundations of Geometry*, 2nd ed. Chicago: Open Court, 1980 (1899).

[30] R. Honsberger, *Mathematical Gems III*, Mathematical Association of America, 1996.

[31] M. Horblit and K. Nielsen, *Plane Geometry Problems with Solutions*, Barnes & Noble, 1947.

[32] Y. Ionin and L. Kurlyandchik, Some things never change, *Quantum*, September/October 1993.

[33] V.M. Klopsky, Z.A. Skopetz, M.I. Yagodovsky, *Geometry, 9–10*, Prosveschenie, Moscow, 1977. In Russian.

[34] A.I. Kostrikin, Yu. I. Manin. *Linear Algebra and Geometry (Algebra, Logic and Applications)*, Gordon and Breach Science Publishers, 1989.

[35] A.A. Leman, *Collection of Moscow mathematical competitions problems*, Moscow, Prosveshchenie, 1965. In Russian.

[36] E. Maor, *Trigonometric Delights*, Princeton University Press, 1998.

[37] B. Mazur, *Imagining Numbers*, Farrar, Straus and Giroux, 2003.

[38] Z.A. Melzak, *BYPASSES: A Simple Approach to Complexity*, John Wiley & Sons, 1982.

[39] P.S. Modenov, *Geometric Problems*, Nauka, Moscow, 1979. In Russian.

[40] P.S. Modenov, *Geometric Transformations*, Vol. 1, Academic Press Inc., 1965.

[41] P.S. Modenov, *Problems for a Special Course of Elementary Mathematics*, Vysshaya Shkola, Moscow, 1960. In Russian.

[42] P.J. Nahin *An Imaginary Tale. The Story of $\sqrt{-1}$*, Princeton University Press, 1998.

[43] J. Needham, *Science and Civilization in China*, Vol. 3, Cambridge University Press, 1959.

[44] D. Pedoe, *A Course of Geometry For Colleges and Universities*, Cambridge University Press, 1970.

[45] D. Pedou, "Thinking Geometrically," *The American Mathematical Monthly*, Vol. 77, No. 7, 1970, pp. 711–721.

[46] I. Peterson, "Squaring Circles," *Mathematical Association of America Online: Ivars Peterson's MathTrek*, 2004, http://www.maa.org/mathland/mathtrek_11_01_04.html.

[47] V.V. Prasolov, *Problems in Plane Geometry*, Nauka, Moscow, 1986. In Russian.

[48] V. Priebe and E.A. Ramos, Proof without Words, *Mathematics Magazine*, Mathematical Association of America, December 2000.

[49] H. Rademacher, O. Toeplitz, *The Enjoyment of Mathematics*, Princeton University Press, 1957.

[50] B.A. Rosenfeld. *Multidimensional Spaces*, Nauka, Moscow, 1966. In Russian.

[51] B. Russell, "The Teaching of Euclid," *The Mathematical Gazette* *2* (33), 1902, pp. 165–167.

[52] I.J. Schoenberg, *Mathematical Time Exposures*, Mathematical Association of America, 1982.

[53] H. Schwerdtfeger, *Geometry of Complex Numbers*, Dover, New York, 1979.

[54] A. Seidenberg, "The ritual origin of geometry," *Archive for History of Exact Sciences 1*, 1962, pp. 488–527.

[55] I.H. Sivashinsky, *Inequalities and Problems*, Nauka, Moscow, 1967. In Russian.

[56] J.R. Smart, *Modern Geometries*, Brooks/Cole Publishing Company, 1998.

[57] F. Swetz, J. Fauvel, O. Bekken, B. Johansson, and V. Katz, *Learn from the Masters!*, Mathematical Association of America, 1997.

[58] T. Tao, *Solving Mathematical Problems: a personal perspective*, Oxford University Press, 2006.

[59] P.J. Taylor, *International Mathematics: Tournament of the Towns, Questions, and Solutions, Tournament 6 to 10 (1984 to 1988)*, Australian Mathematical Foundation Ltd, Belconnen, ACT.

[60] O.N. Tsuberbiller, *Problems and exercises in Analytical Geometry,* 9th ed., Nauka, Moscow 1968. In Russian.

[61] G. Van Brummelen, *The Mathematics of the Heavens and the Earth: The Early History of Trigonometry*, Princeton University Press, 2009.

[62] Glen Van Brummelen, personal communication, 2009.

[63] N.B. Vasiliev, S.A. Molchanov, A.L. Rosental, A.P. Savin, *Mathematical Competitions. Geometry*, Nauka, Moscow, 1974. In Russian.

[64] E.C. Wallance and S.W. West, *Roads to Geometry*, Prentice Hall, Upper Saddle River, N.J., 2003.

[65] H. Weyl. *Space,Time, Matter*, Springer, 1923.

[66] Wikipedia, "Area of a Disk," *Wikipedia, The Free Encyclopedia*, 2008, `http://en.wikipedia.org/w/index.php?title=Area_of_a_disk&oldid=227321257`.

[67] Wikipedia, "Hilbert's Third Problem," *Wikipedia, The Free Encyclopedia*, 2008, `http://en.wikipedia.org/w/index.php?title=Hilbert's_third_problem&oldid=228763618`.

[68] Wikipedia, "Laurent Cassegrain," *Wikipedia, The Free Encyclopedia*, 2008, `http://en.wikipedia.org/w/index.php?title=Laurent_Cassegrain&oldid=214812789`.

[69] I. M. Yaglom, *Geometric Transformations I*, New Mathematical Library **8**, Mathematical Association of America, 1962.

[70] I. M. Yaglom, *Complex Numbers*, FM, Moscow, 1963. In Russian.

[71] I. M. Yaglom, *Geometric Transformations II*, New Mathematical Library **21**, Mathematical Association of America, 1968.

[72] I. M. Yaglom, *Geometric Transformations III*, New Mathematical Library **24**, Mathematical Association of America, 1973.

Index

About the Authors

Owen Byer graduated with a B.A. in mathematics, with secondary education certification, from Messiah College, Grantham, PA in 1989. He received his M.S. (1991) and Ph.D. (1996) in mathematics from University of Delaware. Owen taught for three years at Northwestern College, Orange City, IA. He is currently Professor of Mathematics at Eastern Mennonite University. Owen is a member of MAA and ACMS.

Felix Lazebnik was born in Kiev, the former USSR. Felix received his MA (1975) from Kiev State University and the Ph.D. in mathematics from the University of Pennsylvania (1987). Felix has taught mathematics for 35 years at different levels, including four years in a high school. Since 1987, he has been with the Department of Mathematical Sciences at University of Delaware. As a Professor of Mathematics there, he teaches mathematics and does research with graduate and undergraduate students. He served for five years as the Managing Editor of *The Electronic Journal of Combinatorics* and is a member of their Editorial Board. He is a member of the AMS, MAA, and the ICA.

Deirdre Longacher Smeltzer graduated with a B.A. in mathematics from Eastern Mennonite University, Harrisonburg, VA in 1987. She received her M.S. (1989) and Ph.D. (1994) in mathematics from University of Virginia. Deirdre taught for four years at University of St. Thomas, St. Paul, MN. She is currently a Professor of Mathematics and the chair of the Mathematical Sciences department at Eastern Mennonite University. Deirdre is a member of MAA (and former officer of MD-DC-VA section) and ACMS.